AF844775

Advances in Biochemistry in Health and Disease

Volume 23

Series Editor

Naranjan S. Dhalla, Institute of Cardiovascular Sciences, St. Boniface Hospital, Winnipeg, MB, Canada

Editorial Board

Roberto Bolli, Medicine & Cardiology Department, 3rd Floor, University of Louisville, Louisville, KY, USA

Ramesh Goyal, Delhi Pharmaceutical Sciences & Research, New Delhi, India

Chandrasekharan Kartha, Cardiovascular Diseases and Diabetes Biology, Kerala Institute of Medical Sciences, Thiruvananthapuram, Kerala, India

Lorrie Kirshenbaum, St. Boniface General Hospital, Winnipeg, MB, Canada

Naoki Makino, Kyushu University, Fukuoka, Japan

Jawahar L. L. Mehta, Division of Cardiology, University of Arkansas for Medical Sciences, Little Rock, AR, USA

Bohuslav Ostadal, Institute of Physiology, Czech Academy of Sciences, Prague, Czech Republic

Grant N. Pierce, St. Boniface General Hospital, Winnipeg, MB, Canada

Jan Slezak, Institute for Heart Research, Slovak Academy of Sciences, Karlova Ves, Slovakia

Andras Varro, Pharmacology & Pharmacotherapy, University of Szeged, Szeged, Hungary

Karl Werdan, Martin Luther University Halle-Wittenber, Halle (Saale), Sachsen-Anhalt, Germany

William B. Weglicki, School of Medicine and Health Sciences, George Washington University, Washington, USA

Advances in Biochemistry in Health and Disease focus on the latest developments in biochemical research with implications for health and disease. This book series consists of original edited volumes and monographs, presented by leading experts in the field and provides an up to date and unique source of information for all those interested in the fundamental, biochemical processes of the latest and emerging topics and techniques.

Covering a wide variety of topics, this book series is a valuable source of information from those at the lab bench through to the Health Care workers.

More information about this series at https://link.springer.com/bookseries/7064

Paramjit S. Tappia · Bram Ramjiawan ·
Naranjan S. Dhalla
Editors

Cellular and Biochemical Mechanisms of Obesity

Springer

Editors
Paramjit S. Tappia
Asper Clinical Research Institute
St. Boniface Hospital Research
Winnipeg, MB, Canada

Bram Ramjiawan
Asper Clinical Research Institute
St. Boniface Hospital
Winnipeg, MB, Canada

Naranjan S. Dhalla
Institute of Cardiovascular Science
St. Boniface Hospital Research
Winnipeg, MB, Canada

ISSN 2512-2142 ISSN 2512-2150 (electronic)
Advances in Biochemistry in Health and Disease
ISBN 978-3-030-84762-3 ISBN 978-3-030-84763-0 (eBook)
https://doi.org/10.1007/978-3-030-84763-0

© Springer Nature Switzerland AG 2021
This work is subject to copyright. All rights are reserved by the Publisher, whether the whole or part of the material is concerned, specifically the rights of translation, reprinting, reuse of illustrations, recitation, broadcasting, reproduction on microfilms or in any other physical way, and transmission or information storage and retrieval, electronic adaptation, computer software, or by similar or dissimilar methodology now known or hereafter developed.
The use of general descriptive names, registered names, trademarks, service marks, etc. in this publication does not imply, even in the absence of a specific statement, that such names are exempt from the relevant protective laws and regulations and therefore free for general use.
The publisher, the authors and the editors are safe to assume that the advice and information in this book are believed to be true and accurate at the date of publication. Neither the publisher nor the authors or the editors give a warranty, expressed or implied, with respect to the material contained herein or for any errors or omissions that may have been made. The publisher remains neutral with regard to jurisdictional claims in published maps and institutional affiliations.

This Springer imprint is published by the registered company Springer Nature Switzerland AG
The registered company address is: Gewerbestrasse 11, 6330 Cham, Switzerland

Preface

The World Health Organization's long-standing definition of obesity is abnormal or excessive fat accumulation that may impair health. The United Nations goals on diet-related disease will fail to be met because of the mounting numbers of people becoming obese or overweight, with almost 1 billion of the world's adults projected to be obese by 2025. Thus, with the prominence of the obesity epidemic, different strategies to reduce obesity have become public health priorities. In this regard, according to the 2019 United Nations Sustainable Development Goals (SDG) Report, the 2030 agenda has provided a blueprint for shared prosperity in a sustainable world—a world where all people can live productive, vibrant and peaceful lives on a healthy planet; obesity is one of the factors identified as a concern for achieving the SDG.

It is evident that obesity is a serious health problem with an increased risk of several common diseases including diabetes, cardiovascular disease, renal disease liver disease, infertility, cancer and neurological complications. Although the primary basis of obesity and overweight is an imbalance between calorie intake and calorie expenditure, the underlying biochemical and metabolic processes that cause obesity are not fully understood. Women, compared to men, have higher percent body fat and deposit it in a different pattern, with relatively more adipose tissue in the hips and thighs in women and central obesity typical of men. In fact, more women are obese than men. These differences may be due to different mechanisms.

Our earlier two volumes on this topic, published in this series "Advances in Biochemistry in Health and Disease", have focused on the pathophysiology of obesity-induced health complications and the biochemistry of cardiovascular dysfunction in obesity. This book examines the origins, etiology and complexities of obesity and is intended to bring together a comprehensive overview of the current status of the knowledge on the pertinent mechanisms that are associated with the development of obesity including inflammation/cytokines, hormonal deregulation, activation of the sympathetic nervous system, oxidative stress, metabolic derangements and impairment of signal transduction pathways. While such mechanisms are responsible for the development of obesity, they also contribute to the occurrence of co-morbidities, and thus this book also provides important mechanistic information on obesity-related diseases. The intent of this volume is to emphasize that obesity itself is a major health hazard and that it should be considered as a multifactorial metabolic disease in its own right other than as a risk factor for different diseases.

There are 21 chapters in two different parts in this book, comprising of Part I: Pathophysiologic Mechanisms of Obesity (11 chapters) and Part II: Therapeutic Mechanisms of Obesity (10 chapters). Each chapter in the book will hopefully further advance the understanding of the cellular and biochemical mechanisms of obesity as well as stimulate and motivate biomedical researchers and scientists to develop innovative approaches to prevent obesity as well as obesity-induced adverse complications. Furthermore, this book will serve as a highly useful resource for health professionals, medical students, fellows, residents and graduate students, particularly in the field of nutritional sciences. The contributors to this book are international leading experts on the etiology and pathophysiology of obesity. The book will underscore the multifactorial nature of the mechanisms of obesity from basic to advanced aspects of our biochemical understanding. Although the genetics of obesity is an extensive topic in itself, introduction into the topic is provided by two specific chapters that describe the genetics and epigenetics and obesity.

This monograph covers a broad range of mechanisms involved in the pathophysiology of obesity as well as examines current and advanced concepts in the development of obesity and their relationship to the development of other concomitant problems that impact on overall well-being of an individual. Despite pharmacological interventions, exercise, calorie intake restriction, surgery and behavioural approaches currently undertaken to treat obesity, we hope that the reader will understand that the cause of obesity is multifactorial and that understanding the complexity and comprehensive nature of the mechanisms involved in the etiology of obesity may contribute to the development of novel and effective strategies for the prevention and treatment of obesity. Furthermore, understanding the cellular and biochemical mechanisms associated with the pathophysiology of obesity will also help in further understanding the development of obesity-related health complications.

We are grateful to the St. Boniface Hospital Albrechtsen Research Centre, for their infrastructural support in this endeavour. We are thankful for the efficient and professional assistance provided by Mr. Muruga Prashanth Rajendran in the production of the high quality chapters in this book. We also express our gratitude to Mr. Rajan Muthu for his excellent assistance in developing not only this monograph, but, overall, in the production of the trilogy of books in the very important topic of obesity that is now being considered by some experts in the field, as a major disease. We also thank Dr. Gonzalo Cordova for his understanding and efforts in improving the quality of the chapters presented in these books and post-production assistance from Mr. Madanagopal Deenadayalan. Finally, we thank Springer Nature Switzerland AG in their willingness to undertake this project and for the extraordinary support for the preparation of the series on "Advances in Biochemistry in Health and Disease" and the volumes therein.

Winnipeg, Canada

Paramjit S. Tappia
Bram Ramjiawan
Naranjan S. Dhalla

Contents

Contributors

Salah Abohelaika Clinical Pharmacology Department, Qatif Central Hospital, Ministry of Health, Qatif, Saudi Arabia

Ernest Adeghate Department of Anatomy, College of Medicine and Health Sciences, United Arab Emirates University, Al Ain, United Arab Emirates

Concepción María Aguilera Faculty of Pharmacy, Department of Biochemistry and Molecular Biology II, University of Granada, Granada, Spain;
IBS.GRANADA, Complejo Hospitalario Universitario de Granada, Granada, Spain;
Institute of Nutrition and Food Technology "José Mataix", Center of Biomedical Research, University of Granada, Granada, Spain;
CIBER Physiopathology of Obesity and Nutrition (CIBERobn), Institute of Health Carlos III, Madrid, Spain

Saeeda Al Jaberi Department of Anatomy, College of Medicine & Health Sciences, United Arab Emirates University, Al Ain, United Arab Emirates

Fahad Al-Muhanna Department of Internal Medicine, King Fahd Hospital of the University, College of Medicine, Imam Abdulrahman Bin Faisal University, Dammam, Saudi Arabia

Moayed S. Alavian Internal Medicine-Research Center, Middle East Liver Disease Center, Tehran, Iran

Luigi Aloe Fondazione Iret Tecnopolo R. Levi-Montalcini, Rome, Italy

Fadhel A. Alomar Department of Pharmacology and Toxicology, College of Clinical Pharmacy, Imam Abdulrahman Bin Faisal University, Dammam, Saudi Arabia

Mohammed A. Alshabeed Department of Development Medicine, King Adbullah International Medical Research Center, Riyadh, Saudi Arabia

Augusto Anguita-Ruiz Faculty of Pharmacy, Department of Biochemistry and Molecular Biology II, University of Granada, Granada, Spain;
IBS.GRANADA, Complejo Hospitalario Universitario de Granada, Granada, Spain;

Institute of Nutrition and Food Technology "José Mataix", Center of Biomedical Research, University of Granada, Granada, Spain;
CIBER Physiopathology of Obesity and Nutrition (CIBERobn), Institute of Health Carlos III, Madrid, Spain

Abdullah Adil Ansari Department of Biology, Faculty of Natural Sciences, University of Guyana, Georgetown, Guyana

Shalini Behl School of Life Sciences, Manipal Academy of Higher Education, Dubai, UAE

Keshore R. Bidasee Department of Pharmacology and Experimental Neuroscience, and Environment and Occupational Health, University of Nebraska Medical Center, Omaha, NE, US;
Nebraska Redox Biology Center, Lincoln NE, US

Deniz Billur Faculty of Medicine, Department of Histology and Embryology, Ankara University, Ankara, Turkey

George N. Chaldakov Department of Anatomy and Cell Biology, Medical University, Varna, Bulgaria;
Institute for Advanced Study, Varna, Bulgaria

Sukhinder Kaur Cheema Department of Biochemistry, Memorial University, St. John's, NL, Canada

Athena Cohen Department of Anatomy, College of Medicine & Health Sciences, United Arab Emirates University, Al Ain, United Arab Emirates;
Department of Chemistry, Emory University, Atlanta, GA, USA

Octavian M. Crețu Department of Surgery - Surgical Semiotics I and Centre for Hepato-Biliaryand Pancreatic Surgery, Faculty of Medicine, "Victor Babeș" University of Medicine and Pharmacy, Timișoara, Romania

Emanuel Cummings School of Medicine, College of Medical Sciences, University of Guyana, Georgetown, Guyana

Seema Dangwal Cardiovascular Institute, Department of Medicine, School of Medicine, Stanford University, Palo Alto, CA, USA

Cyrus DeSouza Department of Internal Medicine, Division of Diabetes Endocrinology & Metabolism, University of Nebraska Medical Center, Omaha, NE, US

Naranjan S. Dhalla Institute of Cardiovascular Sciences, St. Boniface Hospital Albrechtsen Research Centre, Department of Physiology and Pathophysiology, Max Rady College of Medicine, University of Manitoba, Winnipeg, Canada;
Department of Physiology and Pathophysiology, Rady Faculty of Health Sciences, St. Boniface Hospital Albrechtsen Research Centre, Institute of Cardiovascular Sciences, Max Rady College of Medicine, University of Manitoba, Winnipeg, Canada

Chinedum Ogbonnaya Eleazu Department of Physiology, School of Medical Sciences, Universiti Sains Malaysia, Kelantan, Malaysia;
Department of Chemistry, Biochemistry and Molecular Biology, Alex Ekwueme Federal University, Ebonyi State, Ndufu-Alike, Ikwo, Nigeria

Marco Fiore Institute of Biochemistry and Cell Biology, Section of Neurobiology, National Research Council (CNR), Rome, Italy

Asit Ranjan Ghosh Department of Integrative Biology, School of Bio Sciences and Technology, Vellore Institute of Technology, Vellore, Tamil Nadu, India

Frederick Hamel Research Service, Omaha Veterans Affairs Medical Center, Omaha, US

Carlin Hanoman School of Natural Sciences, University of Central Lancashire, Preston, UK;
School of Medicine, College of Medical Sciences, University of Guyana, Georgetown, Guyana

Lourdes Franco Hernandez Department of Physiology, Faculty of Science, University of Extremadura, Badajoz, Spain

Marcia Hiriart Department of Physiology and Cognitive Neuroscience, Autonomous National University of Mexico (UNAM), Mexico City, Mexico

Frank Christopher Howarth Department of Physiology, College of Medicine & Health Sciences, UAE University, Al Ain, UAE

Abla Mohammed Ismail Corniche Hospital, Abu Dhabi, United Arab Emirates

Daniela Jakubowicz Diabetes Unit, Wolfson Medical Center, Sackler Faculty of Medicine, Tel Aviv University, Holon, Israel

Zohre Kiani Department of Internal Medicine, Kerman University of Medical Sciences, Kerman, Iran

Shravanthi S. Kumar Department of Integrative Biology, School of Bio Sciences and Technology, Vellore Institute of Technology, Vellore, Tamil Nadu, India

Zohar Landau Diabetes Unit, Wolfson Medical Center, Sackler Faculty of Medicine, Tel Aviv University, Holon, Israel

Jose Antonio Pariente Llanos Department of Physiology, Faculty of Science, University of Extremadura, Badajoz, Spain

Toyoki Maeda Division of Cardiology and Clinical Gerontology, Department of Internal Medicine, Kyushu University Beppu Hospital, Beppu, Japan

Naoki Makino Division of Cardiology and Clinical Gerontology, Department of Internal Medicine, Kyushu University Beppu Hospital, Beppu, Japan

Jawahar L. Mehta Division of Cardiology, University of Arkansas for Medical Sciences and the Central Arkansas Veterans Healthcare System, Little Rock, AR, USA

Andrea Méndez-Gutiérrez Faculty of Pharmacy, Department of Biochemistry and Molecular Biology II, University of Granada, Granada, Spain;
IBS.GRANADA, Complejo Hospitalario Universitario de Granada, Granada, Spain;
Institute of Nutrition and Food Technology "José Mataix", Center of Biomedical Research, University of Granada, Granada, Spain

Alok Kumar Mishra Department of Integrative Biology, School of Bio Sciences and Technology, Vellore Institute of Technology, Vellore, Tamil Nadu, India

Mahaneem Mohamed Department of Physiology, Unit of Integrative Medicine, School of Medical Sciences, Universiti Sains Malaysia, Kelantan, Malaysia

Caronda J. Moore Kansas State University Innovation Partners Manhattan, Manhattan, KS, US

Ana Beatriz Rodriguez Moratinos Department of Physiology, Faculty of Science, University of Extremadura, Badajoz, Spain

Mahsa Motavaf Department of Genetics, Faculty of Biological Science, Tarbiat Modares University, Tehran, Iran

Danina M. Muntean Department of Functional Sciences – Pathophysiology and Centre for Translational Research and Systems Medicine, Faculty ofMedicine, "Victor Babeș" University of Medicine and Pharmacy, Timișoara, Romania

Victor Udo Nna Department of Physiology, College of Medical Sciences, University of Calabar, Cross River State, Calabar, Nigeria

Shreesh Ojha Department of Pharmacology, College of Medicine & Health Sciences, United Arab Emirates University, Al Ain, United Arab Emirates

Rouzha Z. Pancheva Department of Hygiene and Epidemiology, Faculty of Public Health, Medical University, Varna, Bulgaria

Indrayani Phadtare Department of Biochemistry, Memorial University, St. John's, NL, Canada

Julio Plaza-Díaz Faculty of Pharmacy, Department of Biochemistry and Molecular Biology II, University of Granada, Granada, Spain;
IBS.GRANADA, Complejo Hospitalario Universitario de Granada, Granada, Spain;
Children's Hospital of Eastern Ontario Research Institute, Ottawa, ON, Canada

Matthew Ramjiawan Asper Clinical Research Institute & Office of Clinical Research, St. Boniface Hospital, Winnipeg, MB, Canada

Gorana Rancic Department of Histology, Faculty of Medicine, Nis University, Niš, Serbia

Francisco Javier Ruiz-Ojeda Faculty of Pharmacy, Department of Biochemistry and Molecular Biology II, University of Granada, Granada, Spain;
IBS.GRANADA, Complejo Hospitalario Universitario de Granada, Granada, Spain;
Institute for Diabetes and Obesity, Helmholtz Diabetes Center At Helmholtz Center Munich, Munich, Neuherberg, Germany

Khemraj Rupee School of Medicine, College of Medical Sciences, University of Guyana, Georgetown, Guyana

Sunil Rupee School of Medicine, College of Medical Sciences, University of Guyana, Georgetown, Guyana

Zulqarnain Saeed Department of Psychology and Social Work, Flinders University, Adelaide, Australia

Husam Salah Division of Cardiology, University of Arkansas for Medical Sciences and the Central Arkansas Veterans Healthcare System, Little Rock, AR, USA

Aliyeh Sargazi Razi Hospital, Zahedan University of Medical Sciences, Saravan, Iran;
Imamreza Hospital, Iranshahr University of Medical Sciences, Dalgan, Iran

Mehdi Sepehri Department of Computer, Sanabad Golbahar Institute of Higher Education, Golbahar, Khorasan Razavi, Iran

Zahra Sepehri Internal Medicine, Department of Internal Medicine, Zabol University of Medical Sciences, Zabol, Iran;
Biochemistry and Medical Genetics, Basic Medical Sciences Building, University of Manitoba, Winnipeg, Canada

Anureet K. Shah School of Kinesiology, Nutrition and Food Science, California State University, Los Angeles, CA, USA

Tanya Sharma Division of Cardiology, University of Arkansas for Medical Sciences and the Central Arkansas Veterans Healthcare System, Little Rock, AR, USA

Jaipaul Singh School of Natural Sciences, University of Central Lancashire, Preston, Lancashire, England, UK

Manal M. Smail School of Natural Sciences, University of Central Lancashire, Preston, UK;
Department of Anatomy, College of Medicine and Health Sciences, United Arab Emirates University, Al Ain, United Arab Emirates

Hitesh Soni Flagship Pioneering (FL72 Inc.), Cambridge, MA, USA

Adrian Sturza Department of Functional Sciences – Pathophysiology and Centre for Translational Research and Systems Medicine, Faculty ofMedicine, "Victor Babeș" University of Medicine and Pharmacy, Timișoara, Romania

Joseph Bagi Suleiman Department of Science Laboratory Technology, Akanu Ibiam Federal Polytechnic, Ebonyi State, Unwana, Nigeria

Ahmed Sultan Department of Physiology, College of Medicine & Health Sciences, UAE University, Al Ain, UAE

Paramjit S. Tappia Asper Clinical Research Institute & Office of Clinical Research, St. Boniface Hospital, Winnipeg, MB, Canada

Shani Tsameret Institute of Biochemistry, Food Science and Nutrition, The Robert H. Smith Faculty of Agriculture, Food and Environment, The Hebrew University of Jerusalem, Rehovot, Israel

Naga Sai Shravan Turaga Division of Cardiology, University of Arkansas for Medical Sciences and the Central Arkansas Veterans Healthcare System, Little Rock, AR, USA

Belma Turan Faculty of Medicine, Department of Biophysics, Lokman Hekim University, Ankara, Turkey

A. Cihangir Uguz Department of Biophysics, School of Medicine, Yozgat Bozok Univeristy, Yozgat, Turkey

Hitesh Vaidya Department of Biochemistry, Memorial University, St. John's, NL, Canada

Julio Wainstein Diabetes Unit, Wolfson Medical Center, Sackler Faculty of Medicine, Tel Aviv University, Holon, Israel

Stanislav Yanev Department of Drug Toxicology, Institute of Neurobiology, Bulgarian Academy of Sciences, Sofia, Bulgaria

Part I
Pathophysiologic Mechanisms of Obesity

Chapter 1
Obesity as a Major Health Hazard

Paramjit S. Tappia and Naranjan S. Dhalla

Abstract The well-being of the global population has been challenged with the epidemic of obesity. Obesity is a complex and multifactorial health hazard. While obesity itself can be considered as a metabolic disease that reduces quality of life and life expectancy, the occurrence of a variety of obesity-related complications including cardiovascular diseases, cancer, type 2 diabetes, renal dysfunction, liver defect, mobility limitations and neurological/psychological disorders also contribute to the human and economic burden of obesity. Therefore, advancement of the understanding of the mechanisms as well as causes of obesity can further our knowledge such that appropriate and concerted efforts for prevention of obesity and associated health problems can be undertaken. Since successful therapeutic strategies for obesity are not a one size treatment fits all approach, they involve multifaceted and long-term efforts that can be designed to the individual. Accordingly, this chapter is intended to introduce some of the major causes of obesity as well as the nature and prevalence of obesity-related co-morbidities. In view of the influence of sex and ethnicity in the incidence of obesity and related disorders, some discussion is also provided to identify specific populations at risk for obesity as well as to obesity-induced health complications. Furthermore, some of the therapeutic approaches to prevent or treat obesity that can be tailored to the individual are outlined. From the information provided it is evident that obesity is a major health hazard and appropriate public health measures should be implemented to prevent or reduce or even reverse the impact of this global chronic disease.

Keywords Obesity · Metabolic disease · Obesity-related health complications · Prevention of obesity · Human health and disease

---

P. S. Tappia (✉)
Asper Clinical Research Institute, St. Boniface Hospital, CR3129 – 369 Tache Avenue, Winnipeg, MB R2H 2A6, Canada
e-mail: ptappia@sbrc.ca

N. S. Dhalla
Department of Physiology and Pathophysiology, Rady Faculty of Health Sciences, St. Boniface Hospital Albrechtsen Research Centre, Institute of Cardiovascular Sciences, Max Rady College of Medicine, University of Manitoba, Winnipeg, Canada

© Springer Nature Switzerland AG 2021
P. S. Tappia et al. (eds.), *Cellular and Biochemical Mechanisms of Obesity*,
Advances in Biochemistry in Health and Disease 23,
https://doi.org/10.1007/978-3-030-84763-0_1

Introduction

Although the Canadian and American Medical Associations have stated that obesity is a chronic medical disease [1, 2], this has not been generally accepted [3]. According to the World Health Organization (WHO), the global incidence of obesity has increased almost three-fold since 1975. In 2016, global statistics revealed more than 1.9 billion adults as overweight and 650 million of these as obese. The WHO established the definition of obesity an abnormal or excessive fat accumulation that may impair health [4]; however, a minor change to this definition has been proposed by Sharma and Campbell-Scherer [3] that the word "may" be removed in order to fit the definition of an actual disease. Undoubtedly, obesity is complex and is a multifactorial and challenging chronic disease that is much beyond just excessive ingestion of fat [5].

It is well established that an abnormal or excessive fat accumulation is detrimental and presents an adverse health risk to adults (both men and women) as well as to children. In addition, it is well known that there are differences between men and women in not only their fat mass, but also in the bodily distribution of fat, whereby women have relatively more adipose tissue in the hips and thighs whereas central obesity is typical in men [4]. Severe obesity is more prevalent in women than men, and obesity-related disease risks differ in women and men, which have been attributed to sex hormones [6].

Some of the major influences that can increase an individual's predisposition/risk to developing obesity are summarized in Fig. 1.1. The consequences of obesity is that it can increase the risk of occurrence of a variety of different co-morbid conditions. It is then obvious that if obesity can be prevented or treated (reversed) then the risk of developing obesity-related health complications is diminished. As part of the series on "Advanced in Biochemistry in Health Disease", we embarked on producing 3 books on the topic of obesity. Our 1st book is on the "Pathophysiology of Obesity-Induced Health Complications" [7] outlined the link between obesity and the onset of many adverse health conditions including diabetes, cardiovascular disease, mental health disorders, cancers, sleep disorders, liver and kidney disease and musculoskeletal disorders. The development of these diseases may be attributable to changes in cell function, endocrine disorder, metabolic dysfunction, activation of neurohormonal systems, oxidative stress and modification of different signaling pathways. It should be mentioned that this book was ranked among the top used publications on SpringerLink that concern one or more of the United Nations Sustainable Development Goals.

The 2nd book entitled "Biochemistry of Cardiovascular Dysfunction in Obesity [8] is specific to cardiovascular health and described the impact of obesity on the cardiovascular system and increased predisposition to cardiovascular complications. It described some of the major biochemical mechanisms that lead to the occurrence of myocardial abnormalities and vascular alterations in obesity. The focus of the present book, which is the 3rd part of the trilogy on obesity, examines the origins and etiology of obesity and brings together a comprehensive overview of

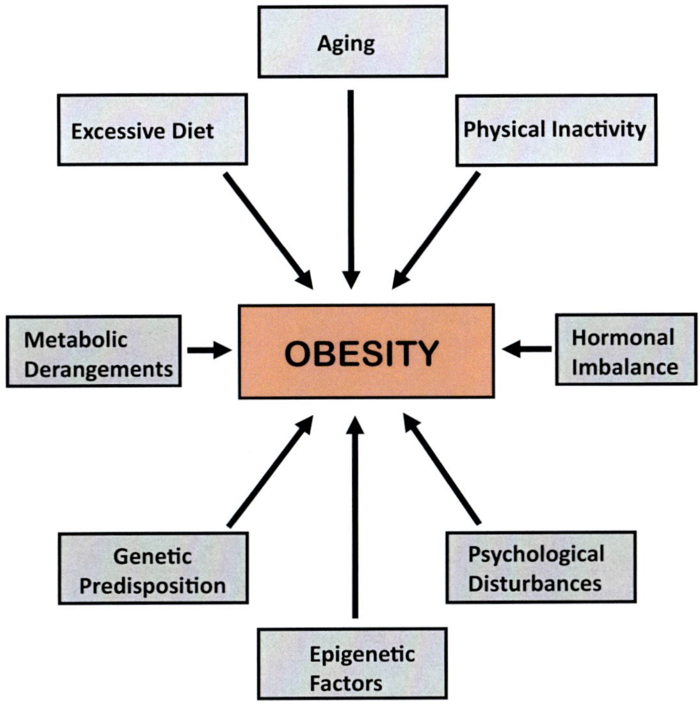

Fig. 1.1 Major causes of obesity

the current status of the knowledge on the pertinent mechanisms that are associated with the development of obesity. Some of the major causes of obesity such as hormonal imbalance, metabolic derangements, excessive diet, physical inactivity, aging, psychological/neurological disturbances and genetic/epigenetic factors are briefly described in this chapter. In addition, this chapter will also discuss some of the major health complications that occur in overweight/obese individuals as well as current approaches in the treatment/prevention of obesity.

Causes of Obesity

Excessive Nutrition and Physical Inactivity

There has been an increase in the intake of high fat content energy dense foods across the world, and with the concomitant sedentary lifestyle due to decreased physical activity, has led to increasing incidence of obesity as well as associated health complications. Interestingly, some current occupations are less physically demanding, as compared to earlier times, with job-related energy expenditure now reported to be lower than in the last 50 years [9].

Processed convenience and fast foods are known to be more energy dense and lack micronutrients [10]. In addition, ultra-processed foods are not only more energy dense, but also have higher content of added sugars, salt, and artificial flavorings, colors, and preservatives [11]. The consumption of ultra-processed foods has markedly increased over the years and has been reported to constitute 60% of the total energy intake in the U.S. adult population [12–14]. Indeed, the higher ingestion of ultra-processed foods has led to a global exponential amplification in the rates of excess weight and abdominal obesity [15–19]. Interestingly, a recent study found that participants consumed more calories when given ad libitum access to ultra-processed diet compared to ad libitum access to an unprocessed food diet [20]. Furthermore, it was revealed that individuals on the ultra-processed diet gained weight, whereas those on the unprocessed diet lost weight, providing additional evidence that high consumption of ultra-processed foods may result in weight gain.

The individual's risk for weight gain is dependent energy intake and energy expenditure [21]. A prolonged period of positive energy balance as a consequence of energy intake being consistently higher than expenditure, increases an individual's the risk for becoming overweight or developing obesity, largely through the expansion of white adipose tissue [22]. An important contributor to the obesogenic environment that has resulted in an increase in the incidence of obesity is the change in the intake in the amount and type of food. Food servings and portion sizes have gradually increased to the extent that such portion-distortion is now perceived as normal [23–25] and has resulted in higher amount of food consumption [25, 26].

Metabolic Derangements

Two metabolic phenotypes of obesity have been proposed, namely metabolically healthy obese (MHO) and metabolically unhealthy obese (MUO) [27]. Individuals that are categorized as MHO have preserved insulin sensitivity, normal adipose function, lower visceral fat mass and lower fat content in the liver [28]. These aspects of MHO is suggestive that a subgroup of obese individuals may not have cardio-metabolic risk factors that occur in obese subjects, emphasizing that adipose tissue function has an important role to play in metabolic health [29]. It has been suggested that central accumulation of body fat is associated with insulin resistance whereas distribution of body fat in a peripheral pattern is metabolically less important [30]. Metabolic obesity occurs when there is a shift in the body weight set point s to an abnormally high level [31]. Although genetic makeup has a role to play in the development of metabolic diseases including metabolic syndrome, type 2 diabetes and obesity, changes in diet, exercise and aging also have a contributory influence [32].

The white adipose tissue has a key role in whole-body energy status and metabolism. It not only stores excess energy, but also secretes various hormones and metabolites to regulate body energy balance [33]. On the other hand, mitochondria are central to ATP production, energy expenditure and the removal of harmful oxyradicals [34]. However, excessive energy substrates can result in mitochondrial dysfunction that negatively impacts lipid and glucose metabolism [34]. It should be mentioned that normal mitochondrial function is essential in the ability of adipocytes to maintain balance between energy storage and expenditure [34]; nutrient overload leads to mitochondrial dysfunction, which in turn leads to obesity.

Hormonal Imbalance and Aging

A adipose tissue is involved in the regulation of hormones and inflammation [35] and adipose tissue dysfunction can lead to low-grade inflammation due to adipocyte production of inflammatory cytokines and extracellular proteins that induce the infiltration and activation of immune cells and promote inflammatory processes [36]. It is pointed out that adipokine and cytokines regulate appetite and energy balance, glucose and lipid metabolism as well as neuroendocrine function [35]. The peripheral energy state is signaled to the central nervous system by several neuropeptides hormones, including leptin [37]. Indeed, leptin acts on the neurons in the hypothalamus and hippocampus as well as on neurons in the brain stem to regulate food intake, glucose and lipid metabolism and energy expenditure [37, 38]. In contrast, glucagon is considered to act as a counter-regulatory hormone that negatively influences energy balance, increases energy expenditure and activates satiety [39].

While the incidence of obesity is higher in adults over the age of 60 years [40], the mortality effect of extreme obesity is greater among younger than older adults [30]. There are several factors that contribute to the occurrence of overweight and obesity in older people and include age-related alterations in life style, depression, modification of endocrine systems, sympathetic tone, oxidative stress as well as other comorbidities [41]. Of note, changes in the body composition due to aging, results in a redistribution of fat from peripheral and subcutaneous sources to a central location and thereby increasing the waist circumference and waist-to-hip ratio. The use of BMI as a classification for weight status in adults has been questioned as it may not be the appropriate tool for assessing the degree of body fat as it underestimates the degree of adiposity [42, 43]. The risk for obesity during the aging process is amplified by the natural loss in muscle mass and strength (sarcopenia) [44]; muscle is metabolically more active than fat, and thus reduces the physical performance [40], which itself exacerbates the risk for becoming overweight or development of obesity. Such age-related changes may be as a consequence of a shift in the hormonal balance due to aging.

Genetic Predisposition and Epigenetic Factors

It has been estimated that genetic factors account for 40–75% of variation in BMI [45]; in fact, more than 300 genes have been determined to be linked to BMI, waist-to-hip ratio, and adiposity [46]. Although genetically predisposed individuals are at greater risk for higher BMI, there is an increase in the risk for obesity when genetic factors are combined with environmental factors, suggesting that environmental factors may have a greater influence on the development of obesity than genetic factors [47]. It should be mentioned that genetic factors have also been suggested to determine the ability to undertake physical activity and thus dictates sedentary behavior [48], creating a vicious cycle of events that may exacerbate obesity.

Several theoretical models have been proposed to explain the genetics of obesity [49–53]. The set point theory proposes that body weight is genetically determined, and that internal genetically determined mechanisms exist, which are designed to protect against alterations in bodyweight [49]. On the other hand, the thrifty gene theory suggests that genetic variants cause some individuals to store energy from food more efficiently and expend less energy at rest and during physical activity [45]. While this proposition may be considered as economical especially when there is a deficiency of energy intake, it may be deemed as a disadvantage particularly as energy is readily available under the current obesogenic environment. Although other theories have also been proposed [50–53], there is an opportunity and scope for advancing investigation in this field in order to fully understanding the genetic-environment interaction with respect to the development of obesity.

Psychological Disturbances

Emotional issues and psychological disturbances are a common feature in overweight individuals and is considered to play a significant role in the development of obesity [54]. On the other hand, in view of the negative perception of obesity, it has been suggested that psychological disturbances may arise due to obesity rather than the cause of obesity itself [54]. Several issues can be seen to regulate appetite and influence weight gain/loss or its maintenance. For example, appetite may be influenced by psychosocial factors [55] and by several other aspects that include modulation of brain neurotransmitters, altered liver metabolism, dysfunction in the nutrient/sensory constituents of the diet, environmental stressors, behavioral changes with respect to dieting as well as use of centrally acting psychotropic substances [55].

It should be mentioned that psychological and emotional distresses linked to weight gain/obesity are not limited to adults. In this regard, children growing up in a dysfunctional family environment or experiencing psycho-emotional imbalance, are prone to weight gain, which can increase the predisposition to further weight gain and eventual obesity [56, 57]. The weight gain/obesity have been suggested to be attributed to prolonged stress, increased appetite and higher energy intake, low-grade inflammation as well as depressed basal metabolism [56, 57].

Obesity-Induced Health Complications

Although the global focus still remains to avert childhood malnutrition and to ensure adequate food supply for the world's population, there is now the additional challenge of managing obesity and obesity-related health complications [58] including diabetes, cardiovascular disease (CVD) and cancer (Fig. 1.2). Some of the major health complications are described below.

Cardiovascular Disease

It is universally accepted that obesity is an independent risk factor for cardiovascular disease (CVD). Indeed, there is a wealth of epidemiological and clinical evidence that

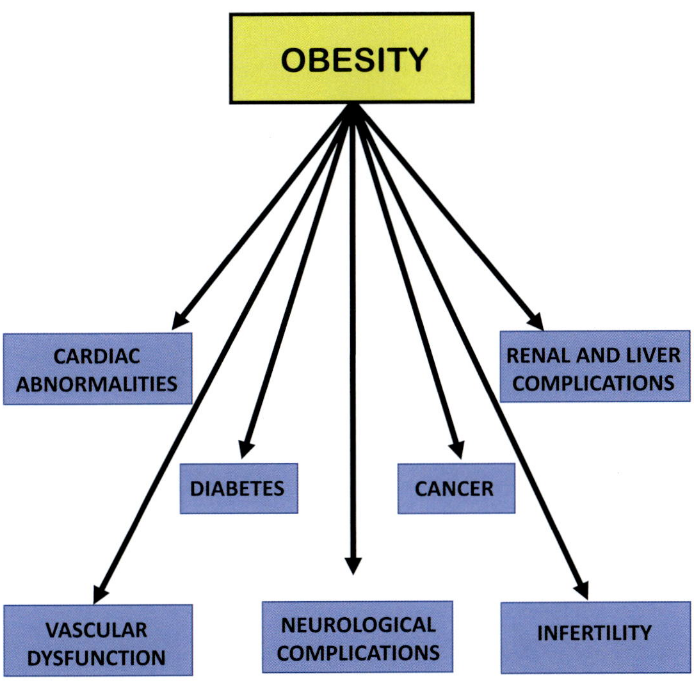

Fig. 1.2 Obesity-induced health complications

have demonstrated a link between obesity and several forms of CVD including coronary heart disease, heart failure, hypertension, stroke, atrial fibrillation and sudden cardiac death [59]. Having obesity can accelerate the progression of atherosclerosis as well as induce adverse ventricular remodeling [60]. Being overweight, with central accumulation of adipose tissues also increases the predisposition to CVD [61, 62]. Structural and functional adaptations of the cardiovascular system occur in response to excess body weight and to adipokine mediated inflammation and vascular abnormalities [59]. Furthermore, the activation of neurohormonal systems i.e. the sympathetic nervous system and the renin angiotensin system results in endothelial dysfunction, coronary calcification and activation of coagulation and play a significant role in the progression of cardiovascular complications [36]; in fact, obesity-induced CVD complications can also occur indirectly as a consequence of hyperglycemia and abnormal blood lipid profile.

It should be noted that obesity-related CVD has also emerged as a major health concern in children mainly due to the increasing incidence of childhood obesity; indeed, obese children have an elevated risk of cardiovascular morbidity and mortality [63]. CVD is also disproportionately higher in racial and ethnic minority populations [64]; in fact, obese African-American women are twice as likely to present with CVD [65]. While the aforementioned has described the relationship between obesity and increased risk for CVD, a better CVD prognosis for some overweight or obese individuals has been reported [66]. This has been referred to as a phenomenon known as the obesity paradox. This paradoxical observation is more evident in overweight individuals and those being classified as class I obese, but less apparent in more severe obesity. However, more exploration is required to fully comprehend the nature and the mechanisms involved in the obesity paradox.

Renal Disease

Obesity is a direct risk factor for chronic kidney disease (CKD) as well as acute kidney injury [67]. Obesity can also increase the risk of CKD indirectly as it increases the risk of diabetes and hypertension, which are considered as the most prominent risk factors for CKD. Obesity increases metabolic demand and thus causes an increase in the workload of the kidneys and subsequent hyperfiltration, which ultimately results in functional decline of the kidneys and CKD [68, 69]. It is interesting that obese patients with end-stage renal disease (ESRD), the obesity itself is associated with better survival in ESRD that has been linked also to race/ethnicity. In this regard, in a study involving African American hemodialysis patients it was found that overweight (BMI >27.5 kg/m^2) hemodialysis patients had higher survival rates than patients with normal BMI [70]. This observation has led some researchers to suggest that this may be related to the obesity paradox [71].

Liver Disease

Obesity increases the risk of nonalcoholic fatty liver disease (NAFLD), which is rapidly emerging as the most common cause of chronic liver disease due to the increase in the incidence of obesity. In fact, the risk of NAFLD increases with increasing BMI [72]. The major characteristic of NAFLD is steatosis (fatty liver) and occurs as a result of an imbalance between higher rates of hepatic fatty acid uptake and de novo fatty acid synthesis relative to the rate of fatty acid oxidation and export [73, 74]. This shift toward a net gain in the level of intrahepatic triglyceride initiates several other metabolic complications including defective glucose, fatty acid and lipoprotein metabolism that ultimately results in insulin resistance, dyslipidemia, and other cardiometabolic complications [75]. It is interesting to note that racial/ethnic background has been suggested to influence the association between BMI and NAFLD [76–78].

Cancer

An increase in the risk of certain types of cancers has been linked to being overweight and obese [79, 80]. For example, obesity is reported to be associated with increased cancers of the colon, breast (post-menopausal women), endometrium (the lining of the uterus), kidney, esophagus, gallbladder, ovaries, and pancreas [81, 82]. In the U.S. about 630,000 people in 2014 were diagnosed with overweight and obesity-related cancer. Several mechanisms have been proposed for the occurrence of obesity-induced cancers. In this regard, low-grade inflammation [83] can cause abnormalities in the chemical structure in DNA causing a dysregulation of cell growth; similarly, excessive amounts of estrogen produced by adipose tissue increase the risk for breast, endometrial and ovarian cancer through the growth of cancer cells, while the increased levels of insulin-like growth factor-1 can cause the development of colon, kidney, prostate and endometrial cancers by promoting a shift toward cell proliferation, suppression of cellular apoptosis and inducing angiogenesis that supports abnormal cell growth and tumor development [84–87]. Leptin, which is increased in obesity, can also induce abnormal cell proliferation [88], while on the other hand, adiponectin has been reported to exert an antiproliferative effect [89]. Both these hormones are altered in obesity such that the balance between them may induce abhorrent cell growth.

Male and Female Infertility

Infertility can occur in both obese men and women. A three-fold increase in the risk of infertility has been reported in obese women that has been suggested to be due to obesity-related disruption of normal ovulation, endometrial and embryo development [90, 91]. Leptin has also been implicated in reducing fertility in obese women [92], where high levels of leptin cause disruption of endometrial maturation and embryo implantation into the endometrium. A contributory factor is also that leptin can inhibit the production of progesterone [92], which is required for endometrial development. Polycystic ovary syndrome (PCOS) is associated with obesity and has also been linked to infertility [90, 91, 93]. Obesity and infertility is not restricted to women because some studies have shown risk of infertility is higher in obese men [94] due to a decrease in circulating free testosterone levels that results in a low sperm count [95]. Furthermore, obesity-associated alterations in male hormones can also lead to erectile dysfunction, which is considered as the major cause of infertility in obese men [96].

Mobility and Skeletal Muscle Dysfunction

Obesity causes a chronic overload stress on the muscles of the legs that leads to functional difficulties including loss of muscle strength, postural and dynamic imbalance and mobility problems [97]. Low-grade inflammation in obesity is considered to cause musculoskeletal complications [98]. Depressed skeletal muscle contractile activity and reduced mobility in obesity have been linked to defective myocyte Ca^{2+}-signaling that impairs the excitation–contraction coupling as well as excitation-transcription coupling [99]. Age-related muscle loss (sarcopenia) is associated with obesity has been termed as sarcopenic obesity [100] and is related to mobility limitations. Overall, although obese, elderly ($\geq$60 years of age) are at higher risk for mobility disability, obese women are at greater risk for mobility limitations as compared to men [101].

Osteoarthritis is a prevalent joint condition that mostly affects the joints of the knee, hip and back and is associated with pain and disability [102]. It is the foremost reason for disability and its incidence is rising due to increasing obesity and the aging population [103]. Being overweight or obese the pressure to the joints is increased which causes a degeneration of the cartilage at a greater rate than what would otherwise occur [104]. The pathogenesis of osteoarthritis is thought to involve the production of pro-inflammatory cytokines (interleukin-1, interleukin-6 and tumor necrosis factor-α) [105]. Interestingly, this debilitating condition is more likely to develop in African-American women and has been linked to lifestyle and increased BMI in this population cohort [106]. Overall, osteoarthritis and consequent mobility issues have emerged as a global health concern as this disease is reported to affect more than half of the 65 years and older aged population [106].

Strategies for Prevention and Treatment of Obesity

Nutrition and Lifestyle

The management of obesity as a chronic disease has to put the emphasis on improving overall well-being that includes both physical and mental health [107]. A number of different approaches that can be tailored to the individual for the prevention/treatment (reversal) of obesity can be undertaken and are summarized in Fig. 1.3. Weight reduction, or at least the prevention of additional weight gain should be considered as the first line approach of obesity management. Such a regimen is aimed to reduce energy intake and increasing physical activity to increase energy expenditure to facilitate fat loss [108]. For this approach to be effective, a concerted effort is required; however, there are several factors that can delay or restrict benefits such as having available time, low motivation, other commitments and physical incapacity [109]. It is interesting to note that older aged males tend to be more compliant with lifestyle interventions in order to achieve weight reduction [110]. There are several components to the regimen of lifestyle/behavioral change in order to achieve a sustained loss in weight. This includes one hour of physical activity/day, consistently consuming a low calorie, low fat diet and monitoring of body weight [111, 112]. In this regard,

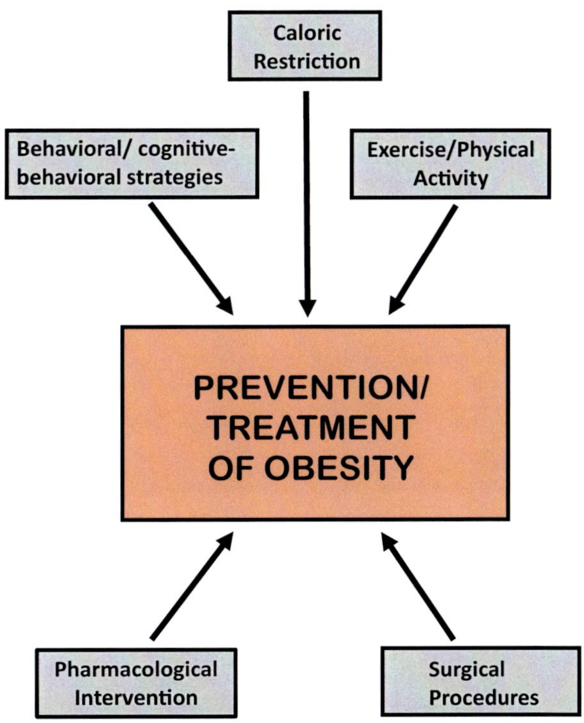

Fig. 1.3 Strategies for the prevention and treatment of obesity

if weight loss can be maintained for two to five years, the likelihood of longer-term success is increased [111]. Caloric restriction (450–800 kcal/day) has been used in obesity management for weight loss; however, the effectiveness and challenges of this approach for long-term sustained weight loss have been questioned [113–115].

Pharmacological Interventions

If behavioral and lifestyle changes alone do not result in weight loss, alternative and/or complementary therapies should be instituted. Pharmacological interventions for obesity management are generally only recommended for individuals with a BMI of $\geq$30 kg/m^2, or $\geq$27 kg/m^2 with one or more co-existing diseases such as hypertension, dyslipidemia, or type 2 diabetes. The Food and Drugs Administration of the US currently approves five anti-obesity medications, each exhibiting different mechanisms of action [36]. All of these medications induce weight loss and are associated with improved cardiometabolic risk factors [116]. Orlistat, is the earliest approved obesity drug and inhibits pancreatic lipase and the digestion of dietary triglycerides (TG). Glucagon like peptide-1 (GLP-1) is an incretin secreted by the intestinal tract, which reduces feeding and improves metabolic parameters linked to obesity. Thus, GLP-1 receptors have emerged as a promising target in the treatment of metabolic disorders [39]. In fact. Liraglutide, a long-acting GLP-1 analogue receptor agonist is used in obese individuals with a BMI >27.0 K/m^2 and with weight related comorbid conditions for the management of their chronic weight [36].

The 5-HT 2C receptor agonist, Lorcaserin, is used in overweight or obesity as an adjunct to a reduced calorie intake and increased physical activity and if the individual has at least one weight-related comorbid disease such as hypertension, diabetes and dyslipidemia. It acts on the 5-HT receptors in the hypothalamus to stimulate the satiety centres as well as by activating anorexigenic pathways to reduce appetite [36]. The combination therapy of naltrexone-bupropion also acts centrally (in the hypothalamus) and is used for long-term weight management in obese individuals that also have obesity-related comorbidities as it is a sustained release formulation [36]. The combination of immediate-release phentermine with extended-release topiramate is an amphetamine analog stimulant for the short-term treatment of obesity in overweight or obese adults with at least one concomitant adverse condition and is used in combination with a low-calorie diet and increased physical activity. Phentermine antagonizes the α-adrenoceptors in the hypothalamus, resulting in an increase in leptin levels, which, in turn, suppresses appetite; while topiramate increases the activity of γ-aminobutyric acid to suppress appetite [36].

Several other potential candidate targets for treatment of obesity have been proposed. Despite the development of resistance to leptin, it can still be explored as a potential target for the management of obesity [117]. Hypothalamic AMP-activated protein kinase (AMPK) directly influences feeding behaviour and thus has been suggested as a target for the treatment of obesity as it regulates energy intake as well as energy expenditure [118]. The sympathetic nervous system activity is also associated with energy balance. In this regard, sympathomimetic agents can decrease food intake and increase resting metabolic rate [119]. Beneficial shits in the composition of the gut microflora through the use of prebiotics and probiotics could provide a strategy for the management of obesity [120]. However, this notion is controversial as studies involving the gut microbiota composition of obese individuals suggest that a simple modification of the gut microflora may not be effective in the treatment of obesity [121].

Surgical Approaches

Generally, obese individuals with a BMI >40 kg/m^2 or >35 kg/m^2 with co-morbidities such as type 2 diabetes, dyslipidemia, hypertension and renal disease are eligible candidates for bariatric surgery for sustained weight loss [122]. The most common bariatric procedures are gastric banding, sleeve gastrectomy, and gastric bypass and are performed to reduce the size of the stomach such that less food and therefore calories are consumed; However, changes in the circulating levels of gastrointestinal hormones as well as the gut microbiome may also contribute to energy homeostasis and weight loss [123, 124].

In severe cases of obesity, bariatric surgeries maybe performed; however, they are not without risk of several complications [125]. In this regard, malabsorption issues following such type of surgeries, nutrient deficiencies may occur and thus supplementation would be necessary [126, 127]. In addition, the bioavailability of oral drugs can also be impacted, which would require dosage monitoring following surgery [128]. It is interesting to note that while surgical procedures exert an initial loss in weight, a subsequent plateau or even weight gain can occur after surgery [129–131]. In these cases, lifestyle interventions would need to be implemented as an adjunct to surgery if weight loss is to be sustained.

Conclusion

Obesity as a chronic metabolic disease and major health hazard is much more than being a condition that arises as consequence of increased nutrient intake [132]. Both adults and children can be afflicted with being overweight or obese; indeed, obesity is a global public health problem of great magnitude [133] as it can also lead to the development of several complications, as described in this chapter, that in turn exacerbates the threat to global health. For an effective strategy for the prevention, treatment and reversal of obesity and associated co-morbidities, sex, ethnicity and education/socioeconomic status must be included in the discussion. From the aforementioned while weight loss through a combination of different lifestyle factors including reduction in food intake, regular physical activity and stress management can be seen to be appropriate initial approaches, which if necessary can evolve to include pharmacological agents, and in the worst-case scenario surgical procedures can be undertaken in extreme cases and if the individual is a suitable candidate. Understanding the pathogenesis of obesity as well as the pathophysiological mechanisms that cause obesity-related health complications will assist in the advancement specific treatment/prevention modalities. It is evident that obesity is largely a preventable condition; however, with the stigma of obesity and its impact on society, economy and the individual, a concerted and meaningful effort needs to be undertaken to reduce the global burden of obesity.

Acknowledgements The infrastructural support for this work was provided by the St. Boniface Hospital Albrechtsen Research Centre, Winnipeg, Canada.

References

1. Pollack A (2013) AMA recognizes obesity as a disease. The New York Times, 18 June 2013
2. Rich P (2015) CMA recognizes obesity as a disease. Canadian Medical Association, 9 Oct 2015
3. World Health Organization (2021) Obesity and overweight. https://www.who.int/en/news-room/fact-sheets/detail/obesity-and-overweight. Accessed 3 Feb 2021
4. Sharma AM, Campbell-Scherer DL (2017) Redefining obesity: beyond the numbers. Obesity 25:660–661
5. Campfield LA, Smith FJ (1999) The pathogenesis of obesity. Baillieres Best Pract Res Clin Endocrinol Metab 13:13–30
6. Dugani S, Gaziano TA (2016) 25 by 25: achieving global reduction in cardiovascular mortality. Curr Cardiol Res 18:10
7. Tappia PS, Ramjiawan B, Dhalla NS (eds) (2020) Pathophysiology of obesity-induced health complications. Springer Nature Switzerland AG
8. Tappia PS, Bhullar SK, Dhalla NS (eds) (2020) Biochemistry of cardiovascular disease. Springer Nature Switzerland AG
9. Church TS, Thomas DM, Tudor-Locke C et al (2011) Trends over 5 decades in U.S. occupation-related physical activity and their associations with obesity. PLoS One 6:e19657
10. Wellard L, Havill M, Hughes C et al (2015) Energy-dense fast food products cost less: an observational study of the energy density and energy cost of Australian fast foods. Aust N Z J Public Health 39:544–545
11. Moubarac JC, Parra DC, Cannon G, Monteiro CA (2014) Food classification systems based on food processing: significance and implications for policies and actions: a systematic literature review and assessment. Curr Obes Rep 3:256–272
12. Marrón-Ponce JA, Tolentino-Mayo L, Hernández-F M, Batis C (2018) Trends in ultra-processed food purchases from 1984 to 2016 in Mexican households. Nutrients 11
13. Solberg SL, Terragni L, Granheim SI (2016) Ultra-processed food purchases in Norway: a quantitative study on a representative sample of food retailers. Ultra-processed food purchases in Norway: a quantitative study on a representative sample of food retailers. Public Health Nutr 19:1990–2001
14. Martínez Steele E, Baraldi LG, Louzada ML et al (2016) Ultra-processed foods and added sugars in the US diet: evidence from a nationally representative cross-sectional study. BMJ Open 6:e009892
15. Adams J, White M (2015) Characterization of UK diets according to degree of food processing and associations with socio-demographics and obesity: cross-sectional analysis of UK National Diet and Nutrition Survey (2008–12). Int J Behav Nutr Phys Act 12:160

16. Julia C, Martinez L, Allès B et al (2018) Contribution of ultra-processed foods in the diet of adults from the French NutriNet-Santé study. Public Health Nutr 21:27–37
17. Louzada ML, Baraldi LG, Steele EM et al (2015) Consumption of ultra-processed foods and obesity in Brazilian adolescents and adults. Prev Med 81:9–15
18. Mendonça RD, Pimenta AM, Gea A, de la Fuente-Arrillaga C et al (2016) Ultraprocessed food consumption and risk of overweight and obesity: the University of Navarra Follow-Up (SUN) cohort study. Am J Clin Nutr 104:1433–1440
19. Canella DS, Levy RB, Martins AP et al (2014) Ultra-processed food products and obesity in Brazilian households (2008–2009). PLoS One 9:e92752
20. Hall KD, Ayuketah A, Brychta R et al (2019) Ultra-processed diets cause excess calorie intake and weight gain: an inpatient randomized controlled trial of ad libitum food intake. Cell Metab 30:67-77.e3
21. Piaggi P (2019) Metabolic determinants of weight gain in humans. Obesity 27:691–699
22. Cuthbertson DJ, Steele T, Wilding JP, Halford JC, Harrold JA, Hamer M, Karpe F (2017) What have human experimental overfeeding studies taught us about adipose tissue expansion and susceptibility to obesity and metabolic complications? Int J Obes 41:853–865
23. Livingstone MB, Pourshahidi LK (2014) Portion size and obesity. Adv Nutr 5:829–834
24. Nielsen SJ, Popkin BM (2003) Patterns and trends in food portion sizes, 1977–1998. JAMA 289:450–453
25. Zheng M, Rangan A, Meertens B, Wu JHY (2017) Changes in typical portion sizes of commonly consumed discretionary foods among Australian adults from 1995 to 2011–2012. Nutrients 9
26. Ello-Martin JA, Ledikwe JH, Rolls BJ (2005) The influence of food portion size and energy density on energy intake: implications for weight management. Am J Clin Nutr 82:236S-241S
27. Iacobini C, Pugliese G, Fantauzzi CB et al (2019) Metabolically healthy versus metabolically unhealthy obesity. Metabolism 92:51–60
28. Bluher M (2014) Predisposition—obesity phenotype. Dtsch Med Wochenschr 139:116–120
29. Tsatsoulis A, Paschou SA (2020) Metabolically health obesity: criteria, epidemiology, controversies, and consequences. Curr Obes Rep 9:109–120
30. Engin A (2017) The definition and prevalence of obesity and metabolic syndrome. Adv Exp Med Biol 960:1–17
31. Yu Y-H (2017) Making sense of metabolic obesity and hedonic obesity. Diabetes 9:656–666
32. Heindel JJ, Blumberg B, Cave M et al (2017) Metabolism disrupting chemicals and metabolic disorders. Reprod Toxicol 68:3–33
33. Heinonen S, Jokinen R, Rissanen A, Pietilainen KH (2020) White adipose tissue mitochondrial metabolism in health and in obesity. Obes Rev 21:e12958
34. Bournat JC, Brown CW (2010) Mitochondrial dysfunction in obesity. Curr Opin Endocrinol Diabetes Care 17:446–452
35. Pataky Z, Bobbioni-Harsch E, Golay A (2010) Obesity: a complex growing challenge. Exp Clin Endocrinol Diabetes 118:427–433
36. Cercato C, Fonseca FA (2019) Cardiovascular risk and obesity. Diabetol Metab Syndr 11:74
37. van der Klaauw AA (2018) Neuropeptides in obesity and metabolic disease. Clin Chem 64:173–182
38. Liu J, Yang X, Yu S, Zheng R (2018) The leptin resistance. Adv Exp Med Biol 1090:145–163
39. Gonzalez-Garcia I, Milbank E, Dieguez C et al (2019) Glucagon, GLP-1 and thermogenesis. Int J Mol Sci 20:3445
40. Batsis JA, Zagaria AB (2018) Addressing obesity in aging patients. Med Clin North Am 102:65–85
41. Apostolopoulou M, Savopoulos C, Michalakis K et al (2012) Age, weight and obesity. Maturitas 71:115–119
42. JafariNasabian P, Inglis JE, Reilly W et al (2017) Aging human body: changes in bone, muscle and body fat with consequent changes in nutrient intake. J Endocrinol 234:R37–R51
43. Palermo A, Tuccinardi D, Defeudis G et al (2016) BMI and BMD: the potential interplay between obesity and bone fragility. Int J Environ Res Public Health 13:544

44. Studenski SA, Peters KW, Alley DE et al (2014) The FNIH sarcopenia project: rationale, study description, conference recommendations, and final estimates. J Gerontol A Biol Sci Med Sci 69:547–558
45. Reddon H, Patel Y, Turcotte M et al (2018) Revisiting the evolutionary origins of obesity: lazy versus peppy-thrifty genotype hypothesis. Obes Rev 19:1525–1543
46. Goodarzi MO (2018) Genetics of obesity: what genetic association studies have taught us about the biology of obesity and its complications. Lancet Diabetes Endocrinol 6:223–236
47. Brandkvist M, Bjørngaard JH, Ødegård RA et al (2019) Quantifying the impact of genes on body mass index during the obesity epidemic: longitudinal findings from the HUNT Study. BMJ 366:l4067
48. den Hoed M, Brage S, Zhao JH et al (2013) Heritability of objectively assessed daily physical activity and sedentary behavior. Am J Clin Nutr 98:1317–1325
49. Farias MM, Cuevas AM, Rodriguez F (2011) Set-point theory and obesity. Metab Syndr Relat Disord 9:85–89
50. Stöger R (2008) The thrifty epigenotype: an acquired and heritable predisposition for obesity and diabetes? BioEssays 30:156–166
51. Speakman JR (2007) A nonadaptive scenario explaining the genetic predisposition to obesity: the "predation release" hypothesis. Cell Metab 6:5–12
52. Sellayah D, Cagampang FR, Cox RD (2014) On the evolutionary origins of obesity: a new hypothesis. Endocrinology 155:1573–1588
53. Belsare PV, Watve MG, Ghaskadbi SS, Bhat DS, Yajnik CS, Jog M (2010) Metabolic syndrome: aggression control mechanisms gone out of control. Med Hypotheses 74:578–589
54. Wadden TA, Stunkard AJ (1987) Psychopathology and obesity. Ann N Y Acad Sci 499:55–65
55. Blundell JE (1990) Appetite disturbance and the problems of overweight. Drugs 39(Suppl. 3):1–19
56. Hemmingsson E (2014) A new model of the role of psycholical and emotional distress in promoting obesity: conceptual review with implications for treatment and prevention. Obes Rev 15:769–779
57. Jauch-Chara K, Oltmanns KM (2014) Obesity—a neuropsychological disease? Systematic review and neuropsychological model. Prog Neurobiol 114:84–101
58. Wagner KH, Brath H (2012) A global view on the development of non communicable diseases. Prev Med 54(Suppl):S38–S41
59. Koliaki C, Liatis S, Kokkrinos A (2019) Obesity and cardiovascular disease revisiting an old relationship. Metabolism 92:98–107
60. Kachur S, Lavie CJ, de Schutter A et al (2017) Obesity and cardiovascular diseases. Minerva Med 108:212–228
61. Van Gaal LF, Mertens IL, De Block CE (2006) Mechanisms linking obesity with cardiovascular disease. Nature 444:875–880
62. Din-Dzietham R, Liu Y, Bielo MV et al (2007) High blood pressure trends in children and adolescents in national surveys, 1963 to 2002. Circulation 116:1488–1496
63. Cote AT, Harris KC, Panagiotopoulos C et al (2013) Childhood obesity and cardiovascular dysfunction. J Am Coll Cardiol 62:1309–1319
64. Kurian AK, Cardarelli KM (2007) Racial and ethnic differences in cardiovascular disease risk factors: a systematic review. Ethn Dis 17:143–152
65. Franklin NC, Arena R (2016) Personalized weight management interventions for cardiovascular risk reduction: a viable option for African-American women. Prog Cardiovasc Dis 58:595–604
66. Elagizi A, Kachur S, Lavie CJ (2018) An overview and update on obesity and the obesity paradox in cardiovascular diseases. Prog Cardiovasc Dis 61:142–150
67. Danziger J, Chen KP, Lee J et al (2016) Obesity, acute kidney injury, and mortality in critical illness. Crit Care Med 44:328–334
68. Garcio-Carro C, Vergara A, Bermejo S et al (2021) A nephrologist perspective on obesity: from kidney injury to clinical management. Front Med 8:655871

69. Martin-Taboada M, Vila-Bedmar R, Medina-Gomez G (2021) From obesity to chronic kidney disease: how can adipose tissue affect renal function? Nephron 21
70. Fleischmann E, Teal N, Dudley J et al (1999) Influence of excess weight in mortality and hospital stay in 1346 hemodialysis patients. Kidney Int 55:1560–1567
71. Kleine C-E, Moradi H, Streja E, Kalantar-Zadeh K (2018) Racial and ethnic disparities in the obesity paradox. Am J Kidney Dis 72(Suppl. 1):S26–S32
72. Wang S, Zhang J, Zhang J et al (2020) A cohort study on the correlation between body mass index trajectories and new-onset non-alcoholic fatty liver disease. Zhonghua Gan Zang Bing Za Zhi 28:597–602
73. Ibrahim SH, Kohli R, Gores GJ (2011) Mechanisms of lipotoxicity in NAFLD and clinical implications. J Pediatr Gastroenterol Nutr 53:131–140
74. Alves-Bezerra M, Cohen DE (2017) Trglyceride metabolism in the liver. Compr Physiol 8:1–8
75. Fabbrini E, Sullivan S, Klein S (2010) Obesity and nonalcoholic fatty liver disease: biochemical, metabolic, and clinical implications. Hepatology 51:679–689
76. Petersen KF, Dufour S, Feng J et al (2006) Increased prevalence of insulin resistance and nonalcoholic fatty liver disease in Asian-Indian men. Proc Natl Acad Sci USA 103:18273–18277
77. Romeo S, Kozlitina J, Xing C et al (2008) Genetic variation in PNPLA3 confers susceptibility to nonalcoholic fatty liver disease. Nat Genet 40:1461–1465
78. Browning JD, Szczepaniak LS, Dobbins R et al (2004) Prevalence of hepatic steatosis in an urban population in the United States: impact of ethnicity. Hepatology 40:1387–1395
79. Nieman KM, Romero IL, Van Houten B, Lengyel E (2013) Adipose tissue and adipocytes support tumorigenesis and metastasis. Biochim Biophys Acta 183:1533–1541
80. Smith KB, Smith MS (2016) Obesity statistics. Prim Care 43:121–135
81. De Pergola G, Silvestris F (2013) Obesity as a major risk factor for cancer. J Obes 2013:291546
82. Pischon T, Nimptsch K (2016) Obesity and risk of cancer: an introductory overview. Recent Results Cancer Res 208:1–15
83. Iyengar NM, Gucalp A, Dannenberg AJ, Hudis CA (2016) Obesity and cancer mechanisms: tumor microenvironment and inflammation. J Clin Oncol 34:4270–4276
84. Markowska A, Stanisławiak-Rudowicz J, Jaszczyńska-Nowinka K (2017) The role of overweight and obesity in selected gynecological malignancies. Eur J Gynaecol Oncol 38:335–341
85. O'Flanagan CH, Bowers LW, Hursting SD (2015) A weighty problem: metabolic perturbations and the obesity-cancer link. Horm Mol Biol Clin Investig 23:47–57
86. Kaaks R (2001) Plasma insulin, IGF-I and breast cancer. Gynecol Obstet Fertil 29:185–191
87. Stoll BA (1998) Teenage obesity in relation to breast cancer risk. Int J Obes Relat Metab Disord 22:1035–1040
88. Xu Y, Tan M, Tian X et al (2020) Leptin receptor mediates the proliferation and glucose metabolism of pancreatic cancer cells via AKT pathway activation. Mol Med Rep 21:945–952
89. Ramzan AA, Bitler BG, Hicks D et al (2019) Adiponectin receptor agonist AdipoRon induces apoptotic cell death and suppresses proliferation in human ovarian cancer cells. Mol Cell Biochem 461:37–46
90. Silvestris E, de Pergola G, Rosania R, Loverro G (2018) Obesity as disruptor of the female fertility. Reprod Biol Endocrinol 16:22
91. Broughton DE, Moley KH (2017) Obesity and female infertility: potential mediators of obesity's impact. Fertil Steril 107:840–847
92. Lin Q, Poon SL, Chen J et al (2009) Leptin interferes with 3′,5′-cyclic adenosine monophosphate (cAMP) signaling to inhibit steroidogenesis in human granulosa cells. Reprod Biol Endocrinol 7:115
93. Escobar-Morreale HF (2018) Polycystic ovary syndrome: definition, aetiology, diagnosis and treatment. Nat Rev Endocrinol 14:270–284
94. Mah PM, Wittert GA (2010) Obesity and testicular function. Mol Cell Endocrinol 316:180–186

95. Salas-Huetos A, Maghsoumi-Norouzabad L, James ER et al (2021) Male adiposity, sperm parameters and reproductive hormones: an updated systematic review and collaborative meta-analysis. Obes Rev 22:e13082
96. Rastrelli G, Lotti F, Reisman Y et al (2019) Metabolically healthy and unhealthy obesity in erectile dysfunction and male infertility. Expert Rev Endocrinol Metab 14:321–334
97. Tomlinson DJ, Erskine RM, Morse CI et al (2016) The impact of obesity on skeletal muscle strength and structure through adolescence to old age. Biogerontology 17:467–483
98. Collins KH, Herzog W, MacDonald GZ et al (2018) Obesity, metabolic syndrome, and musculoskeletal disease: common inflammatory pathways suggest a central role for loss of muscle integrity. Front Physiol 9:112
99. Tallis J, James RS, Seebacher F (2018) The effects of obesity on skeletal muscle contractile function. J Exp Biol 22:jeb163840
100. Goisser S, Kemmler W, Porzel S et al (2015) Sarcopenic obesity and complex interventions with nutrition and exercise in community-dwelling older persons—a narrative review. Clin Interv Aging 10:1267–1282
101. Vincent HK, Vincent KR, Lamb KR (2010) Obesity and mobility disability in the older adult. Obes Rev 11:568–579
102. Chen D, Shen J, Zhao W et al (2017) Osteoarthritis: toward a comprehensive understanding of pathological mechanism. Bone Res 5:16044
103. Johnson VL, Hunter DJ (2014) The epidemiology of osteoarthritis. Best Pract Res Clin Rheumatol 28:5–15
104. Kulkarni K, Karssiens T, Kumar V, Pandit H (2016) Obesity and osteoarthritis. Maturitas 89:22–28
105. Wang T, He C (2018) Pro-inflammatory cytokines: the link between obesity and osteoarthritis. Cytokine Growth Factor Rev 44:38–50
106. Musumeci G, Aiello FC, Szychlinska MA et al (2015) Osteoarthritis in the XXIst century: risk factors and behaviours that influence disease onset and progression. Int J Mol Sci 16:6093–6112
107. Obesity Canada (2019) https://obesitycanada.ca/managing-obesity/. Accessed 10 July 2019
108. Semlitsch T, Stigler FL, Jeitler K et al (2019) Management of overweight and obesity in primary care—a systematic overview of international evidence-based guidelines. Obes Rev. https://doi.org/10.1111/obr.12889 (Epub ahead of print)
109. Sweeting AN, Caterson ID (2017) Approaches to obesity management. Intern Med J 47:734–739
110. Burgess E, Hassmén P, Pumpa KL (2017) Determinants of adherence to lifestyle intervention in adults with obesity: a systematic review. Clin Obes 7:123–135
111. Wing RR, Phelan S (2005) Long-term weight loss maintenance. Am J Clin Nutr 82:222S-225S
112. Thomas JG, Bond DS, Phelan S et al (2014) Weight-loss maintenance for 10 years in the National Weight Control Registry. Am J Prev Med 46:17–23
113. Parretti HM, Jebb SA, Johns DJ et al (2016) Clinical effectiveness of very-low-energy diets in the management of weight loss: a systematic review and meta-analysis of randomized controlled trials. Obes Rev 17:225–234
114. Fothergill E, Guo J, Howard L et al (2016) Persistent metabolic adaptation 6 years after "The Biggest Loser" competition. Obesity (Silver Spring) 24:1612–1619
115. Hansen TT, Hjorth MF, Sandby K et al (2019) Predictors of successful weight loss with relative maintenance of fat-free mass in individuals with overweight and obesity on an 8-week low-energy diet. Br J Nutr 27:1–12
116. Daneschvar HL, Aronson MD, Smetana GW (2016) FDA-approved anti-obesity drugs in the United States. Am J Med 129(879):e1–e6
117. Cui H, Lopez M, Rahmouni K (2017) The cellular and molecular bases of leptin and ghrelin resistance in obesity. Nat Rev Endocrinol 13:338–351
118. Lopez M (2018) Hypothalamic AMPK and energy balance. Eur J Clin Invest 48:e12996
119. Tentolouris N, Liatis S, Katsilambros N (2006) Sympathetic system activity in obesity and metabolic syndrome. Ann N Y Acad Sci 1083:129–152

120. DiBaise JK, Zhang H, Crowell MD et al (2008) Gut microbiota and its possible relationship with obesity. Mayo Clin Proc 83:460–469
121. Bianchi F, Duque ALRF, Saad SMI, Sivieri K (2019) Gut microbiome approaches to treat obesity in humans. Appl Microbiol Biotechnol 103:1081–1094
122. International Federation for the Surgery of Obesity and Metabolic Disorders (2019) Are you a candidate. https://www.ifso.com/are-you-a-candidate/. Accessed 11 July 2019
123. Madsbad S, Dirksen C, Holst JJ (2014) Mechanisms of changes in glucose metabolism and bodyweight after bariatric surgery. Lancet Diabetes Endocrinol 2:152–164
124. Tremaroli V, Karlsson F, Werling M et al (2015) Gastroplasty induce long-term changes on the human gut microbiome contributing to fat mass regulation. Cell Metab 22:228–238
125. Ahmad A, Kornrich DB, Krasner H et al (2019) Prevalence of dumping syndrome after laparoscopic sleeve gastrectomy and comparison with laparoscopic Roux-en-Y gastric bypass. Obes Surg 29:1506–1513
126. Stroh C, Manger T, Benedix F (2017) Metabolic surgery and nutritional deficiencies. Minerva Chir 72:432–441
127. Gregory NS (2017) The effects of bariatric surgery on bone metabolism. Endocrinol Metab Clin North Am 46:105–116
128. Angeles PC, Robertsen I, Seeberg LT et al (2019) The influence of bariatric surgery on oral drug bioavailability in patients with obesity: a systematic review. Obes Rev. https://doi.org/10.1111/obr.12869 (Epub ahead of print)
129. Lamarca F, Melendez-Araújo MS, Porto de Toledo I et al (2019) Relative energy expenditure decreases during the first year after bariatric surgery: a systematic review and meta-analysis. Obes Surg. https://doi.org/10.1007/s11695-019-03934-0 (Epub ahead of print)
130. Chang SH, Stoll CR, Song J et al (2014) The effectiveness and risks of bariatric surgery: an updated systematic review and meta-analysis, 2003–2012. JAMA Surg 149:275–287
131. Gloy VL, Briel M, Bhatt DL et al (2013) Bariatric surgery versus non-surgical treatment for obesity: a systematic review and meta-analysis of randomised controlled trials. BMJ 347:f5934
132. Symonds ME, Budge H, Frazier-Wood AC (2013) Epigenetics and obesity: a relationship waiting to be explained. Hum Hered 75:90–97
133. Aller EEJG, Abete I, Astrup A et al (2011) Starches, sugars and obesity. Nutrients 3:341–369

Chapter 2
Endocrine Role of Adipose Tissue in Obesity and Related Disorders

Shravanthi S. Kumar, Alok Kumar Mishra, and Asit Ranjan Ghosh

Abstract Adipose tissue is a complex network of connective tissue constituting many cells types. It is a tissue whose role was assumed to be limited to storage and metabolism of energy and insulation. However the discovery of leptin followed by further elucidation of numerous hormones and factors secreted by adipose tissue and roles in metabolism and cell regulation it is regarded as an endocrine organ. Adipose tissue regulates the energy balance being influenced by either ligand-mediated transcription factors (C/EBP, PPARγ, RXR) or with other types of transcription factors with the secretion of several different signalling molecules (adiponectin, leptin etc.). The receptors are expressed onto different organs. Thus, these signalling molecules are involved in multiple function of the body. While there is a break of the regulatory discipline of adipogenesis, the subject becomes obese. Due to these phenomenal changes, fat cross the site of adipose tissue to get accumulated in non-adipose multiple tissues of different organs which finally results fat-associated toxic environment consequently leading to lipotoxicity; erratic glucose and fat metabolism in different organs. Energy regulation by the endocrine adipose tissue therefore has so far been identified as the key factor to combat obesity and associated diseases.

Keywords Adipose tissue · Adipogenesis · Endocrine organ · Leptin · Adipokine

Abbreviations

WAT	White adipose tissue
TAG	Triacylglycerol
MSC	Mesenchymal precursor cells
PPARγ	Peroxisome proliferator-activated receptor
C/EBP	CCAAT/enhancer-binding proteins

S. S. Kumar · A. K. Mishra · A. R. Ghosh (✉)
Department of Integrative Biology, School of Bio Sciences and Technology, Vellore Institute of Technology, Vellore 632014, Tamil Nadu, India
e-mail: asitranjanghosh@vit.ac.in

© Springer Nature Switzerland AG 2021

P. S. Tappia et al. (eds.), *Cellular and Biochemical Mechanisms of Obesity*, Advances in Biochemistry in Health and Disease 23,
https://doi.org/10.1007/978-3-030-84763-0_2

AMPK AMP activated protein kinase

Introduction

Adipose tissue is a complex network of connective tissue composing of different types of cells which include adipocytes, endothelial cells, fibroblasts, macrophages. It is also commonly known as fat. Anatomically fat or adipose tissue is distributed throughout our body and bears different naming like pericardial (heart), subcutaneous (skin), perirenal (kidney), gonadal (gonad), visceral (viscera), intramuscular (muscles), structural (joints, breast) etc. Before the discovery of its secretory function the role of adipose tissue was thought to be limited to storage of excess energy as fat and insulation to the body [1]. However it does not only contain adipocytes but other cells types which release many hormones, growth factors, proteins, enzymes, cytokines, complement factors into circulation and these are collectively called adipokines. The adipokines help in cellular regulation and metabolic functions in different target tissues (Fig. 2.1). Their roles are well-reported in regulating energy homeostasis and adipocyte differentiation, regulating vascularization and blood flow,

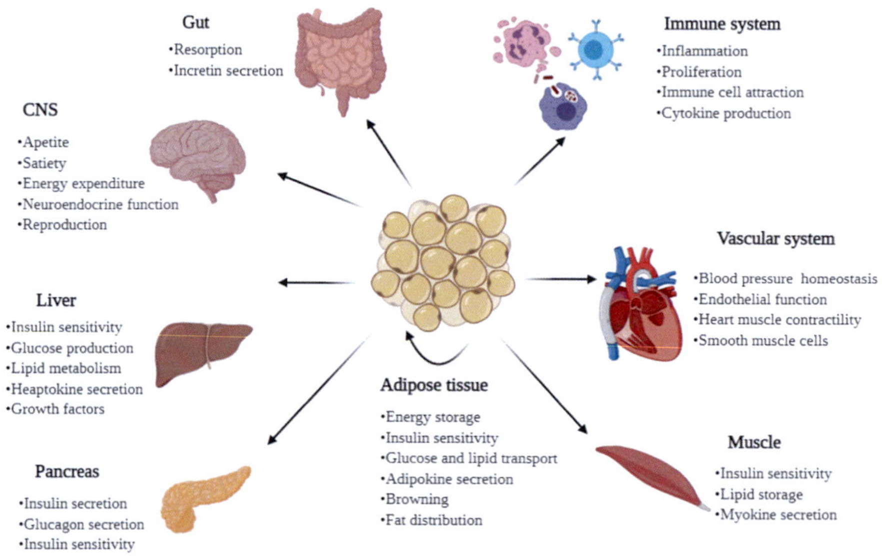

Fig. 2.1 Adipokine secretion helps maintain homeostasis and help in the regulation of biological processes in target organs like brain, liver, gut, muscle, vascular system, immune system, and pancreas. Hence adipose secretion help in maintenance of overall health. Impaired adipose tissue function leading to adverse secretion causes metabolic diseases like diabetes, cardiovascular diseases, inflammatory diseases and cancer

metabolism of lipid and cholesterol and regulation of immune system. Such an important biologics, Adipokines have a good grip in regulation of many metabolic functions in the body and thus its irregular secretions imparts in development and progression of diseases such as obesity, diabetes mellitus, cardiovascular diseases and other metabolic conditions. Hence it is considered as a dynamic organ which helps in regulation of diverse body functions [2].

Adipose tissue is a storehouse of energy which shows differences with morphology and physiology and tissue type and thus classified differently. Two major classes of Adipose tissue are brown adipose tissue (BAT) and white adipose tissue (WAT). BAT have multiple mitochondria and are multilocular cells and its function is energy dissipation to maintain thermal homeostasis whereas unilocular WAT have a central lipid droplet and it function is the storage of triglycerides in the body. The central role of adipose tissue was thought to be limited to energy storage and regulation as well as maintaining thermal homeostasis. However adipose tissue secretes many hormones and growth factors which help in cell regulations in different tissues and organs [3]. With the support of increasing reported insight into metabolic role of adipocytes points to the fact that it functions like an endocrine organ comprising different cell types, and regulate functions in many target tissue and organ through its secretions of paracrine factors and plays a central role in energy storage and cellular differentiation along with regulated immune response throughout the body [4]. The location of adipose tissue in the body determines the factors it secrets and the composition of cell types, number of adipocytes in the tissue and blood vessels and immune cells [5].

In this chapter, with the published evidence we demonstrate the multivariate functions of adipose tissue with particular discussion of its endocrinal potentials in health and diseases.

Adipogenesis

Mesenchymal precursor cells (MSCs) have the ability to commit to form different cell types. Both white and brown adipocyte are formed by MSCs present in the mesoderm but from different lineages. Adipogenic lineage which is composed of Myf5 negative cells produce white adipocytes while brown adipocytes are originated from Myf5 positive cells from myogenic lineage. Myf5 is the key myogenic regulatory factor. The WAT and BAT may have originate from different lineages but have common transcriptional cascades mainly PPARγ and C/EBPs leading to its differentiation [6].

Mature adipocytes are formed by the differentiation of pre adipocyte precursor cells by the sequential activation of cascade of transcription factors. The activation of transcription factors of the activation protein-1 (AP-1) family kick starts the maturation of adipocytes in WAT and induces PPARγ which is a critical transcription factor in promoting adipocyte differentiation. The mammalian target of rapamycin (mTOR) is also involved in adipocyte differentiation via nutrient signalling. Differentiation of pre-adipocytes is also influenced by STATs (signal transducer and activator of transcription), KLF (kruppel like factor) group of proteins, sterol response element–binding protein-1 (SREBP-1), and factors of the C/EBP (CCAAT-enhancer binding proteins) family. Peroxisome proliferator activated receptor γ (PPAR γ) is the key regulator of adipogenesis and is also necessary for maintenance of differentiation [7].

The mature white adipocytes are produced from Myf5 negative precursor cells with the influence of bone morphogenetic protein 2 (BMP2), bones morphogenetic protein 4 (BMP4), PPAR and CCAAT/enhancer-binding proteins. In case of BAT bone morphogenetic protein 7 (BMP7), PPAR and CCAAT/enhancer-binding proteins and also transcriptional co regulators PRDM116 (PR domain containing 16) and PGC-1α help in the differentiation from precursor cells to active brown adipocytes [8] (Fig. 2.2).

Differentiation of adipocytes is influenced by extracellular signals which can be signalling molecules or physiological factors. Insulin, glucocorticoids and bone

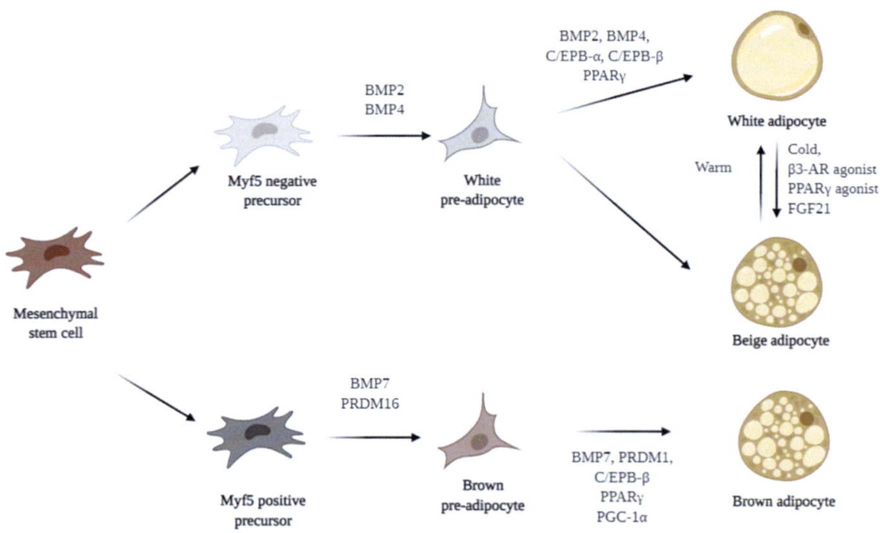

Fig. 2.2 Differentiation of Mesenchymal stem cells into mature adipocytes occurs in 2 steps. Firstly, commitment of MSCs into pre-adipocytes and secondly maturation into active adipocytes by terminal differentiation. The process involves numerous signalling pathways, transcription factors and genes. WAT and BAT originate from different lineage of MSCs namely adipogenic lineage (Myf5 negative) and myogenic lineage (Myf5 positive) respectively but are influenced by same transcriptional cascades of which, Peroxisome proliferator-activated receptor (PPARγ) and CCAAT/enhancer-binding proteins (C/EBPs) play the most important role. Bone morphogenetic protein 2 and bone morphogenetic protein help in commitment of Myf5 negative precursor cells into white pre adipocytes. Bone morphogenetic factor 7 with transcriptional co regulators PRDM1 (PR domain-containing 16) and PGC-1α (peroxisome proliferator activate receptor gamma coactivator 1 alpha) influence the conversion of Myf5 negative precursor cells into mature brown adipocytes. WAT also can be converted into beige adipocytes by cold exposure, β adrenergic agonist, or a PPARγ agonist, FGF21 (Fibroblast growth factor 21)

morphogenetic proteins directly influence PPARγ and C/EBP to induce adipogenesis. Some signalling molecules like the ones from WNT and hedgehog family supress adipogenesis by directly supressing PPARγ and C/EBP pathways or by supressing other adipogenesis inducing signalling pathways such as insulin signalling. Physiological states like oxidative stress, ROS (reactive oxygen species), inflammation, and cold temperature also influence adipogenesis and metabolism in a more complex manner (Fig. 2.1).

Adipogenesis requires numerous transcription factors for the differentiation of precursor cells into mature adipocytes among which CCAAT/enhancer-binding protein β (C/EBPβ) is one such factor from the leucine zipper family. Knockdown of the C/EBPβ in 3T3-L1 has been shown to stop adipogenesis hence it is determined that C/EBPβ imparts an important role in adipocyte differentiation. In the presence of adipogenic stimuli C/EBPβ by bind to promoter which in turn induces C/EPBα and PPARγ [9].

Regulation of energy balance is very critical in a subject that leads to a person obese and simultaneously may lead to be affected with many obesity associated diseases. Therefore energy regulation is so important. PPARs are ligand-activated group of transcription factors that control the storage and dissipation of energy and maintain energy homeostasis. There are several different types of PPAR, like PPARα, PPARβ/δ and PPARγ. PPARγ is important in adipogenesis. Adipocytes are differentiated with the influence of PPARγ essentially. PPARγ heterodimerizes with multifunctional vitamin A receptor, RXR (retinoid X receptor) followed by binding to the specific DNA sequence of the gene at the promoter region, called PPRE (peroxisome proliferator response element) and activate the promoters of the downstream genes (Fig. 2.3).

Types of Adipose Tissue

Fat cells or adipose cells are called adipocytes and they consist of three types white, brown and beige (brite/beige) with different morphology as well as expression. White adipose tissue help in storage of triglycerides of dietary fat and release them as free fatty acids during energy consumption [10]. Brown adipocytes converts energy released into heat to help in thermogenesis whereas beige adipocytes are immature cells with thermogenic potential. The difference between adipocytes are describes in Table 2.1.

Key Hormones Secreted by Adipose Tissue

The adipose tissue, composed of different types of adipocytes which secret hormones may be peptides like adipokines, lipids like lipokine or in other forms which are not fully understood. Some of them function being receptor—specific, like leptin,

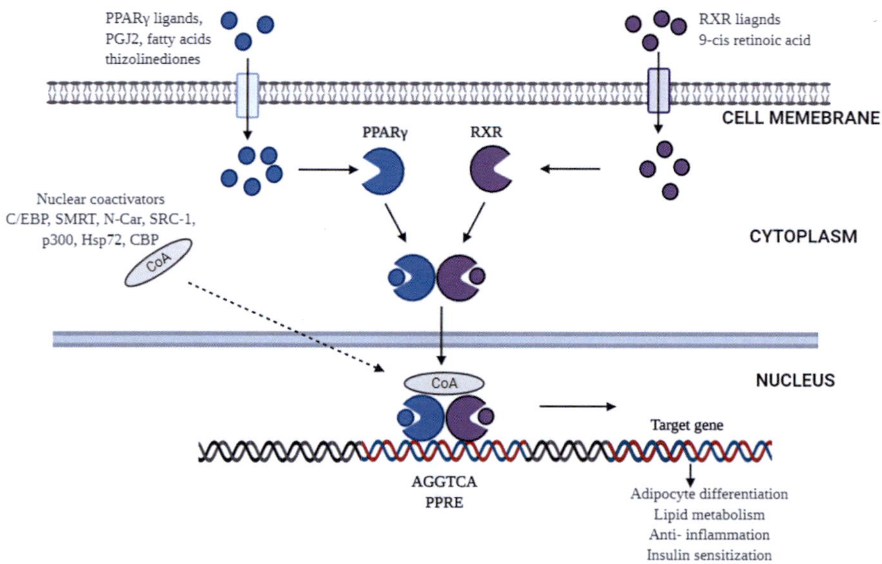

Fig. 2.3 PPARγ is nuclear transcriptional factor which upon binding with RXR forming a heterodimer translocates to the nucleus and binds to particular DNA segment and regulates gene expression. The genes expressed have been linked to adipocyte differentiation and lipid storage

Table 2.1 Differences between White adipose tissue and brown adipose tissue

	White adipose tissue	Brown adipose tissue
Location	Subcutaneous, intra-abdominal, epicardial, gonadal, inguinal, omental	Intrascapular, paravertebral, perirenal, cervical, supraclavicular
Morphology	Spherical (20–200 μm)	Elliptical
Cell composition	• Single lipid droplet • Fewer mitochondria • Flattened peripheral nucleus • Little Endoplasmic reticulum • Low iron content	• Multiple small lipid droplets • Many number of mitochondria • Central oval nucleus • Abundant iron content
Function	Energy storage in the form of triglycerides and endocrine function. Large amount correlates to increased risk of obesity disorders	Dissipate energy for thermogenesis. Large amount correlates to decreased risk of obesity disorder
Origin	Myf5-negative progenitor cells	From Myf5-positive progenitor cells (but there are also Myf5-negative brown fat cells which are derived from other lineages)

adiponectin while others are either non-specific or unknown. Again, receptors are expressed onto different organs. Thus, these signalling molecules are involved in multiple function of the body [11].

Adipocytokines are hormones and these are secreted by adipose tissue and found to be very important in inflammation regulation. Therefore the irregular secretions of these hormones have been known to contribute to inflammation and the development of metabolic syndrome. Adipocytokines secreted by adipose tissue regulates many cell functions eliciting response from tissues like skeletal muscle, endothelium, hypothalamus, pancreas, immune system and many more using different types of signal transductions (autocrine, paracrine, or/and endocrine type). Such important and vital regulatory mechanisms fostered by the WAT secretagogues concluded it as an active endocrine organ [12].

Leptin was the first adipocytokine discovered in 1994 and it garnered interest in adipose secreted factors, its functions, the role it plays in metabolic activities and its influence on other tissues of the body. The factors secreted by adipose tissue identified since leptin has been increasing rapidly forming a new field of research and study.

Adiponectin

Adiponectin is mainly produced by white adipose tissue (WAT. It is a 30 kDa multimeric protein. Many other tissues also express low levels of adiponectin but in different amount depending upon the site and requirement. The peptide adiponectin show diverse nature when compared to human (244 amino acid residues) and other mammals such as mouse (247 amino acid residues). In case of full length human adiponectin, an NH2-terminal hyper-variable region is present. This region generally stretches amino acids from 1 to 18, followed by a collagenous domain comprising 22 Gly-XY repeats. Besides, presence of a COOH-terminal C1q-like globular domain with amino acids from 108 to 244 is also distinct feature. Now it is a well-established fact that adipocytes secreted adiponectin is released directly into the bloodstream. It comprises tri-oligomeric complexes, including trimer (67 kDa), hexamer (140 kDa), and a high molecular weight (300 kDa) multimer. Several molecular chaperones like ERp44 (Endoplasmic Reticulum resident protein 44), Ero1-La (ER oxidoreductase 1-La), and DsbA-L (disulfide-bond A oxidoreductase-like protein) are involved in regulation of biosynthesis and secretion of adiponectin in adipocytes [13].

There are mainly two main adiponectin receptors of 7-pass transmembrane receptors family, AdipoR1 and AdipoR2. Though they fall in the same receptor family, adiponectin receptors (AdipoR1 and AdipoR2) are different from classical G-protein coupled receptors (GPCR) in structure and function with an inverted membrane topology representing a cytoplasmic NH2 terminus and a short, extracellular COOH terminal domain (25 amino acids). Both the receptors AdipoR1 and AdipoR2 are encoded by genes situated on the 1p36.13-q41 and 12p13.31 chromosomal regions, respectively. These receptors have various affinity indices towards adiponectin such as AdipoR1 has a high affinity for globular adiponectin whereas it behaves as a low affinity receptor for full length adiponectin. In terms of distribution it is expressed ubiquitously, with more abundance in skeletal muscles. Similarly, AdipoR2 mainly recognizes full length adiponectin with predominant expression in the liver [14].

Many downstream signalling events are initiated by adiponectin. It involves various adaptor proteins such as APPL1 (Adaptor protein with multiple functional domains) which mediates signal transduction between adiponectin and insulin, interacting directly with insulin receptors. On activation, insulin receptor substrate proteins helps in production of phosphatidylinositol 3,4,5-triphosphate at the plasma membrane by acting as docking platforms for the p85 regulatory subunit of the phosphatidylinositol 3-kinase (PI3K). Furthermore the activation of the PI3K pathway displays a variety of biological response via activation of Akt and its downstream targets. Apart from Akt pathway previous studies also suggest role of APPL1 in the activation of AMP activated protein kinase (AMPK) [15]. This activation is performed by binding and activating protein phosphatase 2A, which result in dephosphorylation and inactivation of protein kinase Cz (PKCz). Furthermore it leads to dephosphorylation of liver kinase B1 (LKB1) at its Ser307, conceding LKB1 to translocate from nucleus to cytoplasm, further activating AMPK for various cellular functions. Thus at cellular level activation of AMPK is a prominent step in mediating the most of the effects of adiponectin. AMPK, has some major functions associated with it and is an important fuel-sensing enzyme. AMPK is found responsible in decreasing cellular energy state by stimulating various pathways. It may correspond via oxidation of fats or inhibition of energy consuming pathways. However, AMPK is not keenly required for survival and its role in fatty acid, triglyceride, and protein synthesis is nominal. Again, released adiponectin significantly increases the expression and activity of PPAR-α, a key transcription factor that regulates various metabolic processes. This is further carried out by up regulation of acetyl CoA oxidase (ACO) and uncoupling proteins (UCPs) which promotes fatty acid oxidation and enhancement of energy expenditure. It has also been suggested that APPL1 and adiponectin improve glucose metabolism in various metabolic tissues via activation of p38 MAPK (Mitogen-activated Protein Kinase) and Rab5, a GTPase effector of APPL1. Activated AMPK further activate eNOS and helps in production of nitric oxide in response to adiponectin which results in vasodilation. In addition, activated AMPK may also trigger apoptosis by inhibiting IKK/NFκB/PTEN [16].

Many previous report suggest that 3T3L1 adipocytes show high expression of adiponectin, which helps in the adipocytes cell differentiation through its autocrine activity being regulated by factors including C/EBPα, PPARγ, and SREBP-1c, respectively which further enhance lipid content and insulin mediated glucose transport. In an animal experiment, it has been found that transgene-mediated overexpression of adiponectin lead to decreased energy expenditure thus promoting morbid obesity. However, down-regulation of TNFα in fat pads, improved glucose metabolism, with reduced macrophage numbers is also visible in ob/ob mice. In addition, augmented vascularisation and expansion of the subcutaneous fat pad were also prominent in over expressed condition. However, continuous over expression of adiponectin for a long time results increase in subcutaneous fat with safeguard against diet induced insulin resistance [17].

Leptin

Leptin and its gene was discovered in Jackson laboratories by cloning of ob/ob mice which comprises of 161 amino acids. The mutation of the leptin gene and its deficiency are involved in various forms of etilogies such as obesity, hyperphagia, diabetes along with infertility and neuroendocrine disorders. In general White adipose tissues (WAT) are involved in secretion of Leptin and its physiological concentration is generally dependent on amount of energy stored in adipocytes and calorific intake along with proportion of fat accumulated. Interestingly, secretion of leptin is also influenced by day duration with higher concentration during evening and early morning [18].

The effect of Leptin can be seen in brain as well as peripheral tissues when it binds to its specific receptors (ObRs). There are different isoforms of Leptin receptors due to alternative splicing which are generally named as ObRa (short leptin receptor isoform) and ObRb (long leptin receptor). Both the isoforms have varied functions; generally ObRa helps in leptin transport across the blood brain barrier whereas the ObRb regulates homeostasis and neuroendocrine function via signal transduction due to higher presence in hypothalamus. Leptin has a major role in suppression as well as stimulation of appetite through its receptors present in the hypothalamus. This action is performed by counteracting the effects of neuropeptide Y which is a feeding stimulant released in the gut as well as regulating a cannabinoid neurotransmitter, adandamide which stimulates appetite. Similar action is performed by promoting the synthesis of an appetite suppressant called α-melanocyte-stimulating hormone which inhibits appetite [19].

The role of leptin and its receptors are also involved in activation of signalling pathways like Janus Kinase-Signal Transducer and Activator of Transcription-3 (JAK-STAT3) which plays an important role in maintaining energy homeostasis. This may further lead to activation of Phosphatidylinositol 3-Kinase (PI3K) which regulates food intake and glucose homeostasis. Apart from PI3K, there are multiple investigations where involvement of many other pathways such as MAPK, AMPK, and mTOR, showed influence on biological role of leptin.

In case of obesity the tolerance to the effects of leptin in circulation may prove vital for disease progression however its complete deficiency due to is extremely rare homozygous mutation is very rare. Among various causes for leptin resistance the downstream leptin receptors are crucial. This may lead to inhibition of leptin signalling or inhibition of transport of leptin across blood brain barriers. Considering both leptin signalling and resistance its advance studies and further research may lead to a breakthrough in development of therapeutic options to manage obesity and metabolic disorders [20].

Resistin

Resistin is an another endocrine product from adipose tissue with 12.5 kDa peptide identified first in mice and contains 108 amino acids with an abundance of cysteine. It was first reported in 2001 and is reported to be involved in resistance to insulin which is a characteristic symptom of type 2 diabetes and hence associates obesity to the development of type 2 diabetes. The antidiabetic class of drugs named thiazolidinediones which lowers insulin resistance is mediated by PPARγ which is present abundantly in adipose tissue hence validating the relationship between obesity and risk for type 2 diabetes [21].

Retinol Binding Protein 4

Retinol binding protein is a factor secreted by adipose tissue which helps in the transport of retinol to surrounding tissue from the liver and is known to influence development of insulin resistance. Obesity causes inflammation causing irregular adipose secretions leading to excessive RBP4 secretion into blood stream. Increased serum RBP4 levels has been found associated with increasing severity of atherosclerosis, risk of cardiovascular conditions as well as development of type 2 diabetes. Insulin stimulates signalling pathways in the skeletal muscles which is inhibited by RBP4 in circulation leading to insulin resistance. Thiazolidineiones which are a class of drugs used to reduce insulin resistance by stimulating PPARγ help in inhibition of excessive RPB4 secretion and hence enhance tissue sensitivity to insulin. Excessive RPB4 in circulation has also been reported to be the cause of the development of atherosclerosis and arterial hypertension by increasing adhesion molecules in endothelial cells. Functional polymorphism in RPB4 gene has been reported to increase risk of developing vascular injury a complication from obesity [22].

Vistafin

Vistafin was first reported in 2005 and is secreted by visceral adipose tissue and functions like insulin. Vistafin is a 52 kDa protein having 491 amino acids (human) and functions as phosphoribosyltransferase and cytokine. Vistafin was reported earlier as Pre-B colony enhancing factor (PBEF) which is a growth factor responsible for B lymphocyte development and secreted in liver tissue, bone marrow, skeletal muscle, neutrophils and foetal membrane. During the differentiation of adipocytes from pre adipocytes vistafin transcription is significantly increased.

Vistafin influences the innate immune system by acting as a pro inflammatory adipocytokine and enhances production of TNF-α and IL-6 to induce leukocyte activation. Vistafin functions as nicotinamide phosphoribosyltransferase involved in

NAD^+ synthesis pathway and hence plays a role in energy metabolism by regulating level of NAD^+ in the cell [23].

Vistafin is involved with many signalling pathways. The phosphatidylinositol 3 kinase (PI3-kinase) signalling involves vistafin which in turn activates MAPK and p70 ribosomal S6 kinase which are recruited in protein synthesis. Protein kinase B (Akt) pathways block the release of visfatin and hence inhibition of PI3kinase. In addition, PI3 kinase is found to be important for insulin-stimulated glucose transport. Besides, Visfatin plays a role in activation of extracellular signal-regulated kinases (ERK)-dependent pathway thereby stimulates angiogenesis [24].

TNF-α

TNF-α is a 23 kDa transmembrane protein which is found as a 17 kDa soluble molecule in circulation after undergoing cleavage. It is an important regulatory and immunomodulatory biomolecule. Thus it regulates immune function, cell differentiation, proliferation, apoptosis and energy metabolism. Macrophages are the primary source of inflammatory cytokines and these macrophages infiltrate into adipocytes and hence lead to increased levels of TNF-α in obese condition. The expression level of mRNA of TNF-α in adipose tissue has shown correlation with insulin resistance and some studies demonstrate that TNF-α can impair insulin signalling. Tumour necrosis factor-alpha receptor 1 (TNFR1) and tumour necrosis factor-alpha receptor 2 (TNFR2) are the cell surface receptors of TNF-α. In adipose tissue TNF-α also stimulates the activation of NFκB pathway (nuclear factor kappa B), MAPK (mitogen-activated protein kinase) pathway, ERK (extracellular signal-regulated protein kinase) pathway, p38 and JNK (c-Jun N-terminal kinase) pathways [12].

Interleukin-6

Adipose tissue is known to secrete 30% of the IL-6 in the body. Increase in obesity leads to increase in IL6 which is known to lead to cardiac issues such as coronary artery disease and atherosclerosis. Many physiological processes are influenced by IL6 and plays a role in metabolism in the body. Increase in IL-6 secretion up regulates SOCS3 which hinders the insulin induced insulin receptors and IRS-1 phosphorylation leading to inhibition of insulin signalling pathway. IL6 receptor α (IL6-Rα) stimulates IL6 signalling through signalling protein IL6ST.

The IL6 signalling is mediated by IL6 receptor α (IL6-Rα) through the signalling protein IL6ST (also known as gp130). The signalling can be mediated by IL6 binding to IL6-Rα/IL6ST in the plasma membrane or by interaction of IL6ST with IL6Rα bound to IL6. IL6 signalling activation in turn activates the JAK1-STAT3 pathway which targets Il4ra promoter leading to increased IL4-Rα expression in macrophages and supresses insulin resistance caused by obesity. Leptin secretion by adipocytes increases in presence of IL6 in circulation and suppresses satiety. Again, IL6 causes to increase adipose tissue lipolysis resulting enhanced hepatic gluconeogenesis and hepatic insulin resistance. It increases the damage of adipose tissue following exercise and in response to cancer cachexia. Furthermore, IL6 signalling induces paraventricular nucleus of the hypothalamus and thus improves energy and glucose homeostasis in response to obesity. Again the role of IL6 to increase insulin secretion by an incretin-based mechanism has been elucidated. IL6 thus displays its potentials acting at key sites of metabolic regulation in several tissues [25].

Angiotensin

Angiotensin is an important part of the renin-angiotensin system (RAS) which supports in the regulation of blood pressure as well as the sodium ion regulation for maintenance of extracellular fluid balance. Presence of RAS system in white and brown adipose tissue has been reported. Adipose tissue is the second major source of angiotensin after liver. Angiotensin is the precursor molecule for angiotensin II which in the main hormone in RAS system is known to play a role organogenesis. Angiotensin II in turn is known to influence the adipocyte growth, lipid metabolism and adipokine regulation. Long chain fatty acids (LCFA) and non-metabolized fatty acids which are present in pre adipocytes behave as activators of the angiotensin gene [26].

Adipokines in Health and Diseases

Obesity is known to be the main cause of many metabolic disorders which includes type 2 diabetes, cardiovascular diseases (like hypertension, coronary heart disease, stroke), fatty liver disease, diseases of central nervous system (dementia), as well as cancer. Interestingly, Adipose tissue is a depot of secretagogue, known to secrete over 600 known secretory proteins which help in maintenance of body mechanisms such as appetite regulation, insulin secretion, fat distribution, inflammation and homeostasis [27]. The adipokine influence of some of the organs like brain, liver, pancreas, gut, immune system, vascular system and adipose tissue in itself has been illustrated in Fig. 2.2.

Obesity leads to alteration in adipokine secretion which results in diseases as it influences many biological processes in the body. Genetic and environmental factors in obese individual can lead to change in chronic positive energy balance leading to a state of obesity. As it has been discussed, overweight leads to irregular distribution of adipose tissue. This brings change in size, distribution, composition and its function. As a consequence, adipocyte hypertrophy, ectopic fat deposition, hypoxia and stress are reported which are the results of adverse changes in adipokine secretions. It is reported that in adipocyte hypertrophy insulin sensitivity is impaired while large adipocytes secrete higher amount of proinflammatory factors [28].

Adipocytes influences immune system by the secretion of factors like adipsin, ASP (acylation stimulating protein). Adipokines like IL1β, IL6, Crp, MCP-1(monocyte chemotactic protein-1), progranulin and chemerin play a role in modulation of inflammation. Leptin, FG21, adiponectin, resistin, vaspin help in the glucose metabolism. Insulin sensitivity is influenced by adipokines like leptin, adiponectin, chemerin; hypertension by angiotensinogen; cell adhesion by PAI-1; vascular function by VEGF; adipogenesis by BMP7; lipid metabolism by CD36; and other functions. Adipokines like leptin, vaspin influence eating disorder like anorexia nervosa and cathepsin, apelin influence atherosclerosis development (Table 2.2) [29]. Adipokines can also function as a biomarker for status of adipose tissue and underlying disorders. Change in adipokine in circulation may point to adipocyte hypertrophy, apoptosis and stress on the tissue. For context an obese individual with insulin sensitivity can be differentiated from an obese individual with insulin resistance who is otherwise metabolically healthy by the adipokine pattern in the serum. The obese insulin sensitive individual will have higher adiponectin concentration and a lower concentration of RBP4, chemerin, progranulin and fetuin-A [30].

Therapeutic Aspect of Adipokines

Over the recent years many novel adipokines have been discovered which influence many biological processes in the body. Elucidating the function and targets can help facilitate use of these adipokines in therapies in obesity and its related disorders.

Leptin Based Therapies

As discussed in the present review leptin plays a major role in energy metabolism and also has a major endocrine function. Hence for the treatment of hypoleptinaemic condition which causes irregular endocrine secretions, recombinant leptin can be used. Hypoleptinaemic conditions are usually seen in extremely lean women having amenorrhoea and also in lipodystrophy patients. Recombinant leptin can also be used to regulate glucose levels of plasma in rodent Type 1 diabetes mellitus models having low leptin levels. Leptin efficiency is however not reported much in human T1DM and hence recombinant leptin administration may not be effective in human patients with T1DM [31].

Leptin increases insulin sensitivity as well as lowers weight and these characteristics of leptin makes it a useful therapeutic option to treat many metabolic diseases. The therapeutic approach of leptin in reducing obesity has been in case of obesity caused by loss of function due to leptin mutation in humans. However this therapeutic approach of using leptin as antidiabetic drug has a major drawback which is that obese

Table 2.2 Factors secrete by adipocytes and its role in health

	Function	References
Leptin	• Energy homeostasis • Regulation of immune function • Regulate food intake • Glucose and fat metabolism	[41]
Adiponectin	• Glucose and lipid metabolism • Anti-inflammatory effect • Angiogenic and vasodilatory properties • Role in insulin resistance/type 2 diabetes and CVD	[7]
Resistin	• Role in insulin resistance and diabetes • Linked to development of atherosclerosis and cardiovascular disease (CVD) • Linked to non-alcoholic fatty liver disease, autoimmune disease, malignancy, asthma, inflammatory bowel disease and chronic kidney disease	[42]
Vistafin	Regulate energy metabolism during stress responses and immune activation	[43]
TNF-α	Pathogenesis of obesity and insulin resistance	[27]
IL-6	Correlated with obesity, impaired glucose tolerance, and insulin resistance	[27]
Retinol binding protein 4	RBP4 is a contributing factor for insulin resistance in obesity	[44]
Angiotensin	It helps in formation of Angiotensin II which is known to regulate blood pressure and electrolyte balance	[45]
Plasminogen activator inhibitor 1 (PAI-1)	Inhibits plasminogen activation which in turn inhibits fibrinolytic system	[46]
Acyclation stimulating protein	Regulation of lipogenesis	[7]
FGF21	• FGF21 is induced by cold exposure and stimulates conversion of WAT into beige adipocytes • Promotes thermogenic gene expression in adipose tissue	[47]
Neuregulin	Counteract chronic inflammation and obesity associated metabolic changes in the liver	[48]
MCP-1	• Contribute to the metabolic abnormalities associated with obesity and insulin resistance • MCP-1 inhibits adipocyte growth and differentiation • Plays a role in development of atherosclerosis	[49]

(continued)

Table 2.2 (continued)

	Function	References
Adipsin (complement factor D)	• It is a component of the alternative complement pathway which is associated with CVD • Potential to modulate inflammatory events	[31]
Apelin	Apelin promotes insulin sensitivity and glucose utilization in adipose and muscle tissues and hence confers anti-obesity and anti-diabetic effect	[50]
Chemerin	It is an inflammatory chemokine and plays a role in immunomodulation	[51]
Omentin-1	Anti-inflammatory effect	[52]
Angiopoietin-like protein 8	Role in pancreatic β cell proliferation	[27]
Cathepsins S, L, K	Regulate glucose and lipid metabolism	[53]
Clusterin	Stimulates tumour progression as well as angiogenesis	[53]
Fetuin-A	Involved in lipid induced inflammation and cancer progression	[54]
Lipocalin 2	Insulin resistance and inflammation	[55]
Nestafin-1	Glucose dependent insulinotropic effect on β cells	[55]
Proragranulin	Chemoattractant protein, inflammation of adipose tissue	[43]
RBP4	Visceral fat distribution, dislipidemia	[56]
TGFβ	Regulates cell proliferation, differentiation and apoptosis	[56]

patients develop resistance to leptin. Furthermore obese rodents in preclinical studies have been reported to have increased level of leptin in their plasma limiting its usage as antiobesity drug.

In the case of leptin resistance there is a hindrance in the leptin uptake in the cerebrospinal fluid and hence there is an impedance in leptin signalling in central nervous system. The therapeutic solution for leptin resistance in obese patients is managed by the use of islet hormone amylin which has been reported to sensitise leptin signalling in central neurons by reducing intracellular stress. However before the use of leptin as a therapeutic drug to treat metabolic diseases its adverse effects must be analysed such as liver fibrosis [32].

Adiponectin Based Therapies

Adiponectin signalling can be regulated and used as a therapeutic approach to help in cardiometabolic diseases. Circulating adiponectin in plasma can be increased by using PPARγ agonists present in abundance in adipose tissue. This therapeutic approach has been reported to lower insulin resistance as well as fatty liver condition in mice in comparison to wild type mice as well as adiponectin deficient mice. Adiponectin receptors can be targeted to be stimulated to secrete more circulating adiponectin by use of agonist AdipoRon which consists of two agonists ADIPOR1 and ADIPOR2. This therapeutic approach was reported to lower insulin resistance an also helped in reducing progression of inflammatory diseases like fibrosis validating its use.

Even though increased concentration of circulating adiponectin in the blood is involved in lowering of insulin resistance and reducing risk of developing type 2 diabetes mellitus, however the lacuna is that these benefits have not been confirmed in large population studies. The therapeutic advantages reported in rodent models have not been confirmed in human clinical studies an require much more exploration to bring its use in the market [33].

Other Adipokine-Based Therapies

Leptin has made large leaps in its use as therapeutic molecule to treat metabolic diseases, however other adipokines and factors secreted by adipose tissue have not been validated and require far more promising reports and research. Adipokines as therapeutic approach to treat metabolic diseases is not ideal as cardiometabolic diseases require long term treatment for its efficacy but long term sustained dosing with adipokines has higher risk of adverse events and maybe life threatening. One such adipokine is neuregulin which is proven effective to help in management of metabolic diseases however due to stimulation of epidermal growth factor receptors (ERBB3 and ERBB4) poses a risk of developing breast cancer. Another factor which limits the usage of adipokines as therapeutics is the fact that developing oral drugs of adipokines poses a challenge [31]. Despite these challenges many adipokines are candidates for use for treatment of metabolic diseases such as FGF21 (fibroblast growth factor 21) which regularises serum glucose levels in rodent animal studies, and diabetes mellitus patients. Furthermore a long acting analogue of FGF21 has been reported to decrease liver steatosis in patients suffering from non-alcoholic steatohepatitis. Growth differentiation factor 15 (GDF15) has also been proven to be beneficial adipokine in reducing obesity in mice and monkey animal models. Omentin is beneficial on reducing cardiac hypertrophy and atherosclerosis and has been validated in various animal models as therapeutic solution for cardiovascular diseases. The beneficial effects of recombinant vaspin against myocardial ischemia and pulmonary hypertension has been reported in rats. Some adipokines cause negative effects on metabolism and cause diseases such as fatty acid binding protein 4 (FABP4) which can be counteracted with antibodies against them. Treatment with antibodies against extracellular FABP4 in mice models has reduced insulin resistance and decreased obesity.

In conclusion adipokines as well as antibodies against adipokines have been proven beneficial as therapeutics in the treatment of metabolic and cardiovascular diseases in rodent and other animal models however their use in human patients is still ambiguous and needs validation [34].

Discussion

Obesity is a definite outcome of significant energy imbalance and is found to be root cause of several diseases. Adipose tissue do the solemn function to regulate the energy balance being influenced by either ligand-mediated transcription factors or with other types of transcription factors with the secretion of several different signalling molecules with endocrinal nature. Adipogenesis is the journey of the developments of adipocytes of different category where differentiation is again monitored by several factors. Hormonal secretions are important in discussion of adipocytes associated physiological functions. Where we observe adiponectin and leptin are well studied.

While there is a breach of the regulatory discipline of adipogenesis, the subject gains body mass and volume and becomes obese. Due to these phenomenal changes, fat cross the site of adipose tissue to get accumulated in non-adipose tissues like brain, heart, gut, liver, muscle and pancreas (Fig. 2.1). This results fat-associated toxic environment and consequently leads to lipotoxicity. Lipotoxic condition influence erratic glucose and fat metabolism in different organs resulting diabetes, insulin resistance, fatty liver, complications associated with cardiovascular diseases, cancers and many more. Energy regulation therefore has so far been identified as the key factor to combat obesity and associated diseases. Several natural ligands to some transcription factors are investigated to be used to control obesity. In our study, probiotic *Pediococcus pentosaceus* GS4 (MTCC 12683) has the ability to biohydrogenate linoleic acid to conjugated linoleic acid (CLA) which is a natural ligand for PPARγ [35]. In another study, we could demonstrate that this CLA producing property can mitigate the azoxymethane-induced toxicity and carcinogenesis in mice [36]. Structural deformities in the intestine was reduced by GS4 by increasing disaccharides associated with brush border membrane, increasing intestinal alkaline phosphatase activity as well as enhancing kidney and liver functions. In another study, we also demonstrated and reported that CLA biohydrogenated by GS4 (MTCC 12683) reduced the proliferation of HCT-116 in vitro by apoptosis and abrogated NF-kB and p-Akt with demonstrable mitigation of colon cancer in mice with triggered apoptosis in colonocytes, PARP [poly(ADP-ribose) polymerase] cleavage, caspase 3 activity, DNA fragmentation and iHDAC (histone deacetylase inhibitors) activity [37]. In separate study, it was demonstrated by another efficient probiotic strain, *Enterococcus faecalis* AG5 assimilates cholesterol and produces free fatty acids including propionate [38] with the ability of controlling obesity among high-fat diet induced mice [39]. This study is an evidence of the role of propionic acid produced by AG5 diminishes the inflammatory response by reduced level of TNF-α; induces apoptosis in 3T3-L1 pre-adipocytes monitoring expression of PPARγ, 5-LOX, NF-kB, p-AKT, caspase-10 and caspase 3/7 [39]. Besides, many hypolipidemic compounds have been formulated along with thiozolidinediones, popularly known as glitazones to treat type 2 diabetes as well as fat regulation.

Development of adipose tissue occurs and response to nutritional changes occur in the foetus. Other than regulating lipid storage and release adipose tissue secretes hormones and factors which control metabolism by regulating cellular function in diverse target cells an organs. After the discovery of leptin in 1994 and its role in metabolism adipose tissue has been considered as an endocrine organ and further research began on other hormones and factors secreted by adipose tissue and its role. As numerous adipokines were discovered its effect on several systems in the body as well as its link to obesity and metabolic disorders were elucidated and a new field of study was established [40].

Subsequently in the years to come, many new adipokines will be discovered and their role in different body systems will be elucidated. The role of these adipokines on metabolism can help develop therapeutic solutions to many metabolic diseases with initial studies validated in animal models which further can be translated into clinical studies in humans. The recent insight pointing to the possibility of stem cell derivation from adipose tissue and also the discovery of many cell types and adipokines, which will help recognise the complex role adipose tissue play in the human body which was once regarded as a simple storage tissue regulating energy metabolism

References

1. Gray SL, Vidal-Puig AJ (2007) Adipose tissue expandability in the maintenance of metabolic homeostasis. Nutr Rev
2. Galic S, Oakhill JS, Steinberg GR (2010) Adipose tissue as an endocrine organ. Mol Cell Endocrinol
3. McGown C, Birerdinc A, Younossi ZM (2014) Adipose tissue as an endocrine organ. Clin Liver Dis
4. Ouchi N, Parker JL, Lugus JJ, Walsh K (2011) Adipokines in inflammation and metabolic disease. Nat Rev Immunol
5. Karastergiou K, Mohamed-Ali V (2010) The autocrine and paracrine roles of adipokines. Mol Cell Endocrinol
6. Lee J-E, Schmidt H, Lai B, Ge K (2019) Transcriptional and epigenomic regulation of adipogenesis. Mol Cell Biol
7. Luo L, Liu M (2016) Adipose tissue in control of metabolism. J Endocrinol
8. Miettinen S, Sarkanen JR, Ashammakhi N (2008) Adipose tissue and adipocyte differentiation: molecular and cellular aspects and tissue engineering applications. Top Tissue Eng
9. Tang QQ, Lane MD (2012) Adipogenesis: from stem cell to adipocyte. Annu Rev Biochem

10. Saely CH, Geiger K, Drexel H (2011) Brown versus white adipose tissue: a mini-review. Gerontology
11. Lehr S, Hartwig S, Sell H (2012) Adipokines: a treasure trove for the discovery of biomarkers for metabolic disorders. Proteomics Clin Appl
12. Rosen ED, Spiegelman BM (2006) Adipocytes as regulators of energy balance and glucose homeostasis. Nature
13. Mandal P, Pratt BT, Barnes M, McMullen MR, Nagy LE (2011) Molecular mechanism for adiponectin-dependent M2 macrophage polarization link between the metabolic and innate immune activity of full-length adiponectin. J Biol Chem
14. Okamoto Y et al (2002) Adiponectin reduces atherosclerosis in apolipoprotein E-deficient mice. Circulation
15. Kadowaki T, Yamauchi T (2005) Adiponectin and adiponectin receptors. Endocr Rev
16. Achari AE, Jain SK (2017) Adiponectin, a therapeutic target for obesity, diabetes, and endothelial dysfunction. Int J Mol Sci
17. Fantuzzi G (2005) Adipose tissue, adipokines, and inflammation. J Allergy Clin Immunol
18. Caro JF, Sinha MK, Kolaczynski JW, Zhang PL, Considine RV (1996) Leptin: the tale of an obesity gene. Diabetes
19. Morgan RC, Considine RV (2018) Leptin. In: Encyclopedia of endocrine diseases
20. Zhou Y, Rui L (2013) Leptin signaling and leptin resistance. Front Med
21. Steppan CM et al (2001) The hormone resistin links obesity to diabetes. Nature
22. Graham TE et al (2006) Retinol-binding protein 4 and insulin resistance in lean, obese, and diabetic subjects. N Engl J Med
23. Berezin AE, Samura TA, Kremzer AA, Berezina TA, Martovitskaya YV, Gromenko EA (2016) An association of serum vistafin level and number of circulating endothelial progenitor cells in type 2 diabetes mellitus patients. Diabetes Metab Syndr Clin Res Rev
24. Berezin AE (2017) The role of vistafin in diabetes-induced impairment of endothelial repair system. Transl Biomed
25. Han MS et al (2020) Regulation of adipose tissue inflammation by interleukin 6. Proc Natl Acad Sci U S A
26. Jagroop IA, Mikhailidis DP (2000) Angiotensin II can induce and potentiate shape change in human platelets: effect of losartan. J Hum Hypertens
27. Fasshauer M, Blüher M (2015) Adipokines in health and disease. Trends in Pharmacol Sci
28. Van Gaal LF, Mertens IL, De Block CE (2006) Mechanisms linking obesity with cardiovascular disease. Nature
29. Blüher M, Mantzoros CS (2015) From leptin to other adipokines in health and disease: facts and expectations at the beginning of the 21st century. Metab: Clin Exp
30. Klöting N et al (2010) Insulin-sensitive obesity. Am J Physiol Endocrinol Metab
31. Scheja L, Heeren J (2019) The endocrine function of adipose tissues in health and cardiometabolic disease. Nat Rev Endocrinol
32. Zhao S et al (2019) Partial leptin reduction as an insulin sensitization and weight loss strategy. Cell Metab
33. Liu Y, Vu V, Sweeney G (2019) Examining the potential of developing and implementing use of adiponectin-targeted therapeutics for metabolic and cardiovascular diseases. Front Endocrinol
34. El Husseny MWA et al (2017) Adipokines: potential therapeutic targets for vascular dysfunction in type II diabetes mellitus and obesity. J Diabetes Res
35. Dubey V, Ghosh AR, Mandal BK (2012) Appraisal of conjugated linoleic acid production by probiotic potential of pediococcus spp. GS4. Appl Biochem Biotechnol 168(5):1265–1276
36. Dubey V, Ghosh AR, Bishayee K, Khuda-Bukhsh AR (2015) Probiotic Pediococcus pentosaceus strain GS4 alleviates azoxymethane-induced toxicity in mice. Nutr Res 35(10):921–929
37. Dubey V, Ghosh AR, Bishayee K, Khuda-Bukhsh AR (2016) Appraisal of the anti-cancer potential of probiotic Pediococcus pentosaceus GS4 against colon cancer: in vitro and in vivo approaches. J Funct Foods 23:66–79

38. Mishra AK, Kumar SS, Ghosh AR (2019) Probiotic Enterococcus faecalis AG5 effectively assimilates cholesterol and produces fatty acids including propionate. FEMS Microbiol Lett
39. Mishra AK, Ghosh AR (2020) Probiotic Enterococcus faecalis AG5 mitigated high fat diet induced obesity and produced propionic acid stimulated apoptosis in 3T3-L1 pre-adipocyte. Life Sci
40. Poulos SP, Hausman DB, Hausman GJ (2010) The development and endocrine functions of adipose tissue. Mol Cell Endocrinol
41. Pereira SS, Alvarez-Leite JI (2014) Adipokines: biological functions and metabolically healthy obese profile. J Recept Ligand Channel Res
42. Jamaluddin MS, Weakley SM, Yao Q, Chen C (2012) Resistin: functional roles and therapeutic considerations for cardiovascular disease. Br J Pharmacol
43. Coelho M, Oliveira T, Fernandes R (2013) Biochemistry of adipose tissue: an endocrine organ. Arch Med Sci
44. Noy N, Li L, Abola MV, Berger NA (2015) Is retinol binding protein 4 a link between adiposity and cancer? Horm Mol Biol Clin Investig
45. Luther JM, Brown NJ (2011) The renin-angiotensin-aldosterone system and glucose homeostasis. Trends Pharmacol Sci
46. Ghosh AK, Vaughan DE (2012) PAI-1 in tissue fibrosis. J Cell Physiol
47. Fisher FF et al (2012) FGF21 regulates PGC-1α and browning of white adipose tissues in adaptive thermogenesis. Genes Dev
48. Jiang J et al (2016) Circulating neuregulin 4 levels are inversely associated with subclinical cardiovascular disease in obese adults. Sci Rep
49. Musi N, Guardado-Mendoza R (2014) Adipose tissue as an endocrine organ. In: Cellular endocrinology in health and disease
50. Than A et al (2015) Apelin enhances brown adipogenesis and browning of white adipocytes. J Biol Chem
51. Helfer G, Wu QF (2018) Chemerin: a multifaceted adipokine involved in metabolic disorders. J Endocrinol
52. Kwon H, Pessin JE (2013) Adipokines mediate inflammation and insulin resistance. Front Endocrinol
53. Blüher M (2013) Adipose tissue dysfunction contributes to obesity related metabolic diseases. Best Pract Res: Clin Endocrinol Metab
54. Ahima RS, Flier JS (2000) Adipose tissue as an endocrine organ. Trends Endocrinol Metab
55. Kershaw EE, Flier JS (2004) Adipose tissue as an endocrine organ. J Clin Endocrinol Metabol
56. Ghaben AL, Scherer PE (2019) Adipogenesis and metabolic health. Nat Rev Mol Cell Biol

Chapter 3
Interaction Between Genetics and Epigenetics in Obesity and Their Clinical Significance

Zahra Sepehri, Mahsa Motavaf, Aliyeh Sargazi, Zohre Kiani, Mehdi Sepehri, and Moayed S. Alavian

Abstract Complex interplay of genetics and environmental factors in developing obesity has attracted a lot of attention in recent years. Different approaches in genetic analysis has illustrated numerous genetic loci that contribute in adiposity traits as monogenic obesity, syndromic obesity, and polygenic obesity. However, current studies are mostly on coding genes. Studies propose that epigenetic modifications such as DNA methylation plays a substantial role in the regulation of genes, involved in obesity-related processes. Also, miRNAs determine the adipocyte fate. Mechanistically they have impact on adipogenesis, adipocyte differentiation, lipid metabolism, glucose homeostasis, and insulin resistance. On the other hand, long non-coding RNAs play a protective role in metabolic dysfunction during obesity. Even though there are huge amount of evidence regarding genetic and epigenetic factors involved in development of obesity, there are lots of questions yet to be answered.

Zahra Sepehri and Mahsa Motavaf have been equally contributed as the first author.

Z. Sepehri (✉)
Internal Medicine, Department of Internal Medicine, Zabol University of Medical Sciences, Zabol, Iran

Biochemistry and Medical Genetics, Basic Medical Sciences Building, University of Manitoba, 745 Bannatyne Avenue, Winnipeg MBR30J9, Canada

M. Motavaf
Department of Genetics, Faculty of Biological Science, Tarbiat Modares University, Tehran, Iran

A. Sargazi
Razi Hospital, Zahedan University of Medical Sciences, Saravan, Iran

Imamreza Hospital, Iranshahr University of Medical Sciences, Dalgan, Iran

Z. Kiani
Department of Internal Medicine, Kerman University of Medical Sciences, Kerman, Iran

M. Sepehri
Department of Computer, Sanabad Golbahar Institute of Higher Education, Golbahar, Khorasan Razavi, Iran

M. S. Alavian
Internal Medicine-Research Center, Middle East Liver Disease Center, Tehran, Iran

© Springer Nature Switzerland AG 2021

P. S. Tappia et al. (eds.), *Cellular and Biochemical Mechanisms of Obesity*, Advances in Biochemistry in Health and Disease 23,
https://doi.org/10.1007/978-3-030-84763-0_3

Keywords Obesity · Overweight · Genetics · Epigenetics · miRNA · LncRNA · Metabolism · Mechanism · Monogenic · Polygenic

Abbreviations

ATP	Adenosine triphosphate
α-MSH	Alpha-melanocyte stimulating hormone
AlkB	Alpha-ketoglutarate-dependent dioxygenase
ADINR	Adipogenic differentiation-induced noncoding RNA
ASMER-1	Adipocyte-specific metabolic related-1
BMI	Body mass index
BAT	Brown adipose
BMP	Bone morphogenetic protein
BMSCs	Bone marrow–derived stromal cells
CpGs	Cytosine guanine site
CEBPs	CCAAT/enhancer binding proteins
DNA	Deoxyribonucleic acid
DMRs	Differentially methylated regions
DZ	Dizygotic
EBF2	Early B-cell factor 2
EBPs	Enhancer-binding proteins
FABP4	Fatty acid-binding protein 4
GWAS	Genome-wide association study
GRS	Genetic risk score
GSK3β	Glycogen synthase kinase 3 beta
HIF	Hypoxia inducible transcription factor
HbA1c	Hemoglobin A1c
HDL	High-density lipoprotein
HOTAIR	HOX Transcript Antisense RNA
HOXD	Homeobox D Cluster
IL	Interleukin
IGF1	Insulin-like growth factor 1
Kb	Kilobase pair
Kg	Kilogram
LncRNA	Long noncoding RNA
LPL	Lipoprotein lipase
MSH	Melanocyte stimulating hormone
MZ	Monozygotic
miRNAs	MicroRNAs
MSC	Mesenchymal stem cell
MAPK1	Mitogen-activated protein kinase 1
miR-30	MicroRNA-30
ncRNA	Non-coding RNA

PC1	Prohormone convertase 1
POMC	Pro-opiomelanocortin
PPARγ	Peroxisome proliferator-activated receptor gamma
PGC1α	Peroxisome proliferator-activated receptor gamma coactivator 1 alpha
PRDM16	PR-domain containing protein 16
PRC2	Polycomb repressive complex 2
RNA	Ribonucleic acid
Runx2	Runtrelated transcription factor 2
SNPs	Single nucleotide polymorphisms
SREBPs	Sterol regulatory element-binding protein
SAT	Subcutaneous adipose tissue
SRA	Steroid receptor RNA activator
TNF	Tumor necrosis factor
TRIM3	Tripartite motif-containing 3
T1D	Type 1 diabetes
T2D	Type 2 diabetes
UBASH3A	Ubiquitin-associated and SH3 domain-containing A
UTR	Untranslated region
UCP1	Uncoupling protein 1
VMRs	Variably methylated regions
WHR	Waist to hip ratio
WAT	White adipose
Wnt	Wingless-related integration site

Introduction

Overweight and obesity are defined as the presence of excessive and abnormal body fat accumulation that is a health risk. Obesity rates are rising rapidly worldwide, becoming a public health problem [1]. Understanding molecular mechanisms underlie changes in body weight and obesity has received a lot of attention in biomedical and clinical research. This potentially provides new insight towards in depth pathophysiology of the issue and designing the new strategies for preventing and curing the condition. Obesity is measured by two anthropometric measures called waist to hip ratio (WHR) and body mass index (BMI) [2].

At this point, it is clear that obesity arises from a complex interplay of genetics and environmental factors. At a molecular level, epigenetics offers a tool that mediates the interactions of these factors. This chapter aims to give an overview of the current knowledge on how genetics and epigenetic regulation are involved in the development of obesity [3].

Genetic Contributions to Obesity

Over the years, different approaches in genetic analysis including linkage analysis, candidate gene association screening, and genome-wide association studies (GWAS) have detected numerous genetic loci that are likely to influence adiposity traits [4]. However, the most common genetic variants are thought to account for less than 1.5% of the overall inter-individual variation in BMI [5, 6]. Thus, undoubtedly genetics plays an important role in obesity-related biochemical processes, yet it alone cannot explain the recent increase in worldwide obesity rates [7]. A single gene presentation will not make a significant change in bodyweight. More than 100 various genes that are discovered in obesity make less than 100 g change in the bodyweight [8]. However, some single genes have a huge influence on bodyweight which are categorized as pleiotropic syndromic obesity [9]. Polygenic traits affect bodyweight much more and are more associated with obesity [10, 11]. It is known now that genetic mutations unbalance numerous metabolic hormones such as leptin, ghrelin, anorexigenic and orexigenic neuropeptides [12]. Moreover, other obesity genes interact in the central nervous system [13].

Monogenic Obesity

A monogene is a gene with a direct effect on the phenotype. The GWAS cleared more than 40 monogenes in obesity [14]. These monogenes mostly interfere with the leptin/melanocortin pathway along with the whole body homeostasis by influencing the central nervous system. Leptin deficiency is still the only genetic disorder that can be modified by pharmacological treatment [15].

The autosomal recessive variant with a mutation in single genes for leptin, leptin receptor, prohormone convertase 1 (PC1), or pro-opiomelanocortin (POMC) has a direct correlation with early onset obesity. People with these mutations are distinguished from some typical phenotypes including red hair, impaired fertility, reduced immunity, and adrenal insufficiency [16].

Most monogenes cause obesity by contributing to the neural system. Different neurologic pathways and mediators are discovered which regulate central nervous system cell receptors. Dopamine, proopiomelanocortin, proopiomelanocortin, melanocortin, natriuretic peptide, and proopiomelanocortin are the most significant mediators in neural obesity. Transcription of some brain hormone genes and neurotransmitters are also involved in body mass regulation. These gene expressions mostly affect energy expenditure, appetite, satiety, and other food consumption behavior. The main genes that interfere with modification of dieting behavior mostly encode dopamine, serotonin and, cannabinoid receptors in the CNS. The most influential genes are *MC4R, NEGR1, SDCCAG8, ETV5, GNPDA2, PRL, FAIM2, NRXN3, KCTD15, FTO* which regulate body mass via hypothalamic signaling [3, 17] (Table 3.1).

Table 3.1 Genes associated with the development of obesity based on their genomic location

Gene	Loci	SNPs	Result	Place of Effect	Genetic Investigation of ANthropometric Traits (GIANT)	Additional phenotype	References
Obesity genes located in chromosome 1							
TBX15_WARS2	1p11_12	rs984222	T-box transcription factor TBX15 Transcription factor regulating growth, development, and fat deposition in adipocytes	Adipose tissue	BMI WHR	Craniofacial anomaly, dwarfism	[167, 168]
PTBP2	1p21.3	rs1555543	Controls neuron maturation and upregulates AdipoR1 expression in muscle tissue		BMI		[169]
NEGR1	1p31	rs2815752, rs3101336, rs2568958	Axon outgrowth Growth regulation Expressed in the hypothalamus and brain cortex. Controls neuron growth, participates in synapse formation, regulates adipocyte differentiation		BMI	Reduced insulin sensitivity	[170]
TNN13K	1p31.1	rs1514175	Troponin kinase, involved in troponin phosphorylation		BMI	Ischemic heart disease	[171]

(continued)

Table 3.1 (continued)

Obesity genes located in chromosome 1							
Gene	Loci	SNPs	Result	Place of Effect	Genetic Investigation of ANthropometric Traits (GIANT)	Additional phenotype	References
DNM3_PIGC	1q24.3	rs1011731	Participates in GLUT6 and GLUT8 transport to adipocytes		WHR	Arterial hypertension	[172]
SEC16B	1q25	rs10913469	Participates in the formation of lysosome membrane and peroxisome biogenesis		BMI	Sex_specific	[173]
ZZZ3	1p31.1	rs17381664	ZZ protein of the zinc finger family				[174]
H6PD	1p36.22	rs6662509	Involved in the development of insulin resistance and metabolic syndrome formation			Bone aging	[175]
LYPLAL1/SLC 30A10	1q41	rs2605100	Triglyceride lipase active in subcutaneous fat of obese patients	Adipose tissue	BMI WHR	Sex_specific	[176]
SDCCAG8	1q43	rs12145833	Located in centrosomes. Expressed in the hypothalamus and pituitary	Hypothalamus	BMI	Schizophrenia	[177]

(continued)

Table 3.1 (continued)

Gene	Loci	SNPs	Result	Place of Effect	Genetic Investigation of ANthropometric Traits (GIANT)	Additional phenotype	References
Obesity genes located in chromosome 2							
FANCL	2p16.1	rs887912	Enhances beta catenin activity		BMI		[178]
POMC	2p23.3	rs713586	Involved in pain development, energy homeostasis, melanocyte stimulation, immune system modulation, and obesity development		BMI	POMC mutations cause monogenic obesity	[179]
TMEM18	2p25	rs6548238, rs2867125, rs4854344, rs7561317, rs11127485	Regulates adipocyte differentiation. Controls neuron development. Differential expression of the TMEM18 gene has been shown in the fat tissue of obese and lean people		BMI	Type 2 diabetes mellitus	
THNSL	2p13	rs1659258					[176]
ZNRF3	2q12.1	rs4823006			WHR		[3]
LRP1B	2q22.2	rs2890652	Regulates insulin sensitivity and BMI		BMI	LRP1B deletion is found in tumor cells	[180]

(continued)

Table 3.1 (continued)

Gene	Loci	SNPs	Result	Place of Effect	Genetic Investigation of ANthropometric Traits (GIANT)	Additional phenotype	References
Obesity genes located in chromosome 3							
ADAMTS9	3p14.1/	rs6795735	Regulates embryogenesis. Digests extracellular matrix proteins		WHR	Type 2 diabetes mellitus	[181]
NISCH_STAB1	3p21.1	rs6784615	Mediates the interaction with insulin receptor		WHR		[3]
CADM2	3p21.1	rs13078807	Belongs to the immunoglobulin superfamily. Participates in cell_to_cell adhesion in epithelial and neural tissues		BMI	Psoriasis	[182]
RSRC1/SHOX2	3q25.32	rs2362965	Found in all vertebrates. Activates or represses transcription depending on cell type			Cornelia de Lange syndrome, schizophrenia	[176]
ETV5	3q27	rs7647305	Transcription factor, controls axon growth	Hypothalamus		Sterility in mice	[183]

(continued)

Table 3.1 (continued)

Obesity Genes located in chromosome 4 and 5							
Gene	Loci	SNPs	Result	Place of Effect	Genetic Investigation of ANthropometric Traits (GIANT)	Additional phenotype	References
GNPDA2	4p12	rs10938397	Glucosamine_6_phosphate deaminase. Involved in carbohydrate metabolism	Hypothalamus		Type 2 diabetes mellitus	[183]
SLC39A8	4q24	rs13107325	Zinc and cadmium transporter		BMI	Schizophrenia	[184]
FLJ35779	5q13.3	rs2112347	Participates in the assembly of spindle microtubules		BMI		[185]
ZNF608	5q23.2	rs4836133	Belongs to the zinc finger family		BMI	Kidney tumor Alzheimer	[186]
CPEB4	5q35.2	rs6861681	Regulates polyadenylation. Expressed in the neural tissue		WHR	Crohn's disease	[187]

(continued)

Table 3.1 (continued)

Obesity genes located in chromosome 6							
Gene	Loci	SNPs	Result	Place of Effect	Genetic Investigation of ANthropometric Traits (GIANT)	Additional phenotype	References
TFAP2B	6p12	rs987237	Transcription factor. Controls cell division and apoptosis	Adipose tissue glucose transport, lipid accumulation, and adiponectin expression	BMI WHR	Type 2 diabetes mellitus	[188]
NCR3, AIF1, BAT2	6p21	rs2844479, rs2260000, rs1077393	Cluster of genes involved in the inflammation of pancreatic beta cells. Located nearby TNF	Adipose tissue Low grad inflammation	BMI	Type 2 diabetes mellitus	[189]
VEGFA	6p21.1	rs6905288	Vascular endothelial growth factor, key mediator in adipocyte synthesis		WHR	Type 2 diabetes mellitus	[190]
NUDT3_HMGA1	6p21.31	rs206936	Controls nucleotide metabolism. Prevents incorporation of wrong intermediates, e.g., 8_oxo_dGTP		BMI		[191]
PRL	6p22.2	rs4712652	Prolactin	Hypothalamus		Reproductive diseases	[192]

(continued)

Table 3.1 (continued)

Obesity genes located in chromosome 6

Gene	Loci	SNPs	Result	Place of Effect	Genetic Investigation of ANthropometric Traits (GIANT)	Additional phenotype	References
LY86	6p25.1	rs1294421	Inflammatory gene. Hypermethylation augments inflammation		WHR	Bronchial asthma	[193]
RSPOS	6q22.33	rs9491696	Triggers angiogenesis and vessel development		WHR	Oncogene in mammary epithelial cells	[3]

Obesity genes located in chromosome 7, 8 and 9

Gene	Loci	SNPs	Result	Place of Effect	Genetic Investigation of ANthropometric Traits (GIANT)	Additional phenotype	References
NFE2L3	7p15.2	rs1055144	Transcription factor belonging to the bZIP family		WHR		[194]
PPP2R2A	8p21.2	rs1594829	Protein phosphatase 2. Participates in cell growth suppression			Cancer	[195]
HNF4G	8q21.11	rs4735692	Transcription factor expressed in the pancreas, kidneys, small intestine, and testes			Early_onset diabetes	[196]

(continued)

Table 3.1 (continued)

Obesity genes located in chromosome 7, 8 and 9

Gene	Loci	SNPs	Result	Place of Effect	Genetic Investigation of ANthropometric Traits (GIANT)	Additional phenotype	References
MSRA	8p23.1	rs7826222, rs17150703	Enzyme repairing protein lesions		BMI	Insulin resistance in mice, schizophrenia	[188]
LRRN6C (LINGO2)	9p21.3	rs10968576	Involved in the formation of dopaminergic neuron plasticity and integrity		BMI	Parkinson's disease	[197]

Obesity genes located in chromosome 10 and 11

Gene	Loci	SNPs	Result	Place of Effect	Genetic Investigation of ANthropometric Traits (GIANT)	Additional phenotype	References
PTER	10p12	rs10508503	Involved in catabolic pathways	Adipose tissue Low grade inflammation	BMI	Insulin resistance	[198]
APOA1 *APOC3* *APOA5*	11q23.3	rs6589566	Blood plasma apolipoprotein, high_density lipoprotein carrier			Atherosclerosis, cardiovascular diseases	[199]

(continued)

Table 3.1 (continued)

Gene	Loci	SNPs	Result	Place of Effect	Genetic Investigation of ANthropometric Traits (GIANT)	Additional phenotype	References
Obesity genes located in chromosome 10 and 11							
MTCH2	11p11.2	rs10838738	Encodes proteins located in mitochondria and involved in apoptosis. Expression is elevated in the fat tissue of obese women	Cellular apoptosis	BMI	Anorexia, type 2 diabetes mellitus	[200]
BDNF	11p14	rs4074134, rs4923461, rs925946, rsl0501087, rs6265	The main growth factor in the brain supporting neurogenesis. Reduced activity in the hypothalamus is associated with hyperphagia and obesity		BMI	Type 2 diabetes mellitus; neurological and psychic disorders	[201]
RPL27A	11p15.4	rs4929949	Ribosomal protein		BMI		[3]

(continued)

Table 3.1 (continued)

Obesity genes located in chromosome 12, 13, 14 and 15							
Gene	Loci	SNPs	Result	Place of Effect	Genetic Investigation of ANthropometric Traits (GIANT)	Additional phenotype	References
ITPR2_SSPN	12p21.1	rs718314	ITPR2 deficiency causes hypoglycemia and body mass decrease		WHR	Kidney tumor	[3]
HOXC13	12q13.13	rs1443512	Transcription factor involved in early embryo development		WHR	Sex_specific	[202]
FAIM2	12q13	rs7138803	Apoptosis in adipocytes	Hypothalamus	BMI	Myocardial infarction	[173]
C12orf51	12q24	rs2074356 rs11066280	Ubiquitin ligase		WHR	Alcohol addiction	[3]
MTIF3_GTF3A	13q12.2	rs4771122	Involved in mitochondrial ribosome synthesis. Dysfunction causes oxidative stress		BMI	Neurodegenerative diseases	[191]
HS6ST3	13q32.1	rs7989336	Involved in cell growth regulation, cell differentiation, and lipid metabolism			Diabetic retinopathy	[203]
PRKD1	14q12	rs11847697	Protein kinase		BMI		[178]
NRXN3	14q31	rs10146997	Membrane protein expressed in neurons	Dopamine neurotransmission Reward behaviour Decision making motivation	BMI	Schizophrenia, alcohol addiction, nicotine addiction	[188]

(continued)

Table 3.1 (continued)

Obesity genes located in chromosome 12, 13, 14 and 15							
Gene	Loci	SNPs	Result	Place of Effect	Genetic Investigation of ANthropometric Traits (GIANT)	Additional phenotype	References
MAP2K5	15q23	rs2241423	Key enzyme of the MAPK signaling pathway regulating adipogenesis processes		BMI	Anorexia	[204]

Obesity genes located in chromosome 16 and 17							
Gene	Loci	SNPs	Result	Place of Effect	Genetic Investigation of ANthropoetric Traits (GIANT)	Additional phenotype	References
SH2B1	16p11.2	rs7498665, rs8049439, rs4788102, rs7498665	Involved in the neural regulation of energy homeostasis and appetite. Enhances the JAK_STAT signaling pathway, the central regulator of body mass control	Increased serum leptin	BMI	Obesity and diabetes in Sh2b1_null mice	[205]
GPRC5B	16p12.3	rs12444979	Orphan receptor. Expressed in pancreatic cells. Elevated expression lowers insulin level		BMI	Neurodegenerative diseases	[206]

(continued)

Table 3.1 (continued)

Obesity genes located in chromosome 16 and 17							
Gene	Loci	SNPs	Result	Place of Effect	Genetic Investigation of ANthropoetric Traits (GIANT)	Additional phenotype	References
ADCY9	16p13.3	rs2531995	Catalyzes conversion of ATP to cAMP		BMI	Epilepsy	[207]
MAF	16q22_q23	rs1424233	Transcription factor involved in adipogenesis and regulation of insulin and glucagon production	Adipogenesis = a Transcription factor	BMI		[3]
FTO	16q22.2	rs993960, rs6499640, rs8050136, rs3751812, rs7190492, rs8044769, rsl558902	Mediates appetite regulation by the nervous system. Participates in nucleic acid methyladenine demethylation		BMI	Diabetes mellitus	[25, 208]
IGFBP4	17q12.2	rs584438	Insulin_like growth factor. Present in skeletal muscles. Involved in bone mineralization			Hereditary breast and ovary cancer	[209]
RPTOR	17q25.3	rs7503807	Regulates cell growth via nutrient accumulation			Autism, neurodegenerative diseases	[210]

(continued)

Table 3.1 (continued)

Gene	Loci	SNPs	Result	Place of Effect	Genetic Investigation of ANthropometric Traits (GIANT)	Additional phenotype	References
Obesity genes located in chromosome 18 and 19							
MC4R	18q22	rs17782313, rs12970134, rs17700144	Food consumption control	paraventricular nucleus of the hypothalamus	BMI (body mass index) WC Waist circumference Hyperphagia Hyperglycemia	hyperinsulinemia, elevated growth rate, and higher bone density	[211]
NPC1	18q11.2	rs1805081	Participates in the intracellular cholesterol and lipid transport		BMI	Involved in appetite Stabilization	[212]
KCTD15	19q13.1 1	rs11084753, rs29941	Involved in melatonin activation	Hypothalamus	BMI	Anorexia, diabetes mellitus	[183]
QPTCL_GIPR	19q13.32	rs2287019	Receptor of appetite_related GIP hormone. Encodes incretin receptors		BMI	Diabetes mellitus	[176]
TMEM160	19q13.32	rs3810291			BMI		[213]

On the other hand, visceral obesity is regulated through different pathways from metabolic hormones to cytokines. Genes encoding adiponectin, adiponutrin, apolipoproteins, guanine nucleotide-binding protein, lipase, insulin, androgen, estrogen, progesterone, corticotropin releasing hormone, growth hormone, cholecystokinin, and vitamin D contribution to the development of obesity. The adipogenesis, metabolism, and thermogenesis enzyme receptors are also coded by the genes which are significantly correlated with obesity; such as ATPase, fatty acid synthase, glycogen synthase, adrenergic alpha, and beta receptors, and fatty acid-binding proteins. Cytokines such as IL6, IL10, and TNF as well as renin-angiotensin pathways also influence the development of obesity [18]. *LYPLAL1/SLC30A10, TFAP2B, NCR3, AIF1, BAT2, PTER, MTCH2, SH2B1, MAF* are the most significant genes that control visceral obesity through regulating serum leptin, adipose tissue, and apoptosis [3, 17] (Table 3.1).

Despite shreds of evidence that argued women are more likely to be obese with the same genotype and the presence of some female-specific obesity genes such as *LYPLAL1*, The GRS did not detect any important sex-specific risk variant and its risk is similar in men and women [19].

Although the following genes are expressed more in females in comparison with a male, there is no evidence for sex phenotypical differences related to these genes. The *GRB14/COBLL1, LYPLAL1/SLC30A10, VEGFA, ADAMTS9, MAP3K1, HSD17B4, and PPARG* are expressed specifically in women and affect waist phenotype [20].

Clinical Significance of Genetic Mutation

The genetic risk score (GRS) is established to analyze GWAS by gathering data from multiple risk single nucleotide polymorphisms (SNPs). GRS-BMI and GRS-obesity correlation studies are more significant in European descent but not African Americans, Asians, and Hispanics [19]. Obesity risk assessment using GWAS and GRS data is less valuable in the absence of environmental factors such as lifestyle, geographic, demographic, socioeconomic, and cultural factors. Although estimation of pure genetic influence on anthropometric indices is difficult because of the multifactorial nature of obesity, some significant genes are detected that influence body mass index. The *FTO, INSIG2,* and *MC4R* variants increase body weight for almost 3 kg, 1 kg, and less than 1 kg respectively. The effect of the genes on anthropometric indexes is indicated in Table 3.1.

Polygenic Obesity

Any trait that is inherited by various alleles of a distinct gene is called a polygenic variant. In polygenic traits, each allele phenotype inheritance causes a small quantitative change in the trait. The addition of all alleles causes a certain phenotype. Bodyweight is adjusted with many types of these polygenic variants. Polygenic obesity or common obesity is a set of polygenic variants and the main cause of obesity mostly with the presence of an obesogenic environment. The hereditability of obesity is discussed in twins and vertical familial obesity. The evolutionary hypothesis argued that these genes are mutated and specified for more efficient nutrition storage and fat-rich diet consumption therefore the person will gain extra weight and will have more survival chance [21]. The GWAS showed these genes inherit as gene blocks simultaneously. Different alleles are involved in the development of obesity. The frequency of each allele translation and related proteins will determine the quantitative body mass index and obesity level. Therefore, every single individual presents a specific set of obesity polygenic variants that are different from another obese individual. Some of the most influential correlated genes in polygenic obesity are melanocortin-4 receptor gene (*MC4R), FTO,* and *INSIG2* [11].

*Melanocortin-4 Receptor Gene (*MC4R*)*

A 188 Kb single gene that influences obesity is *MC4R* which is located on the long arm of chromosome 18 and encodes a G-protein receptor. The polymorphism on rs17782313 has the second strongest obesity effect on individuals after *FTO* the C-allele in this polymorphism cause more than 700 g bodyweight gain. The effect of C-allele is twice in children aged 7–11 years old in comparison with adults and it causes more weight gain in this group. The specific expression of *MC4R* in the paraventricular nucleus of the hypothalamus demonstrated its influence on central nervous system obesity [11].

The metabolic hormones including insulin and leptin stimulate G-protein receptor transcription from *MC4R*. This receptor causes a neuron signaling pathway from proopiomelanocortin (POMC) to alpha-melanocyte-stimulating hormone (α-MSH). The final product which is α-MSH sends a negative feedback response to stimulate *MC4R* translation and causes a satisfaction feeling in the individual. On the other hand, it increases energy expenditure. Mutation in *MC4R* decreases satisfaction feeling and energy expenditure which leads to severe obesity. In conclusion the effect of this cascade will increase the food and fat intake especially in younger ages and lead to early-onset obesity [22]. Furthermore, this mutation is associated with hyperinsulinemia, increased growth rate, and higher bone density [23].

Fat Mass and Obesity-Associated Gene (FTO)

The *FTO* is a known gene for diabetes mellitus type 2 and apparently the first cause of obesity. This gene is located on the long arm of chromosome 16 [24]. Almost six SNPs are found in the *FTO* gene which has a strong correlation with obesity. The A-allele in the rs9939609 polymorphism is associated with a more than 30% increased risk of obesity. This homozygous polymorphism is the reason for almost 1% of all obesities causing 2–3 kg extra weight. The carriers of both *FTO* and *INSIG2* are at risk of overweight obesity [11, 25].

Although the *FTO* does not influence fat and glucose metabolism pieces of evidence are in support of lower insulin response. *FTO* apparently belongs to the alpha-ketoglutarate-dependent dioxygenase (AlkB) family and associated with DNA demethylation. The *FTO* expression in hypothalamic nuclei influences energy intake, feeding, and fasting regulation. It is discussed that *FTO* gene defect is associated with postnatal growth retardation, microcephaly, low psychomotor response, and some other skeletal, neural cardiac, and genital defects [26, 27].

This gene expression on brain hypothalamic arcuate nuclei influences individuals' energy intake by a tendency to more energy-dense food, reduced feeling of satisfaction, and loss of overeating control. It interface with appetite and satiety regulation by coding 2-oxoglutarate-dependent nucleic acid demethylase. It also diminishes adipocyte lipolysis by energy expenditure or physical activity reduction [28, 29].

Insulin-Induced Gene 2 (INSIG2)

The association between obesity and insulin-induced gene 2 (*INSIG-2*) is yet to be discussed. The *INSIG2* is a 10 Kb gene located on the long arm of Chromosome 2. The C-allele of rs7566605 SNP is found in almost 10% of individuals with obesity. Several studies showed a significant association while some studies which are mostly conducted in Asian regions showed less significant relations. It is discussed that CC genotype is mostly associated with severe obesity and does not have a significant correlation with BMI. This genotype and increased fat storage were to protect individuals in tough situations such as starvation or heavy physical activities however after industrialization it is the cause of obesity. Its recessive homozygous trait causes a 1 kg weight increase which is mostly related to higher subcutaneous adipose tissue. It is related to blood fatty acids and cholesterol increase. Interaction of *INSIG2* and sterol regulatory element-binding protein (SREBPs) causes increased serum leptin, increased cholesterol, and fatty acid metabolism and adipogenesis [30].

Syndromic Obesity

The Mendelian syndromic form of obesity which is also called pleiotropic syndrome causes a genetic abnormality that leads to an imbalanced energy profile. This is a set of signs and symptoms along with obesity that follow Mendelian autosomal or X-linked inheritance. The syndromic obesity usually presents by an obese individual who carries other genital or dysmorphic abnormalities appearing in mid-childhood. It is mostly known by Prader–Willi and Bardet–Biedl syndromes. The exact genetic cause of most syndromic obesities is not yet defined because of their low prevalence. Out of total of 79 syndromic obesity, only 19 genetic foundations are completely elucidated. Other syndromes are either unnamed because of various phenotypical presentations or have more than one name for example Carpenter syndrome, which is also named acrocephalopolysyndactyly type II [31, 32].

The X-linked syndromes and related genes are as following: Börjeson-Forssman-Lehmann Syndrome associated with *PHF6*, Chudley–Lowry syndrome associated with *XNP/ATR-X*, Coffin–Lowry syndrome associated with *RSK2/RPS6KA3*, Cornelia de Lange syndrome/Brachmann-de Lange-syndrome associated with *SMC1A, HDAC8*, Kabuki syndrome/Niikawa–Kuroki syndrome associated with *KDM6A/UTX/KABUK2*, Kallmann syndrome/ Hypogonadotropic hypogonadism with anosmia associated with KAL1, Prader–Willi-like phenotype associated with *FMR1* [31, 33–38].

The syndromes which follow the autosomal dominant inheritance are as follows: Albright hereditary osteodystrophy/pseudohypoparathyroidism type Ia is associated with *GNAS1/GNAS;* CHOPS syndrome with *AFF4*, Cornelia de Lange syndrome/Brachmann-de Lange-syndrome with *NIPBL* and *RAD21*; Kabuki syndrome/Niikawa–Kuroki syndrome with *KMT2D/MLL2/ALR/KABUK1*; Kallmann syndrome/ Hypogonadotropic hypogonadism with anosmia with *FGFR1*; Kleefstra syndrome/9q34.3 deletion syndrome with *EHMT1*; Prader–Willi-like phenotype with *SIM1* and *MRAP2*; Proximal 16p11.2 deletion syndrome with *SH2B1* and *KCTD13*; Rubinstein–Taybi syndrome with *CREBBP* and *EP300*; Smith–Magenis syndrome with *RAI1;* WAGRO with *BDNF* [31, 39–45].

Autosomal recessive is the inheritance pattern of the following syndrome by the following genes.

Alström syndrome and *ALMS1*, Bardet–Biedl syndrome/ Laurence-Moon-Bardet–Biedl syndrome and *BBS1, BBS2, BBS3/ARL6, BBS4, BBS5, BBS6/MKKS, BBS7, BBS8/TTC8, BBS9/PTHB1, BBS10, BBS11/TRIM32, BBS12, BBS13/MKS1, BBS14/CEP290, BBS15/WDPCP, BBS16/SDCCAG8/NPHP10, BBS17/LZTFL1, BBS18/BBIP1, BBS19/IFT27, BBS20/IFT172,* and *BBS21/C8ORF37*, Carpenter syndrome/acrocephalopolysyndactyly type II and *RAB23*, Cohen syndrome and *VPS13B/COH1*, Kallmann syndrome/ Hypogonadotropic hypogonadism with anosmia and, *PROKR2, SOX10*, Laron syndrome/growth hormone receptor deficiency, and GHR, MORM (mental retardation, truncal obesity, retinal dystrophy, and micropenis) syndrome and *INPP5E* [31, 46–51].

Moreover; the Angelman syndrome is associated with *UBE3A* and Prader–Willi syndrome/ Prader–Labhart–Willi syndrome is associated with *MKRN3/ZNF127 AND MAGEL2* [49].

Epigenetic Contributions to Obesity

Epigenetics refers to persistent, reversible, and mitotically and/or meiotically heritable changes of gene function that do not involve changes in the underlying DNA sequence [52, 53]. The main epigenetic mechanisms include DNA methylation, non-coding RNAs, and histone modifications. Epigenetic processes are considered as an important source of inter-individual variability that might contribute to complex traits.

Impact of DNA Methylation on Obesity

DNA methylation is the most widely studied epigenetic mechanism known to be involved in biological processes. In eukaryotes, DNA methylation occurs when a methyl group is transferred to the 5′ position of a cytosine residue, specifically at cytosine-phosphate-guanine dinucleotides (CpG). This process is catalyzed by three independently encoded DNA methyl transferases (DNMTs), DNMT1, DNMT3A, and DNMT3B, which use S-adenosyl methionine as the methyl donor.

DNA methylation can exert its function through two main modalities, including physical impeding of transcription factor occupancy on CpG-dense promoters, or via recruiting histone-modifying factors and formation of compact, inactive chromatins [54]. Due to its important regulatory functions, the deregulation of DNA methylation is often associated with multiple human diseases.

In somatic cells, DNA methylation patterns are remarkably stable and transmitted with high fidelity across cell division. DNMT1 is responsible for maintaining pre-existing methylation patterns by using hemimethylated DNA as a substrate for restoring the methyl group to a cytosine residue. On the other hand, DNMT3a and DNMT3b are referred to as de novo methyltransferases, which are responsible for methylation of double-stranded DNA that is not methylated. DNA methylation is generally considered to be associated with gene silencing, and DNA demethylation is usually correlated with gene activation [55–57]. DNA methylation can be removed by two distinct passive and active mechanisms. Passive demethylation refers to the dilution of DNA methylation during successive rounds of replication in the absence of functional DNA methylation maintenance machinery. By contrast, active DNA demethylation is an enzymatic process that removes or modifies the methyl group [58, 59].

DNA methylation is a dynamic epigenetic modification, finely tuned by different intrinsic and extrinsic environmental factors during early development, postnatal maturation, puberty, and aging. The establishment of methylation patterns at the post-implantation stage results in interindividual variation termed metastable epialleles. Due to differences in epigenetic modifications, these alleles show a variable expression state between genetically identical individuals [60]. An individual's epigenome also can change throughout a person's lifetime; however, these epigenetic modifications occur in a tissue-specific manner [61].

A growing body of evidence suggests that DNA methylation plays a significant role in the regulation of genes, involved in obesity-related processes, such as adipogenesis, inflammation, appetite, nutrient metabolism, insulin signaling, and thermogenesis [62]. Thus, the study of DNA methylation patterns provides insights into an explanation and understanding of the etiology of obesity that has been missed by conventional GWAS.

To identify the impact of methylation status on the etiology of obesity three main approaches, including epigenome-wide association studies (EWAS), integrated GWAS and EWAS analyses and candidate-gene approach have been undertaken in the present literature. While the candidate gene approach is hypothesis-driven, genome-wide linkage and genome-wide association are hypothesis-free.

Genome-Wide DNA Methylation Studies of Obesity

The recent development of epigenome-wide technologies for DNA methylation analysis has greatly increased our understanding of the genetic basis of human disease, especially complex traits. In contrast to targeted gene approaches, genome-wide DNA methylation studies simultaneously conduct methylation profiling across a large number of genes and CpGs.

With genome-wide DNA methylation approaches several obesity-associated alterations in DNA methylation patterns at specific sites that are enriched both in obesity candidate genes [63] and in or near genes with no known function related to obesity [63–65].

However, unlike genetic variants that normally remain constant over the lifespan of an individual and can be interpreted as causal for the observed phenotype, DNA methylation patterns can either be stable or get modified reversibly due to fluctuating environments [66]. Thus, while the studies are interested to find the causative roles of DNA methylation in disease etiology, it can also be affected by the disease itself. So, in the context of obesity and related conditions, it is of great importance to determine if methylation patterns play causal roles in the development of these traits or if they are a consequence of disease status, or whether the associations are the result of confounding.

In a pioneer study by Feinberg et al. [67] the detailed methylation status of CpG sites from lymphocyte in several dozen individuals at two different time points, over a decade apart, indicated that a combination of genetic determinants and environmental factors are involved in the regulation of DNA methylation. In this study, more than 200 variably methylated regions (VMRs) distributed over the entire genome, were detected. Half of these VMRs were stable within individuals over time and defined as personalized epigenomic signature or epialleles. Four stable VMRs, located in or near genes previously implicated in regulating body weight or diabetes (*PM20D1*, *MMP9*, *PRKG1*, and *RFC5*) were indicated to be correlated with BMI. *PM20D1* and *MMP9* were earlier described to be upregulated in obese individuals [68–71]. *PRKG1* is involved in nutrient absorption, allocation, storage, and energy acquisition [72], potentially playing a role in weight changes. Moreover, *RFC5*, a replication factor C subunit 5 is required for DNA damage checkpoint control, which make it a potential player in well-known but poorly understood DNA damage–related complications of obesity [73]. These early results strongly support the idea to use methylation signatures as a risk determinant for obesity development.

Based on evidence linking obesity to impaired immune function, Wang et al. [64] conducted a study to explore methylation patterns in peripheral blood mononuclear in obese versus lean subjects. Their findings demonstrated an increased methylation level at one CpG site in the *UBASH3A* (ubiquitin-associated and SH3 domain-containing A) gene and decreased methylation level at one CpG site in *TRIM3* (tripartite motif-containing 3) gene in obese subjects compared with lean controls. *UBASH3A* negatively regulates T Cell Receptor (TCR) signaling and T cell activation [74] through multiple mechanisms [74, 75]. Interestingly, GWAS had previously demonstrated the implication of DNA variants proximal to *UBASH3A* in the development of type 1 diabetes (T1D) [76].

Besides, *UBASH3A* deficiency has been reported to accelerate the development of T1D, which was associated with increased accumulation of β-cell autoreactive T cells in the spleen and pancreatic lymph node [77]. Thus, the findings of Wang et al. [64], when placed in context with existing genetic data, suggest that altered methylation patterns of *UBASH3A* may play a role in obesity-induced immune dysfunction. Intriguingly, *TRIM3*, which belongs to the superfamily of TRIM proteins, is also involved in immune response [78], which may explain why differential methylation patterns of these genes were visible in blood leukocytes. However, these results are not easy to interpret, as the functional consequences of hypermethylation at these two genes, even though related to immune function has not yet been clearly defined.

A study by Dick et al. [79] was the first replication cohorts that applied a genome-wide analysis of methylation at CpG sites concerning BMI. The study reported increased methylation at three CpG sites in intron 1 of *HIF3A* to be positively associated with BMI in both subcutaneous adipose tissue (SAT) and whole blood cells. *HIF3A* is a component of the hypoxia-inducible transcription factor (HIF), which plays a crucial role in cellular and physiological responses to hypoxia during normal or pathological processes.

Compelling experimental data suggest that the HIF system is also implicated in body metabolism, energy expenditure, and obesity [80–83]. *HIF3A* plays a role in glucose and amino acid metabolism, adipocyte differentiation as well as a physiological response to insulin [79, 84, 85].

Dick et al. [79] also reported significant independent associations between genotypes at two SNPs—rs8102595 and rs3826795, upstream of *HIF3A* and methylation. However, these variants did not show any association with BMI, suggesting that increased methylation at the *HIF3A* locus is a result of increased BMI rather than being a causative factor.

But the study did not consider potential modifying effects of environmental factors on the genetic associations. Accordingly, the possible association between environmental factors and the state of *HIF3A* methylation in the development of adiposity was further investigated in subsequent studies [86, 87].

In this regard, Pan et al. [87] applied a candidate-gene approach to investigate the association of infant weight with *HIF3A* methylation at birth. The previous results were replicated showing an association between higher methylation levels at three described *HIF3A* CpGs and greater infant weight or adiposity. However, no particular prenatal factor that strongly influences *HIF3A* hypermethylation, was identified [87]. Interestingly, CC genotype for two of the CpGs and rs3826795 has been associated with higher methylation values at the *HIF3A* locus, and birth weight was more strongly correlated with *HIF3A* methylation in this genotype group. Built on previous studies showing that neonatal methylation patterns are often a product of the interaction of the in utero environment and genotype [88], the authors suggested that the CC genotypic group could be more plastic to environmental exposures in utero.

In two other independent cohorts, Huang et al. [86], demonstrated significant interactions between DNA methylation–associated *HIF3A* SNP (rs3826795) and intakes of vitamin B2, vitamin B12, and folate on 10-year changes in BMI [86]. Even though these results did not indicate significant associations of *HIF3A* rs3826795 variant with adiposity measures, but, positive associations between this DNA methylation at this variant and total B-vitamin intake with 10-year BMI changes, supports the hypothesis that DNA methylation may causally affect body adiposity. This highlights the importance of considering environmental factors when assessing the DNA methylation inference regarding causality.

Association between DNA methylation pattern and obesity-related traits have been replicated in several studies; but, failed to be confirmed in others, indicating substantial false positive findings may exist in these genome-wide methylation studies [89–92]. However, these results are a step toward understanding the functional role of associated variation in the pathophysiology of obesity and identifying new molecular targets to avert its negative health consequences.

Interactions Between Genetics and Epigenetics

A growing body of evidence points towards implications of epigenetic mechanisms, including DNA methylation, in the pathogenesis of various metabolic traits including obesity. As patterns of DNA methylation are suggested to be widely genotype-dependent, genetic variations need to be taken into account when examining DNA methylation in the context of disease.

Thus, studies assessing whether individual variation in DNA methylation at metastable epialleles predicts the risk of adult weight gain would likely give new insights to the field. Genetic variants are suggested to be major contributors to methylation variation and evidence is available that SNPs can affect methylation in nearby (*cis*) or distant (*trans*) CpGs. Moreover, SNPs present in the methyltransferase is demonstrated to affect their enzymatic activities [93]. Thus, despite ample evidence for a correlation between methylation status and obesity, the interpretation of their causal relationship is still challenging.

Along with an increased understanding of the interplay between genetic factors and epigenetics, some studies have incorporated data on genetic variants associated with DNA methylation (methylation quantitative trait loci, meQTL) into results from GWAS for complex traits. These approaches provide potential tools for clarification of the unsolved previously identified disease-associated SNPs and mechanism by which methylation patterns can be transmitted across generations.

However, to date, scarce numbers of studies have focused on meQTL analysis for multifactorial obesity. The study by Drong et al. [94] was the first to report many cis-meQTLs in abdominal SAT as well as a significant association between meQTL and two cis-mRNAs (TNFRSF11B and GOT1) that encode for proteins implicated in T2D and metabolic syndrome [95, 96].

Grundberg et al. [97] extended these findings by assessing the genetic contribution to methylation level changes in adipose tissue from female twins. They reported a high degree of sequence dependency of variable CpG sites. However, the majority of these CpG sites did not have a biological mechanism or their functions have not been fully elucidated. They found that one of the top metabolic disease loci overlapping an enhancer metQTL (rs713586:T > C) [5] significantly affects DNA methylation in an enhancer region upstream of adenylate cyclase 3 (*ADCY3*), a metabolic-disease associated gene. Fine-mapping clarified that this is regulated via altered binding of a transcription factor (USF1) involved in lipid and glucose homeostasis.

Further study by Volkov et al. [98] provided a complete overview of genetic loci in both *cis* and *trans* positions that affect genome-wide DNA methylation in adipose tissue as well as metabolic phenotypes, including lipid and glucose traits. In line with previous mQTL analyses, they reported strong interactions between genetic and epigenetic variation. They reported several meQTLs associated with polymorphisms of previously identified GWAS loci for obesity, lipid, and T2D loci, including *ADCY3/POMC, APOA5, CETP, FADS2, GCKR, SORT1, and LEPR*. By using a causal inference test (CIT), they also demonstrated genetic variants interfere with different metabolic traits such as BMI, HbA1c, and HDL-cholesterol (HDL-c) via altered DNA methylation in human adipose tissue.

In other EWAS, Wahl et al. [91] used Mendelian randomization and a weighted genetic risk score, generated by SNPs known to affect BMI, to elucidate whether DNA methylation changes in blood or adipose tissue are cause or consequence of adiposity. Their findings demonstrated that at the majority of the identified CpG sites, DNA methylation alterations are a consequence and not the cause of adiposity. They further showed an association between the BMI risk score and DNA methylation of *ABCG1*, a gene known to be involved in insulin secretion and pancreatic β -cell function. It is consistent with observations that weight loss influences both *ABCG1* expression and protein activity in adipose tissue [99, 100]. It is noteworthy that they also showed the strongest association for the *ABCG1* locus, with incident T2D. Moreover, a single CpG (cg26663590: *NFATC2IP*) showed evidence of a genetic association for a causal role of methylation on BMI. In accordance with the causal role for *NFATC2IP* methylation in adiposity development, in longitudinal population studies, the authors observed baseline levels of methylation at cg26663590 predict weight gain.

Candidate Gene Approaches for DNA Methylation in Obesity

Multiple studies have used a hypothesis-driven, candidate gene approach where methylation sites in, or near, known candidate genes for obesity susceptibility have been the subject of investigation. The candidate gene methylation studies have focused on a range of genes implicated in obesity such as insulin signaling, immunity, metabolism, appetite control, and circadian clock regulation.

In some cases, the choice of genes has been based on prior analysis of gene expression differences in the same subjects.

For instance, expanded on previous evidence in favor of *FTO* causative role in obesity and T2D, early candidate-gene studies evaluated the methylation status of this gene [101]. DNA methylation associations have been shown in replicated T2D and obesity-related variants of *FTO* gene (SNP rs8050136 and rs9939609) [29, 101, 102].

Despite several studies reporting, obesity-associated SNPs and causal variants of *FTO* [29, 103], the heritability of its expression in skeletal muscle and adipose tissue along with their influence on in vivo glucose and fat metabolism was not influenced by *FTO* susceptibility genotype. Besides, no evidence of allele-specific expression in immortalized lymphoblastic cell lines has been established [104]. Thus, identification of stable haplotype-specific methylation within the *FTO* obesity susceptibility locus potentially aids the exploration of genotype–phenotype interactions in obesity traits.

POMC is another important gene that plays a crucial role in the regulation of energy balance and contributes to both monogenic and common obesity. Several studies have undertaken a candidate gene approach to investigate the association between *POMC* methylation status and BMI [105]. The results from human studies have shown that epigenetic marks at *POMC* (especially intron2/exon3), which is associated with a wide range of weight-related and metabolic outcomes, are sensitive to periconceptional and adolescent nutritional programming. Interestingly evidence from post mortem samples demonstrates that *POMC* DNA methylation is highly correlated across tissues originating from different germ cell layers i.e. brain—ectoderm, and kidney or PBC—mesoderm [105]. This suggests the methylation state was set before the separation of the germ layers at gastrulation. These hypermethylations were reported to interfere with the binding of the transcription enhancer P300 and reduce the expression of the *POMC* transcript.

It is of great importance that in obese individuals methylation of *POMC* was also higher in MSH-positive neurons (melanocyte-stimulating hormone) compared to lean counterparts, suggesting a regulatory function of methylation on the downstream pathway of *POMC* [106]. Importantly, assessment of *POMC* methylation in new-born and adolescent blood samples demonstrated that the methylation pattern appears stable, suggesting that associations with postnatal phenotypes are not driven by reverse causation effects [107].

IGF2 gene and its adjacent *H19* non-coding RNA are imprinted genes that correlate directly with growth, obesity, and body composition. In general, *H19* has a growth restraining effect, and *IGF2* acts as a growth-promoting factor. IGF2 triggers a major signal transduction pathway via activation of the IGF1 receptor, which mediates anabolic effects in adults. Huang et al. [108] found that *IGF2/H19* hypermethylation was associated with greater subcutaneous adiposity, but not with BMI, weight, height, waist circumference, or visceral adiposity in young ADULTS. Notably, assessment of methylation of the *H19* and *IGF2* in differentially methylated regions (DMRs) of monozygotic (MZ) twins and dizygotic (DZ) twins revealed that variation in DNA methylation of these loci is mainly determined by heritable factors and SNPs in cis, rather than the cumulative effect of environmental and stochastic factors occurring with age. Interestingly, paternal obesity has been reported to be associated with hypomethylation at the *IGF2* DMR in leukocytes isolated from umbilical cord blood at birth [109]. It has been hypothesized that the molecular mechanism behind these observations might be a hormonal difference between obese and non-obese parents, inducing an incomplete or unstable establishment of methylation at the *IGF2* DMR during gametogenesis. As a result, exposures to adverse lifestyle factors or poor/over-nutrition during spermatogenesis may affect the reprogramming of methylation profiles at imprinted genes.

Impact of Non-coding RNAs in Obesity

While less than 2% of the human genome is made up of genes that code for proteins, it is likely that ~70 to 80% of the human genome can be transcribed into non-coding RNA (ncRNA). ncRNAs are divided into two categories according to their transcript length: small ncRNAs (<200 nt), such as microRNAs (miRNAs) and long ncRNAs (lncRNAs) (>200 nt).

They play a crucial role in transcriptional and post-transcriptional regulation of the gene. There is an increasing body of evidence documenting the involvement of ncRNA in fundamental processes of obesity, including in adipogenesis, lipid metabolism, and inflammation. Among non-coding RNAs, both miRNAs and lncRNAs have been extensively investigated than others for their roles in these processes.

Role of miRNAs in Obesity

The miRNAs are a class of short non-coding RNAs that play important roles in cell transcriptional regulation. They function via base-pairing with complementary sequences within the 3′-untranslated region (3′UTR) of the target gene and promote gene silencing by inhibition of translation and/or by affecting mRNA stability and degradation.

Results from gain- or loss-of-function studies have provided the initial pieces of evidence for the crucial role of miRNAs in preadipocyte differentiation and adipocyte fate determination [110–112]. For example, adipose-tissue-specific knockout of the miRNA-processing enzyme Dicer (AdicerKO) developed a partial lipodystrophy phenotype, characterized by WAT atrophy, BAT hypertrophy whitening, insulin resistance, dyslipidemia, and premature aging [113, 114].

Furthermore, omics approaches provided tremendous insight into the role of miRNAs in the development of obesity and related metabolic complications [115, 116].

The impact of miRNAs in obesity and metabolic diseases has been indicated to be through their regulatory functions in adipogenesis, adipocyte differentiation, lipid metabolism, glucose homeostasis, and insulin resistance.

Adipogenesis is of major relevance to human disease, as dysregulated adipocyte physiology contributes to obesity and associated diseases. In general, adipose tissue is divided into two subtypes, brown adipose (BAT) and white adipose (WAT). WAT is true a metabolically active organ, which plays the main role in the maintenance of systematic energy through metabolism and storage triglycerides [117]. On the other hand, BAT is specialized in the production of heat [118].

Both common and distinct regulatory mechanisms are involved in the regulation of BAT- and WAT-derived adipocytes. For example, PPARγ (peroxisome proliferator-activated receptor-gamma) and CEBPs (CCAAT/enhancer-binding proteins) are critical regulators for BAT- and WAT-adipogenesis, whereas, UCP1 (uncoupling protein 1), PGC1α (peroxisome proliferator-activated receptor-gamma coactivator 1 alpha), and PRDM16 (PR-domain containing protein 16) are master regulators for brown fat differentiation.

Adipogenic commitment involves signaling factors that modulate conversion of multipotent mesenchymal stem cells (MSCs) to preadipocytes, which later differentiate into mature adipocytes. Several factors including insulin-like growth factor 1 (IGF1), glucocorticoid, cyclic AMP (cAMP)] C/EBPs, Sterol 1 regulatory element-binding protein (SREBP1) and mitogen-activated protein kinase 1 (MAPK1) and GSK3β and PPARγ are involved in the development of adipocyte phenotype. The first evidence suggesting a role for miRNAs in adipogenesis was reported in *Drosophila* flies illustrating that miR-14 and miR-278 play critical roles in the regulation of triacylglyceride metabolism [119, 120]. The deletion of miR-14 in *Drosophila* resulted in increased levels of triacylglycerol and diacylglycerol, whereas miR-14 up-regulation exhibited an opposite effect. Today, a large number of studies are focusing on the role of miRNAs in obesity. Deriving from mouse embryonic fibroblast, 3T3-L1 preadipocytes represent a well-defined in vitro cell model that has been extensively applied to study the mechanisms of adipogenesis. In an early study, Kajimoto et al. [121] examined miRNA expression profiles during 3T3-L1 pre-adipocyte differentiation and identified 21 miRNAs differentially expressed during the process. MiR-21 has been shown to promote adipogenesis by targeting TGFBR2, which mediates transforming growth factor-beta (TGF-)-induced inhibition of adipogenesis [122].

Consistent with this notion, microarray analysis has identified two classes of miRNAs, *miR-143* and the *miR-17/92* cluster, the expression of which is increased during adipogenic differentiation [123, 124].

Antisense knockdown of *miR-143* inhibited adipogenesis [123], whereas overexpression of the *miR-17/92* cluster moderately increased adipocyte formation [124]. MiR-143 is now recognized to induce adipogenesis by targeting ERK5 [123] or pleiotrophin [125], while miR-17–92 cluster promotes adipocyte differentiation via negatively regulating tumor-suppressor Rb2/p130 MiR-30 family (miR-30b/c) is another positive regulator of adipocyte differentiation, acting through the downregulation of the Runtrelated transcription factor 2 (Runx2), an osteogenic transcriptional factor stimulated by Wnt and the BMP pathways. The miR-30b/c has been described to be involved in the regulatory network of thermogenic genes, induced in BAT and subcutaneous WAT under cold [126, 127]. Similar to miR-30, miR-204/211 act as Runx2 attenuators to promote adipogenesis in MSCs and bone marrow–derived stromal cells (BMSCs) [128].

Furthermore, miR-17-5p is reported to inhibit preadipocyte differentiation by negatively regulating receptor co-activator 3 (NCOA3), a critical gene involved in regulating fat accumulation, and fat distribution [129]. Interestingly, the expression level of miR-17-5p in adipose tissue is indicated to be inversely correlated with hyperglycemia and insulin resistance in humans subjects [130].

MiRNAs can also promote adipogenesis through inhibition of MSCs myogenic differentiation. For instance, miR-193b-365 cluster is highly expressed in BAT and is involved in the adipogenesis of brown adipocytes by direct suppression of an adipogenesis inhibitor, Runx1t1, and two pro-myogenic genes, Cdon and Igfbp5 [131]. This cluster is induced by PRDM16, a crucial transcriptional regulator for brown/beige adipocyte differentiation [132]. Importantly, miR-193b is down-regulated in subcutaneous adipose tissue of obese individuals [133, 134].

In the late stages of adipogenesis, C/EBPα and PPARγ act synergistically to fully activate adiponectin gene and lipid metabolic genes. MiR-375 has been reported to promote this stage by suppressing ERK1/2 phosphorylation, and consequently up-regulation of C/EBPα and PPAR2 expression [135].

As a metabolically active organ, adipose tissue has a crucial role in the regulation of fat deposition and energy homeostasis. It functions through the production of several proteins that influence the function of many metabolic organs in endocrine, paracrine, and autocrine manner [136, 137].

From a physiopathological point of view, various factors have been associated with impaired adipose tissue function, such as lipotoxicity, low-grade chronic inflammation, oxidative stress, impaired adipogenesis, and insulin resistance. Interestingly, several miRNAs have recently been found to regulate adipose tissue biology (development and metabolism), thus playing a significant role in the obesity-induced metabolic disorder.

For instance, several individual miRNAs (miR-155, miR-221and miR-223) have been reported to play a role in the inflammatory state of adipose tissue. MiR-155 has been indicated to be up-regulated by TNFα in adipocytes and adipose tissue. Importantly, overexpression of miR-155 results in an increased inflammatory state in adipocytes [138] and reduction of insulin-stimulated glucose uptake in 3T3-L1cells [139]. Gain and loss of function experiments showed that miR-155 affects adipocyte function, probably by targeting PPARγ. Similarly, TNFα enhances the expression of miR-221 in 3T3-L1 adipocytes [140, 141]. At the same time, miR-221facilitates inflammation in WAT and reduces insulin sensitivity in obesity, through suppressing Sirtuin1 (SIRT1). It is noteworthy that as an inflammatory miRNA, expression levels of miR-221 are reported to be positively correlated with BMI, suggesting that it plays a role in obesity-related inflammation [142].

MiR-223 is another important inflammation-related miRNA, playing a crucial role in modulating macrophage polarization and protecting diet-induced adipose tissue inflammation as well as systemic insulin resistance. The anti-inflammatory function of miR-223 is described to be through suppression of classical proinflammatory M1 macrophages response in adipose tissue by targeting Pknox1 [143].

In addition to modulating macrophages activity, miRNAs may mediate adipose tissue inflammation by interactions with adipokines, which act in a paracrine and endocrine manner to elicit pro- and anti-inflammatory effects. Deregulation of adipokines is the main feature of adipose tissue low-grade inflammation and contributes to the pathogenesis of obesity related metabolic disorders. Thanks to adipose tissue miRNA profiling, the expression level of several miRNAs including miR883b-5p, miR-222, miR-143, and mir-95 have been demonstrated to be in correlation with concentrations of various adipokines [144, 145]. Taking together, these pieces of evidence have provided that miRNAs play a significant role in the pathophysiology of obesity and obesity-related disorders.

LncRNAs in Obesity

LncRNAs are emerging as important regulators of gene networks by a variety of mechanisms, including altering the recruitment of transcription factors, interacting with ribonucleoprotein particles, and chromatin-modifying complexes, as well as regulating mRNA and microRNA functions/stability [146].

Recently, several studies have addressed the role of lncRNAs in the network of both WAT and BAT adipogenesis and adipocyte biology.

The lncRNA steroid receptor RNA activator (SRA), which was initially identified to function as a transcriptional coactivator of steroid receptors [147], is among the first described functional lncRNA in adipocytes. SRA is described to affect adipogenesis and adipocyte function in multiple ways, including coactivation of PPARγ, promotion of S-phase entry during mitotic clonal expansion, and regulation of expression of inflammatory genes and signal transduction in response to insulin and TNFα [148]. However, current Ref-seq gene annotation shows that this gene locus is transcribed into three different isoforms, and only one of these transcripts behaves like a 'true' lncRNA: the other two isoforms each harbor an ORF that can potentially be translated into proteins [149–152]. As both SRA lncRNA and SRAPs could regulate the transcriptional activity of the estrogen receptor, which in turn can promote PPARγ gene transcription [152], the role of the noncoding Sra1 transcript in adipogenesis remains inconclusive.

LncRNA HOTAIR, which is expressed in gluteal adipose tissue and cultures of preadipocytes, is also reported to be involved in adipocyte differentiation. This lncRNA is known to regulate the transcriptional silencing of genes in the HOXD locus, which is involved in adipocyte differentiation. HOTAIR creates a scaffold to recruitment PRC2, a silencing complex involved in histone methylation to induce a heterochromatic status on the HOXD locus [76, 153]. Ectopic overexpression of HOTAIR has been reported to enhance abdominal preadipocytes differentiation and increase expressions of functional adipogenic markers including PPARγ, LPL (lipoprotein lipase), FABP4, and AdipoQ, with no effect on preadipocyte proliferation rate [154].

Using mRNA-lncRNA-combined microarray, lncRNA ADINR (adipogenic differentiation-induced noncoding RNA) was identified as a regulator of C/EBPα, [155]. In contrast to previous studies that used preadipocytes, Xiao et al. applied multipotent hMSCs for the loss and gain-of-function experiments. They have suggested that ADINR advantages transcription of the C/EBPα in *cis* by affecting the level of H3K4me3 in its promoter region.

By taking advantage of transcriptome profiling in WAT, Gao et al. [156] reported a pivotal role for two lncRNAs—ASMER-1 (adipocyte-specific metabolic related-1), ASMER-2, during adipogenesis. They showed that in vitro silencing of ASMERs via antisense oligonucleotides suppressed adipogenesis, adiponectin production, and lipid mobilization. They suggested that these effects could be attributed to crosstalk between ASMERs and key adipogenic transcription factors, including PPARγ and INSR.

Taking the important role of leptin in physiological regulation of white adipocyte biology and energy consumption into consideration, Dallner et al. [157] showed that a specific lncRNA, called lncOb is involved in the regulation of leptin expression. Diet-induced obese mice lacking this lncOb showed increased fat accumulation together with reduced circulating levels of leptin. Supplementation of these mice with leptin led to their weight loss [157].

Besides, to play significant regulatory roles in white adipogenesis, several lncRNAs have been identified to be involved in brown adipogenesis. For instance, using a global lncRNA profiling approach during thermogenic adipocyte formation Zhao et al., for the first time identified a lncRNA (named as brown fat lncRNA 1 (Blnc1), that promotes brown and beige adipocyte differentiation. Blnc1 exerts its regulatory function by the formation of a ribonucleoprotein complex with transcription factor EBF2 (early B-cell factor 2) to stimulate the thermogenic gene expression. Further, Blnc1 itself is a target of EBF2, thereby forming a feedforward regulatory loop to drive adipogenesis toward the thermogenic phenotype [158].

In line with these findings, by using RNA-seq approach, Alvarez and colleagues identified 127 BAT-restricted lncRNAs loci induced during differentiation and often targeted by key regulators PPARγ, C/EBPα, and C/EBPβ. Among them, lnc-BATE1, was identified to be required for the establishment and maintenance of BAT identity and thermogenic capacity.

As with miRNAs, accumulating pieces of evidence suggest that lncRNAs play an important role in modulating multiple aspects of immune responses [159]. For example, *Mis* is identified as an anti-inflammatory lncRNA that is down-regulated in multiple macrophage populations from obese insulin-resistant mice and stromal vascular fraction from metabolically unhealthy obese humans. This lncRNA functions through inhibition of macrophage inflammatory response and its down-regulation potentially play a protective role in metabolic dysfunction during obesity [160].

One of the fast-growing fields by the feature of their aptitude to considerably assist to investigate the interaction between genes and the environment is using biosensors. Overall, biosensors can be either transducer such as electrochemical biosensors or be bioreceptor [161]. Epigenetic biosensors are a subcategory of bioreceptors that are affinity-based. For example, when you have a model in which coat color variation in mice would be correlated to epigenetic marks, this model with established epigenetic marks in the early stages of development could be used safely to examine the inducement of nutritional and environmental impacts on the fetal epigenome. Using these epigenetic biosensors scientists have shown that simple dietary changes in mothers can protect fetal epigenome from the deleterious effect of environmental toxicants [162].

Furthermore, there are mouse models that are used as biosensors for examining the developmental origins of health and disease [163].

These epigenetic biosensors are different from other biosensors such as wearable biosensors that are used for monitoring obesity and diabetes [161]. Also, they are different from the electrochemical biosensor that investigators have used for detecting leptin in blood plasma from diet-induced obesity (DIO) mouse model [164] or early diagnosis of childhood obesity [165]. There are also sweat based wearable diagnostics biosensors that can screen artificial sweeteners rapidly and sensitively to protect people from obesity and diabetes [166].

Conclusion and Future Direction

One of the shortcomings of GWAS study is the limited sample size in numbers, race, gender, and origin. Future studies are encouraged to be designed with controlled confounding factors on larger sample size. Moreover, current studies are on coding genes, and the effect of noncoding genes is still unclear. Longitudinal cohort studies are suggested to be performed to assess the genetic risk and its evolution throughout the development from childhood to adulthood. The GRS could elucidate the inheritance risk of obesity as expression of the genes is different.

References

1. Haththotuwa RN, Wijeyaratne CN, Senarath U (2020) Chapter 1—worldwide epidemic of obesity. In: Mahmood TA, Arulkumaran S, Chervenak FA (eds) Obesity and obstetrics, 2nd edn. Elsevier, pp 3–8
2. Herrera BM, Lindgren CM (2010) The genetics of obesity. Curr DiabRep 10(6):498–505
3. Herrera BM, Keildson S, Lindgren CM (2011) Genetics and epigenetics of obesity. Maturitas 69(1):41–49
4. Tam V, Turcotte M, Meyre D (2019) Established and emerging strategies to crack the genetic code of obesity. Obes Rev Off J Int Assoc Study Obes 20(2):212–240
5. Speliotes EK, Willer CJ, Berndt SI, Monda KL, Thorleifsson G, Jackson AU et al (2010) Association analyses of 249,796 individuals reveal 18 new loci associated with body mass index. Nat Genet 42(11):937–948
6. Frayling TM, Timpson NJ, Weedon MN, Zeggini E, Freathy RM, Lindgren CM et al (2007) A common variant in the FTO gene is associated with body mass index and predisposes to childhood and adult obesity. Science (New York, NY) 316(5826):889–894
7. Popkin BM (2001) The nutrition transition and obesity in the developing world. J Nutr 131(3):871S-S873
8. Hinney A, Vogel CI, Hebebrand J (2010) From monogenic to polygenic obesity: recent advances. Eur Child Adolesc Psychiatr 19(3):297–310
9. Kaur Y, De Souza R, Gibson W, Meyre D (2017) A systematic review of genetic syndromes with obesity. Obes Rev 18(6):603–634
10. Pischon T, Boeing H, Hoffmann K, Bergmann M, Schulze MB, Overvad K et al (2008) General and abdominal adiposity and risk of death in Europe. N Engl J Med 359(20):2105–2120
11. Hinney A, Hebebrand J (2008) Polygenic obesity in humans. Obes Facts 1(1):35–42
12. Hebebrand J, Hinney A (2009) Environmental and genetic risk factors in obesity. Child Adolesc Psychiatr Clin N Am 18(1):83–94
13. Willer CJ, Speliotes EK, Loos RJ, Li S, Lindgren CM, Heid IM et al (2009) Six new loci associated with body mass index highlight a neuronal influence on body weight regulation. Nat Genet 41(1):25
14. Sabatti C, Service SK, Hartikainen A-L, Pouta A, Ripatti S, Brodsky J et al (2009) Genome-wide association analysis of metabolic traits in a birth cohort from a founder population. Nat Genet 41(1):35
15. Farooqi IS, Jebb SA, Langmack G, Lawrence E, Cheetham CH, Prentice AM et al (1999) Effects of recombinant leptin therapy in a child with congenital leptin deficiency. N Engl J Med 341(12):879–884
16. Farooqi IS, O'Rahilly S (2005) Monogenic obesity in humans. Annu Rev Med 56:443–458
17. Rankinen T, Zuberi A, Chagnon YC, Weisnagel SJ, Argyropoulos G, Walts B et al (2006) The human obesity gene map: the 2005 update. Obesity 14(4):529–644
18. Yang W, Kelly T, He J (2007) Genetic epidemiology of obesity. Epidemiol Rev 29(1):49–61
19. Belsky DW, Moffitt TE, Sugden K, Williams B, Houts R, McCarthy J et al (2013) Development and evaluation of a genetic risk score for obesity. Biodemography Soc Biol 59(1):85–100
20. Heid IM, Jackson AU, Randall JC, Winkler TW, Qi L, Steinthorsdottir V et al (2010) Meta-analysis identifies 13 new loci associated with waist-hip ratio and reveals sexual dimorphism in the genetic basis of fat distribution. Nat Genet 42(11):949–960

21. Eaton SB, Konner M, Shostak M (1988) Stone agers in the fast lane: chronic degenerative diseases in evolutionary perspective. Am J Med 84(4):739–749
22. Farooqi IS, Keogh JM, Yeo GS, Lank EJ, Cheetham T, O'Rahilly S (2003) Clinical spectrum of obesity and mutations in the melanocortin 4 receptor gene. N Engl J Med 348(12):1085–1095
23. Rhee KE, Phelan S, McCaffery J (2012) Early determinants of obesity: genetic, epigenetic, and in utero influences. Int J Pediatr 2012
24. Wardle J, Carnell S, Haworth CM, Farooqi IS, O'Rahilly S, Plomin R (2008) Obesity associated genetic variation in FTO is associated with diminished satiety. J Clin Endocrinol Metab 93(9):3640–3643
25. Dina C, Meyre D, Gallina S, Durand E, Körner A, Jacobson P et al (2007) Variation in FTO contributes to childhood obesity and severe adult obesity. Nat Genet 39(6):724–726
26. Wåhlén K, Sjölin E, Hoffstedt J (2008) The common rs9939609 gene variant of the fat mass- and obesity-associated gene FTO is related to fat cell lipolysis. J Lipid Res 49(3):607–611
27. Jia G, Yang C-G, Yang S, Jian X, Yi C, Zhou Z et al (2008) Oxidative demethylation of 3-methylthymine and 3-methyluracil in single-stranded DNA and RNA by mouse and human FTO. FEBS Lett 582(23–24):3313–3319
28. Gerken T, Girard CA, Tung Y-CL, Webby CJ, Saudek V, Hewitson KS et al (2007) The obesity-associated FTO gene encodes a 2-oxoglutarate-dependent nucleic acid demethylase. Science 318(5855):1469–1472
29. Fischer J, Koch L, Emmerling C, Vierkotten J, Peters T, Brüning JC et al (2009) Inactivation of the Fto gene protects from obesity. Nature 458(7240):894–898
30. Hall DH, Rahman T, Avery PJ, Keavney B (2006) INSIG-2 promoter polymorphism and obesity related phenotypes: association study in 1428 members of 248 families. BMC Med Genet 7(1):83
31. Geets E, Meuwissen ME, Van Hul W (2019) Clinical, molecular genetics and therapeutic aspects of syndromic obesity. Clin Genet 95(1):23–40
32. van der Valk ES, van den Akker EL, Savas M, Kleinendorst L, Visser JA, Van Haelst MM et al (2019) A comprehensive diagnostic approach to detect underlying causes of obesity in adults. Obes Rev 20(6):795–804
33. Abidi FE, Cardoso C, Lossi A-M, Lowry RB, Depetris D, Mattéi M-G et al (2005) Mutation in the 5′ alternatively spliced region of the XNP/ATR-X gene causes Chudley-Lowry syndrome. Eur J Hum Genet 13(2):176–183
34. Chudley AE, Lowry RB, Hoar DI, Opitz JM, Reynolds JF (1988) Mental retardation, distinct facial changes, short stature, obesity, and hypogonadism: a new X-linked mental retardation syndrome. Am J Med Genet 31(4):741–751
35. Delaunoy JP, Abidi F, Zeniou M, Jacquot S, Merienne K, Pannetier S et al (2001) Mutations in the X-linked RSK2 gene (RPS6KA3) in patients with Coffin-Lowry syndrome. Hum Mutat 17(2):103–116
36. Musio A, Selicorni A, Focarelli ML, Gervasini C, Milani D, Russo S et al (2006) X-linked Cornelia de Lange syndrome owing to SMC1L1 mutations. Nat Genet 38(5):528–530
37. Deardorff MA, Kaur M, Yaeger D, Rampuria A, Korolev S, Pie J et al (2007) Mutations in cohesin complex members SMC3 and SMC1A cause a mild variant of cornelia de Lange syndrome with predominant mental retardation. Am J Human Genet 80(3):485–494
38. Hampshire DJ, Ayub M, Springell K, Roberts E, Jafri H, Rashid Y et al (2006) MORM syndrome (m ental retardation, truncal o besity, r etinal dystrophy and m icropenis), a new autosomal recessive disorder, links to 9q34. Eur J Hum Genet 14(5):543–548
39. Hardelin J-P, Levilliers J, del Castillo I, Cohen-Salmon M, Legouis R, Blanchard S et al (1992) X chromosome-linked Kallmann syndrome: stop mutations validate the candidate gene. Proc Natl Acad Sci 89(17):8190–8194
40. Lederer D, Grisart B, Digilio MC, Benoit V, Crespin M, Ghariani SC et al (2012) Deletion of KDM6A, a histone demethylase interacting with MLL2, in three patients with Kabuki syndrome. Am J Human Genet 90(1):119–124
41. Ng SB, Bigham AW, Buckingham KJ, Hannibal MC, McMillin MJ, Gildersleeve HI et al (2010) Exome sequencing identifies MLL2 mutations as a cause of Kabuki syndrome. Nat Genet 42(9):790–793

42. Gillis LA, McCallum J, Kaur M, DeScipio C, Yaeger D, Mariani A et al (2004) NIPBL mutational analysis in 120 individuals with Cornelia de Lange syndrome and evaluation of genotype-phenotype correlations. Am J Human Genet 75(4):610–623
43. Deardorff MA, Bando M, Nakato R, Watrin E, Itoh T, Minamino M et al (2012) HDAC8 mutations in Cornelia de Lange syndrome affect the cohesin acetylation cycle. Nature 489(7415):313–317
44. Cormier-Daire V, Molinari F, Rio M, Raoul O, De Blois M, Romana S et al (2003) Cryptic terminal deletion of chromosome 9q34: a novel cause of syndromic obesity in childhood? J Med Genet 40(4):300–303
45. Thiele S, Werner R, Grötzinger J, Brix B, Staedt P, Struve D et al (2015) A positive genotype–phenotype correlation in a large cohort of patients with Pseudohypoparathyroidism Type Ia and Pseudo-pseudohypoparathyroidism and 33 newly identified mutations in the GNAS gene. Mol Genet Genomic Med 3(2):111–120
46. Hearn T, Renforth GL, Spalluto C, Hanley NA, Piper K, Brickwood S et al (2002) Mutation of ALMS1, a large gene with a tandem repeat encoding 47 amino acids, causes Alström syndrome. Nat Genet 31(1):79–83
47. Chiang AP, Nishimura D, Searby C, Elbedour K, Carmi R, Ferguson AL et al (2004) Comparative genomic analysis identifies an ADP-ribosylation factor–like gene as the cause of Bardet-Biedl syndrome (BBS3). Am J Human Genet 75(3):475–484
48. Mykytyn K, Nishimura DY, Searby CC, Shastri M, Yen H-j, Beck JS et al (2002) Identification of the gene (BBS1) most commonly involved in Bardet-Biedl syndrome, a complex human obesity syndrome. Nat Genet 31(4):435–438
49. M'hamdi O, Redin C, Stoetzel C, Ouertani I, Chaabouni M, Maazoul F et al (2014) Clinical and genetic characterization of Bardet–Biedl syndrome in Tunisia: defining a strategy for molecular diagnosis. Clin Genet85(2):172–177
50. Kolehmainen J, Black GC, Saarinen A, Chandler K, Clayton-Smith J, Träskelin A-L et al (2003) Cohen syndrome is caused by mutations in a novel gene, COH1, encoding a transmembrane protein with a presumed role in vesicle-mediated sorting and intracellular protein transport. Am J Human Genet 72(6):1359–1369
51. Dodé C, Teixeira L, Levilliers J, Fouveaut C, Bouchard P, Kottler M-L et al (2006) Kallmann syndrome: mutations in the genes encoding prokineticin-2 and prokineticin receptor-2. PLoS Genet 2(10):e175
52. Egger G, Liang G, Aparicio A, Jones PA (2004) Epigenetics in human disease and prospects for epigenetic therapy. Nature 429(6990):457–463
53. van Otterdijk SD, Michels KB (2016) Transgenerational epigenetic inheritance in mammals: how good is the evidence? FASEB J Offic Publ Fed Am Soc Exp Biol. 30(7):2457–2465
54. Zhang B, Gu X, Han X, Gao Q, Liu J, Guo T et al (2020) Crosstalk between DNA methylation and histone acetylation triggers GDNF high transcription in glioblastoma cells. Clin Epigenetics 12(1):47
55. Okano M, Bell DW, Haber DA, Li E (1999) DNA methyltransferases Dnmt3a and Dnmt3b are essential for de novo methylation and mammalian development. Cell 99(3):247–257
56. Berger SL (2007) The complex language of chromatin regulation during transcription. Nature 447(7143):407–412
57. Goldberg AD, Allis CD, Bernstein E (2007) Epigenetics: a landscape takes shape. Cell 128(4):635–638
58. Cacabelos R, Tellado I, Cacabelos P (2019) Chapter 1—the epigenetic machinery in the life cycle and pharmacoepigenetics. In: Cacabelos R (ed) Pharmacoepigenetics, vol 10. Academic Press, pp 1–100
59. Larijani L, Rancourt DE (2018) Chapter 29—stem cell epigenetics and human disease. In: Tollefsbol TO (ed) Epigenetics in human disease, vol 6, 2nd edn. Academic Press, pp 877–902
60. Dolinoy D, Das R, Weidman J, Jirtle R (2007) Metastable epialleles, imprinting, and the fetal origins of adult diseases. Pediatr Res 61:30R-R37
61. Kim M, Costello J (2017) DNA methylation: an epigenetic mark of cellular memory. Exp Mol Med 49(4):e322–e

62. Cheng Z, Zheng L, Almeida FA (2018) Epigenetic reprogramming in metabolic disorders: nutritional factors and beyond. J Nutr Biochem 54:1–10
63. Xu X, Su S, Barnes VA, De Miguel C, Pollock J, Ownby D et al (2013) A genome-wide methylation study on obesity: differential variability and differential methylation. Epigenetics 8(5):522–533
64. Wang X, Zhu H, Snieder H, Su S, Munn D, Harshfield G et al (2010) Obesity related methylation changes in DNA of peripheral blood leukocytes. BMC Med 8:87
65. Almén MS, Jacobsson JA, Moschonis G, Benedict C, Chrousos GP, Fredriksson R et al (2012) Genome wide analysis reveals association of a FTO gene variant with epigenetic changes. Genomics 99(3):132–137
66. Bjornsson HT, Sigurdsson MI, Fallin MD, Irizarry RA, Aspelund T, Cui H et al (2008) Intra-individual change over time in DNA methylation with familial clustering. JAMA 299(24):2877–2883
67. Feinberg AP, Irizarry RA, Fradin D, Aryee MJ, Murakami P, Aspelund T et al (2010) Personalized epigenomic signatures that are stable over time and covary with body mass index. Sci Transl Med 2(49):49ra67
68. O'Hara A, Lim FL, Mazzatti DJ, Trayhurn P (2009) Microarray analysis identifies matrix metalloproteinases (MMPs) as key genes whose expression is up-regulated in human adipocytes by macrophage-conditioned medium. Pflugers Arch 458(6):1103–1114
69. Chavey C, Mari B, Monthouel MN, Bonnafous S, Anglard P, Van Obberghen E et al (2003) Matrix metalloproteinases are differentially expressed in adipose tissue during obesity and modulate adipocyte differentiation. J Biol Chem 278(14):11888–11896
70. Derosa G, Ferrari I, D'Angelo A, Tinelli C, Salvadeo SAT, Ciccarelli L et al (2008) Matrix Metalloproteinase-2 and -9 Levels in Obese Patients. Endothelium 15(4):219–224
71. Benson KK, Hu W, Weller AH, Bennett AH, Chen ER, Khetarpal SA et al (2019) Natural human genetic variation determines basal and inducible expression of PM20D1, an obesity-associated gene. Proc Natl Acad Sci USA 116(46):23232–23242
72. Kaun KR, Sokolowski MB (2009) cGMP-dependent protein kinase: linking foraging to energy homeostasis. Genome 52(1):1–7
73. Włodarczyk M, Nowicka G (2019) Obesity, DNA damage, and development of obesity-related diseases. Int J Mol Sci 20(5):1146
74. Ge Y, Paisie TK, Newman JRB, McIntyre LM, Concannon P (2017) UBASH3A mediates risk for type 1 diabetes through inhibition of T-cell receptor-induced NF-κB signaling. Diabetes 66(7):2033–2043
75. Ge Y, Paisie TK, Chen S, Concannon P (2019) UBASH3A regulates the synthesis and dynamics of TCR-CD3 complexes. J Immunol (Baltimore, Md: 1950) 203(11):2827–2836
76. Barrett JC, Clayton DG, Concannon P, Akolkar B, Cooper JD, Erlich HA et al (2009) Genome-wide association study and meta-analysis find that over 40 loci affect risk of type 1 diabetes. Nat Genet 41(6):703–707
77. Chen Y-G, Ciecko AE, Khaja S, Grzybowski M, Geurts AM, Lieberman SM (2020) UBASH3A deficiency accelerates type 1 diabetes development and enhances salivary gland inflammation in NOD mice. Sci Rep 10(1):12019
78. Ozato K, Shin DM, Chang TH, Morse HC 3rd (2008) TRIM family proteins and their emerging roles in innate immunity. Nat Rev Immunol 8(11):849–860
79. Dick KJ, Nelson CP, Tsaprouni L, Sandling JK, Aïssi D, Wahl S et al (2014) DNA methylation and body-mass index: a genome-wide analysis. Lancet (London, England) 383(9933):1990–1998
80. Park YS, David AE, Huang Y, Park J-B, He H, Byun Y et al (2012) In vivo delivery of cell-permeable antisense hypoxia-inducible factor 1α oligonucleotide to adipose tissue reduces adiposity in obese mice. J Control Release 161(1):1–9
81. Zhang H, Zhang G, Gonzalez FJ, Park S-M, Cai D (2011) Hypoxia-inducible factor directs POMC gene to mediate hypothalamic glucose sensing and energy balance regulation. PLoS Biol 9(7):e1001112–e

82. Shin MK, Drager LF, Yao Q, Bevans-Fonti S, Yoo DY, Jun JC et al (2010) Metabolic consequences of high-fat diet are attenuated by suppression of HIF-1α. PLoS One 7(10):e46562
83. Jiang C, Qu A, Matsubara T, Chanturiya T, Jou W, Gavrilova O et al (2011) Disruption of hypoxia-inducible factor 1 in adipocytes improves insulin sensitivity and decreases adiposity in high-fat diet-fed mice. Diabetes 60(10):2484–2495
84. Hatanaka M, Shimba S, Sakaue M, Kondo Y, Kagechika H, Kokame K et al (2009) Hypoxia-inducible factor-3alpha functions as an accelerator of 3T3-L1 adipose differentiation. Biol Pharm Bull 32(7):1166–1172
85. Heidbreder M, Qadri F, Jöhren O, Dendorfer A, Depping R, Fröhlich F et al (2007) Non-hypoxic induction of HIF-3alpha by 2-deoxy-D-glucose and insulin. Biochem Biophys Res Commun 352(2):437–443
86. Huang T, Zheng Y, Qi Q, Xu M, Ley SH, Li Y et al (2015) DNA methylation variants at HIF3A locus, B-vitamin intake, and long-term weight change: gene-diet interactions in two U.S. cohorts. Diabetes 64(9):3146–3154
87. Pan H, Lin X, Wu Y, Chen L, Teh AL, Soh SE et al (2015) HIF3A association with adiposity: the story begins before birth. Epigenomics 7(6):937–950
88. Teh AL, Pan H, Chen L, Ong ML, Dogra S, Wong J et al (2014) The effect of genotype and in utero environment on interindividual variation in neonate DNA methylomes. Genome Res 24(7):1064–1074
89. Aslibekyan S, Demerath EW, Mendelson M, Zhi D, Guan W, Liang L et al (2015) Epigenome-wide study identifies novel methylation loci associated with body mass index and waist circumference. Obesity (Silver Spring, Md) 23(7):1493–1501
90. Rönn T, Volkov P, Gillberg L, Kokosar M, Perfilyev A, Jacobsen AL et al (2015) Impact of age, BMI and HbA1c levels on the genome-wide DNA methylation and mRNA expression patterns in human adipose tissue and identification of epigenetic biomarkers in blood. Hum Mol Genet 24(13):3792–3813
91. Wahl S, Drong A, Lehne B, Loh M, Scott WR, Kunze S et al (2017) Epigenome-wide association study of body mass index, and the adverse outcomes of adiposity. Nature 541(7635):81–86
92. Sayols-Baixeras S, Subirana I, Fernández-Sanlés A, Sentí M, Lluís-Ganella C, Marrugat J et al (2017) DNA methylation and obesity traits: An epigenome-wide association study The REGICOR study. Epigenetics 12(10):909–916
93. Saradalekshmi KR, Neetha NV, Sathyan S, Nair IV, Nair CM, Banerjee M (2014) DNA methyl transferase (DNMT) gene polymorphisms could be a primary event in epigenetic susceptibility to schizophrenia. PLoS One 9(5):e98182
94. Drong AW, Nicholson G, Hedman AK, Meduri E, Grundberg E, Small KS et al (2013) The presence of methylation quantitative trait loci indicates a direct genetic influence on the level of DNA methylation in adipose tissue. PLoS One8(2):e55923
95. Biscetti F, Porreca CF, Bertucci F, Straface G, Santoliquido A, Tondi P et al (2014) TNFRSF11B gene polymorphisms increased risk of peripheral arterial occlusive disease and critical limb ischemia in patients with type 2 diabetes. Acta Diabetol 51(6):1025–1032
96. Shen H, Damcott C, Shuldiner SR, Chai S, Yang R, Hu H et al (2011) Genome-wide association study identifies genetic variants in GOT1 determining serum aspartate aminotransferase levels. J Hum Genet 56(11):801–805
97. Grundberg E, Meduri E, Sandling Johanna K, Hedman Åsa K, Keildson S, Buil A et al (2013) Global analysis of DNA methylation variation in adipose tissue from twins reveals links to disease-associated variants in distal regulatory elements. Am J Human Genet 93(5):876–890
98. Volkov P, Olsson AH, Gillberg L, Jørgensen SW, Brøns C, Eriksson KF et al (2016) A genome-wide mQTL analysis in human adipose tissue identifies genetic variants associated with DNA methylation, gene expression and metabolic traits. PLoS One11(6):e0157776
99. Johansson LE, Danielsson AP, Parikh H, Klintenberg M, Norström F, Groop L et al (2012) Differential gene expression in adipose tissue from obese human subjects during weight loss and weight maintenance. Am J Clin Nutr 96(1):196–207

100. Aron-Wisnewsky J, Julia Z, Poitou C, Bouillot J-L, Basdevant A, Chapman MJ et al (2011) Effect of bariatric surgery-induced weight loss on SR-BI-, ABCG1-, and ABCA1-mediated cellular cholesterol efflux in obese women. J Clin Endocrinol Metab 96(4):1151–1159
101. Bell CG, Finer S, Lindgren CM, Wilson GA, Rakyan VK, Teschendorff AE et al (2010) Integrated genetic and epigenetic analysis identifies haplotype-specific methylation in the FTO type 2 diabetes and obesity susceptibility locus. PLoS One5(11):e14040
102. Zhou Y, Simmons D, Lai D, Hambly BD, McLachlan CS (2017) rs9939609 FTO genotype associations with FTO methylation level influences body mass and telomere length in an Australian rural population. Int J Obes 41(9):1427–1433
103. Boissel S, Reish O, Proulx K, Kawagoe-Takaki H, Sedgwick B, Yeo GS et al (2009) Loss-of-function mutation in the dioxygenase-encoding FTO gene causes severe growth retardation and multiple malformations. Am J Human Genet 85(1):106–111
104. Verlaan DJ, Ge B, Grundberg E, Hoberman R, Lam KC, Koka V et al (2009) Targeted screening of cis-regulatory variation in human haplotypes. Genome Res 19(1):118–127
105. Candler T, Kühnen P, Prentice AM, Silver M (2019) Epigenetic regulation of POMC; implications for nutritional programming, obesity and metabolic disease. Front Neuroendocrinol 54:100773
106. Crujeiras AB, Campion J, Díaz-Lagares A, Milagro FI, Goyenechea E, Abete I et al (2013) Association of weight regain with specific methylation levels in the NPY and POMC promoters in leukocytes of obese men: a translational study. Regul Pept 186:1–6
107. Kühnen P, Handke D, Waterland Robert A, Hennig Branwen J, Silver M, Fulford Anthony J et al (2016) Interindividual variation in DNA methylation at a putative POMC metastable epiallele is associated with obesity. Cell Metab 24(3):502–509
108. Huang RC, Galati JC, Burrows S, Beilin LJ, Li X, Pennell CE et al (2012) DNA methylation of the IGF2/H19 imprinting control region and adiposity distribution in young adults. Clin Epigenetics 4(1):21
109. Soubry A, Schildkraut JM, Murtha A, Wang F, Huang Z, Bernal A et al (2013) Paternal obesity is associated with IGF2hypomethylation in newborns: results from a Newborn Epigenetics Study (NEST) cohort. BMC Med 11(1):29
110. Mudhasani R, Imbalzano AN, Jones SN (2010) An essential role for Dicer in adipocyte differentiation. J Cell Biochem 110(4):812–816
111. Mudhasani R, Puri V, Hoover K, Czech MP, Imbalzano AN, Jones SN (2011) Dicer is required for the formation of white but not brown adipose tissue. J Cell Physiol 226(5):1399–1406
112. Fujimoto Y, Nakagawa Y, Shingyouchi A, Tokushige N, Nakanishi N, Satoh A et al (2012) Dicer has a crucial role in the early stage of adipocyte differentiation, but not in lipid synthesis, in 3T3-L1 cells. Biochem Biophys Res Commun 420(4):931–936
113. Reis FCG, Branquinho JLO, Brandão BB, Guerra BA, Silva ID, Frontini A et al (2016) Fat-specific Dicer deficiency accelerates aging and mitigates several effects of dietary restriction in mice. Aging 8(6):1201–1222
114. Mori MA, Thomou T, Boucher J, Lee KY, Lallukka S, Kim JK et al (2014) Altered miRNA processing disrupts brown/white adipocyte determination and associates with lipodystrophy. J Clin Investig 124(8):3339–3351
115. Arner P, Kulyté A (2015) MicroRNA regulatory networks in human adipose tissue and obesity. Nat Rev Endocrinol 11(5):276–288
116. Iacomino G, Siani A (2017) Role of microRNAs in obesity and obesity-related diseases. Genes Nutr 12:23–
117. Raajendiran A, Ooi G, Bayliss J, O'Brien PE, Schittenhelm RB, Clark AK et al (2019) Identification of metabolically distinct adipocyte progenitor cells in human adipose tissues. Cell Rep 27(5):1528–40.e7
118. Bijland S, Mancini S, Salt I (2013) Role of AMP-activated protein kinase in adipose tissue metabolism and inflammation. Clin Sci (London, England: 1979) 124:491–507
119. Xu P, Vernooy SY, Guo M, Hay BA (2003) The Drosophila microRNA Mir-14 suppresses cell death and is required for normal fat metabolism. Curr Biol CB 13(9):790–795

120. Teleman AA, Maitra S, Cohen SM (2006) Drosophila lacking microRNA miR-278 are defective in energy homeostasis. Genes Dev 20(4):417–422
121. Kajimoto K, Naraba H, Iwai N (2006) MicroRNA and 3T3-L1 pre-adipocyte differentiation. RNA 12(9):1626–1632
122. Kim YJ, Hwang SJ, Bae YC, Jung JS (2009) MiR-21 regulates adipogenic differentiation through the modulation of TGF-beta signaling in mesenchymal stem cells derived from human adipose tissue. Stem cells (Dayton, Ohio) 27(12):3093–3102
123. Esau C, Kang X, Peralta E, Hanson E, Marcusson EG, Ravichandran LV et al (2004) MicroRNA-143 regulates adipocyte differentiation. J Biol Chem 279(50):52361–52365
124. Wang Q, Li YC, Wang J, Kong J, Qi Y, Quigg RJ et al (2008) miR-17-92 cluster accelerates adipocyte differentiation by negatively regulating tumor-suppressor Rb2/p130. Proc Natl Acad Sci USA 105(8):2889–2894
125. Yi C, Xie W-d, Li F, Lv Q, He J, Wu J et al (2011) MiR-143 enhances adipogenic differentiation of 3T3-L1 cells through targeting the coding region of mouse pleiotrophin. FEBS Lett 585(20):3303–3309
126. Hu F, Wang M, Xiao T, Yin B, He L, Meng W et al (2015) miR-30 promotes thermogenesis and the development of beige fat by targeting RIP140. Diabetes 64(6):2056–2068
127. Zaragosi LE, Wdziekonski B, Brigand KL, Villageois P, Mari B, Waldmann R et al (2011) Small RNA sequencing reveals miR-642a-3p as a novel adipocyte-specific microRNA and miR-30 as a key regulator of human adipogenesis. Genome Biol 12(7):R64
128. Huang J, Zhao L, Xing L, Chen D (2010) MicroRNA-204 regulates Runx2 protein expression and mesenchymal progenitor cell differentiation. Stem cells (Dayton, Ohio) 28(2):357–364
129. Han H, Gu S, Chu W, Sun W, Wei W, Dang X et al (2017) miR-17-5p regulates differential expression of NCOA3 in pig intramuscular and subcutaneous adipose tissue. Lipids 52(11):939–949
130. Klöting N, Berthold S, Kovacs P, Schön MR, Fasshauer M, Ruschke K et al (2009) MicroRNA expression in human omental and subcutaneous adipose tissue. PLoS One4(3):e4699
131. Sun L, Xie H, Mori MA, Alexander R, Yuan B, Hattangadi SM et al (2011) Mir193b-365 is essential for brown fat differentiation. Nat Cell Biol 13(8):958–965
132. Seale P, Kajimura S, Yang W, Chin S, Rohas LM, Uldry M et al (2007) Transcriptional control of brown fat determination by PRDM16. Cell Metab 6(1):38–54
133. Meerson A, Traurig M, Ossowski V, Fleming JM, Mullins M, Baier LJ (2013) Human adipose microRNA-221 is upregulated in obesity and affects fat metabolism downstream of leptin and TNF-α. Diabetologia 56(9):1971–1979
134. Arner E, Mejhert N, Kulyté A, Balwierz PJ, Pachkov M, Cormont M et al (2012) Adipose tissue microRNAs as regulators of CCL2 production in human obesity. Diabetes 61(8):1986–1993
135. Ling HY, Wen GB, Feng SD, Tuo QH, Ou HS, Yao CH et al (2011) MicroRNA-375 promotes 3T3-L1 adipocyte differentiation through modulation of extracellular signal-regulated kinase signalling. Clin Exp Pharmacol Physiol 38(4):239–246
136. Scherer PE (2006) Adipose tissue: from lipid storage compartment to endocrine organ. Diabetes 55(6):1537–1545
137. Al-Lahham SH, Roelofsen H, Priebe M, Weening D, Dijkstra M, Hoek A et al (2010) Regulation of adipokine production in human adipose tissue by propionic acid. Eur J Clin Invest 40(5):401–407
138. Karkeni E, Astier J, Tourniaire F, El Abed M, Romier B, Gouranton E et al (2016) Obesity-associated Inflammation Induces microRNA-155 expression in adipocytes and adipose tissue: outcome on adipocyte function. J Clin Endocrinol Metab 101(4):1615–1626
139. Ying W, Riopel M, Bandyopadhyay G, Dong Y, Birmingham A, Seo JB et al (2017) Adipose tissue macrophage-derived exosomal miRNAs can modulate in vivo and in vitro insulin sensitivity. Cell 171(2):372–84.e12
140. Xie H, Lim B, Lodish H (2009) MicroRNAs induced during adipogenesis that accelerate fat cell development are downregulated in obesity. Diabetes 58:1050–1057
141. Chou WW, Wang YT, Liao YC, Chuang SC, Wang SN, Juo SHH (2013) Decreased microRNA-221 is associated with high levels of TNF-α in human adipose tissue-derived mesenchymal stem cells from obese woman. Cell Physiol Biochem 32(1):127–137

142. Peng J, Zhou Y, Deng Z, Zhang H, Wu Y, Song T et al (2018) miR-221 negatively regulates inflammation and insulin sensitivity in white adipose tissue by repression of sirtuin-1 (SIRT1). J Cell Biochem 119(8):6418–6428
143. Zhuang G, Meng C, Guo X, Cheruku PS, Shi L, Xu H et al (2012) A novel regulator of macrophage activation: miR-223 in obesity-associated adipose tissue inflammation. Circulation 125(23):2892–2903
144. Parra P, Serra F, Palou A (2010) Expression of adipose microRNAs is sensitive to dietary conjugated linoleic acid treatment in mice. PLoS One5(9):e13005
145. Ge Q, Gérard J, Noël L, Scroyen I, Brichard SM (2012) MicroRNAs regulated by adiponectin as novel targets for controlling adipose tissue inflammation. Endocrinology 153(11):5285–5296
146. Gebert LFR, MacRae IJ (2019) Regulation of microRNA function in animals. Nat Rev Mol Cell Biol 20(1):21–37
147. Lanz RB, McKenna NJ, Onate SA, Albrecht U, Wong J, Tsai SY et al (1999) A steroid receptor coactivator, SRA, functions as an RNA and is present in an SRC-1 complex. Cell 97(1):17–27
148. Xu B, Gerin I, Miao H, Vu-Phan D, Johnson CN, Xu R et al (2010) Multiple roles for the non-coding RNA SRA in regulation of adipogenesis and insulin sensitivity. PLoS One5(12):e14199
149. Chooniedass-Kothari S, Emberley E, Hamedani MK, Troup S, Wang X, Czosnek A et al (2004) The steroid receptor RNA activator is the first functional RNA encoding a protein. FEBS Lett 566(1–3):43–47
150. Kawashima H, Takano H, Sugita S, Takahara Y, Sugimura K, Nakatani T (2003) A novel steroid receptor co-activator protein (SRAP) as an alternative form of steroid receptor RNA-activator gene: expression in prostate cancer cells and enhancement of androgen receptor activity. Biochem J 369(Pt 1):163–171
151. McKay DB, Xi L, Barthel KKB, Cech TR (2014) Structure and function of steroid receptor RNA activator protein, the proposed partner of SRA noncoding RNA. J Mol Biol 426(8):1766–1785
152. Coleman KM, Lam V, Jaber BM, Lanz RB, Smith CL (2004) SRA coactivation of estrogen receptor-alpha is phosphorylation-independent, and enhances 4-hydroxytamoxifen agonist activity. Biochem Biophys Res Commun 323(1):332–338
153. Gupta RA, Shah N, Wang KC, Kim J, Horlings HM, Wong DJ et al (2010) Long non-coding RNA HOTAIR reprograms chromatin state to promote cancer metastasis. Nature 464(7291):1071–1076
154. Divoux A, Karastergiou K, Xie H, Guo W, Perera RJ, Fried SK et al (2014) Identification of a novel lncRNA in gluteal adipose tissue and evidence for its positive effect on preadipocyte differentiation. Obesity (Silver Spring, Md) 22(8):1781–1785
155. Xiao T, Liu L, Li H, Sun Y, Luo H, Li T et al (2015) Long Noncoding RNA ADINR Regulates Adipogenesis by Transcriptionally Activating C/EBPα. Stem Cell Rep 5(5):856–865
156. Gao H, Kerr A, Jiao H, Hon CC, Rydén M, Dahlman I et al (2018) Long non-coding RNAs associated with metabolic traits in human white adipose tissue. EBioMedicine 30:248–260
157. Dallner OS, Marinis JM, Lu YH, Birsoy K, Werner E, Fayzikhodjaeva G et al (2019) Dysregulation of a long noncoding RNA reduces leptin leading to a leptin-responsive form of obesity. Nat Med 25(3):507–516
158. Zhao XY, Li S, Wang GX, Yu Q, Lin JD (2014) A long noncoding RNA transcriptional regulatory circuit drives thermogenic adipocyte differentiation. Mol Cell 55(3):372–382
159. Heward JA, Lindsay MA (2014) Long non-coding RNAs in the regulation of the immune response. Trends Immunol 35(9):408–419
160. Stapleton K, Das S, Reddy MA, Leung A, Amaram V, Lanting L et al (2020) Novel long noncoding RNA, Macrophage Inflammation-Suppressing Transcript (MIST), regulates macrophage activation during obesity. Arterioscler Thromb Vasc Biol 40(4):914–928
161. Rodrigues D, Barbosa AI, Rebelo R, Kwon IK, Reis RL, Correlo VM (2020) Skin-integrated wearable systems and implantable biosensors: a comprehensive review. Biosensors (Basel) 10(7)

162. Dolinoy DC (2008) The agouti mouse model: an epigenetic biosensor for nutritional and environmental alterations on the fetal epigenome. Nutr Rev 66(Suppl 1):S7-11
163. Jirtle RL (2014) The Agouti mouse: a biosensor for environmental epigenomics studies investigating the developmental origins of health and disease. Epigenomics 6(5):447–450
164. Sung R, Heo YS (2020) Sandwich ELISA-based electrochemical biosensor for leptin in control and diet-induced obesity mouse model. Biosensors (Basel) 11(1)
165. Pirsean C, Negut C, Stefan-van Staden RI, Dinu-Pirvu CE, Armean P, Udeanu DI (2019) The salivary levels of leptin and interleukin-6 as potential inflammatory markers in children obesity. PLoS One 14(1):e0210288
166. Mehrotra P (2016) Biosensors and their applications—a review. J Oral Biol Craniofac Res 6(2):153–159
167. Zhang X, Ehrlich KC, Yu F, Hu X, Meng XH, Deng HW et al (2020) Osteoporosis- and obesity-risk interrelationships: an epigenetic analysis of GWAS-derived SNPs at the developmental gene TBX15. Epigenetics 15(6–7):728–749
168. Song QY, Meng XR, Hinney A, Song JY, Huang T, Ma J et al (2018) Waist-hip ratio related genetic loci are associated with risk of impaired fasting glucose in Chinese children: a case control study. Nutr Metab 15:34
169. D'Angelo CS, Varela MC, de Castro CIE, Otto PA, Perez ABA, Lourenço CM et al (2018) Chromosomal microarray analysis in the genetic evaluation of 279 patients with syndromic obesity. Mol Cytogenet 11(1):14
170. Dennis EL, Jahanshad N, Braskie MN, Warstadt NM, Hibar DP, Kohannim O et al (2014) Obesity gene NEGR1 associated with white matter integrity in healthy young adults. Neuroimage 102:548–557
171. Samaan Z, Lee YK, Gerstein HC, Engert JC, Bosch J, Mohan V et al (2015) Obesity genes and risk of major depressive disorder in a multiethnic population: a cross-sectional study. J Clin Psychiatry 76(12):1611–1618
172. Zhang X, Cupples LA, Liu C-T (2018) A fine-mapping study of central obesity loci incorporating functional annotation and imputation. Eur J Hum Genet 26(9):1369–1377
173. Hotta K, Nakamura M, Nakamura T, Matsuo T, Nakata Y, Kamohara S et al (2009) Association between obesity and polymorphisms in SEC16B, TMEM18, GNPDA2, BDNF, FAIM2 and MC4R in a Japanese population. J Hum Genet 54(12):727–731
174. Albuquerque D, Stice E, Rodríguez-López R, Manco L, Nóbrega C (2015) Current review of genetics of human obesity: from molecular mechanisms to an evolutionary perspective. Mol Genet Genomics 290(4):1191–1221
175. Chedid MF, do Nascimento FV, de Oliveira FS, de Souza BM, Kruel CR, Gurski RR et al (2019) Interaction of HSD11B1 and H6PD polymorphisms in subjects with type 2 diabetes are protective factors against obesity: a cross-sectional study. Diabetol Metab Syndr 11(1):78
176. Kochetova O, Viktorova T (2015) Genetics and epigenetics of obesity. Biol Bull Rev 5(6):538–547
177. Scherag A, Kleber M, Boes T, Kolbe AL, Ruth A, Grallert H et al (2012) SDCCAG8 obesity alleles and reduced weight loss after a lifestyle intervention in overweight children and adolescents. Obesity 20(2):466–470
178. Williams MJ, Almén MS, Fredriksson R, Schiöth HB (2012) What model organisms and interactomics can reveal about the genetics of human obesity. Cell Mol Life Sci 69(22):3819–3834
179. Parton LE, Ye CP, Coppari R, Enriori PJ, Choi B, Zhang C-Y et al (2007) Glucose sensing by POMC neurons regulates glucose homeostasis and is impaired in obesity. Nature 449(7159):228–232
180. Houde A-A, Ruchat S-M, Allard C, Baillargeon J-P, St-Pierre J, Perron P et al (2015) LRP1B, BRD2 and CACNA1D: new candidate genes in fetal metabolic programming of newborns exposed to maternal hyperglycemia. Epigenomics 7(7):1111–1122
181. Grarup N, Andersen G, Krarup NT, Albrechtsen A, Schmitz O, Jørgensen T et al (2008) Association testing of novel type 2 diabetes risk alleles in the JAZF1, CDC123/CAMK1D, TSPAN8, THADA, ADAMTS9, and NOTCH2 loci with insulin release, insulin sensitivity, and

obesity in a population-based sample of 4,516 glucose-tolerant middle-aged Danes. Diabetes 57(9):2534–2540

182. Morris J, Bailey ME, Baldassarre D, Cullen B, de Faire U, Ferguson A et al (2019) Genetic variation in CADM2 as a link between psychological traits and obesity. Sci Rep 9(1):1–14
183. Ng MC, Tam CH, So WY, Ho JS, Chan AW, Lee HM et al (2010) Implication of genetic variants near negr1, sec16b, tmem18, etv5/dgkg, gnpda2, lin7c/bdnf, mtch2, bcdin3d/faim2, sh2b1, fto, mc4r, and kctd15 with obesity and type 2 diabetes in 7705 Chinese. J Clin Endocrinol Metab 95(5):2418–2425
184. Li D, Achkar J-P, Haritunians T, Jacobs JP, Hui KY, D'Amato M et al (2016) A pleiotropic missense variant in SLC39A8 is associated with Crohn's disease and human gut microbiome composition. Gastroenterology 151(4):724–732
185. Lamiquiz-Moneo I, Mateo-Gallego R, Bea AM, Dehesa-García B, Pérez-Calahorra S, Marco-Benedí V et al (2019) Genetic predictors of weight loss in overweight and obese subjects. Sci Rep 9(1):1–9
186. Hinney A, Albayrak Ö, Antel J, Volckmar AL, Sims R, Chapman J et al (2014) Genetic variation at the CELF1 (CUGBP, elav-like family member 1 gene) locus is genome-wide associated with Alzheimer's disease and obesity. Am J Med Genet B Neuropsychiatr Genet 165(4):283–293
187. Waalen J (2014) The genetics of human obesity. Transl Res 164(4):293–301
188. Bille DS, Banasik K, Justesen JM, Sandholt CH, Sandbæk A, Lauritzen T et al (2011) Implications of central obesity-related variants in LYPLAL1, NRXN3, MSRA, and TFAP2B on quantitative metabolic traits in adult Danes. PLoS One 6(6):e20640
189. Yoganathan P, Karunakaran S, Ho MM, Clee SM (2012) Nutritional regulation of genome-wide association obesity genes in a tissue-dependent manner. Nutr Metab 9(1):1–10
190. Elias I, Franckhauser S, Bosch F (2013) New insights into adipose tissue VEGF-A actions in the control of obesity and insulin resistance. Adipocyte 2(2):109–112
191. Zhao J, Bradfield JP, Zhang H, Sleiman PM, Kim CE, Glessner JT et al (2011) Role of BMI-associated loci identified in GWAS meta-analyses in the context of common childhood obesity in European Americans. Obesity 19(12):2436–2439
192. Rouskas K, Kouvatsi A, Paletas K, Papazoglou D, Tsapas A, Lobbens S et al (2012) Common variants in FTO, MC4R, TMEM18, PRL, AIF1, and PCSK1 show evidence of association with adult obesity in the Greek population. Obesity 20(2):389–395
193. Su S, Zhu H, Xu X, Wang X, Dong Y, Kapuku G et al (2014) DNA methylation of the LY86 gene is associated with obesity, insulin resistance, and inflammation. Twin Res Hum Genet 17(3):183–191
194. Chevillard G, Blank V (2011) NFE2L3 (NRF3): the Cinderella of the Cap'n'Collar transcription factors. Cell Mol Life Sci 68(20):3337–3348
195. Withers SB, Dewhurst T, Hammond C, Topham CH (2020) MiRNAs as novel adipokines: obesity-related circulating MiRNAs influence chemosensitivity in cancer patients. Non-coding RNA 6(1):5
196. Selvanayagam T, Walker S, Gazzellone MJ, Kellam B, Cytrynbaum C, Stavropoulos DJ et al (2018) Genome-wide copy number variation analysis identifies novel candidate loci associated with pediatric obesity. Eur J Hum Genet 26(11):1588–1596
197. Rask-Andersen M, Almén MS, Lind L, Schiöth HB (2015) Association of the LINGO2-related SNP rs10968576 with body mass in a cohort of elderly Swedes. Mol Genet Genomics 290(4):1485–1491
198. Bernhard F, Landgraf K, Klöting N, Berthold A, Büttner P, Friebe D et al (2013) Functional relevance of genes implicated by obesity genome-wide association study signals for human adipocyte biology. Diabetologia 56(2):311–322
199. Wu Y, Yu Y, Zhao T, Wang S, Fu Y, Qi Y et al (2016) Interactions of environmental factors and APOA1-APOC3-APOA4-APOA5 gene cluster gene polymorphisms with metabolic syndrome. PLoS One11(1):e0147946
200. Buzaglo-Azriel L, Kuperman Y, Tsoory M, Zaltsman Y, Shachnai L, Zaidman SL et al (2016) Loss of muscle MTCH2 increases whole-body energy utilization and protects from diet-induced obesity. Cell Rep 14(7):1602–1610

201. Unger TJ, Calderon GA, Bradley LC, Sena-Esteves M, Rios M (2007) Selective deletion of Bdnf in the ventromedial and dorsomedial hypothalamus of adult mice results in hyperphagic behavior and obesity. J Neurosci 27(52):14265–14274
202. Tkatchenko AV, Visconti RP, Shang L, Papenbrock T, Pruett ND, Ito T et al (2001) Overexpression of Hoxc13 in differentiating keratinocytes results in downregulation of a novel hair keratin gene cluster and alopecia. Development 128(9):1547–1558
203. Wang K-S, Wang L, Liu X, Zeng M (2013) Association of HS6ST3 gene polymorphisms with obesity and triglycerides: gene× gender interaction. J Genet 92(3):395–402
204. Lv D, Zhang D-D, Wang H, Zhang Y, Liang L, Fu J-F et al (2015) Genetic variations in SEC16B, MC4R, MAP2K5 and KCTD15 were associated with childhood obesity and interacted with dietary behaviors in Chinese school-age population. Gene 560(2):149–155
205. Doche ME, Bochukova EG, Su H-W, Pearce LR, Keogh JM, Henning E et al (2012) Human SH2B1 mutations are associated with maladaptive behaviors and obesity. J Clin Investig 122(12):4732–4736
206. Kim Y-J, Sano T, Nabetani T, Asano Y, Hirabayashi Y (2012) GPRC5B activates obesity-associated inflammatory signaling in adipocytes. Sci Signal 5(251):ra85–ra
207. Nissen SE, Pillai SG, Nicholls SJ, Wolski K, Riesmeyer JS, Weerakkody GJ et al (2018) ADCY9 genetic variants and cardiovascular outcomes with evacetrapib in patients with high-risk vascular disease: a nested case-control study. JAMA Cardiol 3(5):401–408
208. Lugo-Candelas C, Pang Y, Lee S, Cha J, Hong S, Ranzenhofer L et al (2020) Differences in brain structure and function in children with the FTO obesity-risk allele. Obes Sci Pract 6(4):409–424
209. Yang L, Liu Z, Ling W, Wang L, Wang C, Ma J et al (2020) Effect of anthocyanins supplementation on serum IGFBP-4 fragments and glycemic control in patients with fasting hyperglycemia: a randomized controlled trial. Diabetes Metab Syndr Obes Targets Ther 13:3395
210. Nakayama K, Miyashita H, Iwamoto S (2014) Seasonal effects of the UCP3 and the RPTOR gene polymorphisms on obesity traits in Japanese adults. J Physiol Anthropol 33(1):38
211. Chami N, Preuss M, Walker RW, Moscati A, Loos RJ (2020) The role of polygenic susceptibility to obesity among carriers of pathogenic mutations in MC4R in the UK Biobank population. PLoS Med 17(7):e1003196
212. Liu R, Zou Y, Hong J, Cao M, Cui B, Zhang H et al (2017) Rare loss-of-function variants in NPC1 predispose to human obesity. Diabetes 66(4):935–947
213. Hong J, Shi J, Qi L, Cui B, Gu W, Zhang Y et al (2013) Genetic susceptibility, birth weight and obesity risk in young Chinese. Int J Obes 37(5):673–677

Chapter 4
The Epigenetics and Molecular Interplay in Obesity and Associated Complications

Hitesh Soni and Seema Dangwal

Abstract Obesity is a positive energy balance state associated with high morbidity, mortality and complex pathogenesis. More than one third of USA adults are obese and therefore the need for novel and innovative therapeutic approaches to treat obesity is high. Polygenic factors such as certain diseases, including diabetes, hypertension and cancer, etc., presence of inflammation, oxidative stress, glucose metabolism and insulin resistance add more complexity to obesity pathogenesis and strongly correlate with obesity associated metabolic overload. Epigenetic mechanisms are heritable genetic variability in gene functions influencing pathogenesis of various diseases including obesity without DNA sequence modification. There is a huge information gap in epigenetic regulations during obesity onset, progression and associated conditions. In this chapter, we discuss various epigenetic mechanisms such as DNA methylation, Histone modification and non-coding RNA mediated regulation of obesity, which can be exploited to develop early diagnosis and novel therapies in near future.

Keywords Obesity · Calorie imbalance · Histone modification · DNA methylation · Non-coding RNA · microRNA · Obesity mechanisms · Epigenetics · Obesity biomarkers

Introduction

Obesity is a global epidemic and a public health concern due to higher morbidity, mortality, and healthcare costs. Obesity is also associated with eating behavior and chronic disease conditions such as hypertension, diabetes, and cancer [1]. According to the World Health Organization (WHO), obesity is defined based on body mass

H. Soni
Flagship Pioneering (FL72 Inc.), Cambridge, MA 02139, USA

S. Dangwal (✉)
Cardiovascular Institute, Department of Medicine, School of Medicine, Stanford University, Palo Alto, CA 90403, USA
e-mail: sdangwal@stanford.edu

© Springer Nature Switzerland AG 2021

P. S. Tappia et al. (eds.), *Cellular and Biochemical Mechanisms of Obesity*, Advances in Biochemistry in Health and Disease 23,
https://doi.org/10.1007/978-3-030-84763-0_4

index (BMI). An individual with a BMI of 30 kg/m^2 or higher is considered obese, and a BMI between 25–30 kg/m^2 is considered overweight. In year 2017–2018, the age-adjusted prevalence of obesity among the USA adults was 42.4%, and there were no significant differences between men and women [2]. There are minimal options available for the treatment of obesity (calorie intake restrictions and physical exercise to control obesity), and those options have minimal efficacy. Currently, the need for novel and innovative therapeutic approaches to treat obese patients is ever time high. Due to the lack of effective treatment modalities, the global anti-obesity drug market is predicted to grow at the compound annual growth rate (CAGR) of about 8%, according to the forecast period of 2018–2022 [3]. The anti-obesity research market is concentrated, with a few vital pharmaceutical players in the race, rather than being fragmented over many pharmaceutical companies [3, 4].

Downstream pathway modulation of a disease-causing gene using small molecule approach is a conventional way to treat polygenic diseases. The emerging cutting-edge scientific tools made DNA- or RNA-based gene targeting strategy has become a reality in the medical field for difficult to treat diseases such as cancer. It is well known that endogenous coding and noncoding RNAs have important functions in gene regulation and are expressed differently in disease states. Epigenetics is a rapidly evolving field to study the regulation of those specific genes in many disease conditions without changing the DNA sequence. However, there is a big information gap in epigenetic regulations of obesity pathogenesis and associated conditions. In this chapter, we will discuss the epigenetic regulation of obesity, which may help develop innovative molecular therapies.

Mechanisms of Obesity

Positive energy imbalance resulting from calories intake above the total calories expendature is the fundamental cause of obesity that leads to increased calories storage in adipose tissues [5]. Certain diseases, such as cancer, type 2 diabetes (T2D), and hypertension, add more complexity to the obesity pathogenesis and can act as a risk factor for obesity. There is a strong correlation of chronic inflammation, oxidative stress, glucose metabolism, adipokine secretion, insulin resistance, and metabolic overload in obesity, and involvement of other vital organs [6] (Fig. 4.1). Activation of brain pro-opiomelanocortin (POMC) neurons and stimulation of melanocortin 4 receptor (MC4R) are reportedly involved in obesity. Agouti-related peptide (AgRP) is the endogenous antagonist of MC4R that showed reduction in sympathetic nervous system (SNS) activity and enhances orexigenic (appetite stimulation) signals. On the other hand, an increase in neuropeptide Y (NPY) showed antagonistic activity of the POMC-MC4R axis and responsible for anorexigenic signals (loss of appetite) [7, 8].

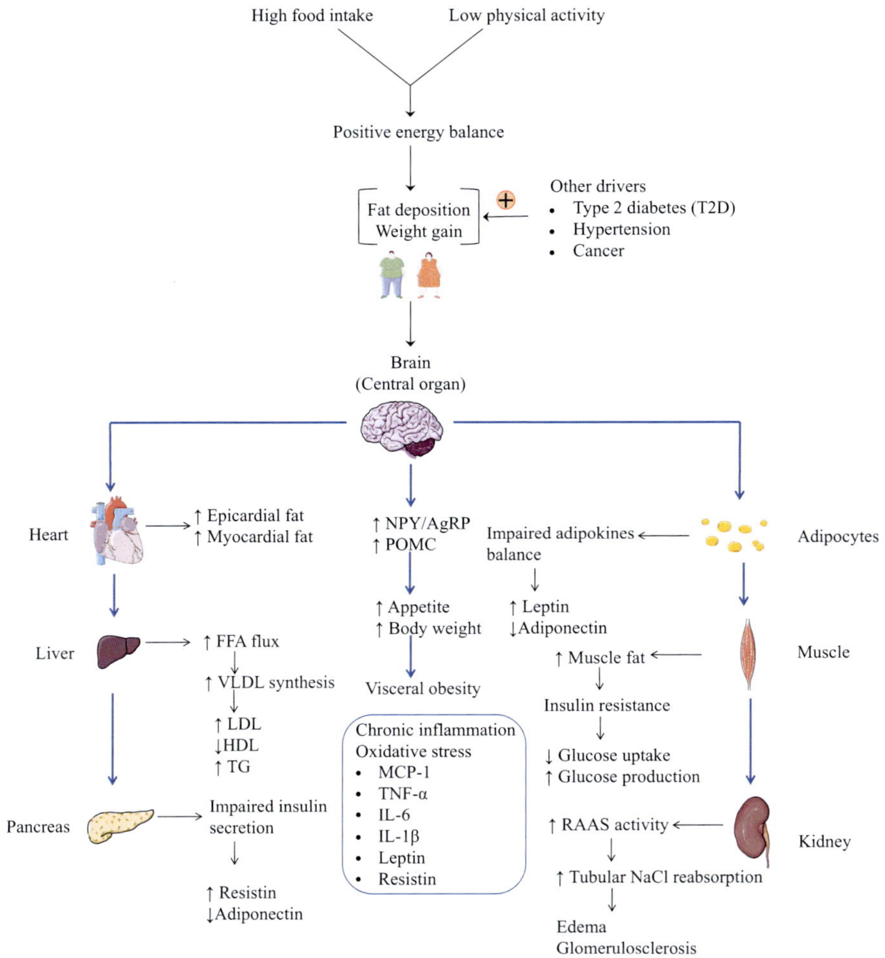

Fig. 4.1 Common mechanism and role of different organs/tissues in obesity. Obesity is the phenomenon of positive energy balance due to high food intake and less physical activity. The brain is the central organ that controls this balance. However, molecular interplay between the brain and other vital organs/tissues such as the heart, liver, pancreas, adipocytes, muscles, and kidneys, is also involved in the polygenic complex disease of obesity. FFA, free fatty acids; VLDL, very-low-density lipoprotein; LDL, low-density lipoprotein; HDL, high-density lipoprotein; TG, triglyceride; NPY, neuropeptide Y; AgRP, agouti-related peptide; POMC, pro-opiomelanocortin; RAAS, renin–angiotensin–aldosterone system; MCP-1, monocyte chemoattractant protein-1; TNF-α, tumour necrosis factor alpha; IL-6, interleukin 6, IL-1β; Interleukin 1 beta

Epigenetic Regulation of Obesity

The Greek prefix "Epi" means "on top of" and the term 'Epigenetics' is the study of heritable phenotypic alteration without affecting DNA sequence. Epigenetic mechanisms in obesity are heritable modulations of genetic variability and gene function that do not involve changes in DNA sequence. Obesity cannot be explained only by monogenic factors; on the contrary, it is considered polygenic in a majority of obese populations. However, single gene mutation studies showed that leptin, leptin receptor, and brain-derived neurotrophic factor (BDNF) are involved in early-onset obesity. Epigenetic modulations such as DNA methylation, histone modification, and noncoding RNA (ncRNA) are now recognized as significant contributors to understanding inter-individual variability in obese individuals and may serve as an essential source of information for future molecular therapies for polygenic complex diseases such as obesity [9].

DNA Methylation Links to Obesity

DNA methylation is an epigenetic regulatory mechanism that controls cell fate and lineage in mammals. Methylated DNA cannot bind to gene expression machinery, and thus the gene expression process is repressed. DNA methylation is carried out by the group of enzymes known as DNA methyltrasferases (DNMTs). There are three types of DNMTs: DNMT1, DNMT3a, and DNMT3b. DNMT1 is responsible for maintenance of methylation during the cell division process, whereas DNMT3a and DNMT3b are important in de novo methylation during the cell differentiation process in early development. DNA methylation leads to the addition of a methyl group to the 5th carbon of a cytosine (C) pyrimidine ring that is located next to a guanidine (G) nucleotide. This type of methyl group addition is commonly known as a CpG residue. The repeat sequence of CpG dinucleotides are located near the gene promoter sequence and, in most cases, is hypomethylated whereas gene bodies and other regions are hypermethylated. Hypomethylation near the promoter region is an important starting point for gene transcription, and it avoids truncated transcription. In normal adult cells, only CpG sites at the promoter region are methylated, which is a regulatory element to control transcription. Whereas DNMTs are important for DNA methylation, ten eleven translocation (TET) enzymes oxidize 5-methyl cytosines and promote location specific demethylation processes [10–12] (Fig. 4.2).

Previous publications support the possibility that DNA methylation modifications are involved in obesity and related comorbidities. However, precise mechanisms related to changes in DNA methylation in obesity have not yet been identified. Preclinical studies suggest that several genes associated with obesity are tumor necrosis factor α (TNF-α), hypoxia inducible factor 3 Subunit α (HIF3 α), CLOCK (a circadian rhythm gene), Insulin-like growth factor binding protein 3 (IGFBP3), and Sterol Regulatory Element Binding Transcription Factor-1 (SREBF1; a group

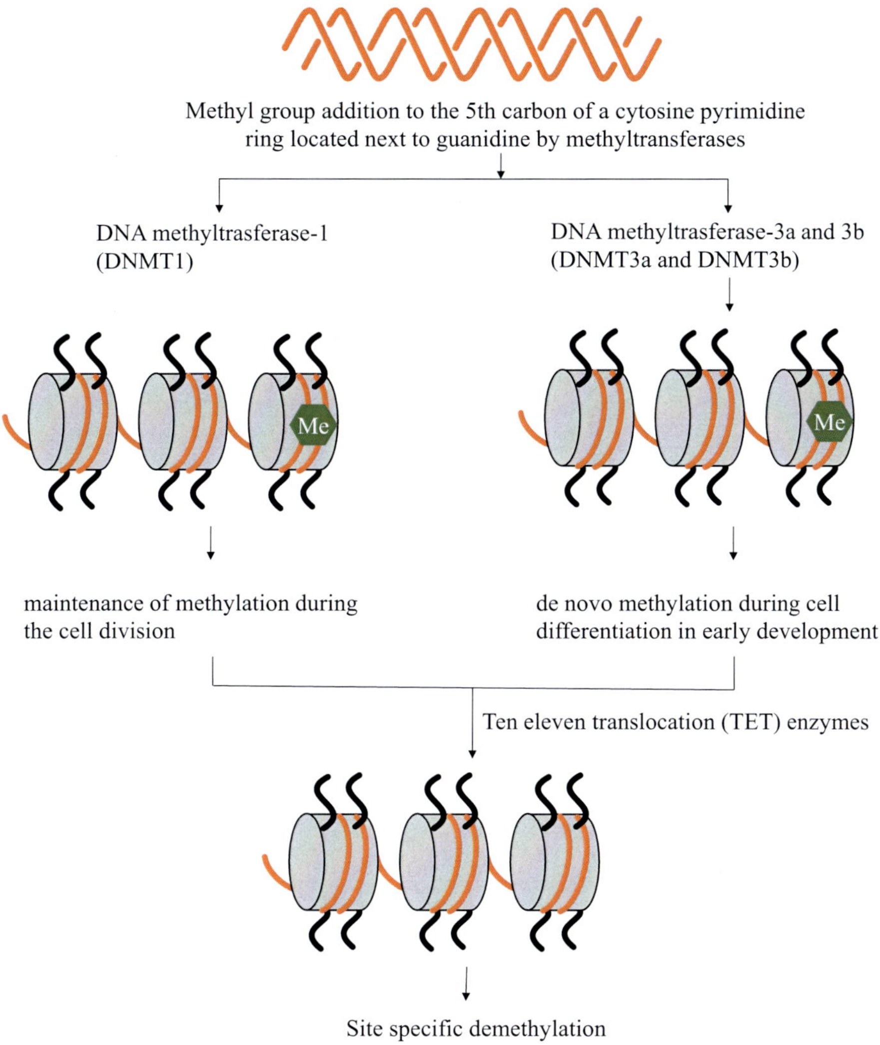

Fig. 4.2 DNA methylation. This is the process of adding a methyl group to the 5th carbon of a cytosine pyrimidine ring next to guanidine by the group of enzymes called methyltransferases (DNMT). DNA demethylation carried out by a group of enzymes known as ten eleven translocation (TET)

of genes involved sterol biosynthesis). The hypoxia-inducible factor (HIF) pathway controls genes related to glycolysis, such as lactate dehydrogenase A (LDHA), pyruvate dehydrogenase kinase 1 (PDK1), and glycogen phosphorylase L (PYGL), which contribute to adipocyte inflammation in white adipose tissue (WAT). These genes either increase glucose production or glucose utilization by stimulating phosphoenolpyruvate carboxykinase (PEPCK) in the liver [13]. It has been demonstrated that

HIF-binding during erythropoietin expression depends on methylation of the 3' UTR region of hypoxia response element (HRE), which suggests that HIF activity is regulated by DNA methylation [14–16]. Mature adipocytes store nutrients and triglycerides for future energy expenditure; the process of converting fibroblast-like progenitor cells to mature adipocytes is called adipogenesis. The peroxisome proliferator-activated receptor gamma (PPAR), co-activator protein CCAAT/enhancer-binding protein-alpha (C/EBP-α), and insulin-like growth factor 1 (IGF1) are mainly involved in process of adipogenesis [17]. In mouse, DNMT1 gene silencing leads to an increase in the adipocyte differentiation rate and alters the gene expression profile as well as lipid accumulation. It has also been observed that genes involved in PPAR, insulin, and adipocytokine pathways showed hypomethylation, which suggests that these genes play a role in regulating adipogenesis [18, 19]. Of the total WAT cells in obese individuals, 40% are characterized by macrophage infiltration, compared to 18% in lean subjects [20]. Macrophage-derived pro-inflammatory cytokines, such as TNF-α , IL-6, and IL-1 stimulate the phosphorylation of insulin receptor substrate 1 (IRS-1), which impairs the insulin signaling process that contributes to obesity [21]. It has been demonstrated that there is a correlation between obesity and change in regular day-light cycles (circadian rhythms). CLOCK is a circadian rhythm gene that has been associated with obesity, and it enhances the glycogen synthase 2 (Gys2) gene transcription to control glycogen synthesis. Other genes that control circadian rhythms include brain and muscle aryl hydrocarbon receptor nuclear translocator (ARNT)-like 1 (BMAL-1) and associated genes such as period (PER) and cryptochrome (CRY) [22, 23]. The CRY gene leads to a decrease in Cyclic adenosine monophosphate (cAMP) activity to affect gluconeogenesis. The PER gene inhibits PPAR signaling and suppresses the adipogenesis process [24]. Conversely, BMAL-1, a transcription factor that is a principal driver of the molecular clock in mammals, upregulates PPAR to promote the process of adipogenesis [25]. There is an intriguing correlation between circadian rhythm genes and up- or down-regulation of PPAR to influence the adipogenesis process. Methylation of the promoter sequence of PPAR gene by enzyme DNMT1 has been studied to elucidate the link between macrophage activation and inflammation in obesity [26]. These studies suggest the importance of abnormal methylation of circadian rhythm regulatory genes in the disruption of metabolic processes linked to obesity. The role of global DNA methylation has been studied in metabolic syndromes, and the role of adipogenesis genes such as PPAR and PPAR, lipid metabolism genes such as retinoid X receptor alpha (RXRA), sterol regulatory element binding transcription factor 2 (SREBF2), SREBF1, stearoyl-CoA desturase-1 (SCD1), lipoprotein lipase (LPL), liver X receptor beta (LXRb), and genes involved in inflammation such as low-density lipoprotein receptor-related protein (LRP), leptin (LEP) and tumor necrosis factor (TNF) has been reported. The group has provided data supporting the hypothesis that global DNA methylation and methylation at specific genes related to adipogenesis, lipid metabolism, and inflammation has strong correlation with metabolic syndromes [27].

Histone Modification and Obesity

There are five types of positively charged histone proteins in the human: H1, H2A, H2B, H3, and H4. The H1 protein acts as a linker protein that keeps nucleosomes together, whereas H2A, H2B, and H3 are packaging material for negatively charged DNA, which makes the core unit of chromatin called nucleosome [28]. Strong interaction among positively charged lysines and arginines of the histone tail and negatively charged phosphate groups of DNA make a packed histone unit that is not capable of performing cellular functions such as transcription, replication, or repair of DNA molecules. The enzyme histone acetyltransferase (HAT) catalyzes the histone acetylation process, which reduces the positive charge on histone tails to destabilize the nucleosome, resulting in a relaxed chromatin structure and activation of the gene transcription process. In contrast, the enzyme histone deacetylase (HDAC) leads to more compact structure of the chromatin that represses the gene transcription process [29] (Fig. 4.3).

The catabolism of carbohydrate, lipid, and protein leads to generation of the key metabolite acetyl Co-A, which enters into the tricarboxylic acid (TCA) cycle for the generation of the energy molecule adenosine triphosphate (ATPs). In excessive nutritional conditions, acetyl Co-A also acts as an important donor of the acetyl group in histone acetylation. Therefore, inhibition of histone acetylation was proposed as a therapeutic target to treat obesity. Preclinical studies in mice treated with ethanolic extract of berry showed a decrease in liver lipid accumulation due to reduction in liver histone acetylation [30]. However, there is a paucity of clinical data on the potential therapeutic benefits of histone acetylase inhibitors in obesity and related conditions [31–33]. Histone deacetylase inhibition showed improvement in insulin sensitivity and higher energy consumption in obese mice [34]. Inhibition of histone deacetylase resulted in reduced levels of acetyl Co-A found in WAT, liver, and pancreas in a high fat diet (HFD) mouse model [35]. The correlation between between acetyl Co-A

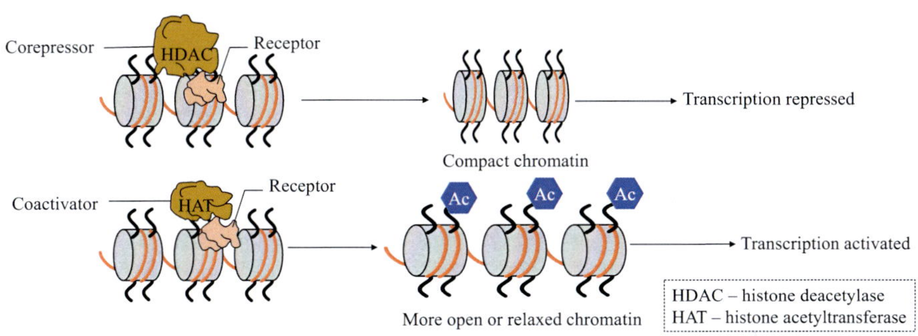

Fig. 4.3 Histone modification process. This process is carried out by either corepressor HDAC or coactivator HAT, which leads to repression or activation of transcription, respectively. HDAC, histone deacetylase; HAT, histone acetyltransferase

levels and histone acetylation may be associated with the suppression of the expression of key enzyme acetyl Co-A synthatase 2 in the adipose tissue [36]. Proteomic analysis in the liver samples of diet-induced obese (DIO) mice also suggested the role of acetylation in obesity and found that about 30% of histones in the mouse liver samples were lysine acetylated [37]. More studies are needed to explore the role of histone acetylation and deacetylation in obesity and related complications.

Noncoding RNAs in Obesity and Related Conditions

Noncoding RNAs (ncRNAs), once considered to be "dark matter of the genome", constitute a major part of the human genome and represent RNAs without coding potential [38, 39]. Several subclasses of ncRNAs (Fig. 4.4), including microRNAs (miRNAs/miRs) and long ncRNAs, which are preferentially being studied. NcRNAs can be broadly classified as short or long ncRNAs based on the transcript length. Short ncRNAs (<200 nucleotides) mainly comprise highly conserved transcripts of approximately twenty nucleotides in length, consisting of 7–8 nucleotide long seed sequences, that imperfectly bind to complementary base-pair sequences in the 3′ untranslated region (UTR) of mRNA transcripts to inhibit translation [40]. To date, approximately two thousand miRNAs, one kind of short ncRNA, have been identified

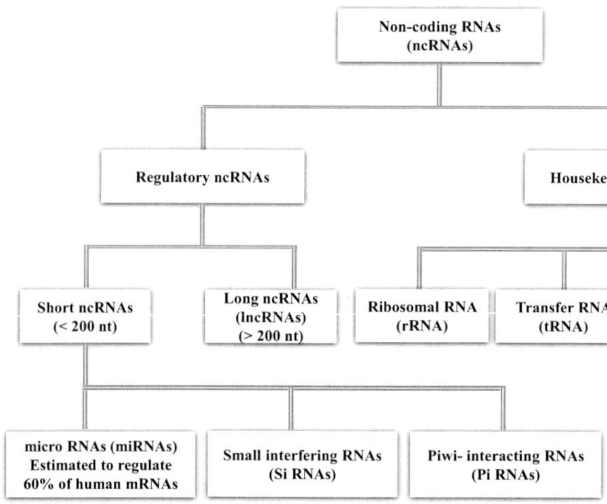

Fig. 4.4 Types of noncoding RNAs (ncRNAs). There are two major types of ncRNAs: regulatory and housekeeping. Regulatory ncRNAs are further classified as short ncRNAs (less than 200 nucleotides) and long-non coding RNAs (lncRNAs, more than 200 nucleotides). Short ncRNAs are further divided into micro RNAs (miRNAs), small interfering RNAs (siRNAs), Piwi- interacting RNAs (piRNAs). miRNAs are estimated to regulate 60% of all human mRNAs. Housekeeping RNAs are ribosomal RNA (rRNA), transfer RNA (tRNA), small nuclear RNA (snRNA), and small nucleolar RNA (snoRNA)

and mapped to the human genome [38, 40]. miRNAs are detected in almost all tissues and body fluids and function as a team with other miRNAs to fine-tune gene expression and regulate cellular functions under stress [41–43]. System wide whole-genome profiling suggests varied pathophysiological conditions result in differential stress responsiveness and spatiotemporal variations in ncRNA expression [41, 44, 45].

Some miRNAs are encoded in the introns of protein-coding genes transcribed in the nucleus. Nuclear RNAse polymerase II and the RNase III enzyme Drosha create pre-miRNA, transcripts that form an RNA hairpin of about 70 nucleotides [40, 46]. Exportin-5 exports the pre-miRNA to the cytoplasm where it is processed by Dicer, a large RNase protein, to produce a double-stranded miRNA duplex. One strand of this miRNA duplex forms a complex with argonaute-2 (Ago-2) protein called the RNA-induced silencing complex (RISC). Mature miRNA binds with the complementary 3′ UTR of target mRNAs, leading to inhibition of the mRNA translation process [38, 40, 46] (Fig. 4.5).

There are nutritional interventions and exercise programs available to control obesity, but despite that the prevalence is increasing with an alarming rate. The key driver for this worldwide epidemic is adipose tissue which is considered a major regulator in energy homeostasis [47]. Two categories of adipose tissues play an important role in energy storage of fat in the form of triglycerides and burning of lipids by thermogenesis, called white adipose tissue (WAT) and brown adipose tissue (BAT) [48]. In addition to sedentary lifestyle and unhealthy food habits there are inter-individual factors such as environment, genetics, and epigenetics plays a significant role in obesity. Alterations in ncRNAs, including miRNAs, are considered major drivers in obesity and metabolic disease, among others [49]. miRNAs are of paramount importance due to their potential role in protein regulation and plasticity [50]. Previous work used high-throughput microarray and RNA-sequencing methods to find that 70 miRNAs were significantly up- or down-regulated in mature adipocytes compared to preadipocyte. 50 of 799 miRNAs significantly differed between fat cells in obese compared to lean individuals. 17 of these 799 miRNAs were well correlated with BMI, glycemic index, and triglyceride parameters [51] (see Table 1 for a list of miRNAs up- or down-regulated in obesity). Subsequently, it was demonstrated that miR-519d expression was significantly changed in scWAT in obese patients [52]. Work done by Arner et al. (2012) provided evidence that 11 miRNAs (miR-26a, 30c, 92a, 126, 143, 145, 193a/b, 652, and let 7 a/d) were altered in scWAT in obese women compared to lean control [53]. Further, Mansego et al. (2017) reported that CpG sites of the coding regions of miR-1203, miR-412, and miR-216a were distinctively methylated in obese children when compared with a non-obese group [54]. It was also shown that weight loss programs resulted in altered expression of miR-197b, let-7d, miR-125b, miR-1309, miR-132-3p, miR-422, miR-935, and miR-4772, indicating that epigenetic regulation of miRNAs is associated with controlled diet [55–58].

As mentioned previously, adipogenesis is highly regulated by several transcription regulators linked to PPAR [59]. miRNAs reported to be upregulated in the adipogenesis process include miR-21, 26b, 30, 103, 143, 148, 181a, 199a, and

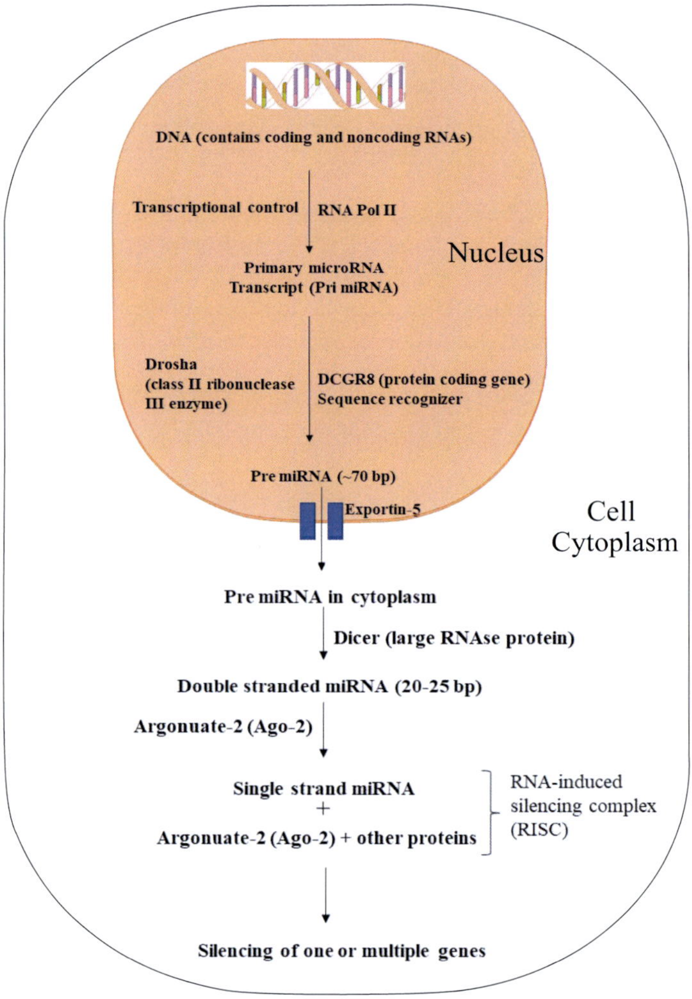

Fig. 4.5 MicroRNA-biogenesis and functions in cells. Some miRNAs are encoded in the introns of protein-coding genes transcribed in the nucleus. Nuclear RNase polymerase II and the RNase III enzyme Drosha generate pre-miRNA. This pre-miRNA enters the cytoplasm by exportin-5 and is processed by Dicer to produce a double-stranded miRNA duplex. One strand of this duplex forms a complex with argonaute-2 (Ago-2) and other proteins to form the complex called RNA-induced silencing complex (RISC). This complex may lead to silencing of one or multiple genes

378. miRNAs such as let-7, miR-22, 125a, 93a/b, and 224 were downregulated in adipogenesis [60]. Overexpression of miR-377, a master mediator of adipogenesis, showed an increase in insulin sensitivity by inhibiting sirt1 and decreasing the process of adipocyte differentiation [61]. Increased expression of miR-30 accelerated the process of thermogenesis and enhanced the process of fat burning [62]. Adipokines, such as monocyte chemoattractant protein-1 (MCP-1), IL-6, and TNF

α, are inflammatory molecules released from expanded (hypertrophy) adipocytes, and considered to be major drivers in obesity. Several miRNAs (miR-193b, 26a, 92a, and 145) are involved in the adipokine production [63–65].

Diabetes, a hyperglycemic condition, is strongly associated with a higher amount of fat mass, and reports suggest a link with expression of miRNAs. In skeletal muscle, miR-133 regulated glucose metabolism, and the condition of insulin resistance and obesity were associated with miR-33 [66–68]. There was a positive correlation between overexpression of miR-24, 30d, and 146a and obesity and hyperglycemia [69].

Lipolysis is a process of releasing energy through the breakdown of triglyceride to free fatty acid (FFA). In human cultured adipocytes, it has been observed that upregulation of miR-30c, 652, 193b, and 145 enhanced lipolysis, whereas upregulation of miR-26a and let-7d resulted in inhibition of lipolysis [64]. Overexpression of other miRNAs, such as miR-124, miR-378, and miR-145, demonstrated a role in adipocyte lipolysis regulation [70–73]. There is also a direct link between obesity and miRNA-mediated regulation of lipid storage. For example, miR-22 has been found to decrease triglyceride storage, whereas upregulation of miR-24 correlated with downregulation of genes associated with the process of lipolysis, such as fatty acid synthase (FASN), adenosine triphosphate citrate lyase (ACLY), and stearoyl-CoA desaturase-1 (SCD-1) [74]. Upregulation of miR-204-5p has been shown to inhibit kruppel-like factor-3 (KLF-3), a negative transcriptional regulator of adipogenesis, leading to the induction of fat accumulation [75]. On the contrary, overexpression of miR-199a-3p inhibited fat accumulation in brown adipose tissue through the mammalian target of rapamycin (mTOR) signaling [76]. Depletion of miR-7a from the liver showed the induction of fat accumulation, indicating that miR-7a has a role in nonalcoholic fatty liver disease (NAFLD) [77]. Interestingly, miR-206 has been associated with stimulation of the insulin signaling pathway through sterol regulatory element-binding protein 1c (SREBP1C) gene, inhibiting the process of lipogenesis and gluconeogenesis [78]. It was also previously shown that miR-139-5p, 30b-5p, 122-5p, 422a, and 146-5p were downregulated in obese patients with NAFLD [79].

Browning of the WAT, a process of diverting excess fat into heat production, has recently evolved as a potential strategy for treating obesity [80]. Upregulation of miR-30 has been observed to enhance the expression of the thermogenic gene in WAT by inhibiting receptor interacting protein-140 (RIP-140), a thermogenic nuclear receptor corepressor protein. In addition, studies in mice also showed the role of miR-32 in stimulating brown fat thermogenesis and activating browning of WAT [81, 82]. Similarly, miR-455 has been reported to activate PPAR and inhibit adipogenic suppressors to stimulate the browning process [83]. Overexpression of miR-27 showed upregulation of browning markers such as mitochondrial uncoupling protein 1 (UCP-1) and PPARγ coactivator 1-alpha (PGC-1α) [84]. Moreover, miR-155, 133, 34a, and 378 have been observed to inhibit browning repressors, thus, stimulating browning [85–88]. It has also been reported that exosomes (extracellular vesicles) from LPS-induced macrophages showed alterations in miR-530, 127, 143, and 486 in adipose tissues, but did not alter adipogenesis or fat accumulation [89]. Exosomes containing miRNAs are present in various biological fluids such as plasma, serum, urine, saliva, sweat, and even breast milk. It has been reported that exosomes derived from adipocytes are involved in the release of inflammatory cytokines in obesity [90]. Given the varied involvement of several miRNAs, they provide an exciting avenue for future work to better understand the association with obesity.

Secondary classification of ncRNAs is based on the linearity of transcripts and can be classified as linear (e.g., miRNAs, lncRNAs, etc.) or circular (e.g., circular RNAs [circRNAs]). In circRNAs, the 3′ and 5′ ends of linear RNA transcripts join together to make a loop structure with very high stability compared to the linear forms [46, 91]. There is a high abundance of circRNAs, which, paired with a high phylogenetic conservation, makes them an excellent area of research for ncRNA-based therapies. However, their functional significance is still widely unexplored [92–94].

Recently, the role of long noncoding RNAs (lncRNAs; > 200 nucleotides) has been reported in adipogenesis and lipid accumulation. LncRNA BATE1was showen to inhibit genes of WAT and maintain the BAT genes [95]. Zhao et al. reported that Blnc1, an lncRNA in brown adipocytes, interacts with transcription factor EBF2 and upregulates thermogenesis genes [96]. Furthermore, lncRNAs are of particular interest due to their role in post-transcriptional modification via polyadenylation (addition of a poly(A) tail to a messenger RNA) and splicing (removal of introns and joining of exons). LncRNA expresion is tissue-specific, therefore creating potential interest in the lncRNA field for targeted diagnosis and treatment for many diseases, including obesity. Circulating miRNAs (post-transcriptional regulators), such as miR-17-5p and miR-132, are not produced in exosomes and had decreased expression in obese versus lean subjects [97] (Table 1). However, more work is needed to understand the role of circulating miRNAs and lncRNAs in obesity.

Clinical Potential of NcRNAs in Therapy and Biomarker Research

Several classes of synthetic oligomers that can modify miRNA expression in cells and organisms [45, 46, 98]. In addition, it is easy to generate miRNA transgenic and knockout small animals or cells to study the functional effects of miRNAs in vitro and in vivo [40, 41, 44, 99, 100]. Noncoding RNAs, such as miRNAs, circulating RNAs, and lncRNAs, are stable and tissue-specific, making them attractive biomarkers for various diseases. For example, functional experiments revealed that circulating miR-122 may act as specific biomarker for type 2 diabetes [101]. In obese subjects, it has been found that there was overexpression of circulating miRNAs miR-140-5p,-142-3p and reduced expression of circulating miRNAs miR-532-5p,-125b,-15a,-520c-3p,-221,-130b, and-423-5p [102]. All of these miRNAs could be considered potential biomarkers for obesity. Endogenous miRNAs act as potential regulator of gene function, which may be upregulated or downregulated in obesity. Synthesized miRNA duplex enters the cell, forms the complex with RISC, and miRNA-guided RISC binds with therapeutic target and inhibits the translation. Therefore, synthetic mimetics or inhibitors of specific miRNAs may be useful as molecular therapeutic intervention in obesity in the future. There are no specific miRNA-based therapies currently available to treat obesity. However, this strategy has reached clinical trials in diseases such as cancer and cardiovascular diseases [103, 104], and miRNAs as potential early biomarkers have been reported in cardio-metabolic diseases [105].

In summary, there is exponential research interest and growth in the field of ncRNAs and their importance as biomarkers and drug targets in obesity. However, selectivity, delivery methods, efficacy, and adverse events related to ncRNAs-based therapies need to be carefully evaluated before becoming a reliable means of treating obesity, or other devastating diseases.

References

1. Wolfenden L, Ezzati M, Larijani B, Dietz W (2019) The challenge for global health systems in preventing and managing obesity. Obes Rev 20:185–193
2. Hales CM, Carroll MD, Fryar CD, Ogden CL (2020) Prevalence of obesity and severe obesity among adults: United States, 2017–2018. NCHS Data Brief, no 360. National Center for Health Statistics, Hyattsville, MD
3. https://www.globaldata.com/six-new-obesity-drugs-set-launch-2026/
4. https://www.businesswire.com/news/home/20181119005718/en/Global-Anti-Obesity-Drugs-Market-2018-2022-Global-Increase
5. Yanovski JA (2018) Trends in underweight and obesity—scale of the problem. Nat Rev Endocrinol 14:5–6
6. Verdile G, Keane KN, Cruzat VF, Medic S, Sabale M, Rowles J, Wijesekara N, Martins RN, Fraser PE, Newsholme P (2015) Inflammation and oxidative stress: the molecular connectivity between insulin resistance, obesity, and alzheimer's disease. Mediators Inflamm 2015:1–17
7. Lee EB, Mattson MP (2014) The neuropathology of obesity: insights from human disease. Acta Neuropathol 127:3–28
8. da Silva AA, do Carmo JM, Wang Z, Hall JE (2014) The brain melanocortin system, sympathetic control, and obesity hypertension. Physiology 29:196–202
9. Holliday R (2006) Epigenetics: a historical overview. Epigenetics 1:76–80
10. Suzuki MM, Bird A (2008) DNA methylation landscapes: provocative insights from epigenomics. Nat Rev Genet 9:465–476
11. Deaton AM, Bird A (2011) CpG islands and the regulation of transcription. Genes Dev 25:1010–1022
12. Jones PA (2012) Functions of DNA methylation: islands, start sites, gene bodies and beyond. Nat Rev Genet 13:484–492
13. Choi JH, Park MJ, Kim KW, Choi YH, Park SH, An WG, Yang US, Cheong J (2005) Molecular mechanism of hypoxia-mediated hepatic gluconeogenesis by transcriptional regulation. FEBS Lett 579:2795–2801
14. Pfeiffer S, Krüger J, Maierhofer A et al (2016) Hypoxia-inducible factor 3A gene expression and methylation in adipose tissue is related to adipose tissue dysfunction. Sci Rep 6:27969
15. Main AM, Gillberg L, Jacobsen AL, Nilsson E, Gjesing AP, Hansen T, Pedersen O, Ribel-Madsen R, Vaag A (2016) DNA methylation and gene expression of HIF3A: cross-tissue validation and associations with BMI and insulin resistance. Clin Epigenet 8:89
16. Watson JA, Watson CJ, McCann A, Baugh J (2010) Epigenetics: the epicenter of the hypoxic response. Epigenetics 5:293–296
17. Jin C, Zhuo Y, Wang J et al (2018) Methyl donors dietary supplementation to gestating sows diet improves the growth rate of offspring and is associating with changes in expression and

DNA methylation of insulin-like growth factor-1 gene. J Anim Physiol Anim Nutr 102:1340–1350
18. Broholm C, Olsson AH, Perfilyev A, Gillberg L, Hansen NS, Ali A, Mortensen B, Ling C, Vaag A (2016) Human adipogenesis is associated with genome-wide DNA methylation and gene-expression changes. Epigenomics 8:1601–1617
19. Londono Gentile T, Lu C, Lodato PM, Tse S, Olejniczak SH, Witze ES, Thompson CB, Wellen KE (2013) DNMT1 is regulated by ATP-citrate lyase and maintains methylation patterns during adipocyte differentiation. Mol Cell Biol 33:3864–3878
20. Lumeng CN, Bodzin JL, Saltiel AR (2007) Obesity induces a phenotypic switch in adipose tissue macrophage polarization. J Clin Invest 117:175–184
21. Rogero M, Calder P (2018) Obesity, inflammation, toll-like receptor 4 and fatty acids. Nutrients 10:432
22. Laermans J, Depoortere I (2016) Chronobesity: role of the circadian system in the obesity epidemic: role of circadian clocks in obesity. Obes Rev 17:108–125
23. Mayeuf-Louchart A, Zecchin M, Staels B, Duez H (2017) Circadian control of metabolism and pathological consequences of clock perturbations. Biochimie 143:42–50
24. Mazzoccoli G, Pazienza V, Vinciguerra M (2012) Clock genes and clock-controlled genes in the regulation of metabolic rhythms. Chronobiol Int 29:227–251
25. Ghaben AL, Scherer PE (2019) Adipogenesis and metabolic health. Nat Rev Mol Cell Biol 20:242–258
26. Wang X, Cao Q, Yu L, Shi H, Xue B, Shi H (2016) Epigenetic regulation of macrophage polarization and inflammation by DNA methylation in obesity. JCI Insight. https://doi.org/10.1172/jci.insight.87748
27. Castellano-Castillo D, Moreno-Indias I, Sanchez-Alcoholado L, Ramos-Molina B, Alcaide-Torres J, Morcillo S, Ocaña-Wilhelmi L, Tinahones F, Queipo-Ortuño M, Cardona F (2019) Altered adipose tissue DNA methylation status in metabolic syndrome: relationships between global DNA methylation and specific methylation at adipogenic, lipid metabolism and inflammatory candidate genes and metabolic variables. JCM 8:87
28. Erler J, Zhang R, Petridis L, Cheng X, Smith JC, Langowski J (2014) The role of histone tails in the nucleosome: a computational study. Biophys J 107:2911–2922
29. Bowman GD, Poirier MG (2015) Post-translational modifications of histones that influence nucleosome dynamics. Chem Rev 115:2274–2295
30. Chung M-Y, Shin EJ, Choi H-K, Kim SH, Sung MJ, Park JH, Hwang J-T (2017) Schisandra chinensis berry extract protects against steatosis by inhibiting histone acetylation in oleic acid–treated HepG2 cells and in the livers of diet-induced obese mice. Nutr Res 46:1–10
31. Shi L, Tu BP (2015) Acetyl-CoA and the regulation of metabolism: mechanisms and consequences. Curr Opin Cell Biol 33:125–131
32. Bender DA (2012) The metabolism of "surplus" amino acids. Br J Nutr 108:S113–S121
33. Henagan TM (2014) A review of mitochondrial-derived fatty acids in epigenetic regulation of obesity and type 2 diabetes. JNHFS. https://doi.org/10.15226/jnhfs.2014.00127
34. Li F, Wu R, Cui X, Zha L, Yu L, Shi H, Xue B (2016) Histone deacetylase 1 (HDAC1) negatively regulates thermogenic program in brown adipocytes via coordinated regulation of histone H3 Lysine 27 (H3K27) deacetylation and methylation. J Biol Chem 291:4523–4536
35. Carrer A, Parris JLD, Trefely S et al (2017) Impact of a high-fat diet on tissue Acyl-CoA and histone acetylation levels. J Biol Chem 292:3312–3322
36. Gao X, Lin S-H, Ren F et al (2016) Acetate functions as an epigenetic metabolite to promote lipid synthesis under hypoxia. Nat Commun 7:11960
37. Nie L, Shuai L, Zhu M et al (2017) The landscape of histone modifications in a high-fat diet-induced obese (DIO) mouse model. Mol Cell Proteomics 16:1324–1334
38. Beermann J, Piccoli M-T, Viereck J, Thum T (2016) Non-coding RNAs in development and disease: background, mechanisms, and therapeutic approaches. Physiol Rev 96:1297–1325
39. The ENCODE Project Consortium (2012) An integrated encyclopedia of DNA elements in the human genome. Nature 489:57–74

40. Dangwal S, Thum T (2014) microRNA therapeutics in cardiovascular disease models. Annu Rev Pharmacol Toxicol 54:185–203
41. Dangwal S, Stratmann B, Bang C, Lorenzen JM, Kumarswamy R, Fiedler J, Falk CS, Scholz CJ, Thum T, Tschoepe D (2015) Impairment of wound healing in patients with type 2 diabetes mellitus influences circulating MicroRNA patterns via inflammatory cytokines. Arterioscler Thromb Vasc Biol 35:1480–1488
42. Osipova J, Fischer D-C, Dangwal S et al (2014) Diabetes-associated MicroRNAs in pediatric patients with type 1 diabetes mellitus: a cross-sectional cohort study. J Clin Endocrinol Metab 99:E1661–E1665
43. Lankisch TO, Voigtländer T, Manns MP, Holzmann A, Dangwal S, Thum T (2014) MicroRNAs in the bile of patients with biliary strictures after liver transplantation: Bile mi RNAS after liver transplantation. Liver Transpl 20:673–678
44. Ucar A, Gupta SK, Fiedler J et al (2012) The miRNA-212/132 family regulates both cardiac hypertrophy and cardiomyocyte autophagy. Nat Commun 3:1078
45. Catanzaro G, Filardi T, Sabato C, Vacca A, Migliaccio S, Morano S, Ferretti E (2020) Tissue and circulating microRNAs as biomarkers of response to obesity treatment strategies. J Endocrinol Invest. https://doi.org/10.1007/s40618-020-01453-9
46. Dangwal S, Schimmel K, Foinquinos A, Xiao K, Thum T (2016) Noncoding RNAs in heart failure. In: Bauersachs J, Butler J, Sandner P (eds) Heart failure. Springer International Publishing, Cham, pp 423–445
47. Spiegelman BM, Flier JS (2001) Obesity and the regulation of energy balance. Cell 104:531–543
48. Bartelt A, Heeren J (2014) Adipose tissue browning and metabolic health. Nat Rev Endocrinol 10:24–36
49. Martínez JA, Milagro FI, Claycombe KJ, Schalinske KL (2014) Epigenetics in adipose tissue, obesity, weight loss, and diabetes. Adv Nutr 5:71–81
50. Arner P, Kulyté A (2015) MicroRNA regulatory networks in human adipose tissue and obesity. Nat Rev Endocrinol 11:276–288
51. Ortega FJ, Moreno-Navarrete JM, Pardo G, et al (2010) MiRNA expression profile of human subcutaneous adipose and during adipocyte differentiation. PLoS ONE 5:e9022
52. Martinelli R, Nardelli C, Pilone V et al (2010) miR-519d overexpression is associated with human obesity. Obesity 18:2170–2176
53. Arner E, Mejhert N, Kulyte A et al (2012) Adipose tissue MicroRNAs as regulators of CCL2 production in human obesity. Diabetes 61:1986–1993
54. Mansego ML, Garcia-Lacarte M, Milagro FI, Marti A, Martinez JA, GENOI members (2017) DNA methylation of miRNA coding sequences putatively associated with childhood obesity: miRNA DNA methylation and childhood obesity. Pediatr Obes 12:19–27
55. Casado-Díaz A, Anter J, Müller S, Winter P, Quesada-Gómez JM, Dorado G (2017) Transcriptomic analyses of adipocyte differentiation from human mesenchymal stromal-cells (MSC): transcriptomic analyses of MSC adipogenesis. J Cell Physiol 232:771–784
56. Garcia-Lacarte M, Martinez JA, Zulet MA, Milagro FI (2018) Implication of miR-612 and miR-1976 in the regulation of TP53 and CD40 and their relationship in the response to specific weight-loss diets. PLoS ONE 13:e0201217
57. Marques-Rocha JL, Milagro FI, Mansego ML, Zulet MA, Bressan J, Martínez JA (2016) Expression of inflammation-related miRNAs in white blood cells from subjects with metabolic syndrome after 8 wk of following a Mediterranean diet–based weight loss program. Nutrition 32:48–55
58. Milagro FI, Miranda J, Portillo MP, Fernandez-Quintela A, Campión J, Martínez JA (2013) High-throughput sequencing of microRNAs in peripheral blood mononuclear cells: identification of potential weight loss biomarkers. PLoS ONE 8:e54319
59. Chen C, Cui Q, Zhang X, Luo X, Liu Y, Zuo J, Peng Y (2018) Long non-coding RNAs regulation in adipogenesis and lipid metabolism: emerging insights in obesity. Cell Signal 51:47–58

60. Engin AB (2017) MicroRNA and adipogenesis. In: Engin AB, Engin A (eds) Obesity and lipotoxicity. Springer International Publishing, Cham, pp 489–509
61. Li X, Yang Y, Yan R et al (2018) miR-377-3p regulates adipogenic differentiation of human bone marrow mesenchymal stem cells by regulating LIFR. Mol Cell Biochem 449:295–303
62. Hu F, Wang M, Xiao T, Yin B, He L, Meng W, Dong M, Liu F (2015) miR-30 promotes thermogenesis and the development of beige fat by targeting RIP140. Diabetes 64:2056–2068
63. Marques-Rocha JL, Samblas M, Milagro FI, Bressan J, Martínez JA, Marti A (2015) Noncoding RNAs, cytokines, and inflammation-related diseases. FASEB J 29:3595–3611
64. Lorente-Cebrián S, Mejhert N, Kulyté A, Laurencikiene J, Åström G, Hedén P, Rydén M, Arner P (2014) MicroRNAs regulate human adipocyte lipolysis: effects of miR-145 are linked to TNF- α. PLoS ONE 9:e86800
65. Belarbi Y, Mejhert N, Lorente-Cebrián S, Dahlman I, Arner P, Rydén M, Kulyté A (2015) MicroRNA-193b controls adiponectin production in human white adipose tissue. J Clin Endocrinol Metab 100:E1084–E1088
66. Granjon A, Gustin M-P, Rieusset J et al (2009) The microRNA signature in response to insulin reveals its implication in the transcriptional action of insulin in human skeletal muscle and the role of a sterol regulatory element-binding protein-1c/Myocyte enhancer factor 2C pathway. Diabetes 58:2555–2564
67. Yin H, Pasut A, Soleimani VD et al (2013) MicroRNA-133 controls brown adipose determination in skeletal muscle satellite cells by targeting Prdm16. Cell Metab 17:210–224
68. Price NL, Singh AK, Rotllan N et al (2018) Genetic ablation of miR-33 increases food intake, enhances adipose tissue expansion, and promotes obesity and insulin resistance. Cell Rep 22:2133–2145
69. Nunez Lopez YO, Garufi G, Pasarica M, Seyhan AA (2018) Elevated and correlated expressions of miR-24, miR-30d, miR-146a, and SFRP-4 in human abdominal adipose tissue play a role in adiposity and insulin resistance. Int J Endocrinol 2018:1–7
70. Das S, Stadelmeyer E, Schauer S, Schwarz A, Strohmaier H, Claudel T, Zechner R, Hoefler G, Vesely P (2015) Micro RNA-124a regulates lipolysis via adipose triglyceride lipase and comparative gene identification 58. IJMS 16:8555–8568
71. Zhang Y, Li C, Li H, Song Y, Zhao Y, Zhai L, Wang H, Zhong R, Tang H, Zhu D (2016) miR-378 activates the Pyruvate-PEP futile cycle and enhances lipolysis to ameliorate obesity in mice. EBioMedicine 5:93–104
72. Kulyté A, Lorente-Cebrián S, Gao H, Mejhert N, Agustsson T, Arner P, Rydén M, Dahlman I (2014) MicroRNA profiling links miR-378 to enhanced adipocyte lipolysis in human cancer cachexia. Am J Physiol Endocrinol Metabol 306:E267–E274
73. Lin Y-Y, Chou C-F, Giovarelli M, Briata P, Gherzi R, Chen C-Y (2014) KSRP and MicroRNA 145 are negative regulators of lipolysis in white adipose tissue. Mol Cell Biol 34:2339–2349
74. Wang M, Li L, Liu R, Song Y, Zhang X, Niu W, Kumar AK, Guo Z, Hu Z (2018) Obesity-induced overexpression of miRNA-24 regulates cholesterol uptake and lipid metabolism by targeting SR-B1. Gene 668:196–203
75. Du J, Zhang P, Gan M et al (2018) MicroRNA-204-5p regulates 3T3-L1 preadipocyte proliferation, apoptosis and differentiation. Gene 668:1–7
76. Gao Y, Cao Y, Cui X et al (2018) miR-199a-3p regulates brown adipocyte differentiation through mTOR signaling pathway. Mol Cell Endocrinol 476:155–164
77. Lai C-Y, Lin C-Y, Hsu C-C, Yeh K-Y, Her GM (2018) Liver-directed microRNA-7a depletion induces nonalcoholic fatty liver disease by stabilizing YY1-mediated lipogenic pathways in zebrafish. Biochimica et Biophysica Acta (BBA)—Molecul Cell Biol Lipids 1863:844–856
78. Wu H, Zhang T, Pan F, Steer CJ, Li Z, Chen X, Song G (2017) MicroRNA-206 prevents hepatosteatosis and hyperglycemia by facilitating insulin signaling and impairing lipogenesis. J Hepatol 66:816–824
79. Latorre J, Moreno-Navarrete JM, Mercader JM, Sabater M, Rovira Ò, Gironès J, Ricart W, Fernández-Real JM, Ortega FJ (2017) Decreased lipid metabolism but increased FA biosynthesis are coupled with changes in liver microRNAs in obese subjects with NAFLD. Int J Obes 41:620–630

80. Goody D, Pfeifer A (2019) MicroRNAs in brown and beige fat. Biochimica et Biophysica Acta (BBA)—Molecul Cell Biol Lipids 1864:29–36
81. Nautiyal J, Christian M, Parker MG (2013) Distinct functions for RIP140 in development, inflammation, and metabolism. Trends Endocrinol Metab 24:451–459
82. Ng R, Hussain NA, Zhang Q, Chang C, Li H, Fu Y, Cao L, Han W, Stunkel W, Xu F (2017) miRNA-32 drives brown fat thermogenesis and trans-activates subcutaneous white fat browning in mice. Cell Rep 19:1229–1246
83. Zhang H, Guan M, Townsend KL et al (2015) Micro RNA-455 regulates brown adipogenesis via a novel HIF 1an- AMPK - PGC 1α signaling network. EMBO Rep 16:1378–1393
84. Sun L, Trajkovski M (2014) MiR-27 orchestrates the transcriptional regulation of brown adipogenesis. Metabolism 63:272–282
85. Trajkovski M, Ahmed K, Esau CC, Stoffel M (2012) MyomiR-133 regulates brown fat differentiation through Prdm16. Nat Cell Biol 14:1330–1335
86. Sun L, Xie H, Mori MA, Alexander R, Yuan B, Hattangadi SM, Liu Q, Kahn CR, Lodish HF (2011) Mir193b–365 is essential for brown fat differentiation. Nat Cell Biol 13:958–965
87. Fu T, Seok S, Choi S, Huang Z, Suino-Powell K, Xu HE, Kemper B, Kemper JK (2014) MicroRNA 34a inhibits beige and brown fat formation in obesity in part by suppressing adipocyte fibroblast growth factor 21 signaling and SIRT1 function. Mol Cell Biol 34:4130–4142
88. Pan D, Mao C, Quattrochi B, Friedline RH, Zhu LJ, Jung DY, Kim JK, Lewis B, Wang Y-X (2014) MicroRNA-378 controls classical brown fat expansion to counteract obesity. Nat Commun 5:4725
89. De Silva N, Samblas M, Martínez JA, Milagro FI (2018) Effects of exosomes from LPS-activated macrophages on adipocyte gene expression, differentiation, and insulin-dependent glucose uptake. J Physiol Biochem 74:559–568
90. Zhao H, Shang Q, Pan Z, Bai Y, Li Z, Zhang H, Zhang Q, Guo C, Zhang L, Wang Q (2018) Exosomes from adipose-derived stem cells attenuate adipose inflammation and obesity through polarizing M2 macrophages and beiging in white adipose tissue. Diabetes 67:235–247
91. Lee RC, Feinbaum RL, Ambros V (1993) The C. elegans heterochronic gene lin-4 encodes small RNAs with antisense complementarity to lin-14. Cell 75:843–854
92. Chen C, Zhang X, Deng Y, Cui Q, Zhu J, Ren H, Liu Y, Hu X, Zuo J, Peng Y (2020) Regulatory roles of circRNAs in adipogenesis and lipid metabolism: emerging insights into lipid-related diseases. FEBS J febs. 15525
93. Zaiou M (2020) The emerging role and promise of circular RNAs in obesity and related metabolic disorders. Cells 9:1473
94. Zaiou M (2020) circRNAs signature as potential diagnostic and prognostic biomarker for diabetes mellitus and related cardiovascular complications. Cells 9:659
95. Alvarez-Dominguez JR, Bai Z, Xu D et al (2015) De Novo reconstruction of adipose tissue transcriptomes reveals long non-coding RNA regulators of brown adipocyte development. Cell Metab 21:764–776
96. Zhao X-Y, Li S, Wang G-X, Yu Q, Lin JD (2014) A long noncoding RNA transcriptional regulatory circuit drives thermogenic adipocyte differentiation. Mol Cell 55:372–382
97. Heneghan HM, Miller N, McAnena OJ, O'Brien T, Kerin MJ (2011) Differential miRNA expression in omental adipose tissue and in the circulation of obese patients identifies novel metabolic biomarkers. J Clin Endocrinol Metab 96:E846–E850
98. Dangwal S, Foinquinos A, Thum T (2018) MicroRNAs: novel therapeutic targets for diabetic wound healing. In: Veves A, Giurini JM, Guzman RJ (eds) The diabetic foot. Springer International Publishing, Cham, pp 237–246
99. Foinquinos A, Batkai S, Genschel C et al (2020) Preclinical development of a miR-132 inhibitor for heart failure treatment. Nat Commun 11:633
100. Schimmel K, Jung M, Foinquinos A et al (2020) Natural compound library screening identifies new molecules for the treatment of cardiac fibrosis and diastolic dysfunction. Circulation 141:751–767

101. Willeit P, Skroblin P, Moschen AR et al (2017) Circulating MicroRNA-122 is associated with the risk of new-onset metabolic syndrome and type 2 diabetes. Diabetes 66:347–357
102. Ortega FJ, Mercader JM, Catalán V et al (2013) Targeting the circulating microRNA signature of obesity. Clin Chem 59:781–792
103. Agostini M, Knight RA (2014) miR-34: from bench to bedside. Oncotarget 5:872–881
104. Huang C-K, Kafert-Kasting S, Thum T (2020) Preclinical and clinical development of noncoding RNA therapeutics for cardiovascular disease. Circ Res 126:663–678
105. Hulsmans M, Holvoet P (2013) MicroRNAs as early biomarkers in obesity and related metabolic and cardiovascular diseases. CPD 19:5704–5717

Chapter 5
Cellular and Biochemical Mechanisms Driving the Susceptibility of Obese Subjects to Covid-19 Infection

Manal M. Smail, Jaipaul Singh, Abla Mohammed Ismail, Emanuel Cummings, Carlin Hanoman, Sunil Rupee, Khemraj Rupee, and Ernest Adeghate

Abstract Overweight is a major global health problem currently affecting almost 2 billion people worldwide. An additional 800 million are obese. These figures showed that 40% of the global adult population aged 18 years, and over are overweight while 14% are obese. What is now worrying is that more than 40 million children worldwide, as young as 5 years of age are either overweight or obese. Individuals with a body mass index (BMI) of 25–29 kg/m^2 are considered to be overweight while obesity is the term used when the BMI is 30 kg/m^2 and over. Obesity is an imbalance between calorie intake and calorie expenditure. In general, obesity can be caused by excessive eating and reduced physical activity. Obesity is a major risk factor for non-communicable diseases such as diabetes mellitus, respiratory and liver dysfunctions, sleep apnea, chronic inflammation, compromised immune system, renal failure, cancer, musculoskeletal disorders, cardiovascular diseases and others. Obesity is also a major risk factor for coronavirus disease 19 (Covid-19), which can induce severe cases of pneumonia and sepsis or acute respiratory distress syndrome. In many cases, Covid-19 causes severe and long-lasting damage to the lungs and other vital organs of the body resulting in death. This review describes the cellular and biochemical mechanism(s) whereby obese patients become susceptible to Covid-19 infection. It also outlines how obesity on its own can affect the lungs, which in turn become more compromised in cases of Covid-19 disease resulting in the imminent death of the patient.

M. M. Smail · J. Singh · C. Hanoman
School of Natural Sciences, University of Central Lancashire, Preston PR1 2HE, UK
e-mail: Jsingh3@uclan.ac.uk

A. M. Ismail
Corniche Hospital, Abu Dhabi, United Arab Emirates

E. Cummings · C. Hanoman · S. Rupee · K. Rupee
School of Medicine, College of Medical Sciences, University of Guyana, Georgetown, Guyana

M. M. Smail · E. Adeghate (✉)
Department of Anatomy, College of Medicine and Health Sciences, United Arab Emirates University, Al Ain, United Arab Emirates
e-mail: eadeghate@uaeu.ac.ae

© Springer Nature Switzerland AG 2021

P. S. Tappia et al. (eds.), *Cellular and Biochemical Mechanisms of Obesity*, Advances in Biochemistry in Health and Disease 23,
https://doi.org/10.1007/978-3-030-84763-0_5

Keywords Obesity · Overweight · Covid-19 · Respiratory dysfunction · BMI · ARDS · Death

Introduction

During the past two decades, the global prevalence of obesity has increased from 30.5 to 42.4% while the prevalence of severe obesity increased from 4.7 to 9.2%. Currently, 2 billion adults, over the age of 18 years, are overweight and more than 800 million are obese globally [1]. We call a person obese if the person has a body mass index (BMI) of 30 kg/m^2 or over. On the other hand, an overweight individual has a BMI of 25–29 kg/m^2. Estimations by the World Health Organization showed that currently about 30% of people worldwide are obese [2–4]. The recent explosion in the number of obese individuals has occurred in the past few decades. For example in European countries, the number of subjects suffering from obesity has risen threefold within the last twenty years. It is also a fact that about one third of the population of USA is either overweight or obese. Likewise, the burden of the obese individual on the healthcare system and the community is tremendous. For example, more than 30,000 lives are lost, because of obesity, every year in the United Kingdom and as high as £3.5 billion ($5.4 billion USD) is spent annually to care for patients with obesity. In the USA, the cost of looking after an obese patient is 36% higher when compared to that of an individual with normal weight [5–7]. Currently, obesity is not only confined to developed countries but it is also prevalent in low-income developing countries as well. Moreover, what is more worrying is that globally, more than 40 million children, as young as 5 years of age are obese [8, 9].

Overweight and obesity are major risk factors of numerous chronic diseases and disorders as well as health conditions. They include diabetes mellitus (DM), heart disease, liver disease, kidney failure, some cancers, respiratory diseases and many others [10–12]. In fact, obesity occurs in conjunction with many metabolic diseases such as DM and hyperlipidemia [13]. Overweight and obese subjects developed respiratory dysfunctions over time due to excess fat deposits and fatty musculature in the body [14]. As such, obese individuals are more susceptible to Covid-19 infection, which is primarily a respiratory infection. Thus, it is of paramount importance to understand how obesity can affect the respiratory system and how obesity makes patients more susceptible to Covid-19 infection. This review attempts to ascertain why obese subjects are more prone to developing Covid-19 infection with emphasis on the cellular and biomedical mechanisms involved in the process. It also addresses the epidemiology of Covid-19 infection, the insulting effect of obesity on the respiratory system, the cellular and biochemical mechanisms of action of Covid-19 on the respiratory system, obesity and other comorbidities as major health risks for Covid-19 infection and acute and chronic inflammation due to a combination of obesity and Covid-19.

Epidemiology of Covid-19

The Covid-19 pandemic, caused by the severe acute respiratory syndrome coronavirus 2 (SARS-CoV-2) swept across the globe, causing tens of thousands of deaths and massive economic disruption since 2020 [15, 16]. Over 80,000 Covid-19 cases were first reported from Wuhan, China in December 2019. A large number of these infected cases were located in the Hubei area of Wuhan [17]. This Covid-19 pandemic actually peaked in China in January–February of the year 2020 and the number of new cases decreased substantially by early March [18]. Thereafter, a substantial number of cases were reported in all continents, except for Antarctica, and the number was rising steadily around the world [19]. The WHO Situation Reports of April 21, 2020 showed that there were 2,431,890 confirmed cases of Covid-19 infections including 169,859 deaths. Subsequently, more than 90% of the global COVID-19 cases were found outside of China. Simultaneously, the available data from WHO COVID-19 Situation Reports indicated a steady daily increase in the incidence globally [18].

On April 21, the European Region reached 1,073,947 of Covid-19 infection cases with a whopping 103,989 deaths. At the time, European countries accounted for almost half of the total global reported cases and two-thirds of the number of people who died from Covid-19. The five countries which reported the largest number of Covid-19 infection cases were Spain, with over 200,000 cases, Italy (181,228), Germany (143,457), UK (124,743) and France (114,657). The highest death rate was in Italy (24,114) and followed by Spain (20,852). In the USA, COVID-19 infection was reported in all 50 States. However, the cumulative incidence of Covid-19 infection varied from State to State. This depends on a number of factors, including population density and demographics, extent of testing and reporting and timing of mitigation strategies [19, 20]. The US Center for Disease Control (CDC) reported 764,265 total cases and 42,906 deaths on April 2020 [21]. Covid-19 infection was the second coronavirus outbreak that affected the Middle East. In 2012, MERS-CoV was reported in Saudi Arabia [22].

In April 2020, Saudi Arabia reported a total 10,484 cases and 109 deaths and the United Arab Emirates announced 7,265 Covid-19 cases and 46 deaths [23]. In addition, African countries confirmed that there were 23,267 Covid-19 cases and the five countries which reported the most number of cases were Egypt (3,333), South Africa (3,300), Morocco (3,046), Algeria (2,718) and Ghana (1,042) [24]. Since April 2020, large number of infected people were counted across the globe as a secondary round of Covid-19 infection. By January 15, 2021, more than 94.7 million cases were recorded with the majority in the USA and UK and over 2.03 million deaths, especially in the USA and Brazil [24–26].

Covid-19 infection affects people differentially. A person can harbor the virus for 1–2 days before showing any signs and symptoms of the disease and at the same time, he or she can spread the virus spontaneously to others without knowing that he or she is infected with Covid-19. This is the reason why many people are infected with the disease in an exponential manner. It is also believed that once the virus is in the body, it switches off the innate immune response of the cells in the respiratory airways by reducing interferon production. Interferon is released from cells in the pulmonary system into circulation as a signal that Covid-19 causing virus has infected the cell. It is also said that Covid-19 and its new strain have evolved by mutation as very smart, new viruses outsmarting the cells in the lungs and other organs of the respiratory system. People with obesity, DM and other comorbidities such as hypertension, cardiac failure, hyperlipidemia and other metabolic conditions are more susceptible to Covid-19 infection even faster and with intense severity [27]. It has been shown that infection is more prevalent in patients with DM, which is a component of metabolic syndrome into which obesity is an important component [28].

Covid-19 inducing virus does not discriminate. It can infect anyone irrespective of ethnicity, color, religion, gender, health and age but it is more likely to cause severe illness in people who are over the age of 60 years, those who have certain illnesses such as obesity, DM with poor glycemic control, cardiovascular diseases, respiratory dysfunction and others. Moreover, health and social care professional workers who come from Black, Asian and Minority Ethnic (BAME) groups, lower socioeconomic status and obese and diabetic subjects are more likely to be infected and possibly die from Covid-19 infection [29–32]. In a recent study, it was reported that people with a combination of either type 1 or type 2 DM, with obesity, are at a higher risk for the infection of Covid-19 and possible dying from the disease [29]. Since obesity predisposes people to type 2 DM [14], this study reviewed and investigated the effect of elevated body mass index (BMI) as a predictor for being hospitalized with Covid-19 infection and possibly dying from the disease, especially since obesity has a direct adverse and insulting effect on lung function and more so, Covid-19 is a respiratory disease. Firstly, it is important to understand how obesity (BMI) $\geq$30 kg/m^2 can induce respiratory dysfunctions.

Obesity-Induced Respiratory Dysfunctions

As major epidemics, both overweight and obesity can cause multiple systemic complications in the body with direct adverse effect on respiratory system [14] (Fig. 5.1). Overweight and obesity increase oxygen consumption and carbon dioxide production [31]. They also stiffen the wall of respiratory tract and increase the physical workload required for ventilation. Indeed, overweight and obesity create mechanical impedance that restrict the expansion and the ability of the diaphragm, lungs and chest cavity to perform their functions regarding oxygen supply to tissues and expulsion of carbon dioxide from the body [33, 34].

The accumulation of fat in the thoracic region reduces the elasticity and flexibility of the respiratory system and weakens the diaphragm and other muscles involved in respiration. With the rise of weight and subsequently BMI, the volume of lung decreases due to restricted air entry [35]. Lung volumes such as forced expiratory volume in one second, functional residual capacity, forced vital capacity, expiratory reserve volume, residual volume and total respiratory capacity are also compromised

Fig. 5.1 Flow diagram showing how obesity can affect the respiratory wellbeing of a person

with increasing BMI. Overweight and obesity around the abdomen further lead to a worsening lung functions and respiratory symptoms. For example, the deposition of fat in the abdominal wall and around the abdominal viscera hampers the movement of the diaphragm and reduces lung expansion during inspiration leading to a reduced lung capacity. On the other hand, lower BMI cause fewer problems with respiration. The prognosis of respiratory conditions like chronic obstructive lung disease, and obstructive sleep apnea, is poorer in obese patients compared to individuals with normal weight. Obesity can also induce exertional dyspnea or severe breathlessness due to minor exertions [36]. Asthmatic crisis and poor progression of the disease is significantly higher in obese patients compared to lean mates. In fact, the prevalence of asthma is 38% more common in overweight individuals and 92% higher in people with obesity [37]. Obese asthma patients require more hospitalizations compared to lean asthma patients because they suffer from many more episodes of asthmatic crisis. This is partly due to poorer ventilation in obese individuals compared to lean subject with asthma. Low ventilation and low residual volume may eventually lead to hypoxia. This phenomenon is called “obesity hypoventilation syndrome”. Obese people are more susceptible to a serious condition called pulmonary embolism [38, 39], a condition that can lead to cardiac failure and imminent death. This is exactly what happens in obese patients who are infected with covid-19. Many obese subjects have short and narrowed airways which can lead to aspiration pneumonia [40]. In this situation, stomach contents move into the lungs causing severe pneumonia due to acids from the stomach regurgitating back towards the pharyngo-laryngeal region. Obese subjects also experience severe breathlessness during physical exercise. Indeed, this symptom is common in obese individuals, with or without overt respiratory disease [41]. This is because obesity increases oxygen demand and consumption with a concomitant increase in carbon dioxide release. In addition, obesity is associated with stiffening of respiratory muscles, chest wall and the lung, making it difficult for the respiratory system to initiate and effect breathing seamlessly [14]. Since asthma is more common in individuals with obesity, it has been postulated that the adverse effect of obesity on the respiratory tract may contribute to airway dysfunction and worsen the signs and symptoms of asthma. In most cases, obese subjects may have prediabetes or fully blown DM, liver dysfunctions, chronic inflammation, compromised immune system, renal failure, cancer, musculoskeletal disorders, cardiovascular diseases and other conditions [37, 39, 42, 43]. As such, it is important to understand first, how Covid-19 infection can affect the lungs and second, how the risk factors are interrelated, thus making the patients more susceptible to Covid-19 infection.

Mechanisms by which Covid-19-Induces Respiratory Complications

Coronaviruses are important human pathogens and research into their behavior is nearly a century old [22]. As outlined earlier, more than 94.7 million individuals are positive for SARS-CoV-2-induced Covid-19 infection. 2.3 million (2.4%) out of 94.7 million infected people have died [26, 27]. The pandemic has caused unmeasurable pain to individuals, communities, health care systems and economies throughout the world [15, 16, 20, 21]. About 80% of individuals with the disease present with the mild version of the disease in the form of high temperature, unproductive cough, shivering, discomfort, sore throat, skeletal muscle ache, and respiratory symptoms. These are due to the fact that the virus reaches the pulmonary airways or the pulmonary tree causing inflammation of the mucosal lining of the airways, which eventually result in cough and fever [29, 30]. Thereafter, the virus causes even more inflammation as it moves into the alveolar sacs of the lungs resulting in the production of fluid and other materials, which accumulate in and occupy alveolar sacs. Once the alveoli become acutely inflamed, especially in highly susceptible patients (5% of the cases), this in turn is manifested as shortness of breath (>30 bpm), poor O_2 saturation (<93%) and a reduced PaO_2/FiO_2 ratio (<300 mm Hg), all of which may lead to even more inflammation, pulmonary failure, septic shock, pneumonia, multiple organ failure and eventually death. The whole process is referred to as acute respiratory distress syndrome (ARDS) [27, 29] (Fig. 5.2).

Covid-19 infection induces severe micro- and macro-vascular damage in the lung, heart, kidney and whole body as a whole. This is because the infection induces a cascade of events, where the pro-thrombin system is activated causing a derangement in blood coagulation. This abnormal coagulation leads to the formation of micro-thrombus [43, 44]. Covid-19 affects all components of the respiratory system, including the neuromuscular breathing apparatus, the conducting and respiratory portions of the pulmonary airways and alveoli in addition to the pulmonary vascular endothelium. Figure 5.3 shows how covid-19 (SARS-CoV2) induces injury to the

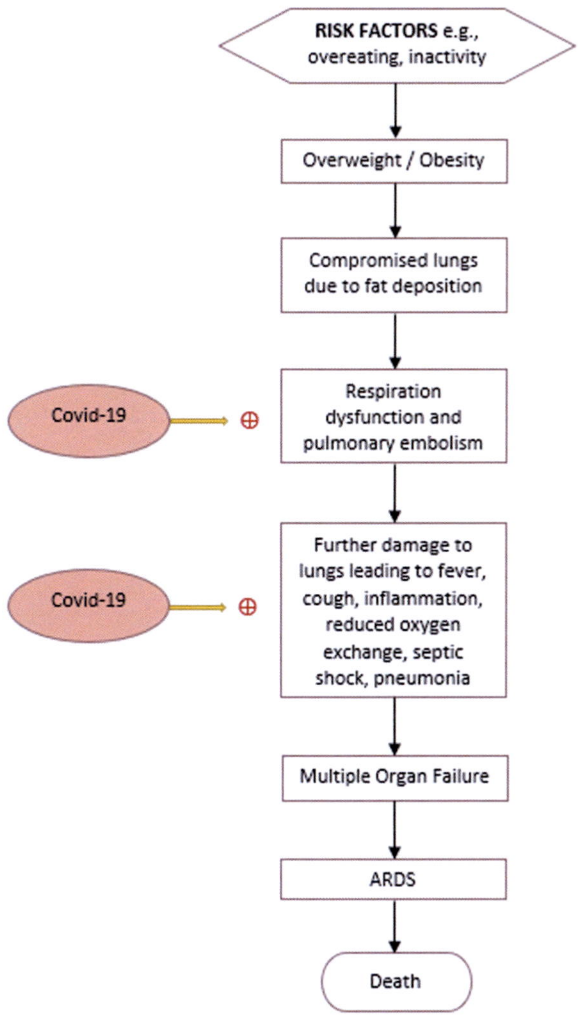

Fig. 5.2 Flow diagram showing the mechanisms by which Covid-19 induces respiratory complications

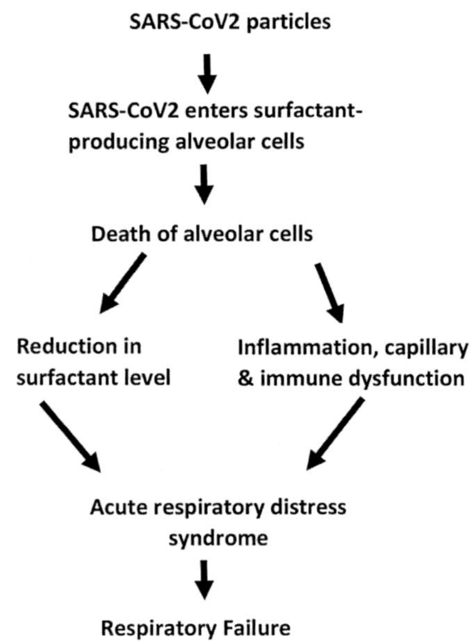

Fig. 5.3 Flow chart of the mechanisms by which SARS-CoV2 causes respiratory failure

alveolar epithelium leading to acute ARDS [43, 45]. Histopathology studies show that venous thromboembolic disease is very common in severe Covid-19 infection. Indeed, microvascular thrombosis has been shown in the alveolar sacs of patients with severe form of Covid-19 [46].

The SARS-CoV2 virus enters the lungs via alveolar epithelial cells expressing ACE2 (angiotensin-converting enzyme-2) and transmembrane serine protease 2 receptors. The coronavirus uses one of its four structural proteins to bind to ACE2 on alveolar type II cells, initiating a fusion between the SARS-CoV2 and the plasma membrane of the host cell. In turn, the transmembrane serine protease 2 receptors separates the ACE2 receptor and the viral glycoprotein S into 2 separate subunits. The virus leaves it coat and its genome is then released into the cytosol of the cell of the respiratory system. This step is followed by replication of the virus using both cytoplasmic organelles of both the viral and host cell. This involves the translation of the viral core proteins S, M, N, and E in the endoplasmic reticulum (ER), followed by assembly of viral particles in the ER-Golgi-intermediate compartment and accompanied by packaging into small wallet granules transferred towards the plasma membrane of the respiratory epithelial cell for eventual exocytosis and release into other compartments of the body. SARS-CoV2 infection-induced AT2 dysfunction or loss is deleterious to the injured lung due to a decrease in surfactant volume. This increases the risk for alveolar collapse and atelectasis. There is also a decrease in AT2 progenitor cells, which causes impaired alveolar type I (AT1) cell replacement, thereby affecting alveolar repair resulting in lung fibrosis. Moreover, ACE2 downregulation enhances over-activity of the ACE/Angiotensin II/AT1 receptor axis thereby worsening the tissue destructive effect of inflammatory response. Finally, viral-induced cytokine release by AT1/AT2 cells results in blood capillary leak and alveolar interstitial immune cell infiltration [43–45].

Diabetes, Obesity and Hypertension as Major Health Risks for Covid-19 Infection

It is well known that ageing is associated with a large variety of comorbidities, including but not limited to DM, obesity, cardiac and renal failures, and hypertension. All of these conditions come with increased risk of dying from Covid-19 [30–33]. Obesity predisposes people to developing DM, a chronic endocrine disease characterized by elevated blood glucose level [10–12]. DM occurs when the body content of insulin is insufficient or the secreted insulin molecules is not effective in helping glucose uptake by insulin sensitive cells. It is projected that the number of people with DM will rise astronomically to more than 650 million worldwide by 2030 from the current number of 470 million people [47]. It has been estimated that more than 80% of all individuals with DM will have complications such as hypertension, atherosclerosis, cardiomyopathy, stroke and chronic kidney disease [11, 12, 48, 49]. According to several studies, severe pneumonia, sepsis, and respiratory failure are the most important causes of the fatal outcomes of Covid-19 infection [30–33]. Moreover, obese subjects are usually prone to developing respiratory and liver dysfunctions, chronic inflammation, compromised immune system, renal failure, cancer, musculoskeletal disorders, CVDs and others due to the obesity making them more compromised to Covid-19 infection. In addition, it has been reported that morbidity and mortality in obese patients are significantly associated with pre-existing cardiovascular conditions [21]. Sudden cardiac death (SCD), for example, is the largest cause of adult deaths each year globally and SCD is accountable for about 50% of all mortality arising from cardiovascular diseases [50]. Indeed, most of the SCDs are the outcome of previously unknown CVDs in the background of obesity and DM [50]. In fact, it has been reported that angiopathies, including coronary artery disease (CAD), peripheral arterial disease and SCD are significantly more prevalent in obese and diabetic individuals compared with non-obese and non-diabetic people [51].

In a study examining the records of 174 patients suffering from COVID-19, individuals who are obese and diabetic develop a more severe form of pneumonia, release more tissue injury-related enzymes, and excessive inflammation and are more likely to die [27]. The degree to which metabolic disease such as DM, obesity or CVD either alone or in association with other co-morbidities affect the prognosis of Covid-19 infection is still elusive but more reports point to the fact that the more severe the metabolic condition, the more fatal a Covid-19 infection will be [30–33]. This review now addresses the involvement of inflammation in infected obese patients with Covid-19.

Covid-19 Combined with Obesity-Induced Acute and Chronic Inflammation

Inflammation, be it acute or chronic, is a mechanism by which the body defends itself [52]. Firstly, acute inflammation starts abruptly, and do not last for more than a few days and involves specific types of cells such as neutrophils. Secondly, chronic inflammation has a longer duration that may range from weeks to months or even years, depending on the type of the causative factor, integrity of the immune system, well-being of the affected person and the overall ability to overcome and repair the inflicted injury [52]. In short, the purpose of an inflammatory reaction is to prevent the harmful sources (biological and physical sources) from inflicting damage to our body systems and repair any tissue damage if that has already occured [53]. However, it may become a problem if an inflammatory response is excessive. In SARS-CoV-2 infection, an excessive and abnormal inflammatory reaction leading to the release of a large quantity of cytokines is a leading cause of death in Covid-19 infected patients, especially those who are obese and diabetic [30–33].

The cytokine storm is a concept derived from the clinical observation between mild infection in obese patients with Covid-19 and those with severe elevation of circulating CXCL10, CCL2 and TNF alpha. Moreover, they require admission into Intensive Care Unit (ICU). In addition, increased level of interleukin-1, interferon-γ, CXCL10, and monocyte chemoattractant protein-1 (MCP-1) was observed in obese individuals with Covid-19 infection. Furthermore, obese patients with Covid-19 and who are in ICU had increased blood levels of Granulocyte colony-stimulating factor (G-CSF), CXCL10, tumor necrosis factor-α, MCP-1, and CCL3 when compared to controls [54, 55]. This SARS-CoV-2-induced cytokine storm stimulates the release of a group of lymphocytes called cytotoxic T cells and Natural killers resulting in the inhibition of the same cell [56].

In the severe form of Covid-19 infection, especially in obese and diabetic patients, this large release of inflammatory cytokines reduces the number of lymphocytes present in the body. This is probably due to exhaustion. The increased activity of macrophages coupled with the release of large amount of inflammatory cytokines can lead to fever, chills and multi-organ failure and eventually the death of the patient [57, 58]

DM and obesity can cause low-grade chronic inflammation, accumulation of bad lipoproteins and initiating the development of labile atherosclerotic plaques that can cause thrombosis. Low grade chronic inflammation has been shown to significantly increase the risk of cardiovascular diseases, especially, stroke and myocardial infarction [59, 60].

Obese people, especially those with comorbidities and over 65 years of age, are more at risk from Covid-19 infection because they have excess fat deposition in their body, especially in the liver, diaphragm, and accessory respiratory muscles around the chest [14, 29–31]. As such, the excess fat seems to disturb the normal homeostasis of the body due to high insulin elevation in the blood. By unknown mechanism(s), the elevated circulating insulin is associated with several abnormalities including the production of inflammatory cytokines and a concomitant reduction of adiponectin, which is known to protect the lungs and the respiratory system from viral insult. This process is directly linked to the excess lipid accumulation in the lungs of the patients due to obesity. As a result, the lungs are unable to handle the invasion of Covid-19 virus leading to a derangement in oxygen absorption. This is due to several factors as discussed earlier in this review. They include reduced activity of the body's immune system due to age and comorbidities such as cardiovascular diseases, DM, lung infection, and even the medication they use regularly, especially anti-inflammatory drugs, which reduce the inflammation induced by the virus [61].

Conclusion

Figure 5.2 is a summary of the relationship between obesity and Covid-19 infection leading to acute respiratory distress syndrome (ARDS) and possible death of the patient. The literature has clearly demonstrated that obesity, due to excess fat deposit in the lungs, liver, and diaphragm has a detrimental effect on the respiratory system and the severity is more intense with the degree of obesity. Obese patients have tremendous problems with breathing and the entry and utilization of oxygen in their body. Moreover, they are susceptible to stroke and also have a compromised immunity system, chronic inflammation, sleep apnea, risk of T2DM, heart attack, hypertension, kidney failure and possibly other comorbidities. Obese people have slim odds against Covid-19 infection, which induces a specific insulting and lethal effects on the respiratory system in the body resulting in detrimental synergistic effects on the pulmonary system in a severe manner making it more difficult for obese subjects to breathe due to a lack of oxygen. In turn and over time, the infection can lead to death of the patients due to oxygen starvation and multiple organ failure.

References

1. So IYH (2020) Obesity and its complications. Adv Biochem Health Dis 19:43–58
2. Chooi Y et al (2019). The epidemiology of obesity. Metab: Clin Exp 92:00006–00010. https://doi.org/10.1016/j.metabol.2018.09.005
3. MacDonald KG Jr et al (2003) Overview of the epidemiology of obesity and the early history of procedures to remedy morbid obesity. Arch Surg 138(4):357–360. https://doi.org/10.1001/archsurg.138.4.357
4. WHO (2018) Non-communicable diseases. Time to deliver
5. Philip W, James T, McPherson K (2017) The cost of overweight. Lancet Public Health 2(5):203–205. https://doi.org/10.1016/S2468-2667(17)30068-3 (published online 5 Apr 2017)
6. Withrow D et al (2011) The economic burden of obesity worldwide: a systematic review of the direct costs of obesity. Obes Rev 12:131–141
7. Arterburn DE et al (2005) Impact of morbid obesity on medical expenditure in adults. Int J Obes 29:334–339
8. Riley JA et al (2005) Early life risk factors for obesity in childhood: cohort study. BMJ 330:1–10

9. Lopez RP (2012) Neighbourhood risk factors for obesity. Obesity 15(8):1903–2161
10. Badwar R et al (2020) Pathophysiology of obesity-related non-communicable chronic diseases and advancement in preventative strategies. Adv Biochem Health Dis 19:317–340
11. Adabag S et al (2015) Obesity related risk of sudden cardiac death in the atherosclerosis risk in communities study. Heart 101:215–221
12. Piernas C et al (2016) Obesity, non-communicable diseases (NCD) risk factors and dietary factors among Chinese school-age children. Asia Pac J Clin Nutr 25(4):828–840
13. Adeghate EA et al (2021) Tackling type 2 diabetes-associated cardiovascular and renal comorbidities: a key challenge for drug development. Expert Opin Investig Drugs 30:85–93. https://doi.org/10.1080/13543784.2021.1865914
14. Mafort T et al (2016) Obesity systemic and pulmonary complications, biochemical abnormalities, and impairment of lung function. Multidiscip Respir Med 11(28):1–11
15. WHO (2020) World Health Organization Director-General's opening remarks at the media briefing on COVID-19
16. WHO (2020) World Health Organization. Director-General's remarks at the media briefing on 2019-nCoV, 11 Feb 2020
17. Zhou P et al (2020) A pneumonia outbreak associated with a new coronavirus of probable bat origin. Nature 579(270):1–10
18. WHO (2020) World Health Organization Director-General's opening remarks at the media briefing on COVID-19, 24 Feb 2020
19. Badr HS et al (2020) Association between mobility patterns and COVID-19 transmission in the USA: a mathematical modelling study. Lancet Infect Dis 20(11):1247–1254
20. Unwin HJT et al (2020) State-level tracking of COVID-19 in the United States. Nat Commun 11(1):1–9
21. CDC (2020) COVID-19 Response Team. Geographic differences in COVID-19 cases, deaths and incidence—United States
22. Perlman S (2020) Another decade, another coronavirus. New Engl J Med 382:760–762
23. Dil S et al (2020) COVID-19 trends and forecast in the Eastern Mediterranean region with a particular focus on Pakistan. Cureus 12(6):e8582. https://doi.org/10.7759/cureus.8582
24. Wells CR et al (2020) COVID-19 on the African continent. Lancet Infect Dis. WHO (2015) Key facts from JMP 2015 report. World Health Organisation
25. WHO (2020) COVID-19 dashboard. Africa Centres for Disease Control and Prevention
26. Worldometers.info (2020) COVID-19 coronavirus pandemic. Dover, Delaware, USA. https://www.worldometers.info/coronavirus/?utm_campaign=homeAdvegas1?. Accessed 24 Jan 2021
27. Mahese E (2020) Covid-19: why are age and obesity are risk factors for serious. BMJ 1:1–11
28. Lotfy M et al (2017) Chronic complications of diabetes mellitus: a mini review. Curr Diabetes Rev 13(1):3–10. https://doi.org/10.2174/1573399812666151016101622
29. Naylor-Wardle J et al (2021) Socioeconomic status and cardiovascular health in the COVID-19 pandemic. Heart (British Cardiac Society). heartjnl-2020-318425. Advance online publication. https://doi.org/10.1136/heartjnl-2020-318425
30. McGurnaghan SJ et al (2020) Risks of and risk factors for Covid-19 disease in people with diabetes: a cohort study of total population of Scotland. The Lancet 20:3405–3408
31. Williamson EJ et al (2020) Factors associated with COVID-19-related death using OpenSAFELY. Nature 584:430–436
32. Bello-Chavolla OY et al (2020) Predicting mortality due to SARS-CoV-2: a mechanistic score relating obesity and diabetes to COVID-19 outcomes in Mexico. J Clin Endocrinol Metab 105:1–11
33. Dixon AE et al (2018) The effect of obesity on lung function. Expert Rev Respir Med 12(9):755–767
34. Brock JM et al (2020) Obesity and the lung: what we know today. Respiration 99(10):856–866
35. Genest M, Pochmalicki G (2004) Diagnostic and therapeutic progress. Venous thromboembolism, cardiac insufficiency and radio contrast agents. Press Med 33(9 Pt 1):623–630

36. He QQ et al (2009) Respiratory health in overweight and obese Chinese children. Pediatr Pulmonol 44(10):997–1002
37. Telenga ED et al (2012) Obesity in asthma: more neutrophilic inflammation as a possible explanation for a reduced treatment response. Allergy 67(8):1060–1068
38. Asghar O, Alam U, Hayat SA, Aghamohammadzadeh R, Heagerty AM, Malik RA (2012) Obesity, diabetes and atrial fibrillation; epidemiology, mechanisms and interventions. Curr Cardiol Rev 8(4):253–264
39. Piper AJ et al (2011) Obesity hypoventilation syndrome: mechanisms and management. Am J Respir Crit Care Med 183(3):292–298
40. Ekici A et al (2020) Pulmonary embolism in obesity-hypoventilation syndrome. Clin Respir J 14(11):1099–1104
41. Goldhaber SZ, Grodstein F, Stampfer MJ, Manson JE, Colditz GA, Speizer FE et al (1997) A prospective study of risk factors for pulmonary embolism in women. JAMA 277(8):642–645
42. Koenig SM (2001) Pulmonary complications of obesity. Am J Med Sci 321(4):249–279
43. Brosnahan S, Jonkman A, Kugler M, Munger SJ, Kaufman D (2020) COVID-19 and respiratory system disorders. Atheroscler Thromb Vasc Biol 40(11):2586–2597
44. Adeghate E, Eid N, Singh J (2021) Mechanisms of COVID 19 induced heart failure: a short review. Heart Fail Rev 2021(26):363–369. https://doi.org/10.1007/s10741-020-10037-x
45. Hassan M, Abbas A, Mohammed MH, Fatima G, Singh RB, Singh J et al (2020) Role of of ARBs and ACEIs in the treatment of SARS-COV2. Eur J Mol Clin Med 7:1–7
46. Ackermann M et al (2020) Pulmonary vascular endothelialitis, thrombosis, and angiogenesis in Covid-19. N Engl J Med 383(2):120–128. https://doi.org/10.1056/NEJMoa2015432
47. Adeghate E et al (2019) An update of SGLT1 and SGLT2 inhibitors in early phase diabetes-type 2 clinical trials. Expert Opin Investig Drugs 28(9):811–820. https://doi.org/10.1080/13543784.2019.1655539
48. Adeghate E et al (2006) An update on the etiology and epidemiology of diabetes mellitus. Ann N Y Acad Sci 1084:1–29. https://doi.org/10.1196/annals.1372.029
49. Zimmet P, Alberti KG, Shaw J (2001) Global and societal implications of the diabetes epidemic. Nature 414(6865):782–787
50. Adabag S et al (2015) Obesity related risk of sudden cardiac death in the atherosclerosis risk in communities study. Heart 101(3):215–221
51. Adeghate E, Singh J (2014) Structural changes in the myocardium during diabetes-induced cardiomyopathy. Heart Fail Rev 19(1):15–23. https://doi.org/10.1007/s10741-013-9388-5
52. Bastard JP, Maachi M, Lagathu C, Kim MJ, Caron M, Vidal H et al (2006) Recent advances in the relationship between obesity, inflammation, and insulin resistance. Eur Cytokine Netw 17(1):4–12
53. Spinas E, Kritas SK, Saggini A, Mobili A, Antinolfi P, Pantalone A (2014) Role of mast cells in atherosclerosis: a classical inflammatory disease. Immunopharmacology 27:517–521
54. Channappanavar R, Fehr AR, Vijay R, Mack M, Zhoa J, Mayerholst DK, Perlman S (2016) Dysregulated type I interferon and inflammatory monocyte-macrophage responses cause lethal pneumonia in SARS-CoV-infected mice. Cell Host Microbe 19(2):181–193
55. Huang C et al (2020) Clinical features of patients infected with 2019 novel coronavirus in Wuhan, China. Lancet 395(10223):497–506
56. Diao B, Wang C (2020) Reduction and functional exhaustion of T cells in patients with Coronavirus disease 2019 (COVID-19). Front Immunol 11:800–827
57. Castelli V, Cimini A, Ferri C (2020) Cytokine storm in COVID-19: "when you come out of the storm, you won't be the same person who walked in." Front Immunol 10:1–11
58. Shaw AC, Goldstein DR, Montgomery RR (2013) Age-dependent dysregulation of innate immunity. Nat Rev Immunol 13(12):875–887
59. Willerson JT, Ridker PM (2004) Inflammation as a cardiovascular risk factor. Circulation 109(21 Suppl-1):1–10
60. Welsh P, Grassia G, Botha S, Sattar N (2017) Targeting inflammation to reduce cardiovascular disease risk: a realistic clinical prospect? Br J Pharmacol 174(22):3898–3913

61. Aoba A, Yamaguchi N, Shibata M, Tsuneizumi T, Sakai T, Chishima T et al (1986) Plasma neuroleptic levels in the elderly patients on propericiazine therapy—possible role of morbidity. Arch Gerontol Geriatr 5(2):147–157

Chapter 6
Dysfunctional Circadian Rhythm Is Associated with Food Consumption, Obesity and Related Metabolic Diseases: Role of Ion Channels

A. Cihangir Uguz, Lourdes Franco Hernandez, Jaipaul Singh, Ana Beatriz Rodriguez Moratinos, and Jose Antonio Pariente Llanos

Abstract Circadian oscillators are the body's biological clocks which exhibited in most of living organisms from bacteria to higher vertebrates. They are responsible for organizing a variety of biochemical and physiological cellular functions with a rhythmic period of a day cycle (24 h, circadian, repeat cycle in a day) even without any timing indicators. Any disruption in synchronization of circadian rhythm (chronodistruption) causes a wide range of complications which can be referred to as metabolic syndrome, obesity or type 2 diabetes mellitus (T2DM). Food intake can be stimulated because of its hedonic properties, although energy need is sufficiently provided. Addiction can be determined as excessive intake of either drug or food. Drug and food addiction shares some similar hedonic neuroadaptative properties in perception reward circuits. That could be as a result of childhood physical or psychological trauma by increasing neurotransmitter hypersensivity or dysregulation. Circadian clocks are key players of hormone synthesis and release, which cause cellular adaptations to the body environment. Ion channels are protein structured gate keepers located in the cell membrane, allowing charged ions to move across the membrane. They contribute and regulate many of cellular functions in the body. Ion channels act as an important player in circadian phases and also subsequent physiological functions by contributing in signaling pathway including homeostasis, gene expression,

---

A. C. Uguz
Department of Biophysics, School of Medicine, Yozgat Bozok Univeristy, Yozgat, Turkey

L. F. Hernandez · A. B. R. Moratinos · J. A. P. Llanos (✉)
Department of Physiology, Faculty of Science, University of Extremadura, 06006 Badajoz, Spain
e-mail: Pariente@unex.es

L. F. Hernandez
e-mail: lourdesfh@unex.es

A. B. R. Moratinos
e-mail: Moratino@unex.es

J. Singh
School of Natural Sciences, University of Central Lancashire, Preston, UK
e-mail: Jsingh3@uclan.ac.uk

© Springer Nature Switzerland AG 2021
P. S. Tappia et al. (eds.), *Cellular and Biochemical Mechanisms of Obesity*,
Advances in Biochemistry in Health and Disease 23,
https://doi.org/10.1007/978-3-030-84763-0_6

etc. Hence, this review focuses on the importance of chronobiology and its role on prevention of obesity, T2DM and regulation of the ion channels by circadian rhythm.

Keywords Circadian rhythm · Chronodistruption · Obesity · Diabetes · Ion channels

Abbreviations

$[Ca^{2+}]_i$	Intracellular calcium (Ca^{2+})
AITC	Allyl isothiocyanate
AP	Action potential
AT	Adipose tissue
BAT	Brown adipose tissue
BK	Big conductance K^+ channels
BMI	Body mass index
Ca^{2+}	Calcium ion
CAPC	Calcium activated K^+ channels
CBT	Core body temperature
CHD	Coronary heart disease
CLOCK	Circadian Kaput cycles
CVDs	Cardiovascular diseases
DD	Constant darkness
EE	Energy expenditure
EPIC	European Prospective Investigation into Cancer
ER	Endoplasmic reticulum
FBG	Fasting blood glucose
GIRK	G protein gated inwardly rectifying K^+ channels
GLP-1	Glucagon-like peptide 1
GLUT-2	Glucose transporter
GOF	Gain of function
GSIS	Glucose-stimulated insulin secretion
HFD	High fat diet
HMGCoA	Hydroxymethylglutaryl-CoA
ICAC	Intermediate-conductance calcium-activated K^+ channels
KK	Knockout mice
LD	Light-dark cycle
LDL	Low-density lipoprotein
LOF	Loss of function
MetS	Metabolic syndrome
MODY	Maturity onset diabetes of the young
MS	Metabolic syndrome
PIP_2	Phosphatidylinositol 4,5-bisphosphate
POMC	Proopiomelanocortin

PSG	Polysomnogrephy
SCCA	Small-conductance calcium-activated
SCN	Suprachiasmatic nucleus
SD	Standard deviation
SNP	Single nucleotide polymorphism
SSTRs	Somatostatin receptors
STZ	Streptozotocin
SUR1	Sulfonylurea receptor 1
T2D	Type 2 diabetes
TRP	Transient receptor potential channels
TT	Sayfa 12 [1]
UCP-1	Uncoupling protein 1
VDCC	Voltage dependent Ca^{2+} channels
WAT	White adipose tissue
WHR	Waist to hip ratio
WT	Wild type
α2-ADRs	Adrenereceptors

Introduction

Nowadays, meals in Western culture revolve around the three meals habit, including breakfast, lunch and dinner. It is widely common (also recommended by some experts) to take a snack in the morning and another snack in the afternoon in order to control appetite throughout the day. As such, the best recommendation is to eat more than four times during the day, but it is not a standard habit, nor in other countries or in past times. Looking to the past, in the times of the Roman Empire, they believed that it was unhealthy to eat more than once per day. Only one big meal was consumed, the "coena", around 4 pm, and two light and quick meals in the morning and at noon [2].

In Medieval times, the term breakfast came from "break the night's fast", because it was the first meal after the evening prayer [3]. Monks also remain silent during meals, hearing the readings of religious texts.

It became necessary to have something to eat before going to work in the Industrial Revolution times, shaping the breakfast we know today. Also, dinner and dining were transformed by the widely use of artificial light, that made possible to eat before dawn and after noon [4]. It is very clear that the habits of eating meals are heavily influenced by social and economics periods, walking along the society changes and necessities.

"Eat like a king in the morning, a prince at noon, and a peasant at dinner" These are the words of Moses ben Maimon or Maimonides 1135–1404.

The total number of people who are suffering from obesity and diabetes mellitus has increased significantly all over the world during the last four decades affecting almost 11% of total population. Diabetes and metabolic disorders develop after the unbalanced homeostatic situation triggered by genetic predispositions and environmental factors. Exceeded energy intake than energy expenditure, prepares the basis for obesity. Disruption of glucose homeostasis in the body can cause high glucose levels to trigger migration, proliferation, apoptosis, necrosis and even promotes tumour progression. Determining the adequate energy expenditure is the best strategy for preventing and treating obesity. Brown adipose tissue (BAT) is defined as the energy-expending part of body which produces heat by inducing uncoupling protein 1 (UCP-1). The temperature of the body varies within a day, as well as depending on the body's activity. With this property, BAT differs from white adipose tissue (WAT). The expansion and activation of BAT can protect individuals from diabetes and obesity. UCP-1 can be found in the inner membrane of mitochondria of BAT. The role of UCP-1 is to uncouple the respiratory chain thereby generating heat. As such, the activation of BAT highlights the energy regulation and fat deposition.

Ion channels have crucial roles in BAT activation and insulin release. In the glucose-stimulated insulin secretion (GSIS) cascade, glucose firstly enters β cells by diffusion via glucose transporter (GLUT-2). Second, glucose is metabolized by enhanced ATP/ADP ratio. Elevation in ATP levels and decreased ADP levels cause closing K_{ATP} channels. Third, membrane voltage levels changes from the K^+ equilibrium state, thereby facilitating Ca^{2+} entry into the cytosol through voltage-dependent Ca^{2+} channels (VDCC). Increased intracellular calcium (Ca^{2+}) ($[Ca^{2+}]_i$) levels trigger insulin secretion form pancreatic β cells. Ion channels are the protein structured gatekeepers of cell membrane allowing ion movement through cell membrane. Among the ion channels, transient receptor potential (TRP) channels are expressed in a variety of tissues and demonstrated that they have a crucial role in regulation of cellular functions, hormone and neurotransmitter release, energy mechanism and cell survival pathways in either diabetic or obese conditions. Moreover, dietary administration of some TRP channel's ligands has been associated with improving effects in diabetic and obese conditions. These findings attract attention of Scientists to focus on ion channels-gating mechanisms. This review discusses the importance of diabetes and obesity and the role of some ion channels contributing the possible mechanisms involved in these metabolic diseases.

Importance of Meals Frequency: When and How Many

There are various studies which associate meal frequency with cholesterol disease, body mass index (BMI), obesity and diabetes. Cultural tradition is believed to be the origin of the concept that the healthier choice is eating three meals per day [5]. Moreover, elevated cholesterol and body weight are two major risk factors for cardiovascular diseases (CVDs [6]. In a study by Paoli et al. [7], they pointed out that a "gorging" diet versus a "nibbling" diet (consumption of frequent smaller meals or snacks) produces worse blood lips in the first case [8]. Another study made by Elestein et al. [9], and confirmed by previous studies, reported that the existence of lower age-adjusted total and low-density lipoprotein (LDL) cholesterol in volunteers who have the habit to eat four or more meals daily, compared to people who reported one or two [9]. Other association was obtained comparing BMI and macronutrient intake with alcohol intake, smoking habit, systolic blood pressure and anthropometric measurements as waist to hip ratio (WHR). But an additional investigation reported a decrease in the risk of obese people who eat less than three meals per day compared to eating more than six meals. Moreover, the Malmo Diet and Cancer study concluded that people who had better lifestyle and eat five or six times per day had lower waist circumference [10].

It is clear that meal frequency is connected to weight control, but the question is how? The big question is how many meals are the best to prevent overweight? Obesity has a big prevalence in modern society and has been doubled since 1980 in more than 70 countries [11]. A big indicator is that the majority of people in the world live in the countries where more people died by overweight and obesity than by underweight [12]. This is a risk factor for type 2 diabetes, (T2DM), dyslipidemia, hypertension and CVDs [13].

In order to prevent overweight and obesity, it is of paramount importance to establish healthy strategies such as eating fewer calories through diet, drugs or bariatric surgery [14] and exercising regularly [15]. The timing of meals and the frequency of meals have a significant influence on weight control [16]. Kahleova et al. [17] showed that eating one or two meals a day was associated with a lower BMI compared to three meals a day. Other authors have seen that there is a relationship between the amount of food ingestions in a day and the amount of snacks eaten (more than three a day) and also the increase in body weight. Changes in weight are related to time of fasting during the night resulting in the longer the overnight fast, the lower the BMI. This reduction is due to the combination of the frequency of meals and night fasting. In addition, there are hormones that help to maintain and also to improve the peripheral circadian clock. These satiety hormones (leptin or ghrelin) will be discussed later in this review and a reduction in oxidative damage, along with increased resistance to stress [17]. There are older [8, 18] and more recent research studies [19] suggesting that more frequent meals can reduce the risk of weight gain. However, recent large prospective studies appear to support the conception that frequent snacking increases the risk of weight gain [20] and T2DM [21]. This is still controversial nowadays.

Some benefits of a lower meal frequency might be supported by investigating acute metabolic responses. One study realized by Taylor and Garrow [22] assessed the effects of isocaloric diets consisting in two or six meals per day on energy expenditure measured in a metabolic chamber. The study carried out by Taylor et al., demonstrated that when two meals are consumed per day compared to six, there are no differences in energy expenditure during the day. However, this habit implies a much higher difference during the night time [22].

Other authors point out that a higher basal energy expenditure in the morning compared to the evening represents an improvement in the health of individuals [23]. On the other hand, there are differences during the day with respect to total energy expenditure [24]. The increase in BMI and its metabolic consequences with a higher frequency of meals are due to the higher intake of energy derived from sugar [25]. However, this is associated metabolic problems, but also to the increase in food stimuli, hunger, and the desire to eat [26]. Thus, a meal habit has a good effect on health and a reduced meal frequency is associated with an irregular eating approach that causes weight gain. Other authors have observed that metabolic alteration can increase cardiovascular risk [27] and decrease the risk of weight gain [17].

In a study, McGrath and Gibney [25] convinced volunteers, who normally ate six times a day to reduce their frequency while persuading those who ate less frequently (three times a day) to increase their frequency to six times. This change led to increased meal frequency and it also induced a significant reduction in blood total low density lipoprotein (LDL) cholesterol level. In addition, this was also combined with a reduction in carbohydrate intake. Due to this interesting observation, it is now possible to consider that the effects of changing meal frequency could also modify the overall macronutrient intake [25].

A lot of research studies were done on Ramadan fasting and energy expenditure because it can be considered a form of time restricted feeding cycle and can be studied (almost) annually [28]. During Ramadan, Muslims cannot eat until dark, creating an opposite fasting window compared to other fasting based on eating during daylight. Even though the duration of Ramadan is short, only four weeks, it usually does not lead into ketosis, but it can stimulate many pathways linked to long-term fasting [29].

On the other hand, when talking about meal frequency and timing, the bigger conflict is breakfast. Is this important or not? Most people believe if they eat breakfast, they will have a lower risk of gaining weight compared to those who skip breakfast compared to those who eat their larger meal at lunch or dinner have a higher risk of gaining weight [17]. Cahill et al. studied the association between the risk of coronary heart disease (CHD) and the frequency of food consumption [30]. Skipping dinner and having breakfast increases energy expenditure for a full day, while skipping breakfast could also trigger a higher concentration of postprandial insulin and greater fat oxidation. As such, there is certain difficulty to change the metabolic habits of our body [31]. Other data suggest that if there are health promoting effects of reducing the frequency of meals then there may be differential effects of skipping breakfast versus dinner. In addition, late intake of meals is associated with an increased risk of obesity and coronary heart disease [32] and snack taking is associated with higher total energy intake and later food consumption in the evening [33].

Meals Frequency and Illness: Cause and Effect

In a study comparing subjects reporting a higher meal frequency (6 times/day) to those who ate once or twice a day, it was observed a lower concentration of total and LDL cholesterol in the group who ate more meals per day, even adjusting age, BMI, physical activity, smoking, total energy intake, and macronutrient distribution. This was discovered by a project within the European Prospective Investigation into Cancer (EPIC) [34]. On the contrary, a recently published analysis within the prospective Seasonal Variation of Blood Cholesterol Study in Worcester County, Massachusetts, proposed an opposite relation. It was demonstrated that more than four meals per day show a lower risk of obesity, compared to a frequency lower than three times per day, even after adjustment for age, sex, physical activity, and total energy intake [35].

Mekary et al. published a study on diabetes. In this article, a 16-year follow-up was carried out that indicated a risk of T2DM in men who ate 1 or 2 times a day when compared with those who ate 3 meals a day. This was the case even when adjusting for age or other factors [21]. Another study, conducted in women during a 6-year follow-up, indicated that there was no correlation between increased frequency of meals and the risk of T2DM [21]. Many studies have examined risk factors compared groups that ate three meals a day after adjusting for total energy intake, diet composition, and other risk factors. Furthermore, as suggested by other authors [36], there is a strong relationship between the frequency of meals and an increased risk of CAD. This relationship is not very strong, and therefore it is difficult to establish this association.

There are several publications that study risk factors, but only one prospective cohort study has investigated the relationship between the frequency of meals and the risk of CAD. Cahill et al. [30] found that men who ate 1–2 meals a day had a fairly thigh risk of CAD of 1.10, while men who ate 4–5 meals a day had a risk of 1.05. In contrast, men who ate 6 times had even a much higher risk of 1.26, compared to those who ate 3 times a day.

McGrath and Gibney observed the reduction of cholesterol, adding some positive proof to the ongoing debate about the relationship between traditional disease markers, such as total cholesterol and LDL cholesterol and CHD [37]. Some authors have observed that high levels of cholesterol in the blood which in turn can increase the risk of both cerebrovascular (CBVDs) and CVDs [38]. Moreover, it is also assumed that the mechanisms involved in lowering cholesterol may be related to the mechanism of cholesterol biosynthesis. With respect to the disease of diabetes, in cholesterol biosynthesis, insulin activates a key enzyme, hydroxymethylglutaryl-CoA (HMGCoA) reductase (the target of statins) [39].

The goal of this review is not to address this mechanism (AMP-activated protein kinase, increased transcription rate, or insulin-induced genes) [40]. However, it is equally important to know that increased blood glucose can lead to an increase in insulin release from beta cells leading to enhanced insulin-induced synthesis of endogenous cholesterol [41]. Thus, an increase in the frequency of meals decreases insulin concentrations if it is compared with the intake of three meals a day, since there is a decrease in cholesterol synthesis [21].

The absence of frequent meals has effects such as the action of insulin, and the elimination of cholesterol (reverse cholesterol transport) after the ingestion of high-fat meals [42] and the inhibitory effects of cholesterol and fats on HMGCoA reductase [43]. Snack intake increases the amount of protein in the diet since the number of macronutrients is affected by the frequency of these meals, blood lipids, and the effects of insulin [44]. Therefore, the data show that there is no correlation between the number of snacks and hunger [44] or at least this relationship is not direct [39]. Therefore it seems like the majority response is related to satiety with a higher protein intake [36]. It should be noted that modifying meal frequencies could also change the energy percentage of particular macronutrients during the day, and also changes in carbohydrates and sugars with proteins [25].

Taking into account the data obtained by several authors, it is possible to assume that the frequency of meals does not seem explain the health effects. Currently there are different data on the frequency of meals and the health of an individual. This can be easily explained since generally, few meals cause the distribution of these effects. Moreover, it will be incorrect to assume that no breakfast, light lunch and a high calorie dinner or a very low number of meals produce a poor metabolic control [21]. Also, the effects of the frequency of meals are related to the timing of the same and the absorption of the macronutrients they contain. Currently, available data on the effects of "bite" (small, frequent meals) compared to "binge" (large and infrequent meals) under isoenergetic conditions [45] are responsible for conflicting results.

Circadian Rhythm and Obesity

Obesity is deeply rooted in our modern society and maybe aggravated by fast food, ultra-processed meals and full-time jobs. All these produces changes in our diet and physical activity, possibly leading to the increase in the prevalence of obese people. However, some research has raised interest in the possibility that changes in our daily behavioral patterns may be a significant contributing factor [46], adding the factor that, in industrial countries, the average of daily sleep has decrease by 1.5 h in the past century, concurrent with a significant increase in obesity [47].

The circadian clock is involved in internal rhythms and therefore metabolic rhythms, including glucose and lipids. Our internal circadian clock is influenced by eating early or late in the day [48]. In the body's circadian timing system, we can find the hypothalamic suprachiasmatic nucleus, which contains a central clock and different peripheral tissue clocks. Although the central clock is in charge of food intake, energy expenditure and insulin sensitivity, others including the peripheral tissue clocks have another additional control. In the intestine, for example, there is a peripheral clock that regulates glucose absorption. There are also other clocks in the liver and adipose tissue that regulate tissue sensitivity to insulin, while another peripheral clock in the pancreas regulates insulin secretion. There is also a molecular clock that regulates lipid metabolism in different tissues [49].

Obesity, T2DM and coronary heart disease (CHD) are promoted by clocks [50]. Animals also have an internal cellular clock mechanism, which helps them to be sensitive to light and therefore to anticipate and adapt to changes in environmental conditions linked to light and dark. The suprachiasmatic nucleus (SCN) was identified in the 1970s as the main biological clock. This mechanism can control not only sleep–wake cycles, but can control hormonal secretion, body temperature, blood pressure, and of course, behavioral variables making the body adapt to the changes that take place in the environment. The hormone melatonin is an important internal clock. This important indolamine is regulated by the light and dark cycles and vertebrates have a high level at night. All hormones have pulses throughout the day. For example, melatonin is highest around 9:00 p.m. and cortisol spikes in the morning, around 8:00 a.m. CVDs may lead to a risk of cardiovascular problems first thing in the morning [51]. The Period (Per1) gene, which has a cyclical rhythm in the pineal gland, sets the rhythm in the endocrine system that gives a signal for the body to control the master clock in the SCN [52]. McHill et al. [53] observed that obese people ingest most of their calories an hour closer to the onset of melatonin, that is, late in the day compared to slimmer people. During the day, glucose tolerance is less well controlled than at night. Diurnal rhythms are influenced by the responsiveness of liver cells, insulin clearance, and peripheral insulin sensitivity, but liver insulin sensitivity is less important. The circadian mechanisms that mark the day and night cycle are affected by light, sleep and wake, physical activity and food intake. In these physiological functions, the timing of meals is an important marker that also influences body weight [54]. This timing of meals influences the central master clock (SCN) and peripheral cell clocks, including genes Bmal1, Clock, Per1/2, Cry1/2, Rev-erb α/ß, Ror α/ß, Dbp, Dec1/2, CK1ε/δ and NPAS2 [52].

Serum ghrelin, insulin resistance (IR) indices and subjective appetite sensation in overweight and obese women are involved in the control of the internal clock and therefore, a low calorie diet, which distributes the calorie intake during the day, influences weight loss [55]. According to Jakubowicz et al., there are positive effects if more calories are ingested during the day, especially at breakfast, and therefore, there is a relationship between meal schedule and body weight [55].

Instead, there is evidence that breakfast is important for body weight. In a study by Dhurandhar et al. [56], they recruited 309 overweight and obese adults. They were then divided into two groups and one group was asked to eat breakfast and others not for 16 weeks. Although all of them complied with the protocol assigned to them, they saw that eating breakfast did not produce weight loss when compared to the group that skipped breakfast. On the other hand, studying cardiovascular health, Uzhova et al. [57] found that skipping breakfast was associated with an increased risk of non-coronary and generalized atherosclerosis independent of conventional CVD risk factors in a sample of asymptomatic middle-aged individuals. Furthermore, Betts et al. [58] found that both lean and obese people expended less energy during the morning when they skipped breakfast than after breakfast. When looking at the effects during the afternoon, they were not as conclusive. A recent study showed that excessive intake during the night is associated with a higher BMI, and thus, there is great heterogeneity, and it is difficult to draw a conclusion [59].

There are some articles on the correlations between night shift workers and the effect of this type of work on night rest [60, 61] and these confirm that poor sleep hygiene can lead to a problem of obesity. The genes involved in circadian Kaput cycles (CLOCK) are essential in metabolic regulation. That is, any alteration of the CLOCK, affects the metabolic pathways [62]. Garaulet et al. observed the associations between the CLOCK 3111T/C (rs1801260) single nucleotide polymorphism (SNP) and baseline body weight and weight reduction in obese patients who participated in a diet based on weight loss based on the Mediterranean diet [1]. Specific alleles in sleep and food intake reinforce the effects of diet [63]. These mechanisms are mediated by the hormones ghrelin and leptin which follow the circadian rhythm and are therefore involved in food intake. Thus, ghrelin is responsible for stimulating appetite and also it affects sleep [64]. This ensures that the neurophysiological and metabolic mechanisms responsible for the control of eating and sleep behaviors are coordinated so that during wakefulness appetite is stimulated and satiety is related to sleep [65].

Garaulet et al. [66] saw the relationship between the SNP CLOCK 3111T/C gene and the decrease in body fat in individuals (n = 500) of a population in which carriers of the (C) allele had more difficulty losing weight than their counterparts TT [1]. The CLOCK locus influences behaviors related to weight loss, and therefore sleep patterns by reducing sleep hours. Treatment of obesity poses a health challenge especially since behavioral therapy is a part of the weight loss program that aims to identify inappropriate behaviors. In turn, this can lead to weight gain, including excessive intake and lack of physical exercise habits [67].

The correct answers are based on the products used for behavioral therapy [68]. The CLOCK 3111T/C SNP gene may contribute to this goal. As mentioned, people who possess the minor C allele have more complications to reduce BMI than TT carriers (difference of 3 kg). The CLOCK 3111T/C gene functions by modifying the half-life of mRNA, which conditions the translation of CLOCK and therefore influences the levels of proteins involved in circadian rhythms and nighttime sleep [69]. Individuals carrying the C allele have a shorter sleep duration and therefore the behavior of these subjects may be affected, and they may even develop mental illnesses [70]. In those people who were carriers of the allele, it was found that they had a higher plasma ghrelin value than non-carriers, and it was more evident in people of 38 years of age.

The correlation between C carriers and the decrease in body fat outweighs the indirect relationship on the duration of sleep and the high concentration of the ghrelin hormone. Peptide Y is activated by this mechanism and is inhibited by proopiomelanocortin (POMC) and cocaine and regulated by amphetamines, and this activation mechanism is a positive feedback [71].

The type of diet is very important when it comes in losing weight and one of the best diets is the Mediterranean Diet, which, as many evidences show, is the best option to lose weight. [72]. Individuals who have the C-gene of the CLOCK3111 T and C SNP have less adherence to the Mediterranean Diet than TT people. It was also observed that there was a higher intake of animal proteins and processed foods, and therefore, a higher intake of trans-fatty acids. This diet made a weight loss program worse [73]. In addition, episodes of chronic insomnia, sleepiness, and fatigue have been reported to produce less physical activity in individuals [74]. It has been proven that the presence of the C allele is associated with a decrease in sleep that produces an increase in daytime fatigue that triggers a lower mental performance.

According to Gómez-Abellán et al. [75], circadian expression patterns in cultured adipose tissue explants were found for all the genes which have been investigated so far, and there exists a correlation analyses between the genetic circadian oscillation (amplitude) and components of the MetS. The parameters affecting each subject were studied and the data obtained by cosinor analysis, defining the circadian rhythms as mesor, amplitude, relative amplitude, acrophase and both per cent of rhythm. There was a greater amount of adipose tissue and BMI. The amplitude of it was revealed that greater adiposity and abdominal obesity (BMI, percentage of body fat, waist, sagittal diameter and conicity index) were correlated with a decrease in the amplitude of the genes ADIPOQ, ADIPOR1 and ADIPOR2 [75]. Moreover, they are related with a lower waist-to-hip ratio and it was also observed that these genes have a circadian rhythmicity and therefore, adiponectin is regulated depending on time. The form of regulation of these genes seems to occur at the level of mRNA. Control of these genes is produced by the suprachiasmatic nucleus in human AT for 24 h. Other genes also follow a circadian mode of action such as the clock genes (PER2, CRY1 and BMAL1) [76]. Likewise, other genes are related to cortisol and the glucocorticoid receptor (GR) and 11β-hydroxysteroid dehydrogenase 1 (11βHSD1) and PPARγ [77].

Physiologically, apart from glucose and lipids, there are other hormones such as insulin, glucagon, GH and cortisol and adiposospecific molecules that have metabolic role to play in the body. Similarly, MetS, such as visfatin, resistin, adipsin, and leptin) also have modes of circadian action [78]. Currently, there are only few studies which are based on the expression of the ADIPOR gene [79]. It is important to note that the circadian rhythm changes in the same way as the amount of adiponectin. Therefore, this relationship indicates that adiponectin has both autocrine and paracrine action. Adiponectin and its receptors are expressed simultaneously and produce an increase in the action of adiponectin and consequently an increase in its action. From a study in experimental animals [80], it is clear that the expression of the adiponectin gene follows a circadian rhythm in adipose tissue deposits with a nadir (minimum expression) during the morning (1000–1100 h). Adiponectin reaches its zenith (maximum) at the same times. This pattern in humans contrasts with the opposite effect in nocturnal animals such as rodents [75].

When the circadian expression of adiponectin is compared with the values found in serum during the day [81], it is observed that the nadir in adipose tissue (AT) anticipates the amount to be found in plasma, Hence, there is a phase delay of 7 h between transcription and secretion of the protein into the blood. Gavrila et al. [81] conducted a study in healthy men, who were compared with morbidly obese women. Their results demonstrated a close relationship to the kinetics of expression, synthesis and secretion of the adiponectin molecule.

The circadian system regulates metabolism in the different organs and tissues involved in the changes that occur in different diseases that lead to the appearance of metabolic syndrome (MetS). Furthermore, it is known that the homeostasis of lipids and glucose has a circadian pattern [82]. Thus, adiponectin is related to glucose metabolism [83]. It is for this reason that adiponectin is called the fat-burning molecule since it has the ability to direct fatty acids to the muscle so that they can be degraded and converted into metabolic energy [83]. This point is important because in this way it decreases the fatty acids in the liver and it also decreases the triglyceride content and therefore there is a greater sensitivity to insulin [83]. Adiponectin has its highest expression point in the morning (10:00 am) and therefore is related to the greater use of fatty acids at this time and therefore, there is a greater tolerance to the hormone insulin [84].

Some studies have shown that disruption of circadian rhythms induces obesity [85]. An adequate circadian rhythm leads to healthy eating habits and reduces obesity, hyperphagia, cancerous diseases and increases life expectancy [60]. Gómez-Abellán et al. [75] found a positive relationship between genetic circadian rhythms (amplitude) and the components of metabolic syndrome (MS) and it was found that biological rhythms are related to adiponectin secretion in women with obesity, leading to the attenuation of obesity. Studies have shown an increase in the markers of adipose tissue and abdominal obesity. They included abdominal diameter, weight, hip waist, body fat, BMI, conicity index and sagittal diameter, all of which were related to less adiponectin synthesis and the amplitude of the ADIPOR1 and ADIPOR2 genes, which characterize a circadian rhythm. This study was performed only on visceral fat and therefore this may explain the characteristics of metabolic syndrome (MetS) and the relationship with visceral fat. Other research previously conducted by Ando et al. [78] showed that the rhythmic expression of the clock genes (Bmal1, Per1, Per2, Cry1, Cry2, and Dbp) and adipocytokines (adiponectin, resistin, and visfatin) was attenuated in visceral adipose tissue of obese knockout mice. Other animal models revealed that mice with a Clock gene alteration are prone to developing obesity and a MetS-like phenotype [84].

In humans, alterations in the circadian rhythms of adiponectin expression appear with the disease of obesity [86]. The disruption of the circadian rhythm is the key in metabolic disorders diseases [87].

Chronodisruption and Diabetes

One of the major health challenges in the world today is the MetS, which affects up to a 25% of the world adult population, a syndrome that derive T2DM. In spite of its widespread presence, its underlying causes are still to be fully determined and understood. Recent studies suggest that some manifestations of MetS [85] might be associated with the disruption of the circadian system (chronodisruption). Therefore, prevalence of adiposity and MetS [78] is increased by sleep deprivation, shift work and exposure to bright light at night. Insulin resistance is an important factor involved in the pathophysiology of MetS. It is characterized by high plasma insulin concentrations that occur due to the failure of insulin function to suppress an increase in plasma glucose. Specifically, IR is defined as the inability of insulin to suppress hepatic glucose uptake and to stimulate glucose uptake in muscle and adipose tissue [88].

It has been recently discovered that circadian clock genes, not only exist in the brain, but also in tissues related to cardiovascular functions and energy metabolism such as the liver, pancreas, heart, and adipose tissue [89].

Adipose tissue (AT) can be found in some peripheral depots and it has relevance in studies concerning obesity and MetS. AT is not just a storage for triglycerides, as adipocytes release free fatty acids and secrete adipokines [90]. Recent studies have confirmed the presence of circadian clock mechanisms in AT and its periodic nature

[90]. The existence of an intracellular circadian clock system that regulates local cell functions might be suggested due to the persistency of gene expression oscillations in vitro. One adipokine that has attracted attention due to its role as a key modulator of insulin sensitivity is adiponectin [90], which can also be considered as a protective factor against MetS alterations [91]. The effects of adiponectin are mediated by binding to two cell membrane receptors: ADIPOR-1 (adiponectin receptor) and ADIPOR2 [92]. The first one is a high-affinity receptor for globular adiponectin, but a low-affinity receptor for full-length adiponectin, and is frequently expressed in muscle. The second one, ADIPOR2, is an intermediate-affinity receptor both for full-length and globular adiponectin and is normally expressed in liver [92]. Both receptors can also be found in adipose tissue, which strongly points to biological effects of adiponectin in AT in a paracrine/autocrine manner [79].

Some studies have demonstrated the diurnal and ultradiurnal dynamics of circulating adiponectin concentrations in humans [81] and in AT from animals [93]. Nevertheless, we still do not know whether this cytokine exerts a circadian oscillation in human AT and whether there are any differences between oscillation in visceral depots and in subcutaneous fat depots.

Adiponectin has been intensively studied in the last few years, due to its protective role in health alterations related to obesity, such as inflammation and insulin resistance [94]. Circulating levels of this hormone exhibit diurnal variations in humans [81]. Nevertheless, specific information regarding circadian rhythmic expression of adiponectin in human adipose tissue has not been reported that we know of. In this study, authors show how adiponectin mRNA levels in AT from obese women fluctuate throughout the day in the same phase as its receptors and how their circadian rhythms are attenuated with MetS features [81].

In a study by Harnandez-Morante et al. [95], they found differences in basal adiponectin expression between adipose tissue depots in women from the present work, being higher in the SC expression. This is in line with a previous study carried out in a similar population and is related to the cardioprotective feature of healthy AT [96]. As regards adiponectin receptors, there was little or no significant differences between both AT location, which is consistent with what has been evidenced in experimental animals [97].

Abnormalities in sleep quantity and quality [98], melatonin regulation [99] and circadian alignment [100] are all associated with an increased risk T2DM. According to several genome-wide association studies, a variation at MTNR1B [101] is associated with diabetes traits, but the mechanism whereby these variants produce elevated T2DM risk still remains unknown. Melatonin, a pineal hormone released at night, has two high-affinity receptors and MTNR1B is one of them. The circadian hormone plays a role in glucose homeostasis [102]. The release of melatonin takes place during overnight fasting while sleeping and high levels of this hormone during an oral glucose load at daytime causes impaired glucose tolerance [103]. Some variants in MTNR1B are related to an increased risk of T2DM, lower glucose-stimulated insulin secretion [104] and fasting blood glucose (FBG) levels in individuals without diabetes [101]. Some studies have determined that MTNR1B rs10830963 is probably the causal variant [105].

There is a relationship between the strange variants of loss of function in MTNR1B and an increase in the risk of diseases such as T2DM that is also related to the MTNR1B gene. Although this MTNR1B gene is important in the maintenance of blood glucose, today it is not known how the mutation in this MTNR1B rs10830963 gene alters normal glucose metabolism [106]. Understanding this mechanism could help to comprehend the mechanisms that influence the risk of glycemia that can help to determine a new route of therapeutic intervention.

Circadian phenotypes were analyzed in 58–96 volunteers by Lane et al. [107] and endogenous circadian physiology parameters were analyzed. Melatonin concentrations were analyzed in different situations, at the beginning of the light, with dim light. The midpoint of the melatonin curve was calculated [108]. Phase was also measured using core body temperature (CBT) nadir, the time when the adjusted circadian curve of CBT was at its minimum [109]. The circadian hormone was also measured in order to ascertain if it was stable and for this, the clearance rate of melatonin in the cytoplasm was measured in addition to the half-life of this hormone. The duration of melatonin secretion was analyzed and measured as the difference of the maximum and the minimum of the curve made with the difference in light. The time to go to bed, the time to wake up, the midpoint of sleep, the hours of sleep were taken into account for 7 days to realize the phenotypes of the sleep schedule. In all the samples analyzed, the rs10830963 gene and 58 other markers were taken into account to avoid stratification of the population [108, 109]. The authors used the Sequenom platform to carry out this study (Broad Institute, Cambridge, MA).

It has been observed that the genotype variant that poses a risk in the suffering of the diabetes-induced disease MTNR1B (rs10830963G) is associated with a late compensation of indole melatonin and the duration in time of the levels of this hormone. The hormone effects take place due to the removal of receptors [110]. On the other hand, it was seen that the increase in T2DM in carriers of the rs10830963G gene is greater in the first sleep compared to late sleep, and that morning melatonin secretion is covered by the increase that occurs later. Thus, the data reveal that MTNR1B rs10830963G causes melatonin production to increase late in the morning and thus early awakening increases the chances of diabetes. Changes in MTNR1B can cause changes in sleep synchronization as it can produce changes in melatonin synthesis [109]. The melatonin receptor 1B (known as Mel1B or MT2) is one of the two transmembrane receptors for melatonin. Research has shown that people carrying the MTNR1B rs10830963G allele develop increased expression of the Mel1B receptor in type b pancreatic cells [104]. The signal produced by melatonin during the night, when diurnal animals are fasting, means that glucose-stimulated basal insulin secretion does not occur [99]. Longer duration of melatonin in people who have the risk allele may result in an increased risk of food intake to coincide with elevated melatonin levels in the morning, resulting in lower glucose tolerance and possibly increased risk diabetes [111–113]. In this way, people who have the laoelo that poses a risk and also have an earlier sleep, have a greater risk of having T2DM, probably because they eat food and also have high levels of melatonin in the morning. In addition, detrimental effects, together with an increase in melatonin levels during the day resulted in a reduction in nocturnal melatonin signaling which also appears to be detrimental. The decrease in nocturnal melatonin signals, due to strange variants of loss of function of the MTNR1B receptor or lower levels of melatonin during darkness [99], is associated with an increased risk of T2DM.

However, it is essential to know the duration and time of night rest since these are cohort studies, and more so, they are probably measured with an error that could be interpreted as true false conclusions. In this sense, it has been concluded that there is a 50% decrease in power associated with a measurement error of one standard deviation (SD) of the trait [114]. This conclusion can be fulfilled in the phenotypes measured by polysomnogrephy (PSG) in which it was observed that the effects of the first night influence the measurements of sleep taken during a single episode of nocturnal PSG without supervision. This study concludes that it is unlikely that the MTNR1B diabetes gene variants also play a role in sleep, and researchers should emphasize in evaluating their role in peripheral tissues of relevance to type 2 diabetes [114].

Taking into account the both beneficial and harmful effects of the disruption of circadian rhythms in metabolism and the suffering of people due to diabetes disease, it has been observed that sleep and the MTNR1B gene relate the circadian rhythm with the alteration of melatonin synthesis. As such this provides clear evidence as to how this influences glucose control and diabetes risk. It is worth highlighting the importance of continuing with the study of the interaction suffered by MTNR1B rs10830963 with the melatonin rhythm phenotypes. These studies should have validity at the level of tissues or phenotypes in individuals. Thus, future investigations may lead to new therapies that lessen the impact of uncontrolled melatonin synthesis. This is most likely to occur through the alterations in insulin secretion mediated by melatonin or the postprandial moment [107].

Role of Ion Channels on Treatment and Prevention of Diabetes and Obesity

Voltage Gated K^+ (Kv) Channels

Almost all of the K_V channels expression was reported in human islets [116]. As mentioned above, their currents can be electrophysiologically recorded from human β-cells. However, their contributions to cellular excitability and insulin secretion offer some controversial results [117, 118]. K_V2.1 and K_V2.2 are responsible from most of K_V currents (65% >) in human β-cells. K_V activity in human β-cells can be inhibited by some toxins called stromatoxin, guangxitoxin-1E and small molecule (RY796) for K_V2.1 and K_V2.2 channels [118]. Moreover, RY976 increase GSIS in human β-cells. Expression levels of K_V2.2 encoding gene *KCNB2* was determined nearly ten times higher, compared to *KCNB1* gene which encodes K_V2 [116]. This indicates the importance of K_V2.2 channels electrical activity in human β-cells. This indicates that K_V2.1 can softly effects electrical changes in human β-cells [117].

However, reduced $K_V2.1$ expression was reported in human islets of T2DM patients [119]. There are some other K_V channels which contribute to electrical excitability, expressed in β-cells of T2DM patients. $K_V1.7$ encoding gene has close interaction with development of T2DM [120]. In an animal study, blockade of $K_V1.7$ channels increased insulin secretion and frequency of action potential firing in islets [121]. $K_V11.1$, and $K_V11.2$ are the other K^+ channels expressed in human β-cells, cardiac cells and neurons. Mutations in $K_V11.1$ cause increased insulin release and its related hypoglycemia [122]. *KCHN2* encodes $K_V11.1$ channels. This is related to KV11.1 polarizing effect on voltage changes in membrane of human β-cells [122]. *KCNH6* encodes $K_V11.2$ channels. Mutations in this gene can result in a rare syndrome which is a different form of either T1DM or T2DM and it is called as maturity onset diabetes of the young (MODY) [123]. It is now known that ascendants of MODY parents with mutation in *KCNH6* gene, will have 50% chance to inherit the gene from their parent(s). Their child will be MODY before his 25 of age. $K_V11.2$ loss of function (LOF) due to mutations in mice is characterized with hyperinsulinemia, similar in adults [123]. This indicates that LOF mutations can induce an increase in Ca^{2+} influx in β-cells which in turn triggers insulin secretion and endoplasmic reticulum (ER) stress. In turn, this causes a failure of pancreatic β-cells to synthesise and secrete insulin which can lead to diabetes [123]. Gain of function (GOF) mutation in K_V11 channels limits voltage-dependent potassium channel activity and thus, reduces insulin secretion which can also result in the development of diabetes [124]. Electrical silencing of $K_V2.1$ channels causes deficiency in free-running of the circadian clock rhythmicity in absolute dark conditions and stops free-running oscillations of PERIOD and TIMELESS proteins [125].

K_{ATP} Channels

It is a well known phenomenon that ion channel activation or inhibition is a response of β cells against elevated blood glucose levels. K_{ATP} channels can be inhibited after inhibition of glucose metabolism. This allows for depolarizing inward cation currents and activate voltage-dependent cation channels in β cells. This in turn, resulted in a change in membrane potential which triggers Ca^{2+} influx into the beta cells to induce insulin secretion. K_{ATP} channels are expressed in β cells and their role is weakly rectifying potassium channels on cell membrane. Their activities can be controlled by intracellular ATP/ADP ratio, which changes according to blood glucose levels. This ratio controls the channel activity as a sensor and tries to balance insulin secretion with correlating to blood glucose levels. Kir6.2 and sulfonylurea receptor 1 (SUR1) are the subunits of K_{ATP} channels. These subunits interact with adenine nucleotides and control the channel response to the ATP/ADP fluctuations. Elevations in ATP/ADP ratio led K_{ATP} channel closure. Kir6.2 is encoded by *KCNJ11* gene and SUR1 is encoded by *ABCC8* gene. Mutations in these genes are related with disorders of insulin secretion [126]. LOF mutations of K_{ATP} channels were related to congenital hyperinsulinism. Up-regulation of ATP-insensitive variant of Kir6.2

was shown in neonatal diabetes in mice [127]. Later, different Kir6.2 and SUR1 mutations were detected and they were related to neonatal diabetes in humans. The damaged insulin secretion in response to elevated glucose levels may occur due to chronic elevations of $[Ca^{2+}]_i$. Sulfonylureas, like glibenclamide, and glinides, like repaglinide and carbamazepine are the inhibitors of K_{ATP} channels. Long term intake of sulfonylureas triggers endoplasmic reticulum stress and hyperexcitability in β cells. Sleep duration can be regulated with many cellular and environmental factors, like seasonal variations of photoperiod and chronotype. It was determined by Allebrandt et al. [128] that sleep disorders are in close relation with genetic factors. Mutations in *ABCC9* gene significantly affect sleep duration variations [128]. Kir6.2 and SUR1, the two subunits of K_{ATP} channels contribute the glucose response in regulation in suprachiasmatic nucleus neurons external light dark cycle [129].

Calcium-Activated Potassium Channels

Calcium-activated potassium channels (CAPC) serve a crucial role in the repolarization period of β cell during action potential (AP). Big conductance K^+ channels (BK), intermediate-conductance calcium-activated (ICAC) K^+ channels and small-conductance calcium-activated (SCCA) K^+ channels are the subtypes of calcium activated (CAPC) K^+ channels expressed in human β cells [116]. Highly expressed CAPC subtype is the Ca^{2+} binding and depolarization activated BK channels. They characterized with quick activation and formation of K^+ currents. For these features, they act important role in repolarizing period of AP of β cell. BK channel inhibition results with increased amplitude of AP and increased depolarization-induced insulin secretion, during inhibited K_{ATP} [130]. LOF mutations in neuronal BK channels induce increased neuronal activity. The same case could happen in β cells which improves GSIS in LOF mutations which is characterized patients with BK channels. Like GOF mutations of BK channels cause faster AP firing rate in neurons, the same situation happens as triggering high frequency of AP in β cells. This high electrical excitability will result in glucose tolerance improvement and Ca^{2+} influx after glucose stimulation. Moreover, SCCA and CAPC K^+ channels limit AP frequency of cells, like they do in neurons. Ca^{2+} influx activates SCCA3 K^+ channels and ICAC and remains open after AP which limits the firing of the other AP. This reduces the frequency of AP. SCCA3 and IK1 encoding genes *KCNN3* and *KCNN4* lowly expressed in human β cells [116]. The insulin secretion is weakly increased by inhibition of IK1 and does not change by inhibition of SCCA3 [131]. Therefore, it is valuable to contribute in Ca^{2+} signaling cascade in T2DM.

G Protein-Gated Inwardly Rectifying K^+ Channels

$G_{\beta\gamma}$ released by Gio/o-coupled receptors activates G protein gated inwardly rectifying K^+ (GIRK) channels. Adrenereceptors (α2-ADRs) and somatostatin receptors (SSTRs) are the two main Gi-coupled receptors expressed in β cells [132]. As such, epinephrine and somatostatin can activate GIRK channels in β cells, by voltage change (hyperpolarization) and reduction in the VDCC activity [133]. Increased Ca^{2+} influx and insulin secretion during Gi receptor normally signals an inhibition by pertusis toxin (PTX; islet activating protein), emphasizing the importance of GIRK channels in controlling Ca^{2+} regulation in β cells. *ADRA2A* (α2A-ADR encoding gene) polymorphism is related with increasing risk factor of developing T2DM. Reduced insulin secretion and highly increased α2A-ADR levels were determined in islets of individuals with *ADRA2A* polymorphism. This cause GIRK activation, hyperpolarization in human β cells and reduction in VDCC activity, respectively as mentioned above. GIRK channel polymorphism and mutation can affect glucose up-regulation. Increased risk factor for developing T2DM was determined with *KCNJ9* mutation (GIRK3 encoding gene) in Indians [134]. GOF mutation was determined with *KCNJ5* mutation (GIRK4 encoding gene) related to familial sinus node dysfunction disease, which can cause sudden unexpected cardiac death [135]. More studies are needed to explore the correlation of glucose homeostasis in these patients. But, overactivity of GIRK4 channels proposed to inhibit insulin secretion. Cholesterol, cytosolic sodium (Na^+), and phosphatidylinositides are the other stimulators of GIRK channels [136]. Changes in cholesterol levels observed in T2DM patients can lead to stimulation of GIRK channels in β cells, insulin secretion and VDCC reduction by activating GIRK conductance. Their activities can be changed under diabetic conditions causing disturbances in human β cells. Phosphatidylinositol 4, 5-bisphosphate (PIP_2) is essential for GIRK channel activation. T2DM depended insulin resistance inhibits insulin-induced PI3K/AKT signaling pathway which in turn induces a significant reduction in GIRK activity [137]. Moreover, increased cytosolic Na^+ levels have been associated with increased GIRK activity [138]. During prolonged electrical excitation, Na^+ influx occurs, like in excitotoxic T2DM conditions. In this situation, Na^+ activates GIRK channels and cause β cell polarization and limits VDCC activation. In contrast to these, excitotoxic conditions may protect and enhance β cell survival by administering Ca^{2+} channel blocker (verapamil) or thioredoxin interacting protein (TXNIP, cellular redox regulator protein) by inhibiting Ca^{2+} influx and apoptosis in T1DM and T2DM [139]. TXNIP expression levels were inhibited by Ca^{2+} channel blockers in human islets and INS-1 cells, inhibiting the calcineurin signaling pathway. Diabetes and glucose jointly up-regulate TXNIP expression levels in β cells. Overexpression of TXNIP triggers apoptotic cascade in β cells. Verapamil reduced expression of TXNIP and apoptosis in β cells. Moreover, verapamil improved glucose levels and saved mice from streptozotocin (STZ) -induced diabetes by promoting β cell survival and its function. Genetic deletion of TXNIP proposes enhanced cell survival and protects obesity and STZ induced diabetes [139].

Ca^{2+} Channels

The main ion channels for Ca^{2+} entry in β-cells are voltage-gated calcium channels (VGCCs). As a universal signaling cation, Ca^{2+} influx is essential for GSIS in β cells and blocking of these channels inhibits insulin secretion [140]. Increased VDCC activation results in elevated insulin secretion and hence, decreased VDCC activity. In turn, this reduces insulin secretion which results as diabetes. The subtypes of VDCCs which has a role in GSIS can be listed as; L ($Ca_V1.2$, $Ca_V1.3$), P/Q ($Ca_V2.1$), and T ($Ca_V3.2$). GOF mutations of VDCCs are characterized with hypoglycemia and hyperinsulinemia [141]. GOF mutations in Cav1.2 (L-type) are related with advanced insulin secretion called as Timothy syndrome which is characterized with intermittent hypoglycemia. Other mutation in *CACNA1D* gene which encodes $Ca_V1.3$, L-type of VDCCs in β cells is characterized with hypoglycemia and hyperinsulinemia. However, VDDC transcription is reduced in T2DM patients. In correlation with insulin reduction, $Ca_V1.3$ encoding gene *CACNA1D* was also reduced in human islets of T2DM patients [142]. This is because of polymorphisms of *CACNA1D* promoter which results in reduction of *CACNA1D* gene. Transcription factor binding promoter of this gene is supposed to reduce $Ca_V1.3$ channel transcription levels [142]. LOF mutation of $Ca_V1.3$ should be predicated as relation between reduced $Ca_V1.3$ levels in T2DM conditions, which cause abnormalities in insulin secretion and as a consequence of this, diabetes would develop [143]. Intracellular signaling cascades are affected in the pathogenesis of diabetic conditions, like regulation of VDCC activity, which plays an important role on Ca^{2+} entry in β cells and insulin secretion. Plasma membrane PIP_2 increase VDCC activity of β cells [144]. It becomes disrupted under insulin resistant of T2DM conditions because of reduced activation of PI3K/AKT signaling pathway in β cells. Decreased PIP_2 levels in β cells are proposed to cause disturbances in VDCC activity in diabetic conditions. Moreover, cytokine exposure to mouse β cells triggers voltage activation of VDCC, which consequently causes insulin secretion [145]. But this could cause significant elevation in intracellular Ca^{2+} levels which in turn can induce mitochondrial dysfunction and endoplasmic reticulum-stress triggered apoptosis in β cells. Verapamil (a calcium channel blocker), administration was reported to inhibit VDCC activity and protect β cells from this stress condition [139]. That could be as a result of reduced stress levels which are in a turn of positive feedback loop with Ca^{2+} influx. Elevated oxidative stress levels trigger Ca^{2+} influx in variety of cells. β cells try to regulate the insulin resistance by increasing the insulin secretion through elevating $[Ca^{2+}]_i$ levels in the early stages of diabetes. This highlights the interaction between β cells and VDCC activity in different stages of diabeters development. Thioredoxin-interacting protein (TXNIP) is increased n β cells during diabetes and in turn, it can triggers β cell apoptosis, whereas down-regulation of TXNIP cause β cell survival [139]. Moreover, the location of Ca^{2+} channels represents importance on insulin secretion in β cells of T2D patients. Ca^{2+} channels closely located to insulin granules triggers higher depolarizations in β cells.

Transient Receptor Potential (TRP) Channels

Regulating glucagon-like peptide 1 (GLP-1) pathway, increasing insulin secretion from β cells (sulfonylureas), increasing tissue sensitivity to insulin (metformin), limit re-absorption of glucose in kidneys (SGLT2 inhibition) or directly injection of insulin can be listed as the pharmacologically treatments of diabetes by targeting end organs. Despite these current treatments, newly investigation of different molecules such as synthetic or natural compounds on glucose or weight handling by various signaling pathways including ion channel activity regulation attracts the attention of numerous researchers. One of these highly received attentions has been classified as TRP family member ion channels. TRP channel history start with the discovery of dysfunctions in phototransduction pathway of *Dropsphila Melanogaster* [146]. Atypical potentials were determined in these mutants. It was determined that the reason was a defect in certain type ion channel, not a failure in photosensitive channel activation or photopigment regeneration. This defect was later named with the transient receptor potentials characterized/caused ion channel family. TRP channels contribute to chemical response, energy balance, thermogenesis, sensory stimuli and temperature control. TRP channel family members have crucial roles in variety of both physiological and pathophysiological situations in living organisms which can be expressed in different types of tissues. TRP channel family has seven subfamily members and generally describe as Ca^{2+} permeable (mostly, but some of them also allows Na^{+} and K^{+} ions to pass) cation channels. Findings on the role of TRP channels on biological rhythms started to exist and most of them have solid findings in correlation with human experiments. Hence, we will summarize findings on the role of TRP channels on diabetic and obese conditions.

TRPA

TRPA1 is the member of TRP ankyrin family, voltage activated channel, which has an increased numbers of ankyrin repeats. Like some of the TRP channels, TRPA1 can also be activated by herbal compound such as cinnamon and its main compound cinnamaldehyde which is the agonist of this channel. Cinnamon is suggested for dietary therapies for diabetes since it controls lipid and glucose levels of T2DM patients. Moreover TRPA1 functions as a temperature regulator of circadian rhythm in neuronal pacemaker [147] and circadian neurons can be activated by TRPA1 [148]. Possible action of cinnamaldehyde on TRPA1 could be due to gastric emptying, ghrelin secretion [149]. Enhanced insulin sensitivity was also determined in this study. In another study, decreased visceral adipose tissue and elevated fatty acid oxidation were determined in mice fed with high-sucrose and high-fat diet (HFD) supplemented with cinnamaldehyde. Additionally, cinnamaldehyde inhibits weight gain in HFD fed mice [150]. As a weight gain marker, leptin/ghrelin ratio was significantly decreased in cinnamaldehyde supplemented HFD fed mice. One of the ingredients of horseradish, mustard and wasabi, allyl isothiocyanate (AITC) is listed among TRPA1 channel agonists. AITC injection (i.v.) and cinnemaldehyde trigger adrenalin secretion in anesthetized rats. However, these reactions totally disappeared in capsaicin administered rats. Moreover, pretreatment with cholinergic blockers (atropine and hexamethonium) decreased AITC and cinnamaldhyde triggered adrenaline secretion which suggest that TRPA1 agonists can activate sensory neurons and trigger adrenalin secretion through central nervous system [151]. In different studies AITC was found to be responsible for protection against free fatty acid induced insulin resistance and enhanced mitochondrial activity in muscle cells [152]. TRPA1 expression was determined in pancreatic β cells, activation of this channel triggers insulin secretion with correlation to inhibition of K_{ATP} [153]. This was supported with the findings that glibenclamide, K_{ATP} inhibitor, acts as TRPA1 agonist [154].

TRPV Channels

TRP Vanilloid 1 (TRPV1) is the most studied TRP subfamily channels on diabetes and obesity. TRPV1 is the firstly described member of TRPV subfamily. TRPV1 is a Ca^{2+} permeable cation channel, and its vanilloid receptor subtype 1 (VR1) was cloned in 1997 [155]. TRPV1 can be activated by both endogenous and exogenous ligands and among of these exogenous ligands, the capsaicin [156], an active component of chili pepper which is the most improving agent for body weight and cellular metabolism. Capsaicin (8-methyl-N-vanillyl-trans-6-nonenamide) causes a burning sensation in target tissue. TRPV1 channels can also activated by noxious heat stimuli and physical abrasion. Capsaicin binding site for TRPV1 is located in intracellular side of the membrane. TRPV1 activation results in depolarization of the membrane and cation influx. Although capsaicin is binding to TRPV1, it also disrupts membrane organization. This implicates the importance of drug discovery and it is better to keep in mind the cellular integrity. Increased energy expenditure (EE) observed in male subjects after having red pepper enriched meal. When the red pepper is included in a meal it increases carbohydrate oxidation and without increasing total EE. This sudden EE increase can be caused by β-adrenergic stimulation of the red pepper. In another study, again capsaicin enriched lunch increased GLP-1 levels while decreasing ghrelin levels, and does not change EE, and satiety [157]. However, two days supplementation of capsaicin increased satiety in a different study [158]. This could be due to extended exposure period. Moreover, daily consumption of chili pepper during four weeks caused a significant resistance of serum lipoproteins to oxidation in healthy individuals. This indicates the long consumption of chili diet can inhibit serum lipoproteins oxidation. Additionally, extended period of chili pepper consumption may reduce resting heart rate and enhance myocardial perfusion in male subjects [159]. It has been reported that activation of TRPV1 channels by capsaicin, in 3T3-L1-preadipocytes and visceral adipose tissue, prevents obesity and adipogenesis in humans and mice [160]. Dose dependently exposure of capsaicin increased $[Ca^{2+}]_i$ levels in 3T3-L1 preadipocytes. Capsaicin induced increased $[Ca^{2+}]_i$ levels were lower in mature adipocytes when comparing with preadipocytes. These findings are consistent with the decreased TRPV1 expression during adipogenesis comparing with knockout mice studies [160]. Additionally, decreased TRPV1 expression levels were found in obese male subjects comparing with lean subjects. This suggests that decreased levels of TRPV1 channels are common associated with diabetes and obesity. Moreover, capsaicin supplemented HFD-fed mice were prevented in developing obesity, which was not observe in TRPV1 knockout mice [160]. Enhanced TRPV1 expression levels and small size adipocytes were associated with the prevention of obesity. In addition to these findings, capsaicin enriched diet prevents weight gain, which did not observed in knockout mice [161]. Obesity prevention was related with the increased expression levels of uncoupling protein 1 (UCP1) and sirtuin-1 [161]. The reason for this is due to the effect of TRPV1 to induce $[Ca^{2+}]_i$ release which in turn triggers the phosphorylation of CaMKII and AMPK. Not only TRPV1, but also TRPV3 and TRPV4 channels contribute to adipocyte functions. TRPV3 activation prevents lipid accumulation and adipogenesis via inhibiting the PPAR-γ expression and phosphorylation of insulin receptor substrate-1 [162]. Administration of diphenylborinic anhydride (DPBA), a TRPV3 agonist, can prevent HFD-induced weight gain and adipogenesis [162]. Increased UCP-1 mRNA expression levels were determined in subcutaneous adipose tissue of TRPV4 knockout mice. Additionally, increased EE levels, related with thermogenic gene programming, were found in TRPV4 knockout mice. Wild type or null mutation characterized mice treated with TRPV4 antagonist showed enhanced thermogenesis in adipose tissue. Moreover, these animals are protected from inflammation of adipose tissue, diet-induced obesity and insulin resistance [163]. This important role of TRPV4 implicates as a target in metabolic diseases and obesity treatment. In some different studies obese mice were fed with capsaicin supplemented HFD. The animals showed decreased leptin, insulin and fasting glucose levels, while adinopectin levels were increased. Similarly, there was a decrease in inflammatory gene expression level [164]. Moreover, mice fed with HFD during three months with supplemented capsaicin result in decreased TNF-α and leptin levels and less weight gain, while thermogenesis of brown adipose tissue was increased [165]. HFD also decreased TRPV1 expression but capsaicin supplementation increased the expression of TRPV1 [165]. Moreover, brain-derived neurotrophic factor and peptide YY were enhanced in capsaicin-supplemented mice [165]. There are some opposite studies in literature in contrast to those findings above. Capsaicin desensitized rats were also protected from weight gain with relation to reduction in food intake and reduce total protein levels. Although many studies implicate the positive impact of activation of TRPV1 channels on diabetes and obesity, some studies present the importance of antagonizing of TRPV1 channels. Neonatally capsaicin administered male rats did not show any significant changes in weight gain, both basal plasma and fasting leptin levels, corticosterone, insulin and adinopectin levels [166]. In another study, neonatally capsaicin-treated adult rats showed decreased corticosterone levels [167]. In a different study, TRPV1 knockout mice became obese wild type (WT) mice kept under same conditions fed with HFD [168]. At the same time, physical activity at night was decreased in TPRV1 knockout mice in correlation with EE rate [168]. Another TRPV1 antagonist, *N*-(tertiarybutylphenyl)-4-(3-chloropyridin-2yl)tetrahydropyrazine-1(*2H*)carbox-amide (BCTC), inhibits TRPV1 currents and decreased neuropathic pain and inflammation in vivo [169]. Plasma glucose and triglyceride levels were significantly decreased after BCTC administration. Additionally, BCTC improved insulin resistance in adipocytes and muscle tissue. These data were achieved as a result of TRPV1 inhibition. BCTC might have two similar effects via increasing insulin secretion and improving insulin resistance in ob/ob mice. WT and knockout mice were kept under light–dark (LD) cycle or constant darkness (DD). Brown adipose tissue (BAT) of WT mice TRPV1 oscillations were highly recorded at scotophase. Period circadian protein homolog encoding gene (PER) 1/2 were observed as same as TRPV1 current. Moreover, BMAL1 (main driver of molecular clock) oscillated in DD. But, oscillations of the clock genes were absent in TRPV1 knockout mice. Additionally, high UCP1 expression of WT was determined in LD when comparing to DD and TRPV1 knockout mice showed high UCP1 oscillations in photophase. TRPV1 knockout mice showed decreased locomotor activity compared to WT when LD applied. TRPV1 knockout mice showed decreased total activity when exposed to only LD. Thus, TRPV1 is a crucial modulator of clock gene oscillations of BAT. This clearly identifies the importance of TRPV1 channels as a novel pharmacological target to treat metabolic disorders [170]. Recently, TRPV1 antagonist was discovered by a private drug company and named AZV1 to this molecule [171]. It was suggested to improve the insulin sensitivity in mice. Eight days of treatment of AZV1 significantly decreased glucose and fructosamine levels comparing control group mice. TRPV1 channels are among the members of thermoreceptors which responsible for entrainment of circadian clock.

TRPM2

TRPM subfamily members show variable levels of permeability to Ca^{2+} and these members are lacking from N-terminal ankyrin repetitions [156]. TRPM2, M3, M4, M5 are the members of this subfamily which plays role in metabolism regulation. TRPM2 can be activated by $[Ca^{2+}]_i$, oxidative stress and related products. The importance of β cells arise from their synthesis and releasing of insulin in response to increased blood glucose levels. Genetic deletion of TRPM2 channels cause protection against STZ-induced deterioration of islets which is a strong finding on the possible role of TRPM2 channels in pathogenesis of diabetes. It has been reported that TRPM2 knockout mice is more insulin-sensitive because of increased glucose metabolism in their heart. These mice show increased EE levels and metabolic genes expression levels. These genes in WAT result in resistance to HFD in mice. TRPM2 knockout mice have increased Akt phosphorylation in the heart and decreased inflammation levels in the liver [172]. TRPM2, TRPM4, and TRPM5 are expressed in human islets, suggesting that they have a modulator role in insulin secretion. In addition to reactive oxygen species, TRPM2 currents can be activated by incretins and glucose [173, 174]. Glucose clearance was enhanced and plasma insulin levels were decreased in TRPM2 knockout mice, which have higher glucose levels. This was related with decreased insulin secretion levels which was triggered by glucose and incretins. Moreover, elevations of $[Ca^{2+}]_i$ was impaired after either insulin or incretin stimulation in β cells of TRPM2 knockout mice [174]. TRPM2 can be activated by incretin-induced PKA phosphorylation. Likewise, $[Ca^{2+}]_i$ levels and adenine nucleotides have more complex activation pathways than others, TRPM2 currents are vital for insulin secretion [175]. TRPM2 was suggested to be a therapeutic target for regulation of inflammation triggered by hyperglycemia-induced oxidative stress in diabetes. GLP-1, at very low concentrations, can activate TRPM2 currents [176]. TRPM2 can be activated by glucose and GLP-1 triggered cADPR pathway in pancreatic islets of mouse. Glucose and GLP-1 induced secretion was inhibited by adrenaline administration, which activates α2A adrenoreceptor. This inhibits cAMP signaling cascade which also inactivates TRPM2 channels. Moreover, ghrelin at low concentrations, inhibits glucose-induced insulin secretion again inhibiting cAMP and reduced TRPM2 signals [177]. Cytoplasmic pH levels also act as a signal marker in TRPM2 activation. It has been reported that acidic pH inhibits TRPM2 channel activity. However, this inhibition disappeared when cytosolic pH was increased [178]. Cytosolic pH increase in glucose stimulated β cells, and this activates TRPM2 currents by ADPR and ADPR triggered elevated levels of $[Ca^{2+}]_i$. Masking and entrainment are two nocturnal or diurnal behaviors which are independent processes occurring in living organisms. Entrainment can be determined as the circadian clock's synchronization to environmental loop, and masking can be determined as direct responses to activity dependent changes. Recently, it was reported that TRPM2 is involved in temperature-dependent masking behavior regulation [179]. TPRM2 is warm-sensitive whereas TRMP8 is cold sensitive cation channels. These two TRP channels are closely related in thermoregulation, which can be changes during a day. It is well known that β cells have increased number of mitochondrion. Glucose stimulations of β cells generate more heat, which is responsible to increase temperature and stimulate TRPM2 channel [180]. This was approved by another study that increased temperature triggered $[Ca^{2+}]_i$ current in β cells were abolished in TRPM2 knockout mice [174]. Both of TRPM2 and TRPM8 channels are involved in temperature-dependent negatively masking behavior in rats [179]. Reduction of body temperature starts with sleep both in diurnal and octurnal period. It is necessary for energy homeostasis and sleep quality. In another study, TRPM2 was shown to limit the fever response and had a role in temperature detection, which protect hypothalamic neurons from high temperature [181].

TRPM3

TRPM3 subfamily has several isoforms. Among of these, TRPM3α2 is more selective to monovalent ions and permeable to Ca^{2+} compared to TRPM3α1, and can be regulated by PIP_2 [182]. Glucose induced elevation of PIP_2 cause activation in β cells. TRPM3 can be inhibited by extracellular monovalent cations. TRPM3 channels are expressed in β cells [183]. TRPM3 can be activated by 3,4-dihydro-*N*-(5-methyl-3-isoxazoyl)-a-phenyl-1(*2H*)-quinolineacetamide (CIM0216) and pregnolone sulphate [184], nifedipine and clotrimazole. Primidone is the antagonist of this channel [185]. Physical stimuli such as hypo-osmolality also activate TRPM3 channels. TRPM3 currents can be inactivated by flavanones [186], progesterone, diclofenac [187], and mefenamic acid [184]. Moreover, TRPM3 currents can be inhibited by G protein βγ [188]. TRPM3 potentiates glutamatergic transmission at cerebellar Purkinje neurons [189] and it is involved in insulin secretion [190]. TRPM3 activation by both of these activators cause elevation in $[Ca^{2+}]_i$ levels and trigger insulin secretion [184] as well as increasing transcription of insulin gene. Moreover, TRPM3 is heat sensitive channel and it can also be activated by glucose-induced heat elevations [180]. This can be an important activation mechanism which can change within a day and in turn affect rhythmiticity of diurnal activities. TRPM3 can contribute $[Ca^{2+}]_i$ signaling in retinal pigment epithelium and sub-retinal space, which contribute regulation of light/dark adaptation. However, TRPM3 is also zinc (Zn^{2+}) selective channel. Fasting glucose levels were determined as normal and subjects of TRPM3 knockout mice were also healthy. TRPM3 can be suddenly activated by heat and its agonist, pregnenolone triggers pain in WT but not in TRPM3 knockout mice. However TRPM3 knockout mice had some deficiencies in heat stimuli response dorsal root ganglions [191].

TRPM4

TRPM4 is impermeable TRP subfamily member to divalent cations and it can be expressed in human β cells. It can be activated by Ca^{2+}, PKC, PIP_2 [192] and inhibited by adenine nucleotides [193]. Some PKC isoforms can regulate β cell function. TRPM4 currents can be inhibited by glibenclamide, cytoplasmic ATP and other nucleotides in a number of cells. Inhibition of this channel results in reduced Ca^{2+} signals, which is related with decreased GSIS in INS-1 cells. It has been reported that TRPM4 plays an important role in the regulation of membrane currents during glucose stimulation [194]. ATP can also inhibit desensitization of Ca^{2+} while decreasing TRPM4 currents. Glucose stimulation increase PIP_2 levels in β cells. PIP_2 shifts voltage activation curve to negative values which protect TRPM4 desensitization by Ca^{2+}. TRPM4 channels are implicated in agonist-induced insulin secretion and glucagon secretion [195]. TRPM4 inhibition by 9-Phenanthrol results in GLP-1 and glucose induced insulin secretion in rats. GLP-1 (even in picomolar concentrations) induced PKC can activate TRPM4 currents and cause Na^+ influx and membrane depolarization. Moreover, low dose GLP-1 stimulation did not observe in islets of TRPM4 knockout mice.

TRPM5

TRPM5 also can be activated by $[Ca^{2+}]_i$ and this type of channels are strongly related with TRPM4 channels. Extracellular low pH can inhibit TRPM5 currents. However, TRPM5 cannot be inhibited by glibenclamide or adenine nucleotides. In addition, TRPM5 can be activated by temperature changes. TRPM5 is characterized with its critical role in the taste signaling. Low levels of mRNA were detected in human β cells. TRPM5 knockout mice do not have type II taste perception and they show reduced levels of GSIS. TRPM5 knockout mice had decreased levels of insulin secretion and reduced levels of glucose induced calcium activity. TRPM5 mutations are related with metabolic syndrome and T2DM. Moreover, TRPM5 expression levels were decreased in lepting signaling deficient and obese mice. TRPM5 might be useful in detecting temperature changes and $[Ca^{2+}]_i$ elevations. An increase in temperature, ranging from 15 to 30 °C was enhanced with sweet stimulus which is missing in TRPM5 knockout mice. Daily consumption of stevioside, which is widely used as a sweetener, increased GSIS and delayed HFD-induced diabetes in wild type mice. However this effect is completely abolished in TRPM5 knockout mice. According to this finding, TRPM5 activity is essential for regulating insulin secretion after food intake and preventing hyperglycemia and moreover, TRPM5 channels are potential target proteins for treating T2D [196]. Quinine, a TRPM5 inhibitor of bitter tasting, supplementation to WT mice diet reduced weight gain compared to control diet. TRPM5 knockout mice have reduced levels of food intake. This might be due to the deficiency of sweet tasting. In a different study, TRPM5 knockout mice gained less weight and they had enhanced glucose tolerance.

TRPM8

TRPM subtype TRPM8 channel is cold-sensing, non-selective cation channel. TRPM8 has the role in temperature and pain sending neurons. It can also be activated by cooling agents, such as icilin and menthol [197, 198]. Studies reveal the importance of TRPM8 on metabolic diseases in last years. TRPM8 knockout mice adipocytes had greater fat droplets density which indicates larger multiocularity comparing with WT mice [199]. These results reveal important an role of TRPM8 in the regulation of molecular clock machinery in BAT [199]. Dieatry menthol supplementation remarkably increased the locomotor activity of WT mice [115]. Hence, the authors assessed the locomotor activity related rhythmic aspects in mice. TRPM8 lacking mice shows delayed onset of locomotor activity in response to light pulses. This could be due to reduction and/or loss variations of PER1 expressions in the eye [199]. This study also highlights the importance of TRPM8 channel activity in integration of environmental light of suprachiasmatic nucleus (SCN) via ocular physiology, which is responsible for keeping the body in the same time region. Findings mentioned in the above study reveal the role of TRPM8 channel contribution with enhanced adaptative thermogenesis in BAT, and associated weight gain [200]. Recently, it was reported that TRPM8 is involved in temperature-dependent masking behavior regulation [179]. Any disruptions in clock machinery can be related with metabolic disorders. In this perspective, pharmacological modulation TRPM8 channels will be one of the main therapeutic targets in regulation of thermogenesis, weight gain and clock gene activation. BCTC is one of the antagonists of TRPM8 channels. BCTC may contribute for targeting the molecular role of inhibiting TRPM8 currents in diabetes and obesity for future studies. Menthol, a TRPM8 agonist, upregulates UCP1 expression and need activation of protein kinase A signaling cascade in adipocytes. TRPM8 plays a vital role of thermogenesis in BAT. Menthol application significantly increased core temperature and locomotor activity in WT mice. However, this effect of menthol was not observed in TRPM8 and UCP1 knockout mice [115].

Conclusion

The schematic diagram in Fig. 6.1 summarises specifically the part of the review relating to the different ion channels activation and inhibition pathways which play major roles in insulin secretion and general view of calcium entry through some TRP channels by various agonists and antagonists. Regarding this review in general, it is now possible to conclude that most of the daily nutrient intake should be ingested first in the morning as breakfast. By having good habits, both in terms of eating times and sleeping times then the body can maintain a healthy and normal circadian rhythms. In turn, this can prevent or delay the development of such non-communicable diseases (NCDs) as obesity, diabetes, heart failure and others. In this way, the disruption of circadian rhythms and imbalances in meal times are a risk factors for certain types of NCDS, despite the existence of innate risk factors such as the genetic load. Additionally, recent advances in ion channel regulation mechanisms and different animal models contribute to discover signaling cascades and in the control of most of them by applying different agonists and antagonists in development of T2DM and obesity. Regulation of molecular clock genes highlights the importance of ion channel activity during physiological processes. One of the main theme of this review is to emphasize the importance of the $[Ca^{2+}]_i$ homeostasis. Hyperexcitability of the cell and alternatively, activated intracellular organelles triggered Ca^{2+} handling varies according to cell type. Different types of cells, expression levels of ion channels and differences in their activation and inhibition pathways are essential in cellular regulation. However, a dysfunction in any one of these can cause different diseases and as such, it is important to undertake therapeutic approach in targeting them. In the view of that information in this review including the cellular cascades, there are still so many scientific studies which are waiting to be explored in the pathophysiology of NCDs including obesity, T2DM, heart diseases, sleep disorders and several others.

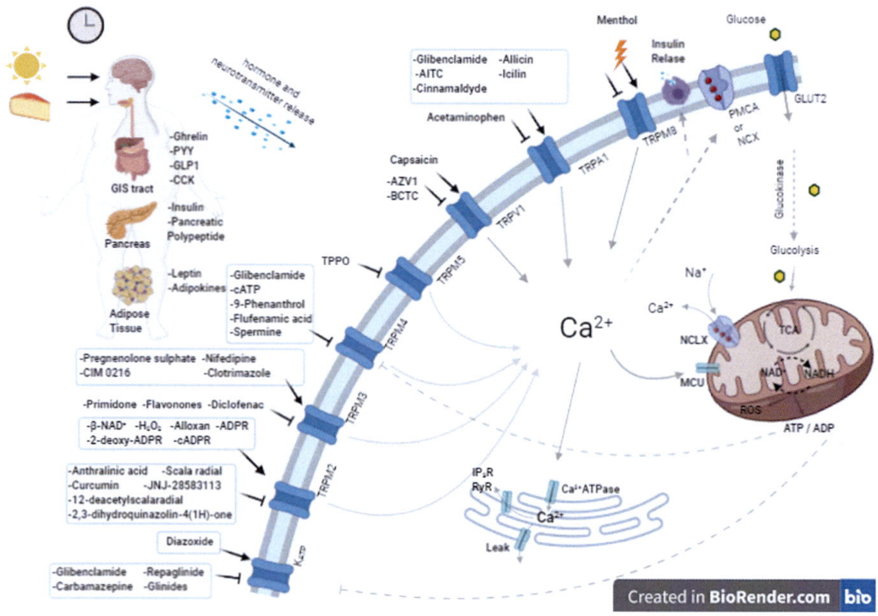

Fig. 6.1 Schematic diagram of the ion channels activation and inhibition pathways outlined above in this review and which play major roles in insulin secretion, and general view of calcium entry through some TRP channels. Their agonists and antagonists are generally listed. Some of them may vary depending on cell type. There are some other molecules which are mentioned in this legend. Some of the agonists and antagonists drawn in figure are also released from intracellular organelles. In addition, K_{ATP} channel is mentioned in the regulation of cellular signaling cascade. Food intake or light exposure triggers hormone release. Membrane voltage can be changed during this process. Glucose entry through GLUT2 induce intracellular signaling cascade. ATP/ADP ratio determines the closure of K_{ATP} channels. Membrane voltage changes or some chemicals result in Ca^{2+} influx through VDCC and TRP channels. Increased $[Ca^{2+}]_i$ levels triggers insulin release. $[Ca^{2+}]_i$ buffered by mitochondria or ER. PMCA and NCX/NCLX are generally used to buffer elevated $[Ca^{2+}]_i$. ATP—adenosine triphosphate, ADP—adenosine diphosphate, Ach—acetylcholine, ACICR—atypical Ca^{2+}-induced Ca^{2+} release, AITC—allyl Isothiocyanate, BCTC—(4-(3-Chloro-2-pyridinyl)-N-[4-(1,1-dimethylethyl)phenyl]-1-piperazinecarboxamide), cADPR—cyclic ADP ribose, $[Ca^{2+}]_i$—free calcium concentration in the cytosol, CCK—cholecystokinin, CIM 0216—3,4-Dihydro-*N*-(5-methyl-3-isoxazolyl)-α-phenyl-1(2*H*)-quinolineacetamide, ER—endoplasmic reticulum, GLP-1—glucagon like peptide-1, GLUT2—glucose transporter 2, K_{ATP} channels—ATP-sensitive K^+ channels, IP_3R inositol—1,4,5 triphosphate receptor, MCU—mitochondrial calcium uniporter, NAD—nicotinamide adenine dinucleotide, NCLX—Na^+–Ca^{2+} exchanger, PMCA—plasma membrane calcium ATP-ase, PYY—peptide YY, RTX—resiniferatoxin, RyR—ryanodine receptor, TCA—tricarboxylic acid cycle, TRP—transient receptor potential, TPPO—TriPhenylPhosphineOxide, VDCC—voltage dependent calcium channels

Acknowledgements This work has supported by Junta de Extremadura grant (GR18040). Authors declare no conflict of interests.

References

1. Garaulet M, Corbalán MD, Madrid JA et al (2010) CLOCK gene is implicated in weight reduction in obese patients participating in a dietary programme based on the Mediterranean diet. Int J Obes (Lond) 34:516–523
2. Flandrin JL, Montanari M (2003) Storia dell'alimentazione, 6th edn. Laterza, Italy
3. Carroll A (2013) Three squares: the invention of the American meal, 1st edn. Basic Books (AZ), New York, USA
4. Affinita A, Catalani L, Cecchetto G et al (2013) Breakfast: a multidisciplinary approach. Ital J Pediatr 39:44
5. Albala K (2003) Food in early modern Europe. Greenwood Publishing Group, Santa Barbara, USA
6. Gwinup G, Byron RC, Roush WH et al (1963) Effect of nibbling versus gorging on serum lipids in man. Am J Clin Nutr 13:209–213
7. Paoli A, Tinsley G, Bianco A et al (2019) The influence of meal frequency and timing on health in humans: the role of fasting. Nutrients 11:719
8. Fabry P, Fodor J, Hejl Z et al (1968) Meal frequency and ischaemic heart-disease. Lancet 2:190–191
9. Edelstein SL, Barrett-Connor EL, Wingard DL, Cohn BA (1992) Increased meal frequency associated with decreased cholesterol concentrations; Rancho Bernardo, CA, 1984–1987. Am J Clin Nutr 55:664–669
10. Holmback I, Ericson U, Gullberg B, Wirfalt E (2010) A high eating frequency is associated with an overall healthy lifestyle in middle-aged men and women and reduced likelihood of general and central obesity in men. Br J Nutr 104:1065–1073
11. Afshin A, Forouzanfar MH, Reitsma MB et al (2017) Health effects of overweight and obesity in 195 countries over 25 years. N Engl J Med 377:13–27
12. Obesity and Overweight, Fact Sheet N 311, Updated Mar 2013. https://www.who.int/newsroom/fact-sheets/detail/obesity-and-overweight. Accessed 15 Jan 2021
13. Koh-Banerjee P, Wang Y, Hu FB et al (2004) Changes in body weight and body fat distribution as risk factors for clinical diabetes in us men. Am J Epidemiol 159:1150–1159
14. Thompson WG, Cook DA, Clark MM et al (2007) Treatment of obesity. Mayo Clin Proc 82:93–101
15. Paoli A, Moro T, Marcolin G et al (2012) High-intensity interval resistance training (HIRT) influences resting energy expenditure and respiratory ratio in non-dieting individuals. J Transl Med 10:237
16. Garaulet M, Gómez-Abellán P (2014) Timing of food intake and obesity: a novel association. Physiol Behav 134:44–50
17. Kahleova H, Lloren JI, Mashchak A et al (2017) Meal frequency and timing are associated with changes in body mass index in adventist health study 2. J Nutr 147:1722–1728
18. Fabry P, Hejl Z, Fodor J et al (1964) The frequency of meals. Its relation to overweight, hypercholesterolaemia, and decreased glucose-tolerance. Lancet 2:614–615
19. Keast DR, Nicklas TA, O'Neil CE (2010) Snacking is associated with reduced risk of overweight and reduced abdominal obesity in adolescents: National Health and Nutrition Examination Survey (NHANES) 1999–2004. Am J Clin Nutr 92:428–435
20. Van der Heijden AA, Hu FB, Rimm EB et al (2007) A prospective study of breakfast consumption and weight gain among U.S. men. Obesity 15:2463–2469

21. Mekary RA, Giovannucci E, Willett WC et al (2012) Eating patterns and type 2 diabetes risk in men: breakfast omission, eating frequency, and snacking. Am J Clin Nutr 95:1182–1189
22. Taylor MA, Garrow JS (2001) Compared with nibbling, neither gorging nor a morning fast affect short-term energy balance in obese patients in a chamber calorimeter. Int J Obes Relat Metab Disord 25:519–528
23. Romon M, Edme JL, Boulenguez C et al (1993) Circadian variation of diet-induced thermogenesis. Am J Clin Nutr 57:476–480
24. Weststrate JA, Weys PJ, Poortvliet EJ et al (1989) Diurnal variation in postabsorptive resting metabolic rate and diet-induced thermogenesis. Am J Clin Nutr 50:908–914
25. Kanaley JA, Heden TD, Liu Y et al (2014) Alteration of postprandial glucose and insulin concentrations with meal frequency and composition. Br J Nutr 112:1484–1493
26. Ohkawara K, Cornier MA, Kohrt WM et al (2013) Effects of increased meal frequency on fat oxidation and perceived hunger. Obesity 21:336–343
27. Alhussain MH, Macdonald IA, Taylor MA (2016) Irregular meal-pattern effects on energy expenditure, metabolism, and appetite regulation: a randomized controlled trial in healthy normal-weight women. Am J Clin Nutr 104:21–32
28. Kul S, Savas E, Ozturk ZA, Karadag G (2014) Does Ramadan fasting alter body weight and blood lipids and fasting blood glucose in a healthy population? A meta-analysis. J Relig Health 53:929–942
29. Mammucari C, Schiaffino S, Sandri M (2008) Downstream of Akt: FoxO3 and mtTOR in the regulation of autophagy in skeletal muscle. Autophagy 4:524–526
30. Cahill LE, Chiuve SE, Mekary RA et al (2013) Prospective study of breakfast eating and incident coronary heart disease in a cohort of male us health professionals. Circulation 128:337–343
31. Nas A, Mirza N, Hagele F et al (2017) Impact of breakfast skipping compared with dinner skipping on regulation of energy balance and metabolic risk. Am J Clin Nutr 105:1351–1361
32. Almoosawi S, Vingeliene S, Karagounis LG, Pot GK (2016) Chrono-nutrition: a review of current evidence from observational studies on global trends in time-of-day of energy intake and its association with obesity. Proc Nutr Soc 75:487–500
33. Leech RM, Timperio A, Livingstone KM et al (2017) Temporal eating patterns: associations with nutrient intakes, diet quality, and measures of adiposity. Am J Clin Nutr 106:1121–1130
34. Titan SM, Bingham S, Welch A et al (2001) Frequency of eating and concentrations of serum cholesterol in the Norfolk population of the European prospective investigation into cancer (epic-Norfolk): cross sectional study. BMJ 323:1286–1288
35. Ma Y, Bertone ER, Stanek EJ et al (2003) Association between eating patterns and obesity in a free-living us adult population. Am J Epidemiol 158:85–92
36. St-Onge MP, Ard J, Baskin ML et al (2017) Meal timing and frequency: Implications for cardiovascular disease prevention: a scientific statement from the American heart association. Circulation 135:e96–e121
37. Lecerf JM, de Lorgeril M (2011) Dietary cholesterol: from physiology to cardiovascular risk. Br J Nutr 106:6–14
38. Nayor M, Vasan RS (2016) Recent update to the US cholesterol treatment guidelines: a comparison with international guidelines. Circulation 133:1795–1806
39. Dietschy JM, Brown MS (1974) Effect of alterations of the specific activity of the intracellular acetyl CoA pool on apparent rates of hepatic cholesterogenesis. J Lipid Res 15:508–516
40. Ness GC, Zhao Z, Wiggins L (1994) Insulin and glucagon modulate hepatic 3-hydroxy-3-methylglutarylcoenzyme a reductase activity by affecting immunoreactive protein levels. J Biol Chem 269:29168–29172
41. Paoli A (2014) Ketogenic diet for obesity: friend or foe? Int J Environ Res Public Health 11:2092–2107
42. Sutherland WH, de Jong SA, Walker RJ (2007) Effect of dietary cholesterol and fat on cell cholesterol transfer to postprandial plasma in hyperlipidemic men. Lipids 42:901–911
43. Paoli A, Mor T, Bosco G, Bianco A et al (2015) Effects of n-3 polyunsaturated fatty acids (omega-3) supplementation on some cardiovascular risk factors with a ketogenic Mediterranean diet. Mar Drugs 13:996–1009

44. Chapelot D (2011) The role of snacking in energy balance: a biobehavioral approach. J Nutr 141:158–162
45. McCrory MA, Campbell WW (2011) Effects of eating frequency, snacking, and breakfast skipping on energy regulation: symposium overview. J Nutr 141:144–147
46. Garaulet M, Madrid JA (2010) Chronobiological aspects of nutrition, metabolic syndrome and obesity. Adv Drug Deliv Rev 62(9–10):967–978
47. Garaulet M, Gómez-Abellán P, Madrid JA (2010) Chronobiology and obesity: the orchestra out of tune. Clin Lipidol 5:181–188
48. Pavlovski I, Evans JA, Mistlberger RE (2018) Feeding time entrains the olfactory bulb circadian clock in anosmic PER2::LUC mice. Neuroscience 393:175–184
49. Hall J (2016) Guyton y Hall. Tratado de fisiología médica, 13th edn. Elsevier, España
50. Albrecht U (2017) The circadian clock, metabolism and obesity. Obes Rev 18:25–33
51. Scheer FA, Kalsbeek A, Buijs RM (2003) Cardiovascular control by the suprachiasmatic nucleus: neural and neuroendocrine mechanisms in human and rat. Biol Chem 384:697–709
52. Kim P, Oster H, Lehnert H et al (2019) Coupling the circadian clock to homeostasis: the role of period in timing physiology. Endocr Rev 40:66–95
53. McHill AW, Phillips AJ, Czeisler CA et al (2017) Later circadian timing of food intake is associated with increased body fat. Am J Clin Nutr 106:1213–1219
54. Poggiogalle E, Jamshed H, Peterson CM (2018) Circadian regulation of glucose, lipid, and energy metabolism in humans. Metabolism 84:11–27
55. Jakubowicz D, Barnea M, Wainstein J et al (2013) High caloric intake at breakfast vs. dinner differentially influences weight loss of overweight and obese women. Obesity 21:2504–2512
56. Dhurandhar EJ, Dawson J, Alcorn A et al (2014) The effectiveness of breakfast recommendations on weight loss: a randomized controlled trial. Am J Clin Nutr 100:507–513
57. Uzhova I, Fuster V, Fernández-Ortiz A et al (2017) The importance of breakfast in atherosclerosis disease: insights from the PESA study. J Am Coll Cardiol 70:1833–1842
58. Betts JA, Chowdhury EA, Gonzalez JT et al (2016) Is breakfast the most important meal of the day? Proc Nutr Soc 75:464–474
59. Fong M, Caterson ID, Madigan CD (2017) Are large dinners associated with excess weight, and does eating a smaller dinner achieve greater weight loss? A systematic review and meta-analysis. Br J Nutr 118:616–628
60. Franco L, Bravo R, Galán C et al (2012) Análisis de la mejora de la calidad del sueño y la ansiedad en estudiantes universitarios, bajo estrés, mediante el consumo de cerveza sin alcohol. Rev Esp Nutr Comunitaria 18(4):218–223
61. Karlsson B, Knutsson A, Lindahl B (2001) Is there an association between shift work and having a metabolic syndrome? Results from a population based study of 27,485 people. Occup Environ Med 58:747–752
62. Von Schantz M (2008) Phenotypic effects of genetic variability in human clock genes on circadian and sleep parameters. J Genet 87:513–519
63. Lamont EW, Legault-Coutu D, Cermakian N, Boivin DB (2009) The role of circadian clock genes in mental disorders. Dialogues Clin Neurosci 9(3):333–342
64. Kotronoulas G, Stamatakis A, Stylianopoulou F (2009) Hormones, hormonal agents, and neuropeptides involved in the neuroendocrine regulation of sleep in humans. Hormones (Athens) 8:232–248
65. Vanitallie TB (2006) Sleep and energy balance, interactive homeostatic systems. Metabolism 55:S30-35
66. Garaulet M, Sánchez-Moreno C, Smith CE et al (2011) Ghrelin, sleep reduction and evening preference: relationships to CLOCK 3111 T/C SNP and wight loss. Plos One 6(2):e17435
67. Garaulet M, Pérez de Heredia F (2010) Behavioural therapy in the treatment of obesity (II): role of the Mediterranean diet. Nutr Hosp 25:9–17
68. Garaulet M, Pérez de Heredia F (2009) Behavioural therapy in the treatment of obesity (I): new directions for clinical practice. Nutr Hosp 24:629–639
69. Monteleone P, Tortorella A, Docimo L et al (2008) Investigation of 3111T/C polymorphism of the CLOCK gene in obese individuals with or without binge eating disorder: association with higher body mass index. Neurosci Lett 435:30–33

70. Benedetti F, Radaelli D, Bernasconi A et al (2008) Clock genes beyond the clock: CLOCK genotype biases neural correlates of moral valence decision in depressed patients. Genes Brain Behav 7:20–25
71. Adamantidis A, de Lecea L (2008) Sleep and metabolism: shared circuits, new connections. Trends Endocrinol Metab 19:362–370
72. Buckland G, Bach A, Serra-Majem L (2008) Obesity and the Mediterranean diet: a systematic review of observational and intervention studies. Obes Rev 9:582–593
73. Van Cauter E, Knutson KL (2008) Sleep and the epidemic of obesity in children and adults. Eur J Endocrinol 159:S59-66
74. Knutson KL (2005) The association between pubertal status and sleep duration and quality among a nationally representative sample of U. S. adolescents. Am J Hum Biol 17:418–424. https://doi.org/10.1002/ajhb.20405
75. Gómez-Abellán P, Gómez-Santos C, Madrid JA et al (2010) Circadian expression of adiponectin and its receptors in human adipose tissue. Endocrinology 151(1):115–122
76. Gómez-Santos C, Gómez-Abellán P, Madrid JA et al (2009) Circadian rhythm of clock genes in human adipose explants. Obesity (Silver Spring) 17:1481–1485
77. Hernandez-Morante JJ, Gomez-Santos C, Milagro F et al (2009) Expression of cortisol metabolism-related genes shows circadian rhythmic patterns in human adipose tissue. Int J Obes (Lond) 33:473–480
78. Garaulet M, Madrid JA (2009) Chronobiology, genetics and metabolic syndrome. Curr Opin Lipidol 20:127–134
79. Rasmussen MS, Lihn AS, Pedersen SB et al (2006) Adiponectin receptors in human adipose tissue: effects of obesity, weight loss, and fat depots. Obesity (Silver Spring) 14:28–35
80. Oliver P, Ribot J, Rodríguez AM et al (2006) Resistin as a putative modulator of insulin action in the daily feeding/fasting rhythm. Pflugers Arch 452:260–267
81. Gavrila A, Peng CK, Chan JL et al (2003) Diurnal and ultradian dynamics of serum adiponectin in healthy men: comparison with leptin, circulating soluble leptin receptor, and cortisol patterns. J Clin Endocrinol Metab 88:2838–2843
82. Seaman GV, Engel R, Swank RL, Hissen W (1965) Circadian periodicity in some physicochemical parameters of circulating blood. Nature 207:833–835
83. Garaulet M, Hernández-Morante JJ, de Heredia FP, Tébar FJ (2007) Adiponectin, the controversial hormone. Public Health Nutr 10:1145–1150
84. Erren TC, Reiter RJ (2009) Defining chronodisruption. J Pineal Res 46:245–247
85. Froy O (2007) The relationship between nutrition and circadian rhythms in mammals. Front Neuroendocrinol 28:61–71
86. Yildiz BO, Suchard MA, Wong ML et al (2004) Alterations in the dynamics of circulating ghrelin, adiponectin, and leptin in human obesity. Proc Natl Acad Sci USA 101:10434–10439
87. Kennaway DJ, Owens JA, Voultsios A et al (2007) Metabolic homeostasis in mice with disrupted Clock gene expression in peripheral tissues. Am J Physiol Regul Integr Comp Physiol 293:R1528–R1537
88. Srinivasan V, Ohta Y, Espino J et al (2012) Metabolic syndrome, its pathophysiology and the role of melatonin. Recent Pat Endocr Metab Immune Drug Discov 7:11–25
89. Kohsaka A, Bass J (2007) A sense of time: how molecular clocks organize metabolism. Trends Endocrinol Metab 18:4–11
90. Cnop M, Havel PJ, Utzschneider KM et al (2003) Relationship of adiponectin to body fat distribution, insulin sensitivity and plasma lipoproteins: evidence for independent roles of age and sex. Diabetología 46:459–469
91. Kobashi C, Urakaze M, Kishida M et al (2005) Adiponectin inhibits endothelial synthesis of interleukin-8. Circ Res 97:1245–1252
92. Yamauchi T, Kamon J, Ito Y et al (2003) Cloning of adiponectin receptors that mediate antidiabetic metabolic effects. Nature 423:762–769
93. Ando H, Yanagihara H, Hayashi Y et al (2005) Rhythmic messenger ribonucleic acid expression of clock genes and adipocytokines in mouse visceral adipose tissue. Endocrinology 146:5631–5636

94. Gil-Campos M, Cañete RR, Gil A (2004) Adiponectin, the missing link in insulin resistance and obesity. Clin Nutr 23:963–974
95. Hernandez-Morante JJ, Milagro FI, Larque E et al (2007) Relationship among adiponectin, adiponectin gene expression and fatty acids composition in morbidly obese patients. Obes Surg 17:516–524
96. Després JP, Lemieux I (2006) Abdominal obesity and metabolic syndrome. Nature 444:881–887
97. Blüher M, Williams CJ, Klöting N et al (2007) Gene expression of adiponectin receptors in human visceral and subcutaneous adipose tissue is related to insulin resistance and metabolic parameters and is altered in response to physical training. Diabetes Care 30:3110–3115
98. Cappuccio FP, D'Elia L, Strazzullo P, Miller MA (2010) Quantity and quality of sleep and incidence of type 2 diabetes: a systematic review and meta-analysis. Diabetes Care 33:414–420
99. McMullan CJ, Schernhammer ES, Rimm EB et al (2013) Melatonin secretion and the incidence of type 2 diabetes. JAMA 309:1388–1396
100. Bass J, Takahashi JS (2010) Circadian integration of metabolism and energetics. Science 330:1349–1354
101. Bouatia-Naji N, Bonnefond A, Cavalcanti-Proença C et al (2009) A variant near MTNR1B is associated with increased fasting plasma glucose levels and type 2 diabetes risk. Nat Genet 41:89–94
102. Karamitri A, Renault N, Clement N et al (2013) Minireview: toward the establishment of a link between melatonin and glucose homeostasis: association of melatonin MT2 receptor variants with type 2 diabetes. Mol Endocrinol 27:1217–1233
103. Cagnacci A, Arangino S, Renzi A et al (2001) Influence of melatonin administration on glucose tolerance and insulin sensitivity of postmenopausal women. Clin Endocrinol (Oxf) 54:339–346
104. Lyssenko V, Nagorny CL, Erdos MR et al (2009) Common variant in MTNR1B associated with increased risk of type 2 diabetes and impaired early insulin secretion. Nat Genet 41:82–88
105. Gaulton KJ, Ferreira T, Lee Y et al (2015) DIAbetes Genetics Replication And Meta-analysis (DIAGRAM) consortium. Genetic fine mapping and genomic annotation defines causal mechanisms at type 2 diabetes susceptibility loci. Nat Genet 47:1415–1425
106. Depner CM, Stothard ER, Wright KP Jr (2014) Metabolic consequences of sleep and circadian disorders. Curr Diabetes Rep 14:507
107. Lane JN, Chang AM, Bjonnes AC et al (2016) Impact of common diabetes risk variant in MTNR1B on sleep, circadian, and melatonin physiology. Diabetes 65:1741–1751
108. St. Hilaire MA, Gronfier C, Zeitzer JM, Klerman EB (2007) A physiologically based mathematical model of melatonin including ocular light suppression and interactions with the circadian pacemaker. J Pineal Res 43:294–304
109. Brown EN, Czeisler CA (1992) The statistical analysis of circadian phase and amplitude in constant-routine core-temperature data. J Biol Rhythms 7:177–202
110. Espino J, Pariente JA, Rodríguez AB (2011) Role of melatonin on diabetes-related metabolic disorders. World J Diabetes 2(6):82–91
111. Srinivasan V, Ohta Y, Espino J, Pariente JA, Rodriguez AB, Mohamed M, Zakaria R (2013) Metabolic syndrome, its pathophysiology and the role of melatonin. Recent Pat Endocr Metab Immune Drug Discov 7(1):11–25
112. Espino J, Rodriguez AB, Pariente JA (2019) Melatonin and oxidative stress in the diabetic state: clinical implications and potential therapeutic applications. Curr Med Chem 26(22):4178–4190
113. Espino J, Pariente JA, Rodríguez AB (2012) Oxidative stress and immunosenescence: therapeutic effects of melatonin. Oxid Med Cell Longev 2012:article 670294
114. Liao J, Li X, Wong TY et al (2014) Impact of measurement error on testing genetic association with quantitative traits. PLoS One 9:article e87044
115. Ma S, Yu H, Zhao Z et al (2012) Activation of the cold-sensing TRPM8 channel triggers UCP1-dependent thermogenesis and prevents obesity. J Mol Cell Biol 4(2):88–96

116. Blodgett DM, Nowosielska A, Afik S et al (2015) Novel observations from next-generation RNA sequencing of highly purified human adult and fetal islet cell subsets. Diabetes 64(9):3172–3181
117. Braun M, Ramracheya R, Bengtsson M et al (2008) Voltage-gated ion channels in human pancreatic beta-cells: electrophysiological characterization and role in insulin secretion. Diabetes 57(6):1618–1628
118. Li XN, Herrington J, Petrov A et al (2013) The role of voltage-gated potassium channels Kv2.1 and Kv2.2 in the regulation of insulin and somatostatin release from pancreatic islets. J Pharmacol Exp Ther 344(2):407–416
119. Fu J, Dai X, Plummer G et al (2017) Kv2.1 clustering contributes to ınsulin exocytosis and rescues human β-cell dysfunction. Diabetes 66(7):1890–1900
120. Kalman K, Nguyen A, Tseng-Crank J et al (1998) Genomic organization, chromosomal localization, tissue distribution, and biophysical characterization of a novel mammalian Shaker-related voltage-gated potassium channel, Kv1.7. J Biol Chem 273(10):5851–5857
121. Finol-Urdaneta RK, Remedi MS, Raasch W et al (2012) Block of Kv1.7 potassium currents increases glucose-stimulated insulin secretion. EMBO Mol Med 4(5):424–434
122. Hyltén-Cavallius L, Iepsen EW, Wewer Albrechtsen NJ et al (2017) Patients with long-QT syndrome caused by ımpaired hERG-encoded Kv11.1 potassium channel have exaggerated endocrine pancreatic and ıncretin function associated with reactive hypoglycemia. Circulation 135(18):1705–1719
123. Yang JK, Lu J, Yuan SS et al (2018) From hyper- to hypoinsulinemia and diabetes: effect of KCNH6 on insulin secretion. Cell Rep 25(13):3800-3810.e6
124. Zhang J, Juhl CR, Hylten-Cavallius L et al (2020) Gain-of-function mutation in the voltage-gated potassium channel gene KCNQ1 and glucose-stimulated hypoinsulinemia—case report. BMC Endocr Disord 20(1):38–43
125. Nitabach MN, Blau J, Holmes TC (2002) Electrical silencing of Drosophila pacemaker neurons stops the free-running circadian clock. Cell 109(4):485–495
126. Ashcroft FM (2005) ATP-sensitive potassium channelopathies: focus on insulin secretion. J Clin Invest 115(8):2047–2058
127. Koster JC, Marshall BA, Ensor N et al (2000) Targeted overactivity of beta cell K(ATP) channels induces profound neonatal diabetes. Cell 100(6):645–654
128. Allebrandt KV, Amin N, Müller-Myhsok B et al (2013) A K(ATP) channel gene effect on sleep duration: from genome-wide association studies to function in Drosophila. Mol Psychiatry 18(1):122–132
129. Yang JJ, Cheng RC, Cheng PC et al (2017) KATP channels mediate differential metabolic responses to glucose shortage of the dorsomedial and ventrolateral oscillators in the central Clock. Sci Rep 7(1):Article 640
130. Raphemot R, Swale DR, Dadi PK et al (2014) Direct activation of β-cell KATP channels with a novel xanthine derivative. Mol Pharmacol 85(6):858–865
131. Henquin JC, Dufrane D, Gmyr V et al (2017) Pharmacological approach to understanding the control of insulin secretion in human islets. Diabetes Obes Metab 19(8):1061–1070
132. Amisten S, Salehi A, Rorsman P et al (2013) An atlas and functional analysis of G-protein coupled receptors in human islets of Langerhans. Pharmacol Ther 139(3):359–391
133. Kailey B, van de Bunt M, Cheley S et al (2012) SSTR2 is the functionally dominant somatostatin receptor in human pancreatic β- and α-cells. Am J Physiol Endocrinol Metab 303(9):E1107-1116
134. Wolford JK, Hanson RL, Kobes S et al (2001) Analysis of linkage disequilibrium between polymorphisms in the KCNJ9 gene with type 2 diabetes mellitus in Pima Indians. Mol Genet Metab 73(1):97–103
135. Kuß J, Stallmeyer B, Goldstein M et al (2019) Familial sinus node disease caused by a gain of GIRK (G-Protein Activated Inwardly Rectifying K+ channel) channel function. Circ Genom Precis Med 12(1):article e002238
136. Bukiya AN, Durdagi S, Noskov S, Rosenhouse-Dantsker A (2017) Cholesterol up-regulates neuronal G protein-gated inwardly rectifying potassium (GIRK) channel activity in the hippocampus. J Biol Chem 292(15):6135–6147

137. Kolic J, Manning Fox JE, Chepurny OG et al (2016) PI3 kinases p110α and PI3K-C2β negatively regulate cAMP via PDE3/8 to control insulin secretion in mouse and human islets. Mol Metab 5(7):459–471
138. Ho IH, Murrell-Lagnado RD (1999) Molecular mechanism for sodium-dependent activation of G protein-gated K+ channels. J Physiol 520 Pt 3(Pt 3):645–651
139. Xu G, Chen J, Jing G, Shalev A (2012) Preventing β-cell loss and diabetes with calcium channel blockers. Diabetes 61(4):848–856
140. Rorsman P, Braun M, Zhang Q (2012) Regulation of calcium in pancreatic α- and β-cells in health and disease. Cell Calcium 51(3–4):300–308
141. Nessa A, Rahman SA, Hussain K (2016) Hyperinsulinemic hypoglycemia—the molecular mechanisms. Front Endocrinol (Lausanne) 7:Article 29. https://doi.org/10.3389/fendo.2016.00029
142. Reinbothe TM, Alkayyali S, Ahlqvist E et al (2013) The human L-type calcium channel Cav1.3 regulates insulin release and polymorphisms in CACNA1D associate with type 2 diabetes. Diabetologia 56(2):340–309
143. Baig SM, Koschak A, Lieb A et al (2011) Loss of Ca(v)1.3 (CACNA1D) function in a human channelopathy with bradycardia and congenital deafness. Nat Neurosci 14(1):77–84
144. de la Cruz L, Puente EI, Reyes-Vaca A et al (2016) PIP2 in pancreatic β-cells regulates voltage-gated calcium channels by a voltage-independent pathway. Am J Physiol Cell Physiol 311(4):C630–C640
145. Wang L, Bhattacharjee A, Zuo Z et al (1999) A low voltage-activated Ca^{2+} current mediates cytokine-induced pancreatic beta-cell death. Endocrinology 140(3):1200–1204
146. Cosens DJ, Manning A (1969) Abnormal electroretinogram from a Drosophila mutant. Nature 224(5216):285–287
147. Lee Y, Montell C (2013) Drosophila TRPA1 functions in temperature control of circadian rhythm in pacemaker neurons. J Neurosci 33(16):6716–6725
148. Sivachenko A, Li Y, Abruzzi KC, Rosbash M (2013) The transcription factor Mef2 links the Drosophila core clock to Fas2, neuronal morphology, and circadian behavior. Neuron 79(2):281–292
149. Camacho S, Michlig S, de Senarclens-Bezençon C et al (2015) Anti-obesity and anti-hyperglycemic effects of cinnamaldehyde via altered ghrelin secretion and functional impact on food intake and gastric emptying. Sci Rep 5:Article 7919
150. Khare P, Jagtap S, Jain Y et al (2016) Cinnamaldehyde supplementation prevents fasting-induced hyperphagia, lipid accumulation, and inflammation in high-fat diet-fed mice. BioFactors 42(2):201–211
151. Iwasaki Y, Tanabe M, Kobata K, Watanabe T (2008) TRPA1 agonists–allyl isothiocyanate and cinnamaldehyde–induce adrenaline secretion. Biosci Biotechnol Biochem 72(10):2608–2614
152. Ahn J, Lee H, Im SW et al (2014) Allyl isothiocyanate ameliorates insulin resistance through the regulation of mitochondrial function. J Nutr Biochem 25(10):1026–1034
153. Cao DS, Zhong L, Hsieh TH et al (2012) Expression of transient receptor potential ankyrin 1 (TRPA1) and its role in insulin release from rat pancreatic beta cells. PLoS One. 7(5):article e38005
154. Kim MJ, Son HJ, Song SH et al (2013) The TRPA1 agonist, methyl syringate suppresses food intake and gastric emptying. PLoS One 8(8):article e71603
155. Caterina MJ, Schumacher MA, Tominaga M et al (1997) The capsaicin receptor: a heat-activated ion channel in the pain pathway. Nature 389(6653):816–824
156. Nilius B, Szallasi A (2014) Transient receptor potential channels as drug targets: from the science of basic research to the art of medicine. Pharmacol Rev 66(3):676–814
157. Smeets AJ, Westerterp-Plantenga MS (2009) The acute effects of a lunch containing capsaicin on energy and substrate utilisation, hormones, and satiety. Eur J Nutr 48(4):229–234
158. Westerterp-Plantenga MS, Smeets A, Lejeune MP (2005) Sensory and gastrointestinal satiety effects of capsaicin on food intake. Int J Obes (Lond) 29(6):682–688
159. Ahuja KD, Robertson IK, Geraghty DP, Ball MJ (2007) The effect of 4-week chilli supplementation on metabolic and arterial function in humans. Eur J Clin Nutr 61(3):326–333

160. Zhang LL, Yan Liu D, Ma LQ et al (2007) Activation of transient receptor potential vanilloid type-1 channel prevents adipogenesis and obesity. Circ Res 100(7):1063–1070
161. Baskaran P, Krishnan V, Ren J, Thyagarajan B (2016) Capsaicin induces browning of white adipose tissue and counters obesity by activating TRPV1 channel-dependent mechanisms. Br J Pharmacol 173(15):2369–2389
162. Cheung SY, Huang Y, Kwan HY et al (2015) Activation of transient receptor potential vanilloid 3 channel suppresses adipogenesis. Endocrinology 156(6):2074–2086
163. Ye L, Kleiner S, Wu J et al (2012) TRPV4 is a regulator of adipose oxidative metabolism, inflammation, and energy homeostasis. Cell 151(1):96–110
164. Wang F, Chen F, Wang G et al (2018) Rapamycin provides anti-epileptogenic effect in a rat model of post-traumatic epilepsy via deactivation of mTOR signaling pathway. Exp Ther Med 15(6):4763–4770
165. Baboota RK, Murtaza N, Jagtap S et al (2014) Capsaicin-induced transcriptional changes in hypothalamus and alterations in gut microbial count in high fat diet fed mice. J Nutr Biochem 25(9):893–902
166. van de Wall EH, Wielinga PY, Strubbe JH, van Dijk G (2006) Neonatal capsaicin causes compensatory adjustments to energy homeostasis in rats. Physiol Behav 89(1):115–121
167. Koopmans SJ, Leighton B, DeFronzo RA (1998) Neonatal de-afferentation of capsaicin-sensitive sensory nerves increases in vivo insulin sensitivity in conscious adult rats. Diabetologia 41(7):813–820
168. Lee E, Jung DY, Kim JH et al (2015) Transient receptor potential vanilloid type-1 channel regulates diet-induced obesity, insulin resistance, and leptin resistance. FASEB J 29(8):3182–3192
169. Pomonis JD, Harrison JE, Mark L et al (2003) N-(4-Tertiarybutylphenyl)-4-(3-cholorphyridin-2-yl)tetrahydropyrazine-1(2H)-carbox-amide (BCTC), a novel, orally effective vanilloid receptor 1 antagonist with analgesic properties: II. In vivo characterization in rat models of inflammatory and neuropathic pain. J Pharmacol Exp Ther 306(1):387–393
170. Moraes MN, Mezzalira N, de Assis LV et al (2017) TRPV1 participates in the activation of clock molecular machinery in the brown adipose tissue in response to light-dark cycle. Biochim Biophys Acta Mol Cell Res 1864(2):324–335
171. Fredin MF, Kjellstedt A, Smith DM, Oakes N (2015) The novel TRPVI antagonist, AZV1, improves insulin sensitivity in ob/ob mice. Diabetologia 58:S289–S289
172. Zhang Z, Zhang W, Jung DY et al (2012) TRPM2 Ca2+ channel regulates energy balance and glucose metabolism. Am J Physiol Endocrinol Metab 302(7):E807–E816
173. Togashi K, Hara Y, Tominaga T et al (2006) TRPM2 activation by cyclic ADP-ribose at body temperature is involved in insulin secretion. EMBO J 25(9):1804–1815
174. Uchida K, Dezaki K, Damdindorj B et al (2011) Lack of TRPM2 impaired insulin secretion and glucose metabolisms in mice. Diabetes 60(1):119–126
175. Uchida K, Tominaga M (2011) TRPM2 modulates insulin secretion in pancreatic β-cells. Islets 3(4):209–211
176. Pang B, Kim S, Li D et al (2017) Glucagon-like peptide-1 potentiates glucose-stimulated insulin secretion via the transient receptor potential melastatin 2 channel. Exp Ther Med 14(5):5219–5227
177. Kurashina T, Dezaki K, Yoshida M et al (2015) The β-cell GHSR and downstream cAMP/TRPM2 signaling account for insulinostatic and glycemic effects of ghrelin. Sci Rep 5:Article 14041
178. Du J, Xie J, Yue L (2009) Modulation of TRPM2 by acidic pH and the underlying mechanisms for pH sensitivity. J Gen Physiol 134(6):471–488
179. Ota W, Nakane Y, Kashio M et al (2019) Involvement of TRPM2 and TRPM8 in temperature-dependent masking behavior. Sci Rep 9(1):Article 3706
180. Silva-Alves JM, Mares-Guia TR, Oliveira JS et al (2008) Glucose-induced heat production, insulin secretion and lactate production in isolated Wistar rat pancreatic islets. Thermochim Acta 474(1–2):67–71

181. Song K, Wang H, Kamm GB et al (2016) The TRPM2 channel is a hypothalamic heat sensor that limits fever and can drive hypothermia. Science 353(6306):1393–1398
182. Badheka D, Borbiro I, Rohacs T (2015) Transient receptor potential melastatin 3 is a phosphoinositide-dependent ion channel. J Gen Physiol 146(1):65–77
183. Oberwinkler J, Philipp SE (2014) TRPM3. Handb Exp Pharmacol 222:427–459
184. Klose C, Straub I, Riehle M et al (2011) Fenamates as TRP channel blockers: mefenamic acid selectively blocks TRPM3. Br J Pharmacol 162(8):1757–1769
185. Krügel U, Straub I, Beckmann H, Schaefer M (2017) Primidone inhibits TRPM3 and attenuates thermal nociception in vivo. Pain 158(5):856–867
186. Straub I, Mohr F, Stab J et al (2013) Citrus fruit and fabacea secondary metabolites potently and selectively block TRPM3. Br J Pharmacol 168(8):183518–183550
187. Suzuki H, Sasaki E, Nakagawa A et al (2016) Diclofenac, a nonsteroidal anti-inflammatory drug, is an antagonist of human TRPM3 isoforms. Pharmacol Res Perspect 4(3):article e00232
188. Badheka D, Yudin Y, Borbiro I et al (2017) Inhibition of transient receptor potential melastatin 3 ion channels by G-protein βγ subunits. Elife 6:article e26147
189. Zamudio-Bulcock PA, Everett J, Harteneck C, Valenzuela CF (2011) Activation of steroid-sensitive TRPM3 channels potentiates glutamatergic transmission at cerebellar Purkinje neurons from developing rats. J Neurochem 119(3):474–485
190. Wagner TF, Loch S, Lambert S et al (2008) Transient receptor potential M3 channels are ionotropic steroid receptors in pancreatic beta cells. Nat Cell Biol 10(12):1421–1430
191. Vriens J, Owsianik G, Hofmann T et al (2011) TRPM3 is a nociceptor channel involved in the detection of noxious heat. Neuron 70(3):482–494
192. Nilius B, Prenen J, Tang J et al (2005) Regulation of the Ca^{2+} sensitivity of the nonselective cation channel TRPM4. J Biol Chem 280(8):6423–6433
193. Leech CA, Habener JF (1998) A role for Ca^{2+}-sensitive nonselective cation channels in regulating the membrane potential of pancreatic beta-cells. Diabetes 47(7):1066–1073
194. Marigo V, Courville K, Hsu WH et al (2009) TRPM4 impacts on Ca^{2+} signals during agonist-induced insulin secretion in pancreatic beta-cells. Mol Cell Endocrinol 299(2):194–203
195. Nelson PL, Zolochevska O, Figueiredo ML et al (2011) Regulation of $Ca(^{2+})$-entry in pancreatic α-cell line by transient receptor potential melastatin 4 plays a vital role in glucagon release. Mol Cell Endocrinol 335(2):126–134
196. Philippaert K, Pironet A, Mesuere M et al (2017) Steviol glycosides enhance pancreatic beta-cell function and taste sensation by potentiation of TRPM5 channel activity. Nat Commun 8:Article 14733
197. Peier AM, Moqrich A, Hergarden AC et al (2002) A TRP channel that senses cold stimuli and menthol. Cell 108(5):705–715
198. Mahieu F, Owsianik G, Verbert L et al (2007) TRPM8-independent menthol-induced Ca^{2+} release from endoplasmic reticulum and Golgi. J Biol Chem 282(5):3325–3336
199. Moraes MN, de Assis LVM, Henriques FDS et al (2017) Cold-sensing TRPM8 channel participates in circadian control of the brown adipose tissue. Biochim Biophys Acta Mol Cell Res 1864(12):2415–2427
200. Yoneshiro T, Aita S, Matsushita M et al (2011) Brown adipose tissue, whole-body energy expenditure, and thermogenesis in healthy adult men. Obesity (Silver Spring) 19(1):13–16

Chapter 7
Hypothalamus-Mediated Actions in the Genesis of Obesity

Matthew Ramjiawan and Paramjit S. Tappia

Abstract Leptin is an anorexigenic hormone produced from the adipose tissue and is involved in the control of hunger and satiety signals. It acts by crossing the blood–brain barrier and transmitting information about energy status to the hypothalamus and consequently suppressing appetite. An adaptive response to acute, short-term exposure to low levels of glucocorticoids in the hypothalamus is to release of anorectic hormones to allow increased energy utilization to counteract a stressful situation, however; chronic exposure to high levels of glucocorticoids due to stress, however, causes degradation of hypothalamic and pituitary protein synthesis, which prevents the generation of anorectic hormones. This subsequently increases the risk for the development of obesity as a consequence of energy overconsumption. Simultaneously, leptin action in the hypothalamus, designed to reduce food intake, also becomes dysregulated through leptin resistance and further contributes to reduced anorectic hormone production resulting in the loss of regulation of food intake. Accordingly, this chapter will briefly describe the role of hypothalamus mediated actions in the predisposition to weight gain and in the pathogenesis of obesity as well as the signal transduction mechanisms that may be targeted for the treatment of obesity.

Keywords Obesity · Hypothalamus · Stress · HPA axis · Leptin · Corticotropin-releasing factor · JAK/STAT pathway · Hunger · Anorectic hormones · Overeating

Introduction

The origins of obesity can be traced back nearly 30,000 years to prehistoric ancestors. Darwinian survival of the fittest commanded that people who stored energy in efficient way, such as through fat, would survive through the winter famine that

---

M. Ramjiawan · P. S. Tappia (✉)
Asper Clinical Research Institute & Office of Clinical Research, St. Boniface Hospital, CR3129 – 369 Tache Avenue, Winnipeg, MB R2H 2A6, Canada
e-mail: ptappia@sbrc.ca

© Springer Nature Switzerland AG 2021

P. S. Tappia et al. (eds.), *Cellular and Biochemical Mechanisms of Obesity*, Advances in Biochemistry in Health and Disease 23,
https://doi.org/10.1007/978-3-030-84763-0_7

followed bountiful summer months where food was less scarce. With ancient Egyptian texts found to publicise recommended diets and ancient Greeks soon discovering the dangers of obesity and its association with disease, the nature of survival has shifted [1]. Natural selection no longer favours such energy storage, instead shifting towards those better able to rid themselves of fat being more suited to survive. In modern times, energy storage is no longer required for success, but instead has shown to be a downward slope to disease.

It is well established that obesity is a global epidemic that is associated with numerous comorbid diseases including diabetes, stroke, cardiovascular disease, cancer, neurodegenerative diseases and many other inflammatory conditions [2–4]. Obesity is categorized as an excessive accumulation of fat, specifically white adipose tissue, that leads to an increased body mass index >30 kg/m^2. With over 650 million adults categorized as obese by the World Health Organization, this preventable disease is easy to overlook as the focus of attention has been on the COVID-19 pandemic [5]. During a time in which physical health is challenged by SARS-CoV-2 viral transmission, psychological stress, due to isolation, is on the rise, and thus it is important to examine the connection between stress and obesity from a physiological perspective to gain better understanding of their interconnected nature [6].

One of the key areas of focus are the neuroendocrine mechanisms mediated by stress pathways in the brain that contribute to the development of obesity. With its powerful endocrine connections, the hypothalamus is directly involved with the mediation of adipose tissue in the body, regardless of how excessive weight gain is caused. Leptin is a hormone predominantly released from adipocytes and acts to alter food intake and control energy expenditure over the long-term [7–9]. In view of the role of the hypothalamus in the regulation of food intake and satiety sensation, this chapter will briefly describe the hypothalamus-leptin connection and provide insight into the processes that are evoked in the hypothalamus-mediated genesis of obesity and potential targets for the treatment of this chronic metabolic disease.

The Hypothalamus

Located in the forebrain, the hypothalamus is a very small, but important brain area formed by various nerve fibers and nuclei. With widespread central nervous system synaptic connections, one of the most principal functions of the hypothalamus is to link the nervous system to the endocrine system via the adjacent pituitary gland [10, 11]. This almond-sized area of the brain has crucial neuroendocrine functions and is chiefly involved in many complex, regulatory homeostatic mechanisms including emotional behavior adjustment, hormone mediation, and hunger. The dysregulation of the hypothalamus plays an integral role in the genesis of obesity through several mechanisms, notably including a negative feedback cycle of stress-induced overeating [12–14], the inability to regulate food intake through hunger sensations

[15, 16], and a downregulation of hormones involved in metabolic homeostasis [16, 17].

HPA Axis

The hypothalamic–pituitary–adrenal (HPA) axis is an elaborate system of bidirectional signalling and feedback interactions between the three component structures that mediate a stress response: the hypothalamus, pituitary gland and adrenal gland [18]. When faced with a stressor, parvocellular neuroendocrine cells found in the paraventricular nucleus (PVN) of the hypothalamus produces corticotropin-releasing factor (CRF), which is released from neurosecretory terminals in the median eminence, at the lower part of the hypothalamus, into a system of microcirculatory blood vessels [11, 19–21]. This is known as the hypothalamo-hypophyseal portal system, which connects the hypothalamus and the anterior pituitary [20, 22]. CRF released by the hypothalamus binds to the corticotropin-releasing factor receptor 1 (CRFR1), located on corticotropes found in the anterior pituitary. The corticotrope cells proceed to synthesize proopiomelanocortin (POMC) protein which, after enzyme-mediated cleavage, secretes several bioactive peptides including adrenocorticotropic hormone (ACTH), β-lipotropin, β-endorphin and melanocyte stimulating hormones (MSH) [11, 23, 24]. ACTH then enters the systemic circulatory system and binds to a melanocortin type 2 receptor (MC2R), which has a highly specific binding affinity for ACTH, found on the surface of adrenal cortical cells to regulate cortisol synthesis. Most MC2R are found in the zona fasciculata, which is where cortisol, the primary endogenous glucocorticoid is synthesized [18, 20, 22, 24]. Cortisol has a wide variety of effects on the body including gluconeogenesis and immune system suppression [24]. Steroid hormones are released and bind to glucocorticoid receptors found in the PVN of the hypothalamus, providing negative feedback through bidirectional signaling [11, 17, 18, 20, 22, 24, 25]. This cycle of activation can be seen in Fig. 7.1.

The HPA axis is designed to regulate diurnal variation in cortisol levels; however, excessive or deficient secretion of cortisol is linked to pathophysiological changes [26]. In fact, studies have shown that central obesity is associated with an increased cortisol level, most likely attributed to an increased activity of the HPA axis [27]. Several lines of evidence indicate that both neuroendocrine dysregulation of the HPA axis as well as altered peripheral cortisol metabolism may play a role in the pathogenesis of abdominal obesity [28]. It should be noted that while ample evidence exists pointing to differential HPA axis in both generalized and abdominal obesity, consistent observations in obesity-related disturbances in the HPA axis need to be established [29, 30].

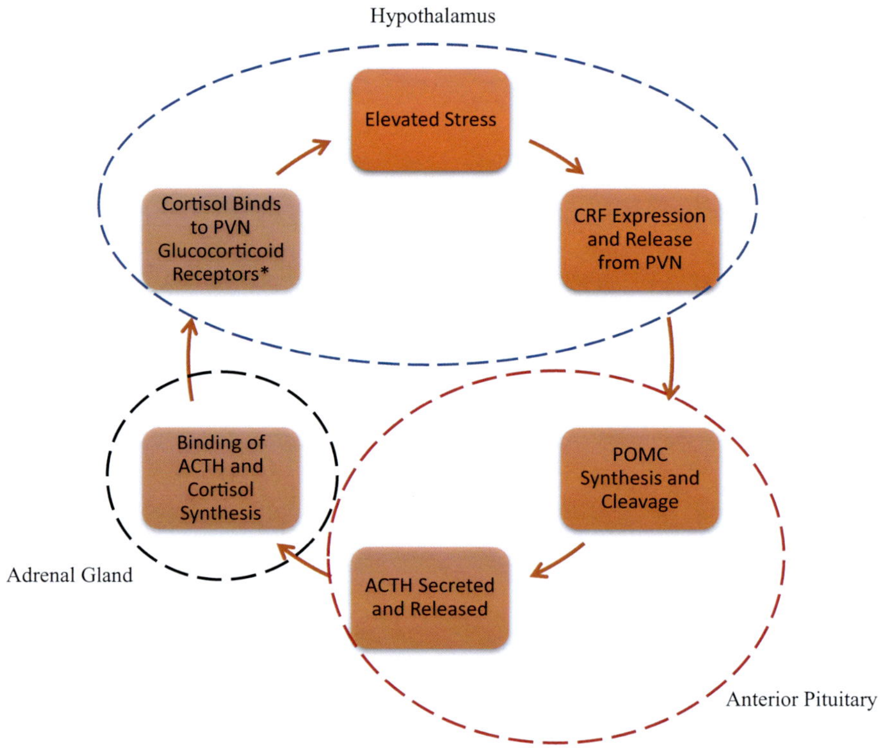

Fig. 7.1 Cycle of HPA Axis Stress-Induced Activation. Chronically elevated stress levels in the hypothalamus result in the activation of the hypothalamus-pituitary-adrenal (HPA) axis. A chain of events originating in the paraventricular nucleus (PVN) produces corticotropin releasing factor (CRF), which is released into a system of microcirculatory blood vessels known as the hypothalamo-hypophyseal portal system that connects the hypothalamus to the anterior pituitary. CRF binds to corticotrope receptors and synthesize proopiomelanocortin (POMC) protein which, after enzymatic cleavage, secretes adrenocorticotropic hormone (ACTH). ACTH enters the systemic circulatory system and binds to receptors found in the zona fasciculata, which initiates cortisol production. Cortisol, a glucocorticoid, binds to receptors found in the PVN of the hypothalamus, causing bidirectional signaling

Stress-Initiated Weight Gain

Chronic stress coupled with a positive energy balance is considered to be a contributory factor to the increased risk for upper body obesity [31, 32]. In 2014, Jauch-Chara and Oltmanns [12] proposed a neuropsychological model for the etiology of obesity as the outcome of a vicious cycle between chronic HPA activation by stressors and physiological and psychological homeostasis. They describe obesity as a by-product of impaired neuropsychological functioning—specifically due to dysfunctional reward system and stress-response functions. Their model states that chronic

HPA axis activation results in neurochemical imbalances and that dysfunction mitigates a negative mood state and causes individuals to consume high-calorie foods that are highly palatable in the attempts to alleviate their stress [25, 33, 34]. It is already well-understood that chronic stress acts as a neuromodulator for dopamine (DA) receptors in the mesolimbic reward pathway, downregulating DA release from the ventral tegmental area (VTA) to the nucleus accumbens [17, 22]. This downregulation of neurological activity combined with chronic HPA axis activation causes a negative feedback cycle of reduced feelings of reward and increased stress, contributing to stress-induced overeating [35]. In addition, individuals who experience severe chronic stress display an increased vulnerability to obsessive–compulsive disorders—contributing to compulsive overindulgence [36, 37]. Consequently, the clinical effect is pervasive and excessive weight gain combined with negative mood states.

Leptin

When examining hypothalamic obesity, it is impossible to overlook the hormone, leptin, as its primary purpose is to regulate fat storage through inhibition of hunger signaling [11]. Its release is directly mediated from signals sent by the hypothalamus, through the pituitary and into the adrenal gland [22]. Discovered by Jeffrey Friedman, in 1994, leptin has a longstanding history of being associated with both obesity and diabetes [38]. In 1949, researchers at the Jackson Laboratory found that the offspring of a non-obese colony were unusually extremely hungry and as a consequence these mice had an unusually high level of food consumption and presented with what researchers named the "ob" gene mutation, on account of the obese nature of the mice. Friedman and others confirmed that the ob gene encoded for a novel hormone that was systemically circulated and suppressed food consumption, which was absent in their obese mice [38–42]. Friedman named the new hormone "leptin" from the Greek word *lepto*, meaning thin.

Leptin is a hormone produced by adipose cells and enterocytes in the small intestine that serves to regulate caloric homeostasis by inhibiting hunger signaling that ultimately diminishes fat storage in adipocytes [43]. Leptin acts on receptors in multiple areas of the brain, including dopaminergic neurons in the VTA and the hypothalamus, but is mostly localized within various areas of the hypothalamus [22, 44]. Within the hypothalamus, leptin notably binds to both the arcuate nucleus and the ventromedial nuclei, which consequently mediates feeding behaviors through both inhibiting hunger signaling, and stimulating sensations of satiety [11, 41, 45]. Leptin acts directly on leptin receptors (Ob-R) found in the cell membrane of a wide range of cell types and is part of a unique class of cytokine receptors as it is a single-transmembrane-domain type 1 cytokine receptor [22, 42, 43, 46]. It also indirectly mediates the effects of glucocorticoids, specifically antagonizing cortisol release from the adrenal glands.

In the lateral hypothalamus, leptin inhibits hunger through degradation of hypothalamic neuropeptide-γ (NPY) mRNA, which encodes for NPY that causes growth of adipose tissue, and is also a potent appetite stimulator [11, 40]. Simultaneously, leptin increases hypothalamic CRF mRNA expression in the PVN [47, 48]. This action shows the opposing nature between leptin and glucocorticoids, as glucocorticoids produce the opposite result: stimulating NPY secretion and inhibiting CRF expression in the hypothalamus.

In the medial hypothalamus however, leptin does not serve to inhibit hunger, but instead stimulates sensations of satiety [43–45]. Thus, the regulation of food intake by leptin is attributable to (a) inhibition of hunger and (b) producing satiety sensations; both of these aspects reduce the total amount of food intake [49–51]. Like CRF in the HPA axis, leptin also stimulates the synthesis of POMC in corticotrope cells [52–54]. Following enzyme-mediated cleavage of POMC, bioactive peptides are synthesized, including ACTH and MSH, with MSH being a potent downstream anorexiant [24]. It is interesting to note that overfeeding of mice can induce leptin resistance [55], suggesting that nutrient excess can eventually lead to an ineffective leptin response that may further augment the occurrence of weight gain and subsequent development of obesity.

Leptin Receptor Signaling

The leptin receptor, Ob-R, has six isoforms (Ob-Ra, Ob-Rb, Ob-Rc, Ob-Rd, Ob-Re, Ob-Rf), which are all encoded by the LEPR gene and is part of the class I cytokine receptor family [22, 39, 43]. All leptin receptor isoforms have identical extracellular parts that consist of six domains: two cytokine receptor homology (CRH) domains, an N-terminal domain, an immunoglobin-like domain, and two membrane-proximal fibronectin type III domains [24, 46]. With regards to obesity, the Ob-Rb has the most impact as the weight-reducing effect of leptin is mediated through Ob-Rb signal transduction in the hypothalamus [41]. The CRH2 domain functions as a high-affinity binding site for leptin. Upon ligand binding, the Ob-Rb activates the JAK/STAT pathway. The JAK/STAT pathway has been examined in depth and is probably the best explored pathway that is activated by leptin [56]. Ligand binding causes the Ob-Rb to undergo homooligomerization which allows for the JAK2 kinase to bind to the Ob-R and results in cross-phosphorylation [56, 57]. The protein Tyr-1138 found in the Ob-R binds and activates transcription factor STAT3, which enters the nucleus and stimulates SOCS3 gene expression [43, 56–58]. The protein, suppressor of cytokine signaling 3 (SOCS3), is a potent negative regulator of the JAK/STAT pathway by negatively inhibiting JAK2, and Tyr-985, ultimately creating a negative feedback loop that limits further leptin signaling. The intracellular signaling mechanisms can be seen in Fig. 7.2. In addition to intracellular action in hypothalamic neurons, STAT3 also binds to corticotropes in the anterior pituitary and supports the production of POMC proteins [59].

Leptin binding to the Ob-R also activates the mitogen-activated protein kinase (MAPK) pathway. Binding to the OB-R, cytoplasmic Src homology 2 domain-containing phosphatase-2 (SHP2) becomes phosphorylated by adjacent JAK2 and

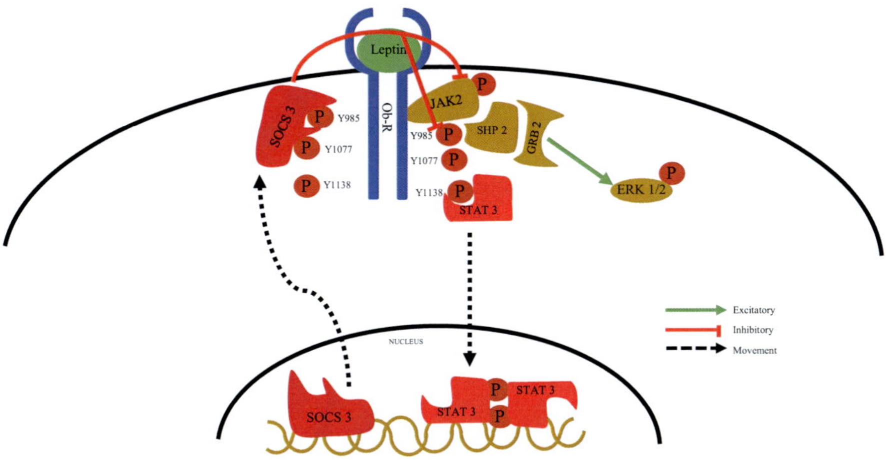

Fig. 7.2 JAK/STAT Hypothalamic Intracellular Signaling Pathway. Upon ligand-binding, the leptin receptor (Ob-R) undergoes homooligomerization, which allows the JAK2 kinase to bind to the Ob-R, causing cross-phosphorylation. Tyr-1138 protein activates STAT3, which enters the nucleus and stimulates SOCS3 production. SOCS leaves the nucleus and negatively inhibits JAK2 and Tyr-985 to create a feedback loop that limits leptin signaling. JAK2 binding also initiates the MAPK pathway, resulting in increased ERK1/2 expression

recruits an adaptor protein, growth factor receptor-bound protein 2, to activate MAPK extracellular signal-regulated kinases 1 and 2 (ERK1/2) [55, 56]. These kinases that are upregulated through high concentrations of leptin are closely associated with obesity and contributes to the anorectic effects of leptin.

While leptin receptors have a self-controlled negative feedback cycle to limit the rate of appetite suppression through the SOCS3 protein, overexpression may lead to increased leptin resistance [17]. One theory of leptin resistance is that chronic, overexpression of leptin through overindulgence causes the intracellular signaling disruption, where increased amounts of STAT3 within the nucleus causes an upregulation of SOCS3 expression, which greatly increases its inhibitory actions—functionally leading to leptin resistance, preventing feelings of satiety and blocking appetite suppression [52].

Conclusion

As the hypothalamus is the initiator for many neuroendocrine processes involved in weight gain, it is expected that any disruption to this brain area will result in changes to homeostatic functioning and rapidly trigger weight changes. It is well known that when the hypothalamus is damaged or lesioned due to surgery, traumatic brain injury, tumors or radiation, metabolic functions are disrupted, and weight gain is often observed. Inadequate management and elevated levels of stress coupled to increased food consumption can result in weight gain and the development of obesity if persistent. From an evolutionary perspective, the suppression of hunger signals would benefit an organism by allowing for increased energy expenditure dedicated towards the removal of the stressor through sympathetic nervous system functioning. Such an adaptive response to acute or short-term exposure to low levels of glucocorticoids in the hypothalamus is attributed to CRF actions in the anterior pituitary and subsequent attenuation of anorectic hormones. On the other hand, the chronic exposure to high levels of glucocorticoids, prevents the formation of anorectic, and satiety-signaling hormones resulting in energy overconsumption leading to weight gain and ultimately, if persistent, to the development of obesity. Obese individuals have elevated levels of leptin due to increased adipocyte production of leptin as a consequence of leptin resistance and are insensitive to the exogenous administration of leptin. Indeed, in obesity, the chronic activation of JAK-STAT3 by increased circulating leptin levels leads to the development of leptin resistance. Since chronic stress leads to altered hypothalamic metabolic functioning, the risk for developing obesity through a deregulation of the Ob-R and JAK/STAT signaling pathway is augmented. Thus, targeting the JAK/STAT pathway may represent another therapeutic approach for the treatment of obesity, particularly under conditions of leptin resistance.

Acknowledgements The infrastructural support for this work was provided by the St. Boniface Hospital Albrechtsen Research Centre, Winnipeg, Manitoba, Canada.

References

1. Haslam D (2007) Obesity: a medical history. Obes Rev 8:31–36
2. World Health Organization (2020) WHO outlines steps to save 7 million lives from cancer. https://www.who.int/news/item/04-02-2020-who-outlines-steps-to-save-7-million-lives-from-cancer. Accessed 04 Feb 2020
3. Janssen I (2013) The public health burden of obesity in Canada. Can J Diabetes 37:90–96
4. Low S, Chin MC, Deurenberg-Yap M (2009) Review on epidemic of obesity. Ann Acad Med Singap 38:57–59
5. WHO (2020) Obesity and Overweight. https://www.who.int/news-room/fact-sheets/detail/obesity-and-overweight. Accessed 04 Feb 2020
6. American Psychological Association (2020) Stress in America 2020. https://www.apa.org/news/press/releases/stress/2020/report. Accessed 04 Feb 2020
7. Izquierdo AG, Crujeiras AB, Casanueva FF, Carreira MC (2019) Leptin, obesity, and leptin resistance: where are we 25 years later? Nutrients 11:2704
8. Triantafyllou GA, Paschou SA, Mantzoros CS (2016) Leptin and hormones: energy homeostasis. Endocrinol Metab Clin North Am 45:633–645
9. Zhang F, Chen Y, Heiman M, Dimarchi R (2005) Leptin structure, function and biology. Vitam Horm 71:345–372
10. Alsop D, Aluru N (2011) In: Farrell FP (ed) The pituitary. Development of the hypothalamus-pituitary-interrenal axis. Academic Press, San Diego, pp 1450–1456
11. Palkovits M (2003) Hypothalamic regulation of food intake. Ideggyogy Sz 56:288–302
12. Jauch-Chara K, Oltmanns KM (2014) Obesity—a neuropsychological disease? Systematic review and neuropsychological model. Prog Neurobiol 114:84–101

13. Pariante CM (2003) Depression, stress and the adrenal axis. J Neuroendocrinol 15:811–812
14. Tsigos C, Chrousos GP (2002) Hypothalamic–pituitary–adrenal axis, neuroendocrine factors and stress J Psychosom Res 53:865–871
15. Gioldasi S, Karvela A, Rojas-Gil AP et al (2019) Metabolic association between leptin and the corticotropin releasing hormone. Endocr Metab Immune Disord Drug Targets 19:458–466
16. Warne JP (2009) Shaping the stress response: interplay of palatable food choices, glucocorticoids, insulin and abdominal obesity. Mol Cell Endocrinol 300:137–146
17. Heiman ML, Ahima RS, Craft LS et al (1997) Leptin inhibition of the hypothalamic-pituitary-adrenal axis in response to stress. Endocrinology 138:3859–3863
18. Aschbacher K, Rodriguez-Fernandez M, van Wietmarschen H et al (2014) The hypothalamic-pituitary-adrenal-leptin axis and metabolic health: a systems approach to resilience, robustness and control. Interface Focus 4:20140020
19. Tache Y, Larauche M, Yuan P-Q, Million M (2018) Brain and gut CRF signaling: biological actions and role in the gastrointestinal tract. Curr Mol Pharmacol 11:51–71
20. Gallagher JP, Orozco-Cabal LF, Liu J, Shinnick-Gallagher P (2008) Synaptic physiology of central CRH system. Eur J Pharmacol 583:215–225
21. Sharma R, Banerji MA (2012) Corticotropin releasing factor (CRF) and obesity. Maturitas 72:1–3
22. Malendowicz LK, Rucinski M, Belloni AS et al (2007) Leptin and the regulation of the hypothalamic-pituitary-adrenal axis. Int Rev Cytol 263:63–102
23. Wei NL, Quan ZF, Zhao T et al (2019) Chronic stress increases susceptibility to food addiction by increasing the levels of DR2 and MOR in the nucleus accumbens. Neuropsychiatr Dis Treat 15:1211–1229
24. Angelousi A, Margioris AN, Tsatsanis C (2020) ACTH action on the adrenals. In: Feingold KR, Anawalt B, Boyce A et al (eds) Endotext. South Dartmouth, MA
25. Ramjiawan M, Tappia PS (2020) Exploration of the bidirectionality of obesity and depression by means of the neuropsychological model of obesity genesis. In: Tappia PS, Ramjiawan B, Dhalla NS (eds) Pathophysiology of obesity-induced health complications. Springer Nature, pp 169–180
26. Chalew S, Nagel H, Shore S (1995) The hypothalamic-pituitary-adrenal axis in obesity. Obes Res 3:371–382
27. Bjorntorp P (1995) Endocrine abnormalities of obesity. Metabolism 44:21–23
28. Pasquali R, Vicennati V, Cacciari M, Pagotto U (2006) The hypothalamic-pituitary-adrenal axis activity in obesity and the metabolic syndrome. Ann N Y Acad Sci 1083:111–128
29. Nieuwenhuizen AG, Rutters F (2008) The hypothalamic-pituitary-adrenal axis in the regulation of energy balance. Physiol Behav 94:169–177
30. Rodriguez ACI, Epel ES, White ML et al (2015) The hypothalamic-pituitary-adrenal axis dysregulation and cortisol activity in obesity: a systematic review. Psychoneuroendocrinology 62:301–318
31. Bose M, Olivan B, Laferrere B (2009) Stress and obesity: the role of the hypothalamic-pituitary-adrenal axis in metabolic disease. Curr Opin Endocrinol Diabetes Obes 16:340–346
32. Sominsky L, Spencer SJ (2014) Eating behavior and stress: a pathway to obesity. Front Physiol 5:434
33. Bruchas RM, Land BB, Chavkin C (2010) The dynorphin/kappa opioid system as a modulator of stress-induced and pro-addictive behaviors. Brain Res 1314:44–55
34. Mcewen BS (2009) Protection and damage from acute and chronic stress: allostasis and allostatic overload and relevance to the pathophysiology of psychiatric disorders. Ann N Y Acad Sci 1032:1–7
35. Berridge KC, Robinson TE (2016) Liking, wanting, and the incentive-sensitization theory of addiction. J Am Psychol 71:670–679
36. Adams TG, B. Kelmendi B, Brake CA et al (2018) The role of stress in the pathogenesis and maintenance of obsessive-compulsive disorder. Chronic Stress 2
37. Sousa-Lima J, Moreira PS, Raposo-Lima C et al (2019) Relationship between obsessive compulsive disorder and cortisol: systematic review and meta-analysis. Eur Neuropsychopharmacol 29:1185–1198

38. Zhang Y, Proenca R, Maffei M et al (1994) Positional cloning of the mouse obese gene and its human homologue. Nature 372:425–432
39. Lee G-H, Proenca R, Montez JM et al (1996) Abnormal splicing of the leptin receptor in diabetic mice. Nature 379:632–635
40. Cohen P, Zhao C, Cai X et al (2001) Selective deletion of leptin receptor in neurons leads to obesity. J Clin Invest 108:1113–1121
41. Fei H, Okano HJ, Li C et al (1997) Anatomic localization of alternatively spliced leptin receptors (Ob-R) in mouse brain and other tissues. Proc Natl Acad Sci 94:7001
42. Vaisse C, Halaas JL, Horvath CM et al (1996) Leptin activation of Stat3 in the hypothalamus of wild–type and ob/ob mice but not db/db mice. Nat Genet 14:95–97
43. Frühbeck G (2005) Intracellular signalling pathways activated by leptin. Biochem J 393:7–20
44. Simonds SE, Cowley MA, Enriori PJ (2012) Leptin increasing sympathetic nerve outflow in obesity: a cure for obesity or a potential contributor to metabolic syndrome? Adipocyte 1:177–181
45. Enriori PJ, Evans AE, Sinnayah P et al (2007) Diet-induced obesity causes severe but reversible leptin resistance in arcuate melanocortin neurons. Cell Metab 5:181–194
46. Ge H, Huang L, Pourbahrami T, Li C (2002) Generation of soluble leptin receptor by ectodomain shedding of membrane-spanning receptors in vitro and in vivo. J Biol Chem 277:45898–45903
47. Huang Q, Rivest R, Richard D (1998) Effects of leptin on corticotropin-releasing factor (CRF) synthesis and CRF neuron activation in the paraventricular hypothalamic nucleus of obese (ob/ob) mice. Endocrinology 139:1524–1532
48. Yamagata S, Kageyama K, Akimoto K et al (2013) Regulation of corticotropin-releasing factor and urocortin 2/3 mRNA by leptin in hypothalamic N39 cells. Peptides 50:1–7
49. Yeung AY, Tadi P (2020) Physiology, obesity neurohormonal appetite and satiety control. StatPearls
50. Ahima RS, Antwi DA (2008) Brain regulation of appetite and satiety. Endocrinol Metab Clin North Am 37:811–823
51. Meek TH, Matson ME, Dorfman MD et al (2013) Leptin action in the ventromedial hypothalamic nucleus is sufficient, but not necessary, to normalize diabetic hyperglycemia. Endocrinology 154:3067–3076
52. Elias CF, Aschkenasi C, Lee C et al (1999) Leptin differentially regulates NPY and POMC neurons projecting to the lateral hypothalamic area. Neuron 23:775–786
53. Balthasar N, Coppari R, Mcminn J et al (2004) Leptin receptor signaling in POMC neurons is required for normal body weight homeostasis. Neuron 42:983–991
54. Cowley MA, Smart JL, Rubinstein M et al (2001) Leptin activates anorexigenic POMC neurons through a neural network in the arcuate nucleus. Nature 411:480–484
55. Wang J, Obici S, Morgan K et al (2001) Overfeeding rapidly induces leptin and insulin resistance. Diabetes 50:2786–2791
56. Harrison DA (2012) The Jak/STAT pathway. Cold Spring Harb Perspect Biol 4:a011205
57. Seif F, Khoshmirsafa M, Aazami H et al (2017) The role of JAK-STAT signaling pathway and its regulators in the fate of T helper cells. Cell Commun Signal 15:23
58. Morton GJ, Blevins JE, Kim F et al (2009) The action of leptin in the ventral tegmental area to decrease food intake is dependent on Jak-2 signaling. Am J Physiol Endocrinol Metab 297:202–210
59. Ernst MB, Wunderlich CM, Hess S et al (2009) Enhanced Stat3 activation in POMC neurons provokes negative feedback inhibition of leptin and insulin signaling in obesity. J Neurosci 29:11582–11593

Chapter 8
Cellular and Molecular Effects of Obesity on the Heart

Ahmed Sultan, Jaipaul Singh, and Frank Christopher Howarth

Abstract Obesity is a serious chronic disease that is responsible for a large number of deaths worldwide. The Body Mass Index (BMI) is widely used to provide definitions of overweight and obesity. A BMI of 20–24 is considered normal, a BMI of 25–29 is considered overweight and a BMI of 30 and over is considered obese. Overweight and obesity are attributed to a variety of risk factors including smoking, genetics, alcohol consumption, high level of stress, physical inactivity, unhealthy diet containing excessive amounts of fat and sugar, food snacking and binging, low calorie expenditure, socio-economic and psychological issues and some medications. Overweight and obesity are also associated with various illnesses including hormonal imbalance, hypothyroidism, insulin resistance (IR), polycystic ovary syndrome and Cushing's syndrome. Obesity is a risk factor for diabetes (diabesity) resulting in hyperglycemia which enhances the generation of reactive carbonyl species (RCS) and reactive oxygen species (ROS) in the myocardium. Both ROS and RCS elicit insults to cardiac muscle which may result in apoptosis, fibrosis, hypertrophy, mitochondrial dysfunction, derangement in cellular calcium homeostasis and electrical signaling. In response to these insults the heart goes through a process of remodeling in order to maintain the demands of the body. Over time there may be development of cardiomyopathy, reduced ejection fraction and perhaps arrhythmias and sudden cardiac death (SCD). In general obesity is a preventable illness. This review describes the cellular and molecular effects of obesity on the heart leading to SCD.

Keywords Risk factors · BMI · Obesity · Diabetes · Heart · Cardiomyopathy

A. Sultan · F. C. Howarth (✉)
Department of Physiology, College of Medicine & Health Sciences, UAE University, Al Ain, UAE
e-mail: chris.howarth@uaeu.ac.ae

A. Sultan
e-mail: 201370241@uaeu.ac.ae

J. Singh
School of Natural Sciences, University of Central Lancashire, Preston, Lancashire, UK
e-mail: JSingh3@uclan.ac.uk

© Springer Nature Switzerland AG 2021

P. S. Tappia et al. (eds.), *Cellular and Biochemical Mechanisms of Obesity*,
Advances in Biochemistry in Health and Disease 23,
https://doi.org/10.1007/978-3-030-84763-0_8

Introduction

Obesity: Epidemiology and Pathophysiology

Obesity is a serious global health problem and it is associated with metabolic syndrome which has been described as the combination of insulin resistance (IR), hypertension, hyperlipidemia and obesity [1, 2]. The World Health Organization (WHO) [3] defines obesity as "an abnormal or excessive fat accumulation in adipose tissue, to the extent that health is impaired". The current gold standard classification of obesity is by means of body mass index (BMI) which is calculated as a person's weight in kilograms divided by the square of height in meters (kg/m^2). For epidemiological purposes, a person is considered overweight if the BMI is greater than 25 kg/m^2 and obese if the BMI $\geq$ 30 kg/m^2 [3]. Excessive adiposity is an important risk factor in the pathophysiology of diabetes mellitus (DM), IR, dyslipidemia, atherosclerosis, hypertension, stroke, psychological, coronary artery disease (CAD), liver disease, osteoarthritis, sleep apnea, some musculoskeletal conditions, gynecological complications, cancer including colon and breast cancer [4]. In recent times, a new method to assess body weight was developed and it was referred to as the body shape index (ABSI). It quantifies the risk associated with abdominal obesity. ABSI is based on waist circumference (WC) and is adjusted for height and weight: ABSI = WC/(BMI (2/3) $\times$ height (1/2)). ABSI might be considered to be a more valuable predictor of mortality than BMI [5]. Obesity has become an important global health concern mainly due to its remarkable rise in prevalence during the last few decades. In 2016, roughly 1.9 billion adults aged 18 years and older globally were pre-obese or overweight and estimates suggest more than 600 million adults were obese. It was estimated that 13% of the global adult population (11% of men and 15% of women) were obese in 2016, meanwhile 39% of adults aged 18 years and above (38% of men and 40% of women) were overweight [3]. The global prevalence of obesity tripled between 1975 and 2016, equivalent to 30% of the world's population. A projected estimate of 340 million children and adolescents aged 5–19 years were either obese or overweight in 2016 and this is of extreme concern to governments worldwide [3]. According to estimates from the latest Lancet Commission report, obesity affected 2 billion people universally in 2015 and the annual costs were approximately $2 trillion US dollars from lost economic productivity and direct health care expenditure [6].

Obesity is a disorder with a multi-factorial basis resulting from a blend of numerous lifestyles, genetic and environmental factors. A sedentary lifestyle, reduction in exercise or physical activity, reduced metabolic rate, thermogenesis and decreased energy expenditure will eventually result in increased energy storage and obesity. Causes of obesity include easy access to addictive appetizing fast foods, leptin resistance, junk foods, assertive marketing of food, sedentary lifestyle, certain medications as well as genetic factors. It is also noteworthy that both overweight and obesity are preventable. Figure 8.1 shows some of the environmental and genetic factors that may contribute to obesity and consequently, leading to metabolic syndrome [7].

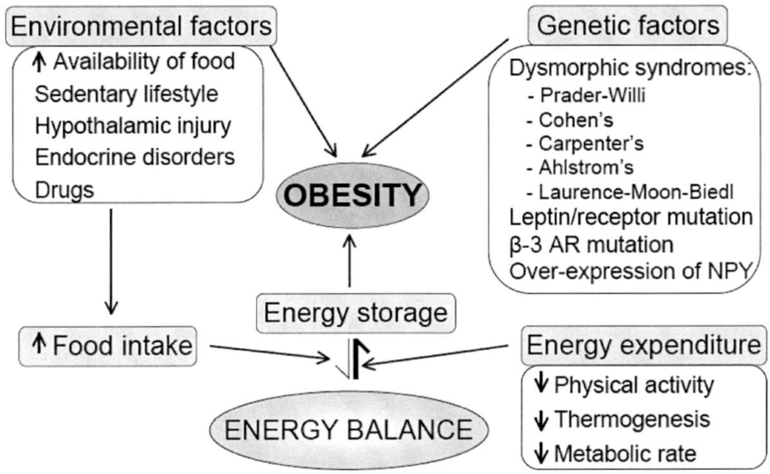

Fig. 8.1 Flow diagram showing the energy equilibrium and etiology of obesity (Adapted from [7])

Obesity is characterized by excess surplus of fat in the body, contributing to harmful metabolic outcomes. Additionally, obesity represents a significant socio-economic and medical global burden. According to a recent published report, the global cost of treating obesity has soared to $2 trillion US dollars [6, 8].

Obesity is the collective result of a harmful increase in adipose tissue resulting from increased fat cell size (hypertrophy) and/or number of cells (hyperplasia) [9, 10]. These two events, together, create a cycle of co-morbidities culminating in obesity. Identifying potential factors and targeting adipocyte biology that can control or normalize these processes may be important in the treatment and prevention of obesity.

Chronic energy imbalance leads to adipocyte hyperplasia and hypertrophy, endoplasmic reticulum stress, plus mitochondrial dysfunction. These processes induce an increase in intracellular and systemic release of adipokines, free fatty acids, and inflammatory mediators that cause adipocyte dysfunction and adverse effects in the liver, pancreatic β-cells, and skeletal muscles and more importantly the heart itself besides vascular beds [11].

The Link Between Obesity and Diabetes

People who have developed obesity are most likely to develop type 2 diabetes mellitus (T2DM) since obesity is a major independent and metabolic risk factor for T2DM. Obesity-induced T2DM, heading towards persistent high level of blood glucose, is normally described as metabolic syndrome or IR or adult pre-diabetes. In this condition, the body can synthesize and release enough insulin in the blood however, the tissues and the cells in the body become somewhat resistant to the actions of insulin [12, 13].

The term "diabesity" was created, due to the strong link between obesity and diabetes. Although the majority of individuals with T2DM are obese, only a relatively small fraction of obese individuals will develop T2DM [14]. The suggested molecular mechanisms underlying these complications remain poorly understood. Figure 8.2 summarizes the three main hypotheses which have been developed in recent years to explain the molecular basis of diabesity [15] and they include the following:

(1) The "inflammation hypothesis" emphasizes that obesity represents a state of chronic inflammation where inflammatory molecules released by infiltrating macrophages in adipose tissue apply pathological changes in β-cells and insulin-sensitive tissues such as the endocrine glands, the muscles, the endocrine pancreas, and the liver.
(2) The "lipid overflow hypothesis" highlights that obesity may result in increased ectopic lipid stores because obese subjects have limited capacity of adipose tissue fat storage. These potentially harmful lipid components and metabolites may utilize cytotoxic effects on peripheral cells from the pancreas, muscles, and liver.
(3) The "adipokine hypothesis" suggests that the main role of white adipose cells is to function as an endocrine organ, and to secrete a range of hormones with autocrine and paracrine activity. Diabesity enlarging fat stores can cause dysfunctional secretion of such endocrine factors, thus resulting in metabolic impairment of insulin target tissues and ultimately failure of insulin producing β-cells. The adipokines have direct effects on the muscles, the central nervous system (CNS) and the liver.

Obesity is associated with many comorbidities such as hypertension, sleep apnea syndrome, atherosclerosis, and CAD. Obesity precedes structural and functional changes in the heart, which ultimately leads to heart failure (HF). Obesity can induce alterations in myocardial structure which increases the risk of atrial fibrillation or arrhythmias and SCD [16]. That is why it is important to investigate the effects of obesity on the heart.

Obesity and the Heart

The heart is actually a muscle pump composed of four-chambers that function to deliver blood and nutrients to all the other organs of the body [17]. The heart is composed of left and right atria and left and right ventricles. The left side of the heart pumps blood through the systemic circulation and the right side of the heart pumps blood through the pulmonary circulation. The orderly contraction of the chambers of the heart are controlled by a specialized electrical conduction system comprising the sinoatrial node (SAN), the atrioventricular node (AVN), right and left bundle branches, besides the Purkinje fibers [17]. Electrical activity of the heart is initiated

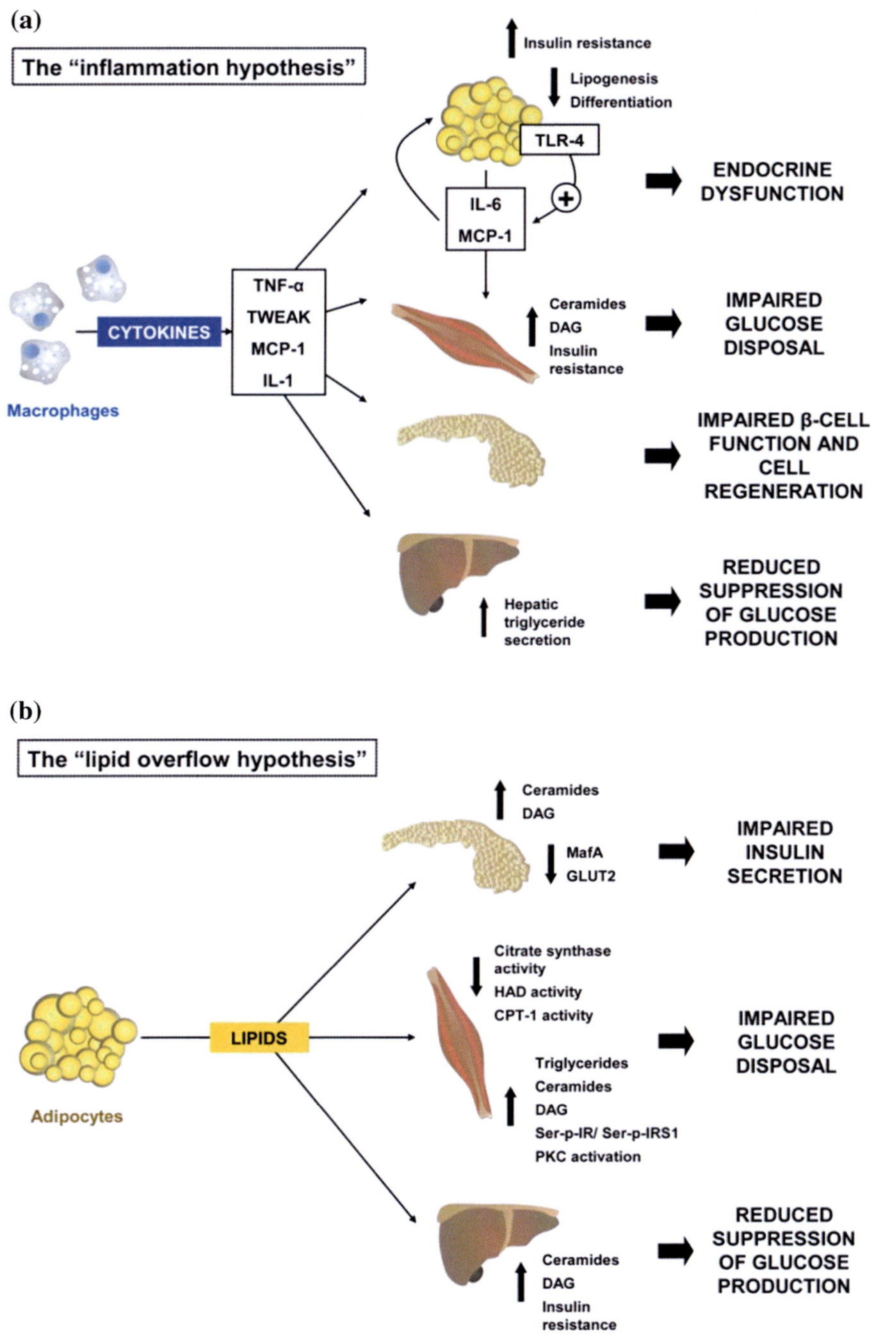

Fig. 8.2 Three main hypotheses involving **a** inflammation, **b** lipid overflow and **c** adipokines which have been developed in recent years that link obesity and diabetes (Adapted from reference [15])

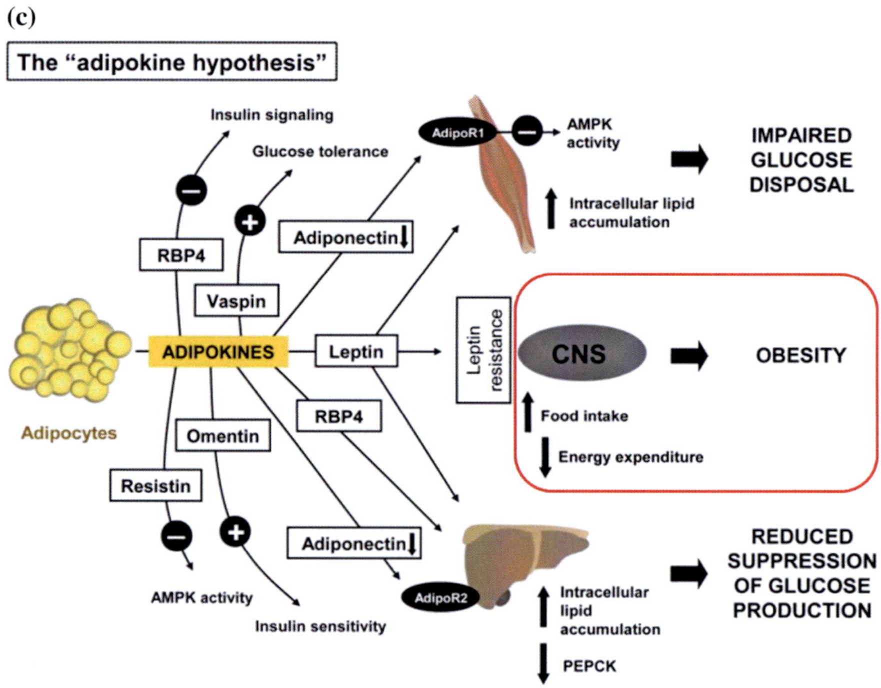

Fig. 8.2 (continued)

by specialized cells in the SAN that are located in the upper part of the wall of the right atrium. SAN cells are able to generate action potentials (APs) spontaneously [18].

APs, produced by the SAN cells, are rapidly conducted across the atria and, after a brief delay in the AVN, are then rapidly conducted through the ventricles. The AP is a brief change in voltage (membrane potential) caused by selective changes in ion channel conductance [17].

According to the Framingham study, overweight and obesity were both associated with an increased relative risk for the development of cardiovascular risk factors and cardiovascular diseases (CVDs) and people with T2DM are 2/3 times more likely to develop CVDs [19, 20].

Animal Models of Obesity and Their Relevance to the Heart

In order to investigate the mechanisms of cardiac remodeling in the context of obesity, several animal models have been used over the years to investigate the function and dysfunction of the heart. These models resemble overweight and/or obesity in humans in addition to comorbidities such as IR, impaired glucose tolerance, hypertension, and diabetes. Zucker diabetic fatty (ZDF) rats were isolated and derived from the Zucker fatty (ZF) rat and are extensively used for obesity and T2DM research. Furthermore, the ZF rat genetic animal model inherits obesity as a Mendelian autosomal recessive trait. The ZF rat, which develops obesity without DM, has a subtle missense mutation (fatty, fa) in the leptin receptor gene (Lepr). This missense mutation (a nucleotide A to C switch or conversion at nucleotide position 806) was discovered in the extracellular specific domain of all the isoforms in ZF (fa/fa) rats and resulted in an amino acid change from glycine (Gln) to proline (Pro) at +269. In the brains of ZF (fa/fa) rats Ob-R isoform messenger ribonucleic acids (mRNAs) remained at similar levels to those found in their lean littermate phenotype named Zucker Lean (ZL) rats [21]. These ZF rats become obese, hyperphagic and hyper-insulinemic. Nevertheless, blood glucose remains at normal levels. The ZDF strain came from a ZF colony [22], which is a relatively new strain of fa/fa homozygous male rats with reproductive capability. They were obtained by mating fa/fa males and fa/+ heterozygous females. The ZDF strain of fa/fa male rats displayed DM-selective breeding and exhibited comparatively high blood glucose levels at a very young age of about 10 weeks and this diabetic strain was referred to as ZDF-Lepr *fa* rat. Whilst none of the ZF fa/+ male rats developed diabetes, the ZDF homozygous fa/fa male rats became diabetic, achieving 100% incidence by the age of 21 weeks. The phenotypic attributes of this diabetic specific strain are different from those of the normo-glycemic ZF rats. The ZDF rat strain with high reproductive capacity provides a valuable animal model of T2DM [22–25].

The storage of fat, especially that within heart muscle, has been coupled with a decrease in cardiac force production, which has adverse hemodynamic consequences and ultimately leading to HF. Furthermore, fats from the blood can be taken up by heart muscle cells where they can be utilized for generating energy which can be stored or used within the mitochondria. Experiments in the ZDF rat have established that fat does in fact accumulate in the heart due to increased transport across the different membranes. Current experiments also suggest that lipids accumulate in the heart in spite of normal mitochondrial content, morphology and long-chain fatty acid oxidation [26]. In addition, increased sarcolemma long-chain fatty acid (LCFA) transport rates and proteins of LCFA result in either a bigger or a larger number of lipid droplets stored within cardiac muscle [26].

Obesity is linked to high blood pressure (BP), which in turn increases the risk of CVDs. Simonds et al. [27] observed that the surge in leptin levels noticed in diet-induced obesity (DIO) promotes an increase in BP in rodents, an outcome that was not previously seen in deficient in leptin or LepR animals. Likewise, humans, with a loss in function of mutations in leptin and LepR, have reduced BP despite severe obesity. This interesting observation suggests that leptin is associated with changes in body weight leading to an alteration in BP in both human and mammalian species. Therefore, the effects of leptin and leptin receptors may be crucial for diabesity and may provide useful therapeutic targets for the prevention of obesity-associated CVDs and/or for the treatment of obesity-associated hypertension [27].

In a study in Sprague–Dawley, ZL, and ZF rats where Lin et al. [28] have examined the SAN to test for the role of explicit actions of leptin on ventricular repolarization and heart rate (HR) in which they found out that adipocytes and leptin receptors appeared in the myocardium and as a result, it was speculated that leptin could directly modify cardiac electrical properties such as HR and QT interval duration and that it can control cardiac electrical activity via the activation of β-adrenergic receptor. These findings indicate that leptin, at high doses (150–300 μg/kg), can stimulate a biphasic effect (decrease and then increase) in HR while at low doses (0.1–30 μg/kg), it can decrease resting HR. The leptin-induced resting HR inhibition was abolished by the presence of a leptin antagonist. Leptin also corrected QTc interval time and increased HR. However, leptin antagonist did not change these two parameters. It was also reported that in isolated ventricular myocytes, leptin (0.03–0.3 μg/ml) increased the action potential duration (APD) and the effects were reversible. Leptin has a direct effect in decreasing HR and in increasing QT interval via its own receptor and independent of β-adrenergic receptor stimulation. Moreover, high concentrations of leptin in the myocardium can produce prolonged QT interval, deep bradycardia and ventricular arrhythmias through inhibition of β-adrenergic receptor activity [28].

Obesity and Ion Channels of the Heart

Propagation of APs depends on voltage-gated ion channels and for this reason it is helpful to discuss these channels in a little more detail, as their dysfunction can result in detrimental CVDs. Voltage-gated ion channels are known as transmembrane proteins that regulate physiological activities such as cardiac muscle cell contraction and relaxation. Obese Wistar rats fed a high fat diet display significant changes in gene expression of ion channels in the left ventricle, which may predispose to arrhythmias. The heart of these obese rats displays significant upregulation of various transport proteins including Cav1.2, Kir2.1, HCN4, NCX1, SERCA2a, RYR2 mRNA and downregulation of ERG mRNA. A significant increase in protein expression of HCN4 was also demonstrated in these obese rats [29].

Axelsen et al. [30] demonstrated that the hearts of fructose-fat fed Sprague–Dawley rats (FFFRs) displayed QRS prolongation *in-vivo*, along with decreased conduction velocity (CV) *ex-vivo*. Similarly, the heart of FFFRs rats showed numerous premature ventricular contractions [30] under relaxed conditions *in-vivo*. Likewise, the isolated hearts from FFFRs displayed an increase in susceptibility to VF following cardiac ischemia–reperfusion. There were no changes in either gap junctional coupling, K^+ or Na^+ current density, and or differences in cell size or fibrosis. This indicates that the mechanism of conductance disturbances in the pre-diabetic heart varies from the disturbances seen in cardiac ischemia and HF [30]. In another study, a high-energy (HE) diet consisting of 33% kcal as fat was fed to a Sprague–Dawley rat model which is a model of obesity. These rats did not develop cardiac hypertrophy, either at the organ level or at the cellular level. In addition, kinetics and densities of the four major ionic membrane currents that control the cardiac AP were found to be similar in both obese and control rats. These include the L-type Ca^{2+} current, the sodium-calcium exchange current (NCX), the transient outward potassium current and the delayed rectifier potassium current. Even though leptin receptor expression was downregulated in the hearts of the HE obese group, no leptin modification was observed in the current densities except for the NCX current densities in the control group which were fed standard chow [31].

In another study, Huang et al. [32] found evidence of decreased protein expression of voltage-gated potassium channels in the obese heart, consequently leading to long QT interval. Their experiments in DIO C57BL/6 J wild-type mice have discovered that diminished protein and mRNA levels of the potassium channel Kv1.5 were triggered by a decline of the transcription factor cyclic AMP response element binding protein (CREB) in heart. CREB knock-down by siRNA diminished Kv1.5. CREB binds to the Kv1.5 promoter located in the heart and increases transcription of human and mouse Kv1.5 promoters. The decline in CREB protein during lipo-toxicity can be salvaged by inhibiting protein kinase D (PKD) [32].

Howarth et al. [33] have previously reported age-dependent changes in myocyte contractility in the hearts of young and ageing ZDF rats. These alterations in contractility are accompanied by changes in Ca^{2+} transport, including a decline in L-type Ca^{2+} current and subtle changes in expression of a variety of genes including the Ca^{2+} channel, membrane transporters, sarcoplasmic reticulum (SR) Ca^{2+} and cardiac muscle proteins [33]. Nevertheless, DIO causes cardiac upregulation of Ca^{2+} transport–related genes in the SR. While obesity also causes an increase in the levels of SR Ca^{2+} ATPase (SERCA2a), ryanodine receptor (RyR2) and phospholamban (PLB) mRNA, it, nevertheless, did not modify the mRNA levels of L-type Ca^{2+} channel (CACNA1C) and NCX in the hearts of Wistar rat [34]. However, there were no changes in the L-type Ca^{2+} channel protein levels and SERCA2a behavior (expression and activity) [35]. Surprisingly, a short period of high fat diet (16 weeks) in Psammomys obesus, a polygenic rodent model, developed for research into obesity, T2DM and cardiovascular diseases [36], resulted in severe changes of cardiac structure, activation of inflammatory and apoptotic mechanisms, and altered expression of calcium-cycling elements [37]. Aromolaran et al. [38] reviewed the literature for altered functional expression of ion channels in various animal models associated with obesity and the data are presented in Table 8.1.

Electrical Conduction System of the Heart and Obesity

Electrocardiography is the process of producing an electrocardiogram (ECG) which is a graph of voltage versus time of the electrical activity of the heart using electrodes placed on the skin. The first interval known as the P wave of the ECG signifies the depolarization of the heart atria. Electrical activity of the heart spreads from the atria to the AVN where there is a short delay prior to the rapid spread of electrical activity across the ventricular myocardium. The second major interval known as the QRS wave of the ECG reflects ventricular depolarization and the final interval called the T wave reflects ventricular repolarization [39–41].

Obesity in humans is associated with prolonged QT, an increase in the frequency of premature ventricular complexes and SCD. Pericardial fat is associated with

Table 8.1 Altered functional expression of ion channels in animal models (Adapted from reference [38])

Current	Gene	mRNA	Protein	Current density	Obese model	Cardiac tissue	QT_c	References
I_{Na}	SCNA5	NR	NR	↔	Rat (SD)	Ventricle	↑	Axelsen et at., 2015
		↑	NR	↑*	Rat [WR)	Ventricle	NR	Ashrafi et al., 2016
$I_{Ca,L}$	CACNA1c	NR	NR	↔	Rat [SDCD)	Ventricle	↑	Ricci et al., 2006
		↔	NR	NR	Rat [WR)	WH	NR	Lima-Lecpoldo et al., 2008
		↑	NR	↓	Rat (ZDF)	Ventricle	NR	Howarth et al., 2012
		NR	↔	NR	Rat (WR)	Ventricle	NR	Leopoldo et al., 2011
		NR	↓	↓	Rat [OZR)	Ventricle	↑	Lirr et al., 2012
		↓	NR	NR	Rat (WR)	WH	NR	Lima-Leopoldo et al., 2013
		↑	NR	↑*	Rat [WR)	Ventricle	NR	Ashrafi et al., 2016
		I↓	NR	NR	Gerbils	WH	NR	Sahraoui et al., 2016
		NR	NR	↓	Rabbit	Ventricle	NR	Luo et al., 2004
		NR	↓	↓	Mice (C57BL/6 J/db/db)	Ventricle	NR	Pereira et al., 2006
I_{to}	$K_V4.2/K_V4.3$	NR	NR	↔	Rat [SDCD)	Ventricle	NR	Ricci et al., 2006
		NR	↑	NR	Mice (ICR)	Atria	NR	Ricci et al., 2006
		NR	↔	NR	Mice (C57BL/6 J)	Ventricle	↑	Huang et al., 2013
		NR	NR	↔	Rat [SD)	Ventricle	↑	Axelsen et al., 2015
		↑	NR	↑*	Rat [WR)	Ventricle	NR	Ashrafi et al., 2016
	$K_V1.4$	↑	NR	↑*	Rat (WR)	Ventricle	NR	Ashrafi et al., 2016
I_{Kur}	$K_V1,5$	↓	↓	NR	Mice (C57BL/BJ)	Ventricle	↑	Huang et al. 2013
		↑	NR	↑*	Rat [WR)	Ventricle	NR	Ashrafi et al., 2016
		NR	↑	NR	Mice (ICR)	Atria	NR	Ricci et al., 2006

(continued)

Table 8.1 (continued)

Current	Gene	mRNA	Protein	Current density	Obese model	Cardiac tissue	QT_c	References
I_K	I_{Kr}:*ERG*	↓	NR	NR	Rat (WH)	Ventricle	NR	Ashrafi et al., 2016
	I_{Ks}:*KCNQ1*	↑	NR	NR	Rat (WR)	Ventricle	NR	Ashrafi et al., 2016
	I_K	NR	NR	↔	Rat [SDCD)	Ventricle	NR	Ricci et al., 2006
		NR	NR	↑	Guinea pig	Atria	NR	Aromolaran et al., 2016
I_{K1}	$K_{ir}2.1$	↑	NR	↑*	Rat [WR)	Ventricle	NR	Ashrafi et al., 2016

↑ Increased; ↓ decreased; ↔ no change; * predicted from computer simulations; *NR* no reported; *SD* Sprague Dawley; *WR* Wister Rats; *SDCD* Sprague Dawley Cesarean Dewed; *ZDE* Zuckec Diabetic fatty rat; *OZR* Obese Zucker Rat; *ICR* imprinting control region; QT_c QT interval corrected for heart rate; *WH* whole heart atrial conduction as measured by P wave indices. The relation is independent of associations with adipose depots outside of the pericardium (thoracic and visceral fat) which is consistent with the hypothesis that pericardial fat induced changes in atrial conduction represents an important mechanism by which pericardial fat predisposes towards atrial fibrillation (AF) [42]. It was found that Sprague–Dawley rats fed high-fat and high-fructose/cholesterol diets for 15 weeks were susceptible to AF. Chan et al. [43] found that IR can promote endogenous factors which can lead to abnormal calcium homeostasis and interstitial fibrosis in atria, modifies the CV and increases ectopic events in the atria, leading to AF. IR may also have a rapid effect on the expression of transforming growth factor (TGF-β1) in fibroblasts and myocytes, contributing to atrial fibrosis. These results indicated that upstream therapy, targeting Ca^{2+}/calmodulin-dependent protein kinase II (CaMKII), TGF-β1 and reducing oxidative stress, is a potentially effective strategy for avoiding AF caused by a diet high in sugar, fat, and cholesterol [43].

Lin et al. [44] discovered that defective calcium inactivation causes long QT interval in obese insulin-resistant Zucker rats (OZR) also known as ZF rats. The heart of these ZF rats, aged 16–17 weeks, demonstrated an increase in APD, prolonged QTc interval and an increase in cell capacitance. Additionally, the L-type Ca^{2+} current in cardiac myocytes showed defective inactivation, leading to increased Ca^{2+} influx. The results indicated that at around 16–17 weeks of age, ZF rats developed (1) altered electrophysiology as evidenced by the prolonged QTc interval, (2) cardiac hypertrophy, (3) increased APD in isolated ventricular myocytes, 4) defective inactivation of L-type Ba^{2+} and Ca^{2+} currents and 5) decreased protein expression of $Ca_{v1.2}$ and calmodulin [44].

The positive effects of aerobic exercise training (AET) in delaying and reversing cardiac dysfunction have been widely studied [45, 46]. Currently, limited research exists on ECG measurement for the evaluation of AET in the ZDF rat. A previous study demonstrated alterations in HRV, R wave amplitudes, QT and QTc intervals in the hearts of ZDF rats and AET was able to rectify R wave amplitude alterations alongside ECG correlates of left ventricular mass [45]. Another study demonstrated that miRNA played a role in cardiac revascularization of AET-induced obese animals. In ZF rats, miRNA-16 in the myocardium was enhanced with a simultaneous reduction in its target VEGF, which could be linked with cardiac microvascular rarefaction in obesity. AET not only normalized levels of these miRNAs, but also promoted angiogenesis, including improvement of angiogenic factors such as CD31 and VEGFR2. Taken together, these data indicate how AET may rebuild the balance between injury and repair of vascular systems in obese animals. Moreover, miRNAs may have great potential as therapeutic targets for treatment of various cardiovascular diseases [46].

Recent in vivo biotelemetry experiments have demonstrated that HR was significantly reduced in ZDF compared to ZF and ZL control rats in the absence of changes in body temperature and physical activity [47]. Age related changes in PQ intervals, QRS intervals, QT and QTc intervals between different groups were also reported. Surprisingly, the QTc interval was significantly increased in the ZF compared to the ZDF rats after correction for the HR.

Obesity and Sudden Cardiac Death

SCD is defined as "death due to cardiac causes occurring within 1 h of the onset of symptoms". The yearly incidence of SCD rises as age advances. CAD is the most common cause while other cardiomyopathies and genetic channelopathies are responsible for most of the remaining cases. In young adults (age < 35 years), the most common cause of SCD are arrhythmias, largely in the context of an apparently normal heart [48]. In general, obesity is associated with an increased risk of SCD especially in non-smoking middle-aged adults. In addition, SCD is aggravated by traditional cardiovascular risk factors such as BP, lipids, DM and CHD. In contrast, abdominal obesity seems to be directly associated with an increased risk of SCD [49].

Narayanan et al. [50] investigated the association between QRS fragmentation on the 12-lead ECG and SCD, in obese/overweight subjects. Fragmentation was defined as "the presence of RSR' patterns and/or notching of the R/S wave in at least 2 contiguous leads". It was reported that QRS fragmentation, particularly in the lateral territory, is a probable risk marker for SCD independent of the ejection fraction [50].

As mentioned previously, mitochondrial reactive oxygen species (mROS) is a primary source of oxidative stress in HF. The mROS can exert both acute events, such as electrical instability responsible for SCD, and chronic HF remodeling events. These are illustrated by either suppression or altered phosphorylation of metabolic, antioxidant and ion transport protein networks. *In-vivo* reduction of mROS reverses and prevents electrical instability, SCD and HF. *In-vivo* scavenging of mROS, not only prevented the progression of HF and eliminated SCD, but also reversed impaired contractility in failing hearts. Moreover, mROS is a key up-stream influencer driving both acute electrophysiological instability (leading to SCD) and chronic proteome remodeling (protein expression and phosphorylation) throughout cardiac decompensation. Comprehensive analysis of mROS-dependent HF suggests that ROS-dependent signal transduction pathways, perhaps in the mitogen-activated protein kinase family, can disrupt the coupling of cytosolic signals to the metabolic, antioxidant, and ion transport sub-proteomes as HF develops [51].

Conclusion

In summary, there are a number of risk factors which are associated with obesity leading to the development of T2DM. Figure 8.3 is a summary of the events leading to diabesity-induced SCD. Both obesity and diabetes are associated with an enhanced risk of developing CVDs, particularly HF and CHD which gives rise to the term "Diabesity". Both obesity and diabetes, via hyperglycemia, elicit the generation of RCS and ROS which induce structural changes in heart muscle leading to apoptosis, fibrosis, hypertrophy, mitochondrial dysfunction, derangement in cellular calcium

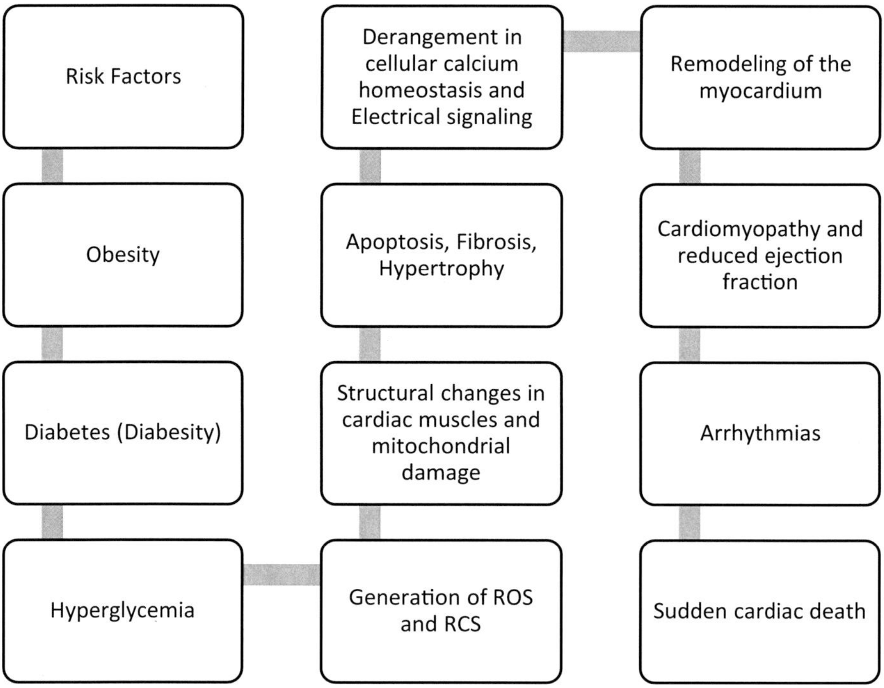

Fig. 8.3 Flow diagram illustrating the various events, including the risk factors, obesity and diabetes, in the development of cardiomyopathy and subsequently, in SCD following. ROS=Reactive oxygen species, RCS = Reactive carbonyl species

homeostasis and electrical signaling and subsequently, remodeling of the heart, all leading to reduced ejection fraction. If left untreated, the heart becomes weaker and unable to meet the constant demands of the body and as a result, arrhythmia develops leading to SCD. The cellular, subcellular and molecular changes that occur in the heart of obese individuals and give rise to electrical conduction defects are complex and need further investigation.

References

1. Reaven GM (1988) Role of insulin resistance in human disease. Diabetes 37:1595–1607. https://doi.org/10.2337/diab.37.12.1595
2. Fellmann L, Nascimento AR, Tibiriça E, Bousquet P (2013) Murine models for pharmacological studies of the metabolic syndrome. Pharmacol Ther 137:331–340. https://doi.org/10.1016/j.pharmthera.2012.11.004
3. World Health Organization (2018) Fact sheets. Obesity and overweight. In: WHO. https://www.who.int/en/news-room/fact-sheets/detail/obesity-and-overweight
4. NHLBI Obesity Education Initiative Expert Panel. National Institutes of Health. National Heart Lung and Blood Institute. The Practical Guide: Identification, Evaluation, and Treatment of

Overweight and Obesity in Adults. NIH Publ No. 00-4084
5. Krakauer NY, Krakauer JC (2012) A new body shape index predicts mortality hazard independently of body mass index. PLoS ONE 7:1–10. https://doi.org/10.1371/journal.pone.0039504
6. Swinburn BA, Kraak VI, Allender S, Atkins VJ, Baker PI, Bogard JR, Brinsden H, Calvillo A, De Schutter O, Devarajan R, Ezzati M, Friel S, Goenka S, Hammond RA, Hastings G et al (2019) The global syndemic of obesity, undernutrition, and climate change: The Lancet Commission report. Lancet 393:791–846. https://doi.org/10.1016/S0140-6736(18)32822-8
7. Gurevich-Panigrahi T, Panigrahi S, Wiechec E, Los M (2009) Obesity: pathophysiology and clinical management. Curr Med Chem 16:506–521. https://doi.org/10.2174/092986709787315568
8. Dobbs R, Swinburn B (2015) The global obesity threat. In: Projuct Synd. McKinsey Glob. Inst. https://www.mckinsey.com/mgi/overview/in-the-news/the-global-obesity-threat. Accessed 6 Jun 2019
9. Vasanji Z, Dhalla NS, Netticadan T (2004) Increased inhibition of SERCA2 by phospholamban in the type I diabetic heart. Mol Cell Biochem 261:245–249. https://doi.org/10.1023/B:MCBI.0000028762.97754.26
10. Howarth FC, Parekh K, Jayaprakash P, Inbaraj ES, Oz M, Dobrzynski H, Adrian TE (2017) Altered profile of mRNA expression in atrioventricular node of streptozotocin-induced diabetic rats. Mol Med Rep 16:3720–3730. https://doi.org/10.3892/mmr.2017.7038
11. de Ferranti S, Mozaffarian D (2008) The perfect storm: obesity, adipocyte dysfunction, and metabolic consequences. Clin Chem 54:945–955. https://doi.org/10.1373/clinchem.2007.100156
12. Weyer C, Funahashi T, Tanaka S, Hotta K, Matsuzawa Y, Pratley RE, Tataranni PA (2001) Hypoadiponectinemia in obesity and type 2 diabetes: close association with insulin resistance and hyperinsulinemia. J Clin Endocrinol Metab 86:1930–1935. https://doi.org/10.1210/jcem.86.5.7463
13. Saklayen MG (2018) The global epidemic of the metabolic syndrome. Curr Hypertens Rep 20:12–20. https://doi.org/10.1007/s11906-018-0812-z
14. Zimmet P, Alberti KGMM, Shaw J (2001) Global and societal implications of the diabetes epidemic. Nature 414:782–787. https://doi.org/10.1038/414782a
15. Chadt A, Scherneck S, Joost H-G, Al-Hasani H (2000) Molecular links between obesity and diabetes: "Diabesity." MDText.com, Inc.
16. Csige I, Ujvárosy D, Szabó Z, Lőrincz I, Paragh G, Harangi M, Somodi S (2018) The impact of obesity on the cardiovascular system. J Diabetes Res 2018:1–12. https://doi.org/10.1155/2018/3407306
17. Klabunde RE (2017) Cardiac electrophysiology: normal and ischemic ionic currents and the ECG. Adv Physiol Educ 41:29–37. https://doi.org/10.1152/advan.00105.2016
18. Crawford M, DiMarco J, Paulus W (2009) Cardiology. https://www.elsevier.com/books/cardiology/crawford/978-0-7234-3485-6. Accessed 10 Oct 2018
19. Turpie AGG, Bauer KA, Eriksson BI, Lassen MR (2002) Overweight and obesity as determinants of cardiovascular risk: the Framingham experience. Arch Intern Med 162:1867–1872. https://doi.org/10.1001/archinte.162.16.1867
20. International Diabetes Federation (2019) IDF diabetes Atlas, 9th edn
21. Takaya K, Ogawa Y, Isse N, Okazaki T, Satoh N, Masuzaki H, Mori K, Tamura N, Hosoda K, Nakao K (1996) Molecular cloning of rat leptin receptor isoform complementary DNAs-identification of a missense mutation in Zucker fatty (fa/fa) rats. Biochem Biophys Res Commun 225:75–83. https://doi.org/10.1006/bbrc.1996.1133
22. Yokoi N, Hoshino M, Hidaka S, Yoshida E, Beppu M, Hoshikawa R, Sudo K, Kawada A, Takagi S, Seino S (2013) A novel rat model of type 2 diabetes: the Zucker Fatty Diabetes Mellitus ZFDM rat. J Diabetes Res 2013:103731–103740. https://doi.org/10.1155/2013/103731
23. Wang B, Chandrasekera PC, Pippin JJ (2014) Leptin- and leptin receptor-deficient rodent models: relevance for human type 2 diabetes. Curr Diabetes Rev 10:131–145. https://doi.org/10.2174/1573399810666140508121012

24. Bray GA (1977) The Zucker-fatty rat: a review. Fed Proc 36:148–153
25. Lirk P, Flatz M, Haller I, Hausott B, Blumenthal S, Stevens MF, Suzuki S, Klimaschewski L, Gerner P (2012) In Zucker diabetic fatty rats, subclinical diabetic neuropathy increases in vivo lidocaine block duration but not in vitro neurotoxicity. Reg Anesth Pain Med 37:601–606. https://doi.org/10.1097/AAP.0b013e3182664afb
26. Holloway GP, Snook LA, Harris RJ, Glatz JFCC, Luiken JJFPFP, Bonen A (2011) In obese Zucker rats, lipids accumulate in the heart despite normal mitochondrial content, morphology and long-chain fatty acid oxidation. J Physiol 589:169–180. https://doi.org/10.1113/jphysiol.2010.198663
27. Simonds SE, Pryor JT, Ravussin E, Greenway FL, Dileone R, Allen AM, Bassi J, Elmquist JK, Keogh JM, Henning E, Myers MG, Licinio J, Brown RD, Enriori PJ, O'Rahilly S et al (2014) Leptin mediates the increase in blood pressure associated with obesity. Cell 159:1404–1416. https://doi.org/10.1016/j.cell.2014.10.058
28. Lin Y-CC, Hull R, Huang J, Martin KH, Davis M, Hileman S, Yu H-GG, Martin KH, Hull R, Davis M, Yu H-GG (2015) Leptin decreases heart rate associated with increased ventricular repolarization via its receptor. Am J Physiol Circ Physiol 309:H1731–H1739. https://doi.org/10.1152/ajpheart.00623.2015
29. Ashrafi R, Yon M, Pickavance L, Yanni Gerges J, Davis G, Wilding J, Jian K, Zhang H, Hart G, Boyett M (2016) Altered left ventricular ion channel transcriptome in a high-fat-fed rat model of obesity: insight into obesity-induced arrhythmogenesis. J Obes 2016:1–12. https://doi.org/10.1155/2016/7127898
30. Axelsen LN, Calloe K, Braunstein TH, Riemann M, Hofgaard JP, Liang B, Jensen CF, Olsen KB, Bartels ED, Baandrup U, Jespersen T, Nielsen LB, Holstein-Rathlou NH, Nielsen MS (2015) Diet-induced pre-diabetes slows cardiac conductance and promotes arrhythmogenesis. Cardiovasc Diabetol 14:87–101. https://doi.org/10.1186/s12933-015-0246-8
31. Ricci E, Smallwood S, Chouabe C, Mertani HC, Raccurt M, Morel G, Bonvallet R (2006) Electrophysiological characterization of left ventricular myocytes from obese Srague-Dwley rat. Obesity 14:778–786. https://doi.org/10.1038/oby.2006.90
32. Huang H, Amin V, Gurin M, Wan E, Thorp E, Homma S, Morrow JP (2013) Diet-induced obesity causes long QT and reduces transcription of voltage-gated potassium channels. J Mol Cell Cardiol 59:151–158. https://doi.org/10.1016/j.yjmcc.2013.03.007
33. Howarth FC (2012) Ventricular myocyte contraction, intracellular calcium and expression of genes encoding cardiac muscle proteins in young and aging Zucker diabetic fatty rat heart reviewed. Hamdan Med J 5:165–172. https://doi.org/10.7707/hmj.v5i2.140
34. Lima-Leopoldo AP, Sugizaki MM, Leopoldo AS, Carvalho RF, Nogueira CR, Nascimento AF, Martinez PF, Luvizotto RAM, Padovani CR, Cicogna AC (2008) Obesity induces upregulation of genes involved in myocardial Ca2+ handling. Brazilian J Med Biol Res 41:615–620. https://doi.org/10.1590/S0100-879X2008000700011
35. Leopoldo AS, Lima-Leopoldo AP, Sugizaki MM, do Nascimento AF, de Campos DHS, de Azevedo Melo Luvizotto R, Castardeli E, Alves CAB, Brum PC, Cicogna AC (2011) Involvement of L-type calcium channel and serca2a in myocardial dysfunction induced by obesity. J Cell Physiol 226:2934–2942. https://doi.org/10.1002/jcp.22643
36. Chaudhary R, Walder KR, Hagemeyer CE, Kanwar JR (2018) Psammomys obesus: a natural diet-controlled model for diabetes and cardiovascular diseases. Curr Atheroscler Rep 20:46–56. https://doi.org/10.1007/s11883-018-0746-6
37. Sahraoui A, Dewachter C, De Medina G, Naeije R, Bouguerra SA, Dewachter L (2016) Myocardial structural and biological anomalies induced by high fat diet in Psammomys obesus gerbils. PLoS ONE 11:1–16. https://doi.org/10.1371/journal.pone.0148117
38. Aromolaran AS, Boutjdir M (2017) Cardiac ion channel regulation in obesity and the metabolic syndrome: relevance to long QT syndrome and atrial fibrillation. Front Physiol 8:431–448. https://doi.org/10.3389/fphys.2017.00431
39. Boyett MR (2009) "And the beat goes on" the cardiac conduction system: The wiring system of the heart. Exp Physiol 94:1035–1049. https://doi.org/10.1113/expphysiol.2009.046920

40. Kashou AH, Kashou HE (2018) Physiology, sinoatrial node (SA Node). In: StatPearls. http://www.ncbi.nlm.nih.gov/pubmed/29083608. Accessed 10 Nov 2018
41. Pinnell J, Turner S, Howell S (2007) Cardiac muscle physiology. Contin Educ Anaesth Crit Care Pain 7:85–88. https://doi.org/10.1093/bjaceaccp/mkm013
42. Friedman DJ, Wang N, Meigs JB, Hoffmann U, Massaro JM, Fox CS, Magnani JW (2014) Pericardial fat is associated with atrial conduction: the Framingham heart study. J Am Heart Assoc 3:1–10. https://doi.org/10.1161/JAHA.113.000477
43. Chan Y-H, Chang G-J, Lai Y-J, Chen W-J, Chang S-H, Hung L-M, Kuo C-T, Yeh Y-H (2019) Atrial fibrillation and its arrhythmogenesis associated with insulin resistance. Cardiovasc Diabetol 18:125–139. https://doi.org/10.1186/s12933-019-0928-8
44. Lin YC, Huang J, Kan H, Castranova V, Frisbee JC, Yu HG (2012) Defective calcium inactivation causes long QT in obese insulin-resistant rat. Am J Physiol Hear Circ Physiol 302:1013–1022. https://doi.org/10.1152/ajpheart.00837.2011
45. VanHoose L, Sawers Y, Loganathan R, Vacek JL, Stehno-Bittel L, Novikova L, Al-Jarrah M, Smirnova IV (2010) Electrocardiographic changes with the onset of diabetes and the impact of aerobic exercise training in the Zucker Diabetic Fatty (ZDF) rat. Cardiovasc Diabetol 9:56–66. https://doi.org/10.1186/1475-2840-9-56
46. Fernandes T, Casaes L, Soci Ú, Silveira A, Gomes J, Barretti D, Roque F, Oliveira E (2018) Exercise training restores the cardiac microrna-16 levels preventing microvascular rarefaction in obese Zucker rats. Obes Facts 11:15–24. https://doi.org/10.1159/000454835
47. Sultan A, Jacobson M, Adeghate E, Oulhaj A, Shafiullah M, Qureshi A, Howarth FC (2021) Effects of obesity and diabesity on heart rhythm in the Zucker rat. Clin Exp Pharmacol Physiol 48:735–747. https://doi.org/10.1111/1440-1681.13473
48. Katritsis DG, Gersh BJ, Camm AJ (2016) A clinical perspective on sudden cardiac death. Arrhythm Electrophysiol Rev 5:177–182
49. Adabag S, Huxley RR, Lopez FL, Chen LY, Sotoodehnia N, Siscovick D, Deo R, Konety S, Alonso A, Folsom AR (2015) Obesity related risk of sudden cardiac death in the atherosclerosis risk in communities study. Heart 101:215–221. https://doi.org/10.1136/heartjnl-2014-306238
50. Narayanan K, Zhang L, Kim C, Uy-Evanado A, Teodorescu C, Reinier K, Zheng Z, Gunson K, Jui J, Chugh SS (2015) QRS fragmentation and sudden cardiac death in the obese and overweight. J Am Heart Assoc 4:1–8. https://doi.org/10.1161/JAHA.114.001654
51. Dey S, DeMazumder D, Sidor A, Brian Foster D, O'Rourke B (2018) Mitochondrial ROS drive sudden cardiac death and chronic proteome remodeling in heart failure. Circ Res 123:356–371. https://doi.org/10.1161/CIRCRESAHA.118.312708

Chapter 9
Dietary Advanced Glycation End Products as Mediators of Obesity: Cellular and Molecular Mechanisms of Action

Chinedum Ogbonnaya Eleazu, Victor Udo Nna, Joseph Bagi Suleiman, and Mahaneem Mohamed

Abstract Obesity, a disorder of body weight regulatory systems that is characterized by the accumulation of excess body fat is increasingly becoming a global pandemic. Despite several approaches that have been applied to mitigate obesity, they have not been able to totally reverse the obesity and its mediated complication. Advanced glycation end products (AGEs) refer to a group of prooxidant heterogenous compounds whose formation results from nonenzymatic reactions between reactive sugars and proteins, lipids and nucleic acids. While supporting evidence has suggested that AGEs may have contributory roles in the pathogenesis of obesity, there are indications that increased consumption of dietary AGEs increases the circulating AGEs levels and their deposition in the tissues including the adipose tissue, increasing the risk of development of obesity and its comorbidities. Identification of the underlying mechanism may provide an important strategy for novel therapeutic approaches against obesity. This chapter therefore provided novel insights into the role of dietary AGEs in the pathogenesis of obesity and the purported mechanisms of action.

C. O. Eleazu · M. Mohamed (✉)
Department of Physiology, School of Medical Sciences, Universiti Sains Malaysia, Kubang Kerian, Kelantan, Malaysia
e-mail: mahaneem@usm.my

C. O. Eleazu
Department of Chemistry, Biochemistry and Molecular Biology, Alex Ekwueme Federal University, Ebonyi State, Ndufu-Alike, Ikwo, Nigeria

V. U. Nna
Department of Physiology, College of Medical Sciences, University of Calabar, Cross River State, P.M.B 1115 Calabar, Nigeria

J. B. Suleiman
Department of Science Laboratory Technology, Akanu Ibiam Federal Polytechnic, Ebonyi State, Unwana, Nigeria

M. Mohamed
Unit of Integrative Medicine, School of Medical Sciences, Universiti Sains Malaysia, Kubang Kerian, Kelantan, Malaysia

© Springer Nature Switzerland AG 2021

P. S. Tappia et al. (eds.), *Cellular and Biochemical Mechanisms of Obesity*, Advances in Biochemistry in Health and Disease 23,
https://doi.org/10.1007/978-3-030-84763-0_9

Keywords Obesity · Advanced glycation end products · Metabolism · PI3K/Akt signaling · Food intake

Abbreviations

Akt	Protein kinase B
ALI	Arginine-lysine imidazole
BMI	Body mass index
IRS-1	Insulin receptor substrate 1
IRS-2	Insulin receptor substrate 2
PI3K	Phosphatidylinositol 3-kinase
RAGE	Receptor for Advanced Glycation End products
ROS	Reactive Oxygen Species
sRAGE	Soluble receptors for advanced glycation end products

Introduction

Obesity is a disorder of body weight regulatory systems that is characterized by the accumulation of excess body fat [1]. According to the World Health Organization [2], obesity is an abnormal or excessive fat accumulation that presents a risk to human health. The development of obesity involves both adipocyte hypertrophy and hyperplasia. Obesity in childhood has been reported to involve adipocyte hyperplasia and hypertrophy while adipocyte hypertrophy arising from imbalanced energy intake was generally suggested to be responsible for most adult-onset obesity. However, adipocyte hyperplasia has also been suggested to contribute to the development of adult-onset obesity, especially morbidly obese patients [3].

Advanced glycation end products (AGEs) refer to a group of prooxidant heterogeneous compounds whose formation results from nonenzymatic reactions between reactive sugars and proteins, lipids and nucleic acids in a reaction that is otherwise known as 'the Maillard reaction'. Recent studies have revealed that consumption of foods that are rich in these AGEs could play fundamental roles in the pathogenesis of chronic diseases including obesity [4]. Given the inability of the current therapeutic approaches for obesity to totally reverse it [5], identification of the underlying mechanism involved in the contribution of AGEs to obesity may provide an important strategy for novel therapeutic approaches against obesity. In line with the above, this chapter discussed the role of dietary AGEs in the pathogenesis of obesity. The mechanisms of action deriving from preclinical and clinical studies that were carried out were also reported.

Global Prevalence of Overweight and Obesity

In current report, the global prevalence of overweight and obesity doubled since 1980 to such an extent that about one third of the world's population is now classified as being overweight or obese [6]. Additionally, studies that were carried out revealed that a total of 1.9 billion and 609 million adults were estimated to be overweight and obese in 2015, respectively, representing approximately 39% of the world's population [6]. The prevalence of obesity was generally higher in women than in men in all age groups, with sex differences being maximal between 50 and 65 years old [6]. Obesity adversely affects nearly all the physiological functions of the body and has become a significant public health threat [6].

Measurement of Obesity

The body mass index (BMI) usually expressed as weight (kg) divided by the square of height $(\text{meters})^2$ is the most acceptable index of measurement of obesity. This index measures the relative weight, adjusted for height which makes for comparisons both within and between populations [1]. Based on BMI cut off point, individuals with BMI values of 18.5 and 24.9 are regarded to have normal weight; individuals with a BMI value between 25 and 29.9 are considered overweight, those with a BMI value equal to or greater than 30 are classified as obese, while BMI values above 40 are regarded as extremely/morbidly obese. Measurement of the waist circumference using a tape has also been found to be diagnostic of obesity as it reveals the amount of fat (visceral fat) in the central abdominal region of the body [1]. This approach has also been reported to reduce the risk of developing cardiovascular diseases, independent of BMI [1, 5, 7].

Advanced Glycation End Products—Overview

AGEs are a group of heterogeneous compounds whose formation results from nonenzymatic reactions between reactive sugars and proteins, lipids and nucleic acids in a reaction that is otherwise known as Maillard reaction as earlier stated [4, 8]. In this reaction, the carbonyl group of a reducing sugar reacts with the amino group of a protein, lipid or nucleic acid, generating Schiff bases which rearrange to Amadori products. Since the Amadori products are relatively unstable, further reactions occur, which eventually lead to the formation of irreversible AGEs [9]. Examples of AGEs include: glycated hemoglobin (HbA1c), Nε-carboxymethyllysine (CML), Nε-carboxyethyl-lysine (CEL), ALI, pentosidine, methylglyoxal, pyrraline and imidazolone, N-fructosyl-lysine, Alkyl formyl glycosyl pyrroles, and others [10, 11].

With respect to their chemical structure, AGEs can be classified as: non-fluorescent crosslinked (e.g. glucosepane, imidazolium dilysine), fluorescent crosslinked (e.g. crossline, pentosidine) and non-crosslinked molecules (e.g. pyralline, CML) [11]. These AGEs exert their effects through their receptors (RAGE) (described in detail in our review, [5] and book chapter [7]).

Sources of AGEs

Based on the differences in their origins, AGEs are categorised into two major classes namely: biologically derived AGEs (biological-AGEs) and environmentally derived AGEs. Biologically derived AGEs are produced endogenously (in vivo) during normal body metabolism, aging, prolonged oxidative stress or due to hyperglycemia [11]. On the other hand, environmentally derived AGEs are obtained exogenously (in vitro) while cigarette smoke and diet (food) are two major sources of environmentally derived AGEs.

Cigarette smoke contains glycation products that are highly reactive (AGEs) which may contribute to the increased AGEs accumulation in serum and tissue as observed in cigarette smokers. Cigarette smoke derived AGEs have been shown to increase the risk of developing cancers and cardiovascular diseases in former smokers [10]. Dietary AGEs constitute a class of pro-oxidant foods and their formation depends directly on the protein, lipid and carbohydrate content of the food as well as on the temperature and conditions of cooking, especially time used for cooking and moisture [12].

Heat treatment of food results in the generation of AGEs and other Maillard reaction products that improve the aroma and flavour of food products. Therefore, dietary AGEs are commonly found in processed foods with their levels being increased by food processing at high temperatures [13, 14]. In fact, foods may contain up to 200 times the initial AGEs content after cooking depending on the cooking method [10]. Food processing and cooking techniques that utilise dry heat (frying, roasting, baking, grilling, barbecuing) result in greater AGEs formation compared with techniques that use lower temperatures for longer periods of time with higher water content, such as boiling or steaming [4, 15, 16].

Higher pH levels have also been reported to increase the formation of AGEs, as the alkaline conditions promote the amino groups to be in their basic deprotonated form, increasing their reactivity. Pre-treating foods with acidic solutions before cooking (to lower their pH) such as marinating with vinegar or lemon has been reported to decrease the formation of dietary AGEs in food products [15]. Animal-derived foods cooked at high temperature, for a prolonged time and under dry conditions have the highest content of AGEs [12]. Other examples of such foods that contribute to high amounts of dietary AGEs include processed cereal products such as biscuits, bakery products and extruded breakfast cereals [15].

In contrast, dairy products, fruits, and vegetables have lower AGEs content. Higher humidity, lower temperatures, and low pH also make minor contributions to the

Table 9.1 Factors that affect the levels of AGEs in foods

Type of food		Method of cooking	
High AGEs	Low AGEs	High AGEs	Low AGEs
Dairy eg. red meat	Boiled grains eg. rice and oatmeal	High temperature	Low temperature
Fried foods	Vegetables	Barbecuing	Steaming
Cheese	Fruits	Grilling	Poaching
Processed foods	Dairy soups	Baking	Stewing
Bakery products eg. biscuits		Frying	Braising
Breakfast cereals		Searing	Boiling
Butter		Broiling	Marinating with vinegar before cooking
Magarine		Toasting	Marinating with lemon before cooking
Mayonnaise		Roasting	Poaching
Oils		Sauteing	
Nuts		Prolonged cooking time	Lesser cooking time
		High pH	Low pH
		Low moisture	High moisture

Sources Uribarri et al. [18]; Tessier and Birlouez-Aragon [19]; Barbosa et al. [20]; Ribeiro et al. [4]; Snelson and Coughlan [15]. AGEs: advanced glycation end products

formation of AGEs [17]. Table 9.1 shows the differences between food groups and cooking methods that contribute to low or high levels of AGEs [4, 18–21].

Digestion, Absorption, Distribution and Excretion of Dietary AGEs

Digestion

Protein and peptide bound AGEs are the major AGEs in the diet [8]. It has been suggested that the structural features of AGEs may have a significant effect in the digestion of glycated proteins and peptides. For instance, reports have shown that AGEs with non-crosslinked structures result in a considerable decrease in glycated protein digestibility. In addition, the digestibility of CML-casein (60% modification of target lysine) was reported to be significantly lower than that of native casein [8].

Absorption

AGEs can exist in either free form as a single amino acid or a free low-molecular-weight peptide, or bound to proteins, forming high-molecular-weight compounds. Dietary AGEs are reportedly absorbed in circulation mostly in the form of free AGEs eg. CML (by simple diffusion) and peptide bound AGEs following gastrointestinal digestion [8] and thereafter, they can get deposited in various tissues [17].

Approximately 10–30% of dietary AGEs are absorbed into the systemic circulation [8], one third of which is excreted by the kidneys while two-thirds linger in the body, contributing to the body's AGEs pool and which binds to several tissues [17, 22]. Therefore, decreased intake of dietary AGEs has been associated with approximately 40% decreases in the levels of AGEs in the body [8]. On the other hand, intestinal absorption of pyrraline is considered to happen mostly as a dipeptide rather than a free amino acid, with the dipeptide form of pyrraline absorbed across the intestinal epithelium using peptide transporter 1 [15].

Distribution of Dietary AGEs

After crossing the epithelial cells of the digestive tract, dietary AGEs enter into circulation and merge with the biological AGEs [8]. Dietary AGEs are distributed in most tissues following their absorption and supporting evidence has shown that intake of dietary AGEs can cause AGEs accumulation throughout the body, including the tissues (gastrointestinal tract, liver, kidneys, lungs, adipose tissue, heart and spleen), serum, urine and faeces [8]. The distribution of dietary AGEs in the body is driven by their higher affinity for some tissues on the basis of covalent or noncovalent binding interactions [8].

Metabolism

It has been reported that AGEs may not act as substrates for detoxification by phase I and phase II enzymes of xenobiotic metabolism [8, 23]. According to the hydrophile-lipophile balance, hydrophilic AGEs do not act as substrates for phase I enzymes in the fatty membranes of the endoplasmic reticulum and due to the glycation of side groups, most AGEs have insufficient typical side groups for phase II coupling reactions, except acidic groups for esterification [8].

Excretion

The renal excretion of dietary AGEs has been estimated to be approximately 30% of the absorbed amount in healthy adults, but less than 5% in patients with renal disease [8]. Between 20 and 50% of ingested CML are reportedly excreted in the faeces, suggesting that there is a proportion of ingested AGEs that are not absorbed and not defecated, and may be metabolised intraluminally by the microbiome [15].

Dietary AGEs and Human Health

Consumption of dietary AGEs have been reported to contribute to oxidative stress and inflammation in animal models, although results were less consistent in human studies [15, 24]. Excessive dietary AGEs consumption in mice was implicated in the development of hepatic inflammation in the absence of steatosis. In a rat model of non-alcoholic fatty liver disease, high dietary AGEs were reported to exacerbate liver injury, inflammation and liver fibrosis while chronic dietary AGEs intake was suggested to cause cognitive decline and Alzheimer's disease [15]. Furthermore, regular consumption of dietary AGEs in healthy individuals was reported to promote the accumulation of CML in some organs such as kidneys, heart, liver, tendons and lungs [8]. On the contrary, consumption of low dietary AGEs was reported to improve markers of inflammation and oxidative stress in haemodialysis patients and those with stage 3 chronic kidney disease [15].

Dietary AGEs and Obesity

Supporting evidence from studies that were carried out have implicated dietary AGEs in the pathogenesis of obesity [17, 25, 26]. In fact, it has been suggested that increased consumption of processed foods in the last 50 years may have favoured increased consumption of dietary AGEs, leading to increased development of obesity, and its mediated complications in the world's population [4, 27].

In a study that was conducted in rats, high-AGEs diet reportedly increased serum levels of AGEs and upregulated ovarian RAGE. These effects were reversed following consumption of a low-AGE diet, administration of an AGE blocker and the use of orlistat (obesity suppressing drug) [17, 28, 29]. To examine the relationship between methylglyoxal accumulation and the development of obesity, Jia et al. [3] compared methylglyoxal accumulation in the white fat tissues from Zucker lean and obese rats. The authors found significantly increased methylglyoxal accumulation in the kidney, fat tissue and serum of the obese rats at age of 16 weeks relative to the lean rats.

Earlier studies by Koyama et al. [30] showed that plasma levels of sRAGE were inversely correlated with BMI. Additionally, sRAGE levels were shown to be significantly reduced in obese women compared to their normal weight counterparts. Furthermore, short-term weight loss programs in the obese women were reported to significantly increase their plasma sRAGE levels [31]. Since sRAGE (acts as a decoy receptor for AGEs that decreases their circulating levels) was inversely associated with obesity [26], these studies are further pointers to the role of AGEs in the pathogenesis of obesity.

Later experimental studies that were conducted by Sayej et al. [10] showed that mice that were fed high-AGEs high-fat diet accumulated more fat than those fed low-AGEs high-fat diet and low-AGEs low-fat diet, respectively. This was evidenced by their larger weight gain and larger epididymal fat pad, indicating the role of dietary AGEs in the pathogenesis of obesity [10]. The authors further reported that the mice that were fed high-AGEs low-fat diet had similar outcomes to those that were fed high-AGEs high-fat diet in terms of weight gain and epididymal fat pad but milder development of hepatosteatosis compared to the group fed high-AGEs high-fat diet [10], further affirming the role of dietary AGEs in the pathogenesis of obesity.

Studies conducted by Uribarri et al. [12], showed a positive relationship between visceral fat and elevated serum concentration of AGEs, suggesting a contributory role of exogenous AGEs in the development of obesity and metabolic syndrome. This statement was further buttressed by the decreased circulating and urinary AGE markers and the improved anthropometric indices that were reported in overweight and obese individuals placed on low-AGEs diets [4, 32]. Additionally, a low-AGEs diet was reported to decrease circulating and urinary AGEs markers, and was further associated with improved anthropometric, glycemic, and cardiometabolic indices, as well as overall decrease in body weight compared with a high-AGEs diet [4].

Since these studies revealed that consumption of foods rich in AGEs play a fundamental role in the pathogenesis of obesity, reduced intake of AGEs is suggested to be beneficial in the management of obesity, independent of consumption of standard energy-restricted diets [4].

Evidence from Pre-clinical and Clinical Studies on the Mechanism of AGEs Mediated Pathogenesis of Obesity

AGEs act as appetite stimulating agents that simultaneously stimulate excessive food intake and inflammation, increasing the risk of obesity [4, 27]. The report by these authors was affirmed by the study that was carried out by Sayej et al. [10] in mice which found increased leptin levels in high-AGEs high-fat diet fed mice compared to other groups, suggesting leptin desensitization, leading to excessive food intake as a potential mechanism of obesity development following feeding of diets high in AGEs.

Leptin is an adipose tissue derived protein hormone that plays important roles in regulating food intake and energy expenditure, including appetite and metabolism [10, 33]. Leptin levels control food intake and energy expenditure by acting on receptors in the mediobasal hypothalamus [34]. Therefore, desensitization of this leptin hormone will lead to its increased plasma levels due to decreased negative feedback control mechanism that regulates adipocyte leptin production [35] leading to increased food intake, decreased energy expenditure and obesity.

The PI3K/Akt signaling pathway is another pathway that has been implicated in dietary AGEs mediated pathogenesis of obesity. The pathway is required for normal body metabolism, regulation of cell proliferation, promotion of lipid biosynthesis, inhibition of lipolysis, etc. [36]. PI3Ks refer to a family of lipid kinases composed of regulatory and catalytic subunits that are activated by phosphorylating the 3-hydroxyl group of the inositol ring of phosphatidylinositol lipids in the plasma membrane [37, 38]. PI3K is activated by various stimuli, including growth factors (epidermal growth factor, platelet-derived growth factor and insulin-like growth factor), cytokines and hormones [38].

A number of polymerization targets receive signals generated by the PI3K downstream cascade. However, the most important mediator is Akt [38]. Akt refers to a class of serine/threonine protein kinase B which controls several cellular functions such as modulation of growth, survival, proliferation and metabolism. Akt has 3 isoforms namely Akt 1 (PKBα), Akt2 (PKBβ) and Akt3 (PKBγ). Akt is phosphorylated at two sites: the catalytic domain by phosphatide-dependent kinase 1 and the carboxy domain by the mammalian target of rapamycin complex2 [39]. Akt 1 is the isoform that contributes to cell proliferation and cell growth, Akt2 is involved in the regulation of glucose transport and uptake by fat and muscle cells while Akt3 is critical in neuronal development [3, 39]. The different functions of these Akt isoforms have been suggested to contribute to Akt signaling diversity [39].

A mutation in the catalytic domain of Akt2 causes severe insulin resistance and T2DM in humans. Additionally, targeted deletion of Akt2 in mice but not Akt1 or Akt3, reportedly led to insulin insensitivity, hyperglycemia, hyperinsulinemia, glucose intolerance, and impaired glucose uptake by the muscle and adipose tissue [39], further affirming the role of Akt2 in glucose homeostasis and glucose uptake by the muscle and adipose tissue.

Activation of insulin receptor causes phosphorylation of tyrosine residues in insulin receptor substrate (IRS)-1 and IRS-2, leading to the activation of the PI3K complex. Activation of PI3K further activates Akt and phosphoinositide-dependent kinase-1, which then convey most of the intracellular effects induced by insulin [36, 37], one of which is the regulation of food intake at the hypothalamus through leptin regulation. Therefore, the PI3K/Akt signaling pathway integrates the effect of insulin and leptin in the regulation of food intake and studies have shown that this pathway is required for the acute effects of leptin, such as leptin mediated reduction in food intake [37].

There are indications that interaction between AGEs and their receptors in the adipose tissue, trigger insulin insensitivity in the adipocytes. These could be evidenced in these two studies:

(a) In an in vitro study that was conducted on 3T3-L1 adipocytes, the presence of AGEs inhibited glucose uptake in the presence or absence of insulin. Additionally, it increased the generation of ROS and the expression of monocyte-1-chemoattractive protein [40].
(b) The study by Monden et al. [41] showed that increased RAGE expression increases adipocyte hypertrophy, suppression of glucose transporter type 4 (GLUT-4), attenuation of insulin-stimulated glucose uptake, and reduction of IRS-1 phosphorylation.

These studies and more, reveal that AGEs/RAGE interactions in the adipocytes, inhibit glucose uptake through increased generation of ROS, cytokines, and other inflammatory mediators, decreased phosphorylation of IRS-1, thereby inhibiting the PI3K-Akt signaling pathway [4]. These lead to increased food intake, adipocyte hypertrophy and obesity [26]. Since Akt2 is the Akt isoform that is involved in glucose uptake in the muscle and adipose tissue, the implication is that AGEs/RAGE interactions in the adipocytes inhibit PI3K/Akt2 signaling, thereby leading to adipocyte hypertrophy. Although PI3K is required in the activation of Akt, Akt activation has also been reported to be mediated by a P13K independent mechanism [36].

In the study by Jia et al. [3] that reported the role of methylglyoxal in adipocyte proliferation using Zucker rat models, the authors found that methylglyoxal accumulation in the white fat tissue of obese Zucker rats aged 16 weeks old (relative to lean Zucker rats), stimulated the phosphorylation of Akt1 and its targets including P21 and P27, leading to adipocyte proliferation and adipogenesis, an action that was reversed by the administration of the AGEs breaker, alagebrium and Akt inhibitor.

Since Akt1 is the Akt isoform that contributes to cell proliferation and cell growth, the authors studied the phosphorylation of Akt1 isoforms from the Zucker lean and obese rats at the age of 16 weeks and found significantly elevated levels of phosphorylated Akt1 compared to Zucker lean rats. The authors further found that methylglyoxal promoted faster cell cycle progression in 3T3-LI cells by increasing the cell number in the S and G1 phases of the cell cycle. Their study therefore suggested that phosphorylation of Akt1 by methylglyoxal promotes phosphorylation of the downstream targets (P21 and P27), enhancing cell cycle, leading to adipogenesis. Since the increased Akt1 phosphorylation associated with methylglyoxal accumulation in obese rats was found in the obese rats aged 16 weeks old, these authors suggested that methylglyoxal stimulated adipogenesis by the up-regulation of Akt signaling pathway could contribute to the development of adult-onset morbid obesity [3].

The findings from these studies as reported above, therefore reveal that consumption of dietary AGEs, increases the circulating AGEs levels and the deposition of AGEs in the tissues (including the adipose tissue). AGEs/RAGE interaction in the adipocytes, inhibits glucose uptake through increased generation of ROS, cytokines and inflammatory mediators, decreases phosphorylation of IRS-1, inhibits PI3K-Akt2 signaling, leading to increased food intake, adipocyte hypertrophy and obesity. Additionally, increased deposition of AGEs in the adipocytes can stimulate phosphorylation of Akt1 and its targets including p21 and p27, leading to adipocyte

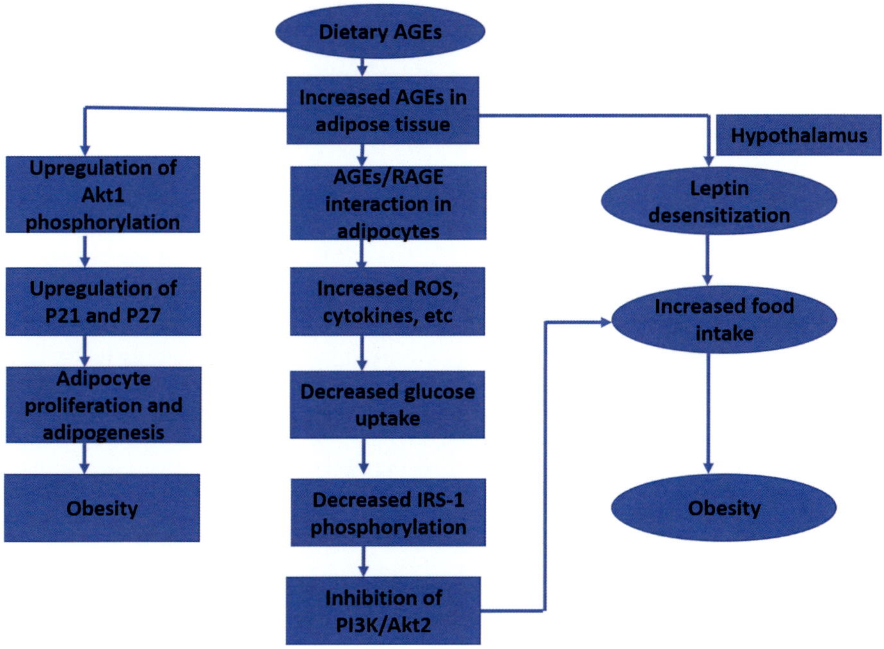

Fig. 9.1 The summary of the mechanisms that explain the role of dietary AGEs in the pathogenesis of obesity. Increased consumption of dietary AGEs leads to increased deposition of AGEs in the tissues including the adipose tissue. This ultimately leads to desensitization of leptin in the hypothalamus, leading to increased food intake and obesity. Increased AGEs in the adipose tissue also increases AGEs/RAGE interaction in the adipocytes leading to increased generation of ROS, cytokines and other inflammatory mediators which causes decreased glucose uptake. Then, this can cause decreased phosphorylation of IRS-1 and inhibition of PI3K-Akt2 signaling, leading to increased food intake and obesity. Additionally, the increased deposition of AGEs in the adipocytes can stimulate phosphorylation of Akt1 and its targets including p21 and p27, leading to adipocyte proliferation and adipogenesis, resulting to obesity

proliferation and adipogenesis. The summary of the mechanisms that explain the role of dietary AGEs in the pathogenesis of obesity is shown in Fig. 9.1.

Conclusions

Preclinical and clinical studies that were conducted have supported the contributory role of dietary AGEs in the pathogenesis of obesity. Three key mechanisms such as: AGEs mediated leptin dysregulation, AGEs/RAGE mediated inhibition of PI3K/Akt2 signaling and AGEs mediated upregulation of Akt1 have been implicated in this pathogenesis. Therefore, decreased consumption of foods that are high in dietary AGEs could be helpful in the prevention of obesity.

References

1. Ferrier DR (2014) Lippincotts biochemistry, 6th edn, pp 634–635
2. World Health Organization (2015) Obesity and overweight. Fact sheet N°311 January 2015 [cited 2016 20 April 2016]. http://www.who.int/mediacentre/factsheets/fs311/en/
3. Jia X, Chang T, Wilson TW, Wu L (2012) Methylglyoxal mediates adipocyte proliferation by increasing phosphorylation of Akt1. PLoS One 7:e36610
4. Ribeiro PVM, Tavares JF, Costa MAC et al (2019) Effect of reducing dietary advanced glycation end products on obesity-associated complications: a systematic review. Nutr Rev 77:725–734
5. Eleazu C, Omar N, Lim OZ et al (2019) Obesity and comorbidity: could simultaneous targeting of esRAGE and sRAGE be the Panacea? Front Physiol 10:787
6. Chooi YC, Ding C, Magkos F (2019) The epidemiology of obesity. Metab 92:6–10
7. Eleazu C, Mohamed M (2020) Targeting advanced glycation end products (esRAGE and sRAGE) for obesity, diabetes, and its associated complications. In: Faintuch J Faintuch S (eds) Obesity and diabetes. Springer Nature Switzerland AG. In Press. https://doi.org/10.1007/978-3-030-53370-0_14
8. Liang Z, Chen X, Li L, Li B, Yang Z et al (2019) The fate of dietary advanced glycation end products in the body: from oral intake to excretion. Critic Rev Food Sci Nutr 2019. https://doi.org/10.1080/10408398.2019.1693958
9. Nowotny K, Jung T, Höhn A et al (2015) Advanced glycation end products and oxidative stress in type 2 diabetes mellitus. Biomolecules 5:194–222
10. Sayej WN, Lii KPR Guo WA et al (2016) Advanced glycation end products induce obesity and hepatosteatosis in CD-1 wild-type mice. Biomed Res Int 2016:12 pages
11. Kosmopoulos M, Drekolias D, Zavras PD et al (2019) Impact of advanced glycation end products (AGEs) signaling in coronary artery disease. Biochimica et Biophysica Acta (BBA)—Molec Basis Dis 1865:611–619
12. Uribarri J, Cai W, Woodward M et al (2015) Elevated serum advanced glycation end products in obese indicate risk for the metabolic syndrome: a link between healthy and unhealthy obesity? J Clin Endocrinol Metab 100:1957–1966
13. Ames JM (2008) Determination of N epsilon-(carboxymethyl) lysine in foods and related systems. Ann N Y Acad Sci 1126:20–24
14. Pouillart P, Mauprivez H, Ait-Ameur L et al (2008) Strategy for the study of the health impact of dietary Maillard products in clinical studies: the example of the ICARE clinical study on healthy adults. Ann N Y Acad Sci 1126:173–176
15. Snelson M, Coughlan MT (2019) Dietary advanced glycation end products: digestion, metabolism and modulation of gut microbial ecology. Nutrients 11:215
16. Delgado-Andrade C, Seiquer I, Haro A et al (2010) Development of the Maillard reaction in foods cooked by different techniques. Food Chem 122:145–153
17. Zhang JJ, Merhi Z (2016) Could advance glycation end products explain the poor response to controlled ovarian hyperstimulation in obese women? J Endocrinol Diab 3:1–9
18. Uribarri J, Woodruff S, Goodman S et al (2010) Advanced glycation end products in foods and a practical guide to their reduction in the diet. J Am Diet Assoc 110:911–916.e12
19. Tessier FJ, Birlouez-Aragon I (2012) Health effects of dietary Maillard reaction products: the results of ICARE and other studies. Amino Acids 42:1119–1131
20. Barbosa JHP, Souza IT, Santana AEG, Goulart MOF (2016) Determination of advanced glycation (AGEs) and lipoxidation (ALEs) end products in foods and biological systems: advances, challenges and perspectives. Quım Nova 39. https://doi.org/10.5935/0100-4042.20160048
21. Kehm R, Rückriemen J, Weber D et al (2019) Endogenous advanced glycation end products in pancreatic islets after short-term carbohydrate intervention in obese, diabetes-prone mice. Nutr and Diab 9:9
22. Uribarri J, Cai W, Sandu O et al (2005) Diet-derived advanced glycation end products are major contributors to the body's age pool and induce inflammation in healthy subjects. Ann New York Acad Sci 1043:461–466

23. Ott C, Jacobs K, Haucke E et al (2014) Role of advanced glycation end products in cellular signaling. Redox Biol 2:411–429
24. Kellow NJ, Coughlan MT (2015) Effect of diet-derived advanced glycation end products on inflammation. Nutr Rev 73:737–759
25. Leuner B, Max M, Thamm K et al (2012) RAGE influences obesity in mice. Effects of the presence of RAGE on weight gain, AGE accumulation, and insulin levels in mice on a high fat diet. Z Gerontol Geriatr 45:102–108
26. Tavares JF, Ribeiro PVM, Coelho OGL et al (2020) Can advanced glycation end-products and their receptors be affected by weight loss? A systematic review. Obes Rev 21:e13000
27. Vlassara H, Striker GE (2011) AGE restriction in diabetes mellitus: a paradigm shift. Nat Rev Endocrinol 7:526–539
28. Diamanti-Kandarakis E, Katsikis I, Piperi C et al (2007) Effect of long-term orlistat treatment on serum levels of advanced glycation end-products in women with polycystic ovary syndrome. Clin Endocrinol (Oxf) 66:103–109
29. Boor P, Celec P, Behuliak M et al (2009) Regular moderate exercise reduces advanced glycation and ameliorates early diabetic nephropathy in obese Zucker rats. Metab 58:1669–1677
30. Koyama H, Shoji T, Yokoyama H et al (2005) Plasma level of endogenous secretory RAGE is associated with components of the metabolic syndrome and atherosclerosis. Arterioscler Thromb Vasc Biol 25:2587–2593
31. Vazzana N, Guagnano MT, Cuccurullo C et al (2012) Endogenous secretory RAGE in obese women: association with platelet activation and oxidative stress. J Clin Endocrinol Metab 97:E1726–E1730
32. Poulsen MW, Bak MJ, Andersen JM et al (2014) Effect of dietary advanced glycation end products on postprandial appetite, inflammation, and endothelial activation in healthy overweight individuals. Eur J Nutr 53:661–672
33. Frago LM (2015) Chowen JAHypothalamic leptin and ghrelin signaling as targets for improvement in metabolic control. Current Pharm Design 21:3596–3605
34. Williams KW, Scott MM, Elmquist JK (2009) From observation to experimentation: leptin action in the mediobasal hypothalamus. Am J Clin Nutr 89:985S–990S
35. Balland E, Cowley MA (2015) New insights in leptin resistance mechanisms in mice. Front Neuroendocrinol 39:59–65
36. Huang X, Liu G, JGuo J, Zhengquan SZ (2018) The PI3K/AKT pathway in obesity and type 2 diabetes. Int J Biol Sci 14:1483–1496
37. Donato J, Frazão R, Elias CF (2010) The PI3K signaling pathway mediates the biological effects of leptin. Arquivos Brasileiros de Endocrinologia & Metabologia 54:591–602
38. Shi X, Wang J, Lei Y et al (2019) Research progress on the PI3K/AKT signaling pathway in gynecological cancer. Mol Med Rep 19:4529–4535
39. Gonzalez E, McGraw TE (2009) The Akt kinases: isoform specificity in metabolism and cancer. Cell Cycle 8:2502–2508
40. Unoki H, Bujo H, Yamagishi S et al (2007) Advanced glycation end products attenuate cellular insulin sensitivity by increasing the generation of intracellular reactive oxygen species in adipocytes. Diabetes Res Clin Pract 76:236–244
41. Monden M, Koyama H, Otsuka Y et al (2013) Receptor for advanced glycation end products regulates adipocyte hypertrophy and insulin sensitivity in mice: involvement of toll-like receptor 2. Diabetes 62:478–489

Chapter 10
Monoamine Oxidase, Obesity and Related Comorbidities: Discovering Bonds

Adrian Sturza, Danina M. Muntean, and Octavian M. Crețu

Abstract Obesity together with diabetes represent nowadays the largest epidemic in human history as well as the heaviest economic burden worldwide. Both conditions are associated with high rates of morbidity and mortality largely due to the association of cardiovascular comorbidities. The current understanding of cardiometabolic pathologies recognize chronic oxidative stress and low-grade inflammation as major pathomechanisms in both adipose tissue and cardiovascular system. The sources of reactive oxygen species (ROS) and the factors enabling the perpetuation of systemic inflammation are far from being elucidated. In the past decade monoamine oxidases (MAO), enzymes at the outer mitochondrial membrane with 2 isoforms, A and B, have emerged as important contributors to the ROS-induced endothelial dysfunction and cardiac injury and more recently, to the dysfunctional adipose tissue. The aim of the present chapter is to summarize information about MAO contribution to obesity and related comorbidities in light of the underlying pathomechanisms and to highlight the potential of MAO inhibitors as candidate molecules for drug repurposing in cardiometabolic pathologies.

Keywords Monoamine oxidase · Obesity · Diabetes · Endothelial dysfunction · Inflammation

A. Sturza · D. M. Muntean (✉)
Department of Functional Sciences – Pathophysiology and Centre for Translational Research and Systems Medicine, Faculty of Medicine, "Victor Babeș" University of Medicine and Pharmacy, Eftimie Murgu Sq. nr. 2, 300041 Timișoara, Romania
e-mail: daninamuntean@umft.ro

O. M. Crețu
Department of Surgery - Surgical Semiotics I and Centre for Hepato-Biliary and Pancreatic Surgery, Faculty of Medicine, "Victor Babeș" University of Medicine and Pharmacy, Eftimie Murgu Sq. nr 2, 300041 Timișoara, Romania

© Springer Nature Switzerland AG 2021
P. S. Tappia et al. (eds.), *Cellular and Biochemical Mechanisms of Obesity*, Advances in Biochemistry in Health and Disease 23,
https://doi.org/10.1007/978-3-030-84763-0_10

Introduction

Obesity has been declared by the World Health Organization as the most threatening chronic disease of the twenty-first century whose prevalence has nearly tripled between 1975 and 2019 and continues to rise, particularly in developing/low-income countries [1]. Defined as abnormal or excessive fat accumulation that increases the body mass index (BMI) and may impair health, overweight (BMI > 25) and obesity (BMI > 30) are the major risk factors for metabolic pathologies, such as metabolic syndrome (MetS) and diabetes mellitus (DM). As for the latter, according to the International Diabetes Federation Atlas, 79% of obese people with DM—hence the term 'diabesity'—currently live in low- and middle income countries [2]. More important, the number of people with DM was predicted to increase to 693 million by 2045 and the most cases (77%) in middle-income countries are below the age of 65, thus posing a huge social and health system burden [3]. Of note, in Romania, the PREDATORR study published back to 2016 reported an overall prevalence of 73.90% for abdominal obesity, 34.7% for overweight, and 38.50%f or MetS [4].

An increased BMI has been also reported as a major risk factor for cardiovascular diseases. According to AHA 2019 Heart Disease and Stroke Statistics, in the past decades the prevalence of obesity has constantly increased in the adult population (mainly among females), ranking on the 2nd place after hypertension [5]. Both overweight and obesity are associated with augmented lifetime risk for hypertension, atherosclerosis, atrial fibrillation, adverse cardiac remodelling, and ultimately, heart failure [6].

Despite the fact that obesity has been classically defined according to the BMI changes as an excess of weight for height, in recent years, the European Association for the Study of Obesity recommended a novel denomination, namely "adiposity-based chronic disease" (ABCD) in order to improve the diagnostic criteria in terms of etiology, adiposity degree and health risks of this ongoing pandemic [7]. The definition is particularly aimed to emphasize the importance of assessing in clinical practice the characteristics [total amount, distribution, and function] of the adipose tissue as major culprit responsible for individual metabolic and cardiovascular risks, beyond the BMI [8–12]. The increased adipose tissue (particularly, the visceral and ectopic fat) is currently view as a dysfunctional, highly active metabolic organ ("adiposopathy") that participates in a constant crosstalk with several organs/tissues and acts as main stage for the two intricate mechanisms that underlie obesity pathogenesis, namely the chronic low-grade inflammation and increased oxidative stress, respectively [9, 12, 13].

Obviously, obesity is associated with high level of systemic oxidative stress and represents an imbalance between antioxidants and pro-oxidants with further disruption of redox signaling pathways and cellular and molecular damage [14]. Among reactive oxygen species [ROS], hydrogen peroxide (H_2O_2) is a the key molecule involved both in physiological oxidative stress ("oxidative eustress") as signalling molecule, as well as in pathological oxidative stress ("oxidative distress") as recently reviewed in Ref. [15]. However, the individual contribution of the numerous ROS sources to the metabolic and cardiovascular complications of obesity is far from being elucidated.

Several sources are responsible for H_2O_2 generation by adipocytes, among which the more important (but not exclusive) are: NADPH oxidases (with Nox4 being the major isoform and source of H_2O_2) and mitochondria where H_2O_2 may result either from the SOD-catalyzed dismutation of superoxide anion occasionally generated by a dysfunctional respiratory chain at the inner mitochondrial membrane or from the constant activity of a dedicated enzyme, monoamine oxidase (MAO) at the outer mitochondrial membrane [16–18]. Indeed, mitochondria have emerged in the past decades as both sources and targets of ROS [19–22] in the vast majority of chronic diseases [23] including obesity [24, 25] and a desirable therapeutic approach should be aimed, at least partly, at improving both mitochondrial function and ROS balance.

We here present a brief overview on the contribution of the MAO-related oxidative stress in experimental and clinical settings of obesity and its major comorbidities and highlight the potential of MAO inhibitors as candidate molecules for drug repurposing in cardiometabolic pathologies.

MAO—An Overview

Monoamine oxidase (EC 1.4.3.4) is a flavin-containing monooxygenase discovered more than 90 years ago [26] and located at the outer mitochondrial membrane in virtual all mammalian tissues, which play a key role in the metabolism of monoaminergic neurotransmitters and biogenic amines according to a reaction that constantly generates an aldehyde, ammonium, and hydrogen peroxide as potential toxic ancillary products; therefore, MAO inhibition is expected to exert beneficial effects (reviewed in [27–30]). Two isoforms, MAO-A and MAO-B, have been identified with variable, species-dependent expression in almost all body tissues, e.g., brain, heart, vasculature, liver, intestine, lung, kidney, thyroid gland, platelets, placenta. In the nervous system, it is mainly MAO-B isoform that protects neurons from exogenous amines, prevents the actions of endogenous neurotransmitters and regulates the intracellular amine content, whereas in the peripheral tissues, in particular MAO-A is involved in the oxidative catabolism of amines and prevention dietary amines (such as tyramine from cheese) penetration into the circulation (see Ref. [31] for a recent review). MAO-A is responsible for degradation of epinephrine, norepinephrine, melatonin, and serotonin; MAO-B degrades phenylethylamine and benzylamine. Both MAO isoforms are responsible for dopamine, tyramine, octopamine and tryptamine metabolization [32, 33].

While the vast majority of research has been focused for more than half century on MAO inhibition in the nervous system as important pharmacological target for neurodegenerative diseases and depression, in the past decades the effects of MAO-A and B inhibition in the cardiovascular system have been systematically addressed by several groups, including ours (for comprehensive reviews see Refs. [34–37]). Accordingly, several studies unequivocally demonstrated that MAOs are constant sources of H_2O_2 in both animal models of disease [38–40] and humans with cardiovascular diseases, obesity and diabetes [20, 22, 25, 35, 41] and will be further detailed.

MAO and Vascular Function

The beneficial effects of MAO inhibition in the setting of endothelial dysfunction were reported in both experimental models and clinical settings. Thus, in spontaneously hypertensive rats (SHR), ex vivo incubation of aortic rings with different MAO inhibitors (organ bath experiments) reversed the impaired vascular function, by improving endothelium-dependent relaxation [42] and reduced the 5-hydroxytryptamine (5-HT)-induced tension in isolated basilar arteries [43]. The latter group reported an increased MAO-A protein expression in basilar arteries harvested from the SHR as compared to the corresponding controls (WKY). More, it was revealed an increased 5-HT-induced contractility effect that was attenuated by clorgyline (irreversible MAO-A inhibitor) and polyethylene glycol-catalase (ROS scavenger).

The implication of MAO in development and evolution of cardiovascular remodeling process and endothelial dysfunction has been studied in the last years.

Thus, in an experimental model of chronic *in vivo* endothelial dysfunction, Sturza et al. reported that after angiotension II (AII) treatment in mice (2 weeks release by minipumps), an upregulation of both MAO isoenzymes expression in aortas occurred; importantly, in this experiment, MAO A and B inhibitors restored the impaired vascular relaxation and also decreased oxidative stress [41]. Of note, the MAO-AII interaction has been already reported (yet overlooked) back to the late 80s: accordingly, in rat neuronal cell cultures of hypothalamus and brainstem, AII in submicromolar ranges increased neuronal norepinephrine (but not dopamine) uptake and stimulated MAO activity via a yet unidentified mechanism [44]. Important, the effect of AII on MAO is not limited to the vascular system since it was also reported in cardiomyocytes where MAO-A activity increased after incubation with AII [45]. Moreover, inhibition of the renin–angiotensin–aldosterone system was associated with a decrease in MAO activity in the heart [46].

Collectively, all these data strongly suggest that MAO-related ROS production contributes to the development of endothelial dysfunction and MAO inhibition can improve the vascular function and also attenuate ROS production (Table 10.1).

MAO and Obesity

Adipose tissue consists of several cell types: adipocytes, fibroblasts, endothelial cells, immune cells with complex physiological and pathological interactions. Besides its

Table 10.1 MAO and endothelial dysfunction—mechanistic evidence

MAO contribution and its inhibition effect	Relevance	References
MAO up-regulation occurred in mice aortas after AII treatment with impaired vascular relaxation (most probably, via NFkB and PI3 kinase pathways); MAO limited cGMP accumulation	MAO A and B are mediators of endothelial dysfunction process	[41]
MAO inhibition improved vascular relaxation in vascular samples from animal models of diabetes (aortas from Zucker diabetic fatty rats and rats with streptozotocin-induced diabetes)	MAO inhibitors are drug candidates to alleviate murine endothelial dysfunction	[47, 48]
MAO-A inhibitor (clorgyline) reduced pulmonary vascular remodeling in rats with pulmonary arterial hypertension	MAO inhibition might be a therapeutic alternative for pulmonary arterial hypertension	[49]
MAO-B inhibitor (selegiline) increased nitric oxide availability	Therapeutic effects in neurodegenerative diseases associated with aging and vascular diseases	[50]
Angiotensin-converting enzyme inhibition increased cardiac catecholamine content and reduced MAO activity	Potential beneficial effects in heart failure	[46]
MAO inhibition improved vascular relaxation in human vascular samples: mammary arteries (isolated from patients subjected to by-pass surgery), mesenteric arteries branches (isolated from patients subjected to abdominal surgery) and brachial artery collaterals (isolated from patients with end-stage renal disease and indication of hemodialysis)	MAOs are mediators of human endothelial dysfunction. MAO inhibitors are drug candidates to alleviate endothelial dysfunction in humans	[17, 42, 43, 48, 51, 52]

classic role of fat deposition, adipose tissue is an important regulator of the redox balance.

MAO is expressed in the adipose tissue being mainly involved in catecholamine clearance. In physiological conditions, hydrogen peroxide, the byproduct of MAO activity, has been reported to exhibit insulin-like effects in adipocytes with glucose uptake and inhibition of lipolysis [53]. Even if there is several evidences (experimental and clinical) showing that in cardiovascular system MAO contributes significantly to ROS production, there is limited information about the implication of this enzyme in pathophysiology of obesity and related complications. What it has been unequivocally demonstrated is that the dysfunctional visceral and ectopic adipose tissues are responsible for generation of high amounts of ROS that are linked with obesity-related pathologies. Obviously, the role of MAO in the setting of obesity is not clear at this moment. Pioneering studies in animal models of obesity reported that administration of non-selective and irreversible MAO inhibitors were able to increase lipolysis [54] and reduce the body weight [55, 56], in contrast with MAO substrates and hydrogen peroxide which displayed antilipolytic effects in vitro [57]. Also, an increased MAO activity was found in the intra-abdominal adipose tissue isolated from dogs with diet-induced obesity [58]. More recent studies confirmed the increase in MAO activity in obese mice [11], dogs [11] and pigs [59] and in mature adipocytes isolated from human subcutaneous adipose tissue [60].

We have initially reported an increased MAO expression in the perivascular adipose tissue from mammary arteries isolated from diabetic and non-diabetic patients with coronary artery disease subjected to the revascularization procedure [51]. More recently, we reported an increased oxidative stress in visceral adipose tissue (VAT) harvested from obese (but not from non-obese) patients with indication of elective abdominal surgery and that MAO-A isoform was predominantly over-expressed; ex vivo inhibition of MAO-A with clorgyline significantly mitigated the oxidative stress in VAT samples from obese patients and had no effect in those from the non-obese group [61].

Important, we have also demonstrated that ex vivo incubation with the same MAO-A inhibitor of mesenteric artery branches isolated from these obese patients significantly improved the endothelium-dependent relaxation and decreased the level of oxidative stress, strongly suggesting that MAO is a druggable target in both adipose and vascular tissue and MAO inhibitors might provide a ‘killing two birds with one stone’ approach in these settings [17].

The increased expression of MAO in the dysfunctional VAT from obese patients points to the important role of this tissue in the control of peripheral catecholamine clearance [62], an observation that clearly requires further investigation in relation with the development of hypertension or insulin resistance. Mechanistic evidence for MAO contribution to the oxidative stress in white adipose tissue and the beneficial effects of its inhibition are depicted in Table 10.2.

MAO and Inflammation

In the setting of obesity, besides chronic oxidative stress, the presence of low-grade inflammation is considered another important risk factor for cardiovascular disease. To date the contribution of MAO to the persistent, chronic inflammation has scarcely been addressed in the literature. Thus far we know that MAO-A is involved in ROS generation in alternatively activated monocytes/macrophages [66, 67]. Both MAO isoforms were reported to be upregulated after ex vivo stimulation and in vivo treatment with lipopolysaccharide in mice [41] and rat [68] aortic rings. This process

Table 10.2 MAO and obesity—mechanistic evidence

MAO contribution and its inhibition effects	Relevance	References
MAO activity is increased in white adipose tissue of obese dogs	MAO is source of ROS in the animal adipose tissue	[58]
MAO non-selective and irreversible inhibitor [phenelzine] reduced body fat in mice and rats; in white adipose tissue phenelzine reduced the lipogenic effect of insulin	MAO inhibition might be a therapeutic alternative in obesity	[56, 63]
MAO non-selective and irreversible inhibitor (pargyline) reduces weight gain and adiposity in obese Zucker rats	MAO inhibition might be a therapeutic alternative in obesity	[55, 64]
MAO A- and B selective inhibitors (moclobemide and selegiline) increased adiponectin production in human bone marrow mesenchymal stem cells	MAO inhibitors can have novel anti-obesity and anti-diabetic properties	[65]
MAO-B selective inhibitor (selegiline) reduced adiposity (both subcutaneous and visceral fat) induced by high-fat, high-sucrose diet in rats	MAO inhibition might be a therapeutic alternative in obesity	[40]
MAO-A expression is upregulated in visceral abdominal adipose tissue isolated from obese patients and its inhibition reduced the ROS generation	MAO is a source of ROS in human adipose tissue MAO inhibition mitigates oxidative stress in adipose tissue	[17]

seems to involve NFkB and PI3 kinase pathways. In this experiment MAO inhibition partially improved the vasomotor function and decrease oxidative stress [41]. Interestingly MAO upregulation triggered by LPS was also reported in a rat model of periodontal disease and treatment with a non-selective irreversible MAO inhibitor phenelzine was able to significantly reduce the amount of H_2O_2 [69]. Another study showed that moclobemide, a selective and reversible MAO-A inhibitor, was able to attenuate vascular inflammation of intra-myocardial arteries in a rat model of ischemia–reperfusion injury [70]. We have recently reported that ex vivo stimulation with interleukin 6 (IL-6, 100 ng/ml, 12 h) of mesenteric arteries branches harvested from patients (both children and adults) subjected to abdominal surgery, was able to elicit MAO upregulation [32].

Increased transcription of the MAO-B isoform has been reported to occur in individuals with HIV encephalitis and macaques with the simian immunodeficiency virus (SIV)-induced lesions of the central nervous system (CNS). It has been postulated that the neuroinflammatory environment of HIV/SIV-associated CNS disease may promote MAO activation, which will be responsible for the decreased dopamine level and increased ROS production, respectively [68]. A list of connections between MAO activity/expression up-regulation and inflammation is presented in Table 10.3.

Table 10.3 MAO and inflammation—mechanistic evidence

MAO contribution and its inhibition effects	Relevance	References
MAO-A upregulation induced by chronic intermittent hypoxia (study in rat hippocampus)	MAO upregulation can lead to neuronal inflammation with subsequent neurodegeneration	[71]
MAO upregulation after LPS administration (mice, rats) most probably via NFkB	MAO is a mediator of endothelial dysfunction in experimental conditions that mimic septic shock	[41, 72]
IL-4/13-induced upregulation of MAO-A expression in the Th2 response (study in human peripheral monocytes)	MAO is a modulator of the immune system	[73]
H2O2 produced by MAO-B elicited mitochondrial dysfunction and NFkB induction; inhibition of MAO-B with rasagiline prevent IL-1β secretion (study in murine bone marrow-derived macrophages and human monocyte-derived macrophages)	Repurposing of MAO inhibitors can be a realistic approach to treat several inflammatory diseases characterized by IL-1β-driven pathology	[74]
MAO-A upregulation in visceral abdominal adipose tissue isolated from obese patients with chronic inflammatory syndrome	MAO is a source of ROS in human adipose tissue MAO inhibition mitigates oxidative stress in adipose tissue	[17]
IL-13-induced MAO-A gene expression in alternatively activated monocytes/macrophage, A549 lung carcinoma cells and in normal human bronchial epithelial cells	Contribution of MAO-A in lung cancer metastasis MAO-A might evolve as a therapeutic target in lung cancer	[75]
TNF-α induces MAO-A expression (cell culture)	MAO-A increased expression as key factor in in neuronal/ cardiovascular diseases associated with inflammatory states	[76]
IL-6 induces MAO-A expression in mesenteric arteries branches isolated from patients subjected to abdominal surgery	Chronic inflammation might potentiate endothelial dysfunction according to a MAO-dependent mechanism	[32]
MAO upregulation triggered by LPS occurred in the rat periodontal disease model	MAO inhibitor, phenelzine, was able to significantly reduce the amount of H_2O_2	[69]
MAO inhibition prevented both mast cell degranulation and altered collagen deposition in the heart (study in streptozotocin induced diabetes in mice)	MAO inhibition might become a therapeutic alternative in diabetes complications	[77]
MAO inhibition with selegiline reduces cigarette smoke-induced oxidative stress and inflammation in airway epithelial cells	MAO inhibition might become a therapeutic alternative in COPD management	[78]

MAO and Diabetes

Few studies have addressed the contribution of MAO-related oxidative stress in the setting of diabetes. Also, a limited number of studies is available with respect to the oxidative stress-mediated endothelial dysfunction in diabetes. Our group firstly reported that MAOs, in particular the MAO-B isoform, are overexpressed (mRNA and protein) in aortas harvested from rats with streptozotocin-induced diabetes and contributed via hydrogen peroxide generation to the endothelial dysfunction. MAO inhibition with clorgyline for MAO-A and selegiline for MAO-B significantly reduced vascular contractility, improved the endothelial-dependent relaxation and decreased by 50% the level of H_2O_2 in diabetic aortic samples [48].

The same observations were recapitulated in an experimental model of type II diabetes, as presented in Table 10.1 [47]**.** We may speculate that upregulation of MAO in diabetes is a direct consequence of hyperglycemia, since incubation of rat aortic samples with high glucose level (400 mg/dl, 12 h) elicited the increase in MAO-A expression [32].

We have also reported that MAO expression is early increased in mammary arteries, again with the predominance of the MAO-B isoform in patients with coronary heart disease and preserved ejection fraction, regardless the presence **of diabetes** [51]. In these samples MAO inhibitors were equally able to improve the vascular relaxation and mitigate the oxidative stress [51].

Elucidation of the pathogenesis of diabetic complications is currently an active field of research. Several clinical trials revealed that normalization of glycemia failed to reduce the rates of major cardiovascular events in the diabetic population, an observation that supports the concept that hyperglycemic environment may be "remembered" in the vasculature [79]. Whether MAOs inhibition might interfere with the hyperglycemic memory has not been investigated so far. As a consequence of hyperglycemia, MAOs become a crucial contributor to vascular oxidative stress and mitochondrial dysfunction and targeted MAO inhibition in the vasculature may provide benefits in the setting of diabetes. Importantly, MAO inhibitors are already available in clinical practice for several brain pathologies, so their utilization in other diseases supports the concept of drug repurposing, highly favored by pharmaceutical industry.

A list of connections between MAO activity/expression and the effect of its inhibition in the setting of diabetes is shown in Table 10.4.

Coda

Obesity has a complex, multifactorial (genetic, environmental, socio-economic, behavioral and cultural) etiopathogenesis and, despite the growing recognition, its pandemic ('globesity') continues unabated worldwide [81]. With the increased global exposure to an obesogenic environment, particularly of children, it is not surprisingly

Table 10.4 MAO and diabetes—mechanistic evidence

MAO contribution and its inhibition effects	Relevance	References
Upregulation of both MAO isoforms (particularly of MAO-B) in aorta isolated from streptozotocin induced diabetic rats with subsequent vascular function impairment; beneficial effects of MAO inhibition	MAO inhibition might be a therapeutic alternative in the management of type 1 diabetes-related vascular complications	[48]
In Zucker diabetic fatty rats MAO inhibitors reduced oxidative stress and improved vascular reactivity	MAO inhibition might be a therapeutic alternative in the management of type 2 diabetes-related vascular complications	[47]
Upregulation of MAO-A in response to in vitro exposure to high glucose in rat aortas	MAO upregulation is a direct effect of hyperglycemia	[32]
Case study—Type 1 diabetic patient improved his glycemic control and hypoglycemic perception using a MAO inhibitor (tranylcypromine)	Impact on glycemic control in diabetic patients	[80]

that it has been predicted, more than a decade ago that one out of five individuals will be obese by 2030 [82, 83].

In light of the epidemic status of overweight/obesity and the numerous related comorbidities (metabolic, cardiovascular, several cancers, chronic venous disease, depression etc.) with negative health and economic outcomes [81], pharmacotherapy is warranted for more than one-third of adults [84]. Importantly, the approved drugs are extremely few in number and efficacy; also, are associated with several side effects [85]. In a recent critical paper, it has been noticed that only 190 drugs were under development to target obesity as compared to 3436 drugs under evaluation for the treatment of cancer [86]. This situation is, at least partially, due to the misconception [shared by some health providers and patients] that obesity is not a disease, but a lifestyle condition, and patients have the primary responsibility for its treatment [85, 86]. Important, the Endocrine Society guidelines on obesity pharmacotherapy cautions "against prescribing medications known to be associated with weight loss if they have no proven beneficial effect on the patient's other identified health issues" [85]. Therefore, efforts to further elucidate the pathophysiology of organ dysfunction in obesity are fully warranted in order to develop novel and effective medications and/or to repurpose the existing ones.

Nowadays, the white adipose tissue is the most investigated active endocrine organ since it is widely accepted that its dysfunction and the ectopic fat accumulation (and not obesity per se) are the major predictors for the development of metabolic diseases [86]. Adipose tissue dysfunction is characterized by predominantly visceral fat accumulation, with maladaptive changes in composition and secretory profile, local inflammation and mitochondrial dysfunction that result in long run in metabolic inflexibility and structural remodeling [87]. In particular, understanding the pathways behind compromised mitochondrial function in white adipose tissue is nowadays an active field of research [88].

Monoamine oxidases are mitochondrial enzymes expressed in both visceral adipose tissue and vasculature being upregulated in the setting of obesity, diabetes and the related cardiovascular comorbidities. MAO inhibition significantly mitigated the oxidative stress in adipose tissue and improved vascular reactivity arteries in both animal models and humans. These data published so far strongly suggest that MAO inhibitors are promising candidate compounds for drug repositioning in cardiometabolic pathologies.

Acknowledgements Work partially suported by the university internal grant code 6POSTDOC/1871/12.02.2020 (A.S.).

References

1. Organization WH (2020) Obesity and overweight. Fact sheet. https://www.whoint/news-room/fact-sheets/detail/obesity-and-overweight. Accessed 10 2020
2. Federation ID (2020) IDF diabetes Atlas, 9th edn. International Diabetes Federation, Brussels, Belgium. https://www.diabetesatlasorg/en/. Accessed 11 Nov 2020
3. Cho NH, Shaw JE, Karuranga S et al (2018) IDF diabetes Atlas: global estimates of diabetes prevalence for 2017 and projections for 2045. Diabetes Res Clin Pract 138:271–281
4. Popa S, Moţa M, Popa A et al (2016) Prevalence of overweight/obesity, abdominal obesity and metabolic syndrome and atypical cardiometabolic phenotypes in the adult Romanian population: PREDATORR study. J Endocrinol Invest 39(9):1045–1053
5. Benjamin EJ, Muntner P, Alonso A et al (2019) Heart disease and stroke statistics-2019 update: a report from the American Heart Association. Circulation 139(10):e56–e528
6. Khan SS, Ning H, Wilkins JT et al (2018) Association of body mass index with lifetime risk of cardiovascular disease and compression of morbidity. JAMA Cardiol 3(4):280–287
7. Fruhbeck G, Busetto L, Dicker D et al (2019) The ABCD of obesity: an EASO position statement on a diagnostic term with clinical and scientific implications. Obes Facts 12(2):131–136
8. Shah RV, Murthy VL, Abbasi SA et al (2014) Visceral adiposity and the risk of metabolic syndrome across body mass index: the MESA Study. JACC Cardiovasc Imaging 7(12):1221–1235
9. Goossens GH (2017) The metabolic phenotype in obesity: fat mass, body fat distribution, and adipose tissue function. Obes Facts 10(3):207–215
10. Rallidis LS, Baroutsi K, Zolindaki M et al (2014) Visceral adipose tissue is a better predictor of subclinical carotid atherosclerosis compared with waist circumference. Ultrasound Med Biol 40(6):1083–1088
11. Abbasi SA, Hundley WG, Bluemke DA et al (2015) Visceral adiposity and left ventricular remodeling: the Multi-Ethnic Study of Atherosclerosis. Nutr Metab Cardiovasc Dis 25(7):667–676
12. Neeland IJ, Poirier P, Despres JP (2018) Cardiovascular and metabolic heterogeneity of obesity: clinical challenges and implications for management. Circulation 137(13):1391–1406
13. Gomez-Hernandez A, Beneit N, Diaz-Castroverde S, Escribano O (2016) Differential role of adipose tissues in obesity and related metabolic and vascular complications. Int J Endocrinol 1216783
14. Sies H, Berndt C, Jones DP (2017) Oxidative stress. Annu Rev Biochem 86:715–748

15. Sies H (2017) Hydrogen peroxide as a central redox signaling molecule in physiological oxidative stress: oxidative eustress. Redox Biol 11:613–619
16. Ursini F, Maiorino M, Forman HJ (2016) Redox homeostasis: the Golden Mean of healthy living. Redox Biol 8:205–215
17. Sturza A, Olariu S, Ionica M et al (2019) Monoamine oxidase is a source of oxidative stress in obese patients with chronic inflammation. Can J Physiol Pharmacol 97(9):844–849
18. Hauck AK, Huang Y, Hertzel AV, Bernlohr DA (2019) Adipose oxidative stress and protein carbonylation. J Biol Chem 294(4):1083–1088
19. Muntean DM, Sturza A, Danila MD et al (2016) The role of mitochondrial reactive oxygen species in cardiovascular injury and protective strategies. Oxid Med Cell Longev 8254942
20. Zorov DB, Juhaszova M, Sollott SJ (2014) Mitochondrial reactive oxygen species (ROS) and ROS-induced ROS release. Physiol Rev 94(3):909–950
21. Murphy MP (2009) How mitochondria produce reactive oxygen species. Biochem J 417(1):1–13
22. Brand MD (2016) Mitochondrial generation of superoxide and hydrogen peroxide as the source of mitochondrial redox signaling. Free Radic Biol Med 100:14–31
23. Murphy E, Ardehali H, Balaban RS et al (2016) Mitochondrial function, biology, and role in disease: a scientific statement from the American Heart Association. Circ Res 118(12):1960–1991
24. McMurray F, Patten DA, Harper ME (2016) Reactive oxygen species and oxidative stress in obesity-recent findings and empirical approaches. Obesity 24(11):2301–2310
25. Woo CY, Jang JE, Lee SE et al (2019) Mitochondrial dysfunction in adipocytes as a primary cause of adipose tissue inflammation. Diabetes Metab J 43(3):247–256
26. Hare ML (1928) Tyramine oxidase: a new enzyme system in liver. Biochem J 22(4):968–979
27. Bortolato M, Chen K, Shih JC (2008) Monoamine oxidase inactivation: from pathophysiology to therapeutics. Adv Drug Deliv Rev 60(13–14):1527–1533
28. Edmondson DE (2014) Hydrogen peroxide produced by mitochondrial monoamine oxidase catalysis: biological implications. Curr Pharm Des 20(2):155–160
29. Youdim MBH (2018) Monoamine oxidase inhibitors, and iron chelators in depressive illness and neurodegenerative diseases. J Neural Transm (Vienna) 125(11):1719–1733
30. Song MS, Matveychuk D, MacKenzie EM et al (2013) An update on amine oxidase inhibitors: multifaceted drugs. Prog Neuropsychopharmacol Biol Psychiatry 44:118–124
31. Tipton KF (2018) 90 years of monoamine oxidase: some progress and some confusion. J Neural Transm (Vienna) 125(11):1519–1551
32. Sturza A, Popoiu CM, Ionica M et al (2019) Monoamine oxidase-related vascular oxidative stress in diseases associated with inflammatory burden. Oxid Med Cell Longev 8954201
33. Youdim MB, Edmondson D, Tipton KF (2006) The therapeutic potential of monoamine oxidase inhibitors. Nat Rev Neurosci 7(4):295–309
34. Deshwal S, Di Sante M, Di Lisa F, Kaludercic N (2017) Emerging role of monoamine oxidase as a therapeutic target for cardiovascular disease. Curr Opin Pharmacol 33:64–69
35. Duicu OM, Lighezan R, Sturza A et al (2016) Assessment of mitochondrial dysfunction and monoamine oxidase contribution to oxidative stress in human diabetic hearts. Oxid Med Cell Longev 8470394
36. Kaludercic N, Carpi A, Menabò R et al (2011) Monoamine oxidases (MAO) in the pathogenesis of heart failure and ischemia/reperfusion injury. Biochim Biophys Acta 1813(7):1323–1332
37. Kaludercic N, Mialet-Perez J, Paolocci N et al (2014) Monoamine oxidases as sources of oxidants in the heart. J Mol Cell Cardiol 73:34–42
38. Kaludercic N, Carpi A, Nagayama T et al (2014) Monoamine oxidase B prompts mitochondrial and cardiac dysfunction in pressure overloaded hearts. Antioxid Redox Signal 20(2):267–280
39. Bianchi P, Kunduzova O, Masini E et al (2005) Oxidative stress by monoamine oxidase mediates receptor-independent cardiomyocyte apoptosis by serotonin and postischemic myocardial injury. Circulation 112(21):3297–3305
40. Nagy CT, Koncsos G, Varga ZV et al (2018) Selegiline reduces adiposity induced by high-fat, high-sucrose diet in male rats. Br J Pharmacol 175(18):3713–3726

41. Sturza A, Leisegang MS, Babelova A et al (2013) Monoamine oxidases are mediators of endothelial dysfunction in the mouse aorta. Hypertension 62(1):140–146
42. Sturza A, Mirica SN, Duicu O et al (2013) Monoamine oxidase—a inhibition reverses endothelial dysfunction in hypertensive rat aortic rings. Rev Med Chir Soc Med Nat Iasi 117(1):165–171
43. Poon CC, Seto SW, Au AL et al (2010) Mitochondrial monoamine oxidase-A-mediated hydrogen peroxide generation enhances 5-hydroxytryptamine-induced contraction of rat basilar artery. Br J Pharmacol 161(5):1086–1098
44. Sumners C, Shalit SL, Kalberg CJ, Raizada MK (1987) Norepinephrine metabolism in neuronal cultures is increased by angiotensin II. Am J Physiol 252(6 Pt 1):C650–C656
45. Manni ME, Zazzeri M, Musilli C et al (2013) Exposure of cardiomyocytes to angiotensin II induces over-activation of monoamine oxidase type A: implications in heart failure. Eur J Pharmacol 718(1–3):271–276
46. Raasch W, Bartels T, Gieselberg A et al (2002) Angiotensin I-converting enzyme inhibition increases cardiac catecholamine content and reduces monoamine oxidase activity via an angiotensin type 1 receptor-mediated mechanism. J Pharmacol Exp Ther 300(2):428–434
47. Sturza A, Noveanu L, Duicu O, Muntean D (2014) P172Monoamine oxidase inhibition corrects endothelial dysfunction in experimental diabetes. Cardiovasc Res 103(suppl 1):S30-S
48. Sturza A, Duicu OM, Vaduva A et al (2015) Monoamine oxidases are novel sources of cardiovascular oxidative stress in experimental diabetes. Can J Physiol Pharmacol 93(7):555–561
49. Sun X-Q, Peters E, Schalij I et al (2018) The effect of Monoamine oxidase A inhibition on experimentally induced pulmonary arterial hypertension. Eur Respir J 52(suppl 62):PA3072
50. Thomas T (2000) Monoamine oxidase-B inhibitors in the treatment of Alzheimer's disease. Neurobiol Aging 21(2):343–348
51. Lighezan R, Sturza A, Duicu OM et al (2016) Monoamine oxidase inhibition improves vascular function in mammary arteries from nondiabetic and diabetic patients with coronary heart disease. Can J Physiol Pharmacol 94(10):1040–1047
52. Utu D, Pantea S, Duicu OM et al (2017) Contribution of monoamine oxidases to vascular oxidative stress in patients with end-stage renal disease requiring hemodialysis. Can J Physiol Pharmacol 95(11):1383–1388
53. Enrique-Tarancón G, Marti L, Morin N et al (1998) Role of semicarbazide-sensitive amine oxidase on glucose transport and GLUT4 recruitment to the cell surface in adipose cells. J Biol Chem 273(14):8025–8032
54. Mattila M, Torsti P (1966) Effect of monoamine oxidase inhibitors and some related compounds on lipid metabolism in rat. Plasma free fatty acids and lipoprotein lipase of the heart and adipose tissue. Ann Med Exp Biol Fenn 44(3):397–400
55. Carpéné C, Iffiú-Soltesz Z, Bour S et al (2007) Reduction of fat deposition by combined inhibition of monoamine oxidases and semicarbazide-sensitive amine oxidases in obese Zucker rats. Pharmacol Res 56(6):522–530
56. Carpéné C, Abello V, Iffiú-Soltész Z et al (2008) Limitation of adipose tissue enlargement in rats chronically treated with semicarbazide-sensitive amine oxidase and monoamine oxidase inhibitors. Pharmacol Res 57(6):426–434
57. Visentin V, Prévot D, Marti L, Carpéné C (2003) Inhibition of rat fat cell lipolysis by monoamine oxidase and semicarbazide-sensitive amine oxidase substrates. Eur J Pharmacol 466(3):235–243
58. Wanecq E, Bour S, Verwaerde P et al (2006) Increased monoamine oxidase and semicarbazide-sensitive amine oxidase activities in white adipose tissue of obese dogs fed a high-fat diet. J Physiol Biochem 62(2):113–123
59. Carter K, Nelson M, Robidoux J et al (2017) Kinetics of neurotransmitter metabolism by monoamine oxidase in porcine heart differs by location and is increased with obesity/metabolic syndrome. FASEB J 31(1_supplement):883.16
60. Bour S, Daviaud D, Gres S et al (2007) Adipogenesis-related increase of semicarbazide-sensitive amine oxidase and monoamine oxidase in human adipocytes. Biochimie 89(8):916–925

61. Ionica M (2020) Novel insights into adipose tissue and vascular dysfunction in obese patients with inflammatory status. PhD thesis defended the 22 of July 2020
62. Rayner JJ, Banerjee R, Francis JM et al (2015) Normalization of visceral fat and complete reversal of cardiovascular remodeling accompany gastric bypass, not banding. J Am Coll Cardiol 66(22):2569–70
63. Carpéné C, Mercader J, Le Gonidec S et al (2018) Body fat reduction without cardiovascular changes in mice after oral treatment with the MAO inhibitor phenelzine. Br J Pharmacol 175(12):2428–2440
64. Carpéné C, Boulet N, Chaplin A, Mercader J (2019) Past, present and future anti-obesity effects of flavin-containing and/or copper-containing amine oxidase inhibitors. Medicines (Basel) 6(1):9
65. Byun Y, Park J, Hong SH et al (2013) The opposite effect of isotype-selective monoamine oxidase inhibitors on adipogenesis in human bone marrow mesenchymal stem cells. Bioorg Med Chem Lett 23(11):3273–3276
66. Cathcart MK, Bhattacharjee A (2014) Monoamine oxidase A (MAO-A): a signature marker of alternatively activated monocytes/macrophages. Inflamm Cell Signal 1(4):e161
67. Bhattacharjee A, Shukla M, Yakubenko VP et al (2013) IL-4 and IL-13 employ discrete signaling pathways for target gene expression in alternatively activated monocytes/macrophages. Free Radic Biol Med 54:1–16
68. Meulendyke KA, Ubaida-Mohien C, Drewes JL et al (2014) Elevated brain monoamine oxidase activity in SIV- and HIV-associated neurological disease. J Infect Dis 210(6):904–912
69. Ekuni D, Firth JD, Nayer T et al (2009) Lipopolysaccharide-induced epithelial monoamine oxidase mediates alveolar bone loss in a rat chronic wound model. Am J Pathol 175(4):1398–1409
70. Vuohelainen V, Hamalainen M, Paavonen T et al (2015) Inhibition of monoamine oxidase A increases recovery after experimental cardiac arrest. Interact Cardiovasc Thorac Surg 21(4):441–449
71. Lam CS, Li JJ, Tipoe GL et al (2017) Monoamine oxidase A upregulated by chronic intermittent hypoxia activates indoleamine 2,3-dioxygenase and neurodegeneration. PLoS One 12(6):e0177940
72. Rațiu C, Uțu D, Petruș A et al (2018) Monoamine oxidase inhibition improves vascular function and reduces oxidative stress in rats with lipopolysaccharide-induced inflammation. Gen Physiol Biophys 37(6):687–694
73. Chaitidis P, Billett EE, O'Donnell VB et al (2004) Th2 response of human peripheral monocytes involves isoform-specific induction of monoamine oxidase-A. J Immunol 173(8):4821–4827
74. Sánchez-Rodríguez R, Munari F, Angioni R et al (2020) Targeting monoamine oxidase to dampen NLRP3 inflammasome activation in inflammation. Cell Mol Immunol
75. Dhabal S, Das P, Biswas P et al (2018) Regulation of monoamine oxidase A (MAO-A) expression, activity, and function in IL-13-stimulated monocytes and A549 lung carcinoma cells. J Biol Chem 293(36):14040–14064
76. Gupta V, Khan AA, Sasi BK, Mahapatra NR (2015) Molecular mechanism of monoamine oxidase A gene regulation under inflammation and ischemia-like conditions: key roles of the transcription factors GATA2, Sp1 and TBP. J Neurochem 134(1):21–38
77. Deshwal S, Forkink M, Hu CH et al (2018) Monoamine oxidase-dependent endoplasmic reticulum-mitochondria dysfunction and mast cell degranulation lead to adverse cardiac remodeling in diabetes. Cell Death Differ 25(9):1671–1685
78. Cui Y, Liu KW, Liang Y et al (2017) Inhibition of monoamine oxidase-B by selegiline reduces cigarette smoke-induced oxidative stress and inflammation in airway epithelial cells. Toxicol Lett 268:44–50
79. Ceriello A (2009) Hypothesis: the "metabolic memory", the new challenge of diabetes. Diabetes Res Clin Pract 86(Suppl 1):S2-6
80. Emory H, Mizrahi N (2017) Glycaemic control by monoamine oxidase inhibition in a patient with type 1 diabetes. Diab Vasc Dis Res 14(2):163–165

81. Hruby A, Hu FB (2015) The epidemiology of obesity: a big picture. Pharmacoeconomics 33(7):673–689
82. Kelly T, Yang W, Chen CS et al (2008) Global burden of obesity in 2005 and projections to 2030. Int J Obes 32(9):1431–1437
83. Meldrum DR, Morris MA, Gambone JC (2017) Obesity pandemic: causes, consequences, and solutions-but do we have the will? Fertil Steril 107(4):833–839
84. Misra M (2013) Obesity pharmacotherapy: current perspectives and future directions. Curr Cardiol Rev 9(1):33–54
85. Bray GA, Heisel WE, Afshin A et al (2018) The Science of obesity management: an endocrine society scientific statement. Endocr Rev 39(2):79–132
86. Saadi E, White G (2014) Rewarding innovation in drug development. Am Health Drug Benefits 7(7):373–374
87. Blüher M (2013) Adipose tissue dysfunction contributes to obesity related metabolic diseases. Best Pract Res Clin Endocrinol Metab 27(2):163–177
88. Heinonen S, Jokinen R, Rissanen A, Pietiläinen KH (2020) White adipose tissue mitochondrial metabolism in health and in obesity. Obes Rev 21(2):e12958

Chapter 11
Adipose Extracellular Matrix Remodeling in Obesity and Insulin Resistance

Francisco Javier Ruiz-Ojeda, Julio Plaza-Díaz, Augusto Anguita-Ruiz, Andrea Méndez-Gutiérrez, and Concepción María Aguilera

Abstract Obesity drives an excessive triglycerides accumulation in adipose tissue, which incites immune cell infiltration, causing fibrosis and inflammation, causing local hypoxia in adipocytes, and ultimately insulin resistance. The extracellular matrix (ECM) complex network of proteins and proteoglycans that offer a scaffold for cells controlling differentiation, migration, repair, survival, and development, and ECM remodeling is required for healthy adipose tissue expansion. To understand the molecular mechanism of this process is a challenge in order to prevent or treat metabolic diseases. This chapter describes the different ECM components and their function related to adipose tissue and their contribution to restore or maintain insulin sensitivity and the whole body metabolism.

Keywords Adipose tissue · Extracellular matrix · Obesity · Remodeling · Insulin signalling · Angiogenesis · Matrix metalloproteinases · Integrins · Collagen

---

F. J. Ruiz-Ojeda · J. Plaza-Díaz · A. Anguita-Ruiz · A. Méndez-Gutiérrez · C. M. Aguilera (✉)
Faculty of Pharmacy, Department of Biochemistry and Molecular Biology II, University of Granada, 18071 Granada, Spain
e-mail: caguiler@ugr.es

IBS.GRANADA, Complejo Hospitalario Universitario de Granada, 18014 Granada, Spain

F. J. Ruiz-Ojeda
Institute for Diabetes and Obesity, Helmholtz Diabetes Center At Helmholtz Center Munich, 85764 Munich, Neuherberg, Germany

J. Plaza-Díaz
Children's Hospital of Eastern Ontario Research Institute, Ottawa, ON K18L1, Canada

A. Anguita-Ruiz · A. Méndez-Gutiérrez · C. M. Aguilera
Institute of Nutrition and Food Technology "José Mataix", Center of Biomedical Research, University of Granada, Avda. del Conocimiento s/n. 18016 Armilla, Granada, Spain

A. Anguita-Ruiz · C. M. Aguilera
CIBER Physiopathology of Obesity and Nutrition (CIBERobn), Institute of Health Carlos III, 28029 Madrid, Spain

© Springer Nature Switzerland AG 2021
P. S. Tappia et al. (eds.), *Cellular and Biochemical Mechanisms of Obesity*, Advances in Biochemistry in Health and Disease 23,
https://doi.org/10.1007/978-3-030-84763-0_11

Abbreviations

MMPs	Matrix Metalloproteinases
TIMPs	Tissue Inhibitors of Metalloproteinases
WAT	White Adipose Tissue
ECM	Extracellular Matrix
CVD	Cardiovascular Disease
T2D	Type-2 Diabetes

Introduction

Adipose tissue expansion requires a suitable extracellular matrix (ECM) remodeling to provide enough space for all individual adipocytes, but also to form new ones. Similar to other tissues, adipose ECM is a network of structural proteins that compromises a scaffolding for cells controlling several biological processes (e.g. migration, survival, cell adhesion, repair, and development) [74]. Adipose ECM is mainly composed of collagens, fibronectin, and a small amount of laminin [48]. Nonetheless, other proteins such as A disintegrin and metalloproteinase domain-containing protein (ADAMs), hyaluronan (HA), osteopontin (OPN), thrombospondins (THBS1), matrix metalloproteases (MMPs), and tissue inhibitor of metalloproteinases (TIMPs), are involved in the reorganization and remodeling of ECM [55]. Moreover, proper ECM remodeling facilitates the angiogenesis that it is vital for healthy adipose tissue expansion in obesity. Thus, the failure of this triggers hypoxia, death of adipocytes, low-grade inflammation, and fibrosis, which is the key player in adipocyte dysfunction, and, finally, local and systemic insulin resistance [68, 66] (Fig. 11.1). In this chapter, we describe the importance of ECM remodeling in adipose tissue to prevent adipocyte dysfunction, as well as the fibrosis, inflammation, and insulin resistance related to obesity and metabolic diseases.

Extracellular Matrix Components in the Adipose Tissue and Obesity

Collagens

Collagen is one of the major ECM component and contributes to the non-cell mass of the adipose tissue with important functions such as differentiation, morphogenesis, cell adhesion and migration, and wound healing in the tissue. Collagen is largely produced by the adipocytes, although other cell populations like stem cells,

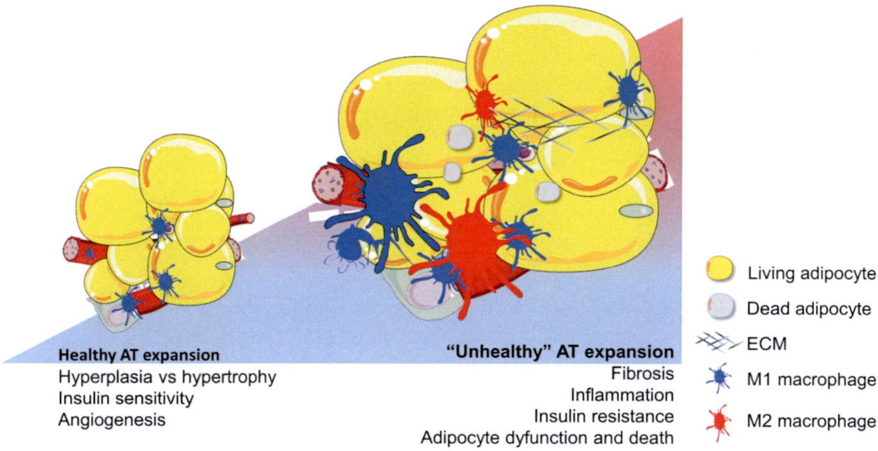

Fig. 11.1 ECM remodeling is required for healthy adipose tissue expansion. In lean and healthy individuals, ECM remodeling allows proper hyperplasia through adipogenesis and the nutrients and oxygen supply into the cell. An "unhealthy" adipose tissue expansion in obesity drives fibrosis (collagen deposition), immune cell infiltration causing inflammation, insulin resistance, and lastly adipocyte dysfunction and death. Abbreviations: AT: adipose tissue

preadipocytes, and endothelial cells, are also producing it. Accumulation of triglycerides in the adipocytes initiative robust mechanical stress, which it is communicated into the intracellular signaling and it is modulated by the external skeleton. In particular, collagen IV, which is the major component of the basement membrane of adipocytes, participates in the adipocyte survival [37]. Nevertheless, collagen I is the most abundant constituent of ECM [38]. In obesity, collagen deposition triggers fibrosis and rigidity, reduces the expandability of the tissue, and causes local insulin resistance [9]. In mice, collagen VI is more present in adipose tissue, and it is able to bind to collagen IV, implicated in adipocyte survival [42]. An increase in adipose tissue content of collagens I,III, V and VI have been reported in high-fat fed (HFD) obese mice [74]. While an increased expression of collagen type VI, α3 (COL6A3) has been observed in adipose tissue in these mice, a suppression of mature adipocyte differentiation was observed in *Col6a3* deficient adipocytes [51]. In addition, IL-6 expression and basal lipolysis are attenuated in these deficient adipocytes. Moreover, Col6a3 knockout mice showed a defect in the adipogenic and lipolytic capacity of adipocytes in normal conditions, supporting the contribution of this collagen type in the adipose tissue to the metabolic homeostasis [51]. In humans, *COL6A3* gene expression is lower in adipose tissue of individuals with obesity, whereas surgery-induced and diet- weight loss rises *COL6A3* gene expression in subcutaneous white adipose tissue (WAT), and leptin regulates its expression [45]. It is pointed out that collagen XVIII is a structural cellular membrane proteoglycan, which is involved in adipocyte differentiation. Reduced adiposity and dyslipidemia have been linked to an attenuation of collagen XVIII, and to the development of glucose intolerance and insulin resistance as well as abnormalities in the circulatory triglyceride concentration and liver lipid content [53]. Such irregularities are related to metabolic disorders. Thus, together together, it can be suggested that accumulation of adipose tissue collagen protects against adipocyte hypertrophy and hyperplasia under conditions of tissue expansion.

Integrins

Integrins are transmembrane receptors composed by α- and β-subunits that are involved in cellular signal transduction. An overlapping feature of all integrins is their role in cell adhesion to ECM as well as cell–matrix interactions. Integrins consist of several different domains that exhibit unique functionalities. In this regard, there is a a large ectodomain that mediates ligand binding, a transmembrane domain and a short intracytoplasmic domain which is associated with actin cytoskeleton. Talins and kindlins are intracellular adaptor proteins that bind to the cytoplasmic domain that resukts in the activation of integrins [67]. Of note, while integrins do not have an intrinsic kinase activity, downstream signal transduction is mediated by an integrin-linked kinase (ILK) and by a focal adhesion kinase (FAK).

Several studies have described an important role and crosstalk between FAK and insulin signaling in adipose tissue, suggesting that integrin signaling is implicated in the insulin action [40]. Indeed, adipose-specific loss of kindlin-2, a focal adhesion protein activating integrins, provokes lipodystrophy and metabolic disturbance in mice [19]. Likewise, Ruiz-Ojeda et al. [59] have revealed that active integrins interact with the insulin receptor in adipocytes modulating WAT insulin sensitivity (Fig. 11.2). Moreover, this study reveals how integrins contribute to brown fat thermogenesis mediated by the basal lamina of brown adipocytes, establishing integrin-extracellular matrix interactions as key regulators of adipose tissue function and whole body metabolism [59].

Matrix Metalloproteinases (MMPs) and Tissue Inhibitors of Metalloproteinases (TIMPs)

The metzincin superfamily of zinc-dependent metalloproteinases comprises the MMP, ADAM, and ADAMTS (ADAM with a Thrombospondin type-1 motif). Principally, MMP is a family of peptidases participating in the degradation of ECM components [7]. Thus, MMPs are indispensable to maintain the ECM remodeling in both normal physiology and diseases. There are different types of MMP1-25 [50].

Adipose tissue expandability is associated with adipogenesis and angiogenesis, and MMP contributes in these processes. In this context, *MMP-9* expression in adipose tissue is associated with the homeostasis model assessment index of insulin resistance (HOMA-IR) in individuals with obesity [71]. In animal models, however, *MMP-3, MMP-11, MMP-12, MMP-13* and *MMP-14* expression are upregulated in visceral WAT, whereas *MMP-7, MMP-9, MMP-16, MMP-24* and *TIMP-4* expression are lower [41]. Besides, MMP-2 and MMP-9 activity are decreased in WAT in high-sucrose-rich diet fed mice with insulin resistance, whereas no changes are reported regarding MMP plasma activity [6]. Indeed, a recent study reported that resistance training might play a crucial role in the maintenance of ECM remodeling in WAT

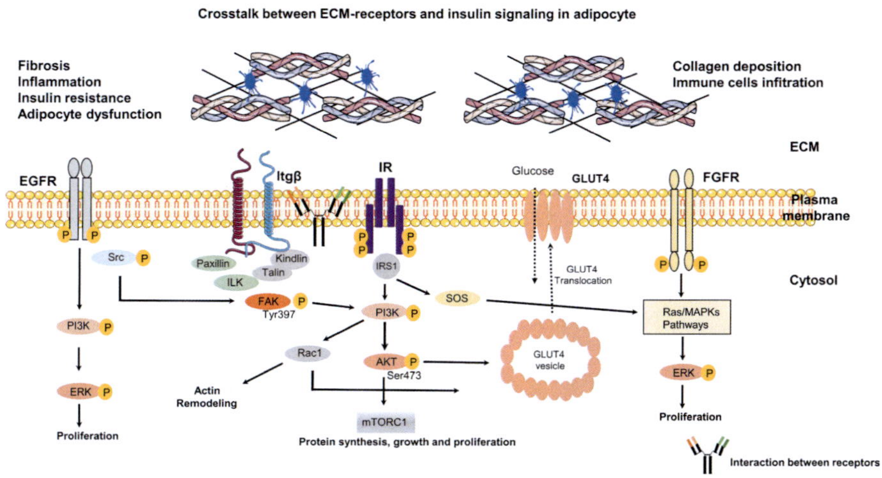

Fig. 11.2 Crosstalk between ECM-receptors and insulin signaling in adipocyte**.** Abbreviations: AKT: protein kinase B; ECM: extracellular matrix; EGFR: epidermal growth factor receptor; ERK: extracellular-signal-regulated kinase; FGF: fibroblast growth factor; FGFR: fibroblast growth factor receptor; GLUT4: glucose transporter type 4; ILK: integrin-linked kinase; IR: insulin receptor; Itgβ: integrin beta subunit; IRS1: Insulin receptor substrate 1; MAPK: mitogen-activated protein kinases; mTOR: mammalian target of Rapamycin; PI3K: phosphatidylinositol 3-kinase; SOS: Son of Sevenless; Src: Proto-oncogene tyrosine-protein kinase

by modulating MMP-2, vascular endothelial growth factor (VEGF)-A, and TIMP-2 activity [16]. An increase in the risk for cardiovascular disease in obesity has been linked to elevated WAT MMP-9 levels [58]. Furthermore, both MMP-2 and MMP-9, secreted from several different cell types including macrophages, endothelial cells, fibroblasts and myofibroblasts have been reported to be increased in the plasma of obese individuals as well as in type 2 diabetics [63]. These two metalloproteinases are involved in the degradation of collagen IV during different pathophysiological conditions such as vascular remodeling, inflammation and atherosclerotic plaque rupture as well as during angiogenesis [26]. MMP-9 gene expression has been reported to be increased in individuals with reduced insulin sensitivity and in those with abnormal BMI [3]. In addition, MMP-14 has recently been identified to be involved in ECM remodeling and dysfunction in obesity [14].

The endogenous TIMPs inhibit the MMPs and there are mainly four: TIMP-1, -2, -3, and -4. The levels of TIMP-1 and TIMP-2 in circulation are higher in individuals with metabolic syndrome and T2D [37]. Indeed, both enzymes are considered as a biomarkers for non-alcoholic fatty liver disease (NAFLD) [1]. In addition, TIMP-1 levels in serum of women with gestational diabetes mellitus and patients with obesity are significantly higher [73]. Nonetheless, TIMP-1 overexpression in pancreatic β-cells protects against diabetes in mice [29], whereas the deletion triggered a higher food intake and adiposity, and serum protein levels are raised in obese pre-diabetic rats [56]. On the other hand, knockout of TIMP-2 increases adiposity in HFD fed mice [27]. Finally, TIMP-3 and -4, play crucial role in insulin sensitivity. Specifically, the knockout of TIMP-3 in mice causes hepatic steatosis and inflammation in WAT [46], while an the overexpression of TIMP-3 prevent it. Regarding TIMP-4, it seems to participate in the deregulation of insulin resistance in mice [60].

Elastin is a ECM component that confers elasticity to the tissue and MMPs degradate it [13]. In particular, MMP-12 is the major MMPs that degradates elastin, and it has been reported that CD11c adipose macrophages (M2) express a lot of levels of MMP-12 under HFD feeding, however, low levels of elastin aggravates insulin resistance [15].

Generally, reorganization and remodeling of ECM drives a bulk of processes in the adipose tissue that further research is required in order to understand the molecular mechanism to develop possible therapies. Accordingly, TIMPs seems to act as endogenous inhibitors of MMPs, which are responsible for degrading excess ECM, but the beneficial effects whether higher TIMP or ADAMTS activities would improve the ECM stability by suppression of MMPs is still unclear [49]. More studies are needed in order to understand the full role of MMPs, TIMPs and ADAMTs in metabolic homeostasis.

Other Components: Osteopontin, Hyaluronan, and Thrombospondin

Osteopontin is another ECM component and is highly expressed in WAT of both HFD fed mice and individuals with obesity [32]. Osteopontin is regularly expressed in WAT macrophages, and the knockout in mice has shown to prevent macrophage infiltration and inflammation in WAT, resulting in improved insulin sensitivity [33]. An up-regulation of osteopontin, Col6, MMP-2 and MMP-9 has been reported to occur in 3T3-L1 pre-adipocyte cells subsequent to treatment with visfatin [18]. While the visfatin-induced increase in Col6 gene expression was determined to be mediated through PI3K, JNK and NF-κB signal transduction, the up-regulation of osteopontin was found to be mediated by signaling via PI3K, JNK, MAPK/ERK and NOTCH1 [18]. Besides, osteopontin plasma levels are higher in T2D patients [5]. With regards to the potential role of osteopontin in insulin resistance, the baseline values of osteopontin might predict 3-year T2D remission in patients undergoing bariatric surgery. Accordingly, circulating levels at baseline of osteopontin were associated with decrease in body weight, BMI, and insulin sensitivity improvements [12].

THBS1 and HA are present in the adipose tissue ECM as well. HA stimulates monocyte adhesion and chemotaxis through the binding to the CD44 [24], and HA is higher in obese mice comparing with their counterparts, and the inhibition of HA improve adipose tissue inflammation and insulin resistance [28]. In humans, HA decreases adipogenesis, but the role in the insulin resistance state is not completely elucidated [75]. Regarding the THBS1 is highly increased in insulin-resistant obese mice and humans [44]. In mice, treatment with recombinant THBS1 decreases insulin signaling in the cultured muscle cell, which could denote an important cross talk between the WAT and skeletal muscle in obesity, indicating a putative therapeutic target against insulin resistance, although further research is needed [43]. Table 11.1 summarizes the main ECM components and its function related to obesity in humans and mice.

Table 11.1 ECM components and function related to obesity in humans and mice

ECM component	Main result	Human/mice	Reference	Main role
Collagens	↑ Collagen type I, III, V and VI expression in WAT of obese mice	Mice	Williams et al. [74]	Main ECM component with functions such as cell adhesion, migration, differentiation, morphogenesis, and wound healing in adipose tissue
	↓ Collagen VI expression in WAT from obese people. Weight loss increased collagen VI expression	Mice	McCulloch et al. [45]	
	Lack of collagen XVIII → reduced adiposity and dyslipidaemia, insulin resistance and glucose intolerance	Human	Petäistö et al. [53]	
	Collagen VI KO mice → metabolic dysfunction	Mice	Oh et al. [51]	
Integrins	Lack of kindling-2 (integrin activator) → metabolic dysfunction	Mice	Gao et al. [19]	Cell adhesion to ECM proteins and cell–matrix interactions
	Lack of kindling-2 (integrin activator) → insulin resistance and BAT altered activity	Mice	Ruiz-Ojeda et al. [59]	
MMPs	↑ MMP-3, -11, -12, -13 and -14 expression in abdominal WAT of obese mice	Mice	Maquoi et al. [41]	Degradation of ECM proteins, regulating ECM remodelling
	↓ MMP-7, -9, -16 and -24 expression in abdominal WAT of obese mice	Mice	Maquoi et al. [41]	
	↓ MMP-2, -9 activity in WAT from insulin resistant mice	Mice	Berg et al. [6]	
	MMP-9 WAT expression positively associated with HOMA in obese adults	Human	Tinahones et al. [71]	
	↑ MMP-2 and -9 plasma levels in obese adults	Human	Tinahones et al. [71]	
	↑ MMP-9 expression in WAT of obese patients and with CVD risk	Human	Ritter et al. [58]	

(continued)

Table 11.1 (continued)

ECM component	Main result	Human/mice	Reference	Main role
	↑ MMP-2 and -9 plasma levels in T2D patients	Human	Signorelli et al. [63]	
	MMP-9 expression in WAT correlated with insulin sensitivity, WC, and BMI	Human	Åkra et al. [3]	
TIMPs	↑ TIMP-1 expression in abdominal WAT of obese mice	Mice	Maquoi et al. [41]	Endogenous MMPs inhibitors, responsible for degrading excess ECM
	↓ TIMP-4 expression in abdominal WAT of obese mice	Mice	Maquoi et al. [41]	
	Overexpression of TIMP-1 in pancreatic β cells protected against diabetes	Mice	Jiang et al. [29]	
	TIMP-3 deletion caused hepatic steatosis and WAR inflammation	Mice	Sakamuri et al. [60]	
	TIMP-4 deletion protected from HFD-induced obesity	Mice	Sakamuri et al. [60]	
	TIMP-1 expression in WAT correlated with insulin sensitivity, WC, and BMI	Human	Akra et al. (2020)	
	↑ TIMP-1 and -2 plasma levels in patients with metabolic syndrome and T2D	Human	Lin et al. [37]	
	↑ TIMP-1 and -2 plasma levels in patients with gestational diabetes and patients with obesity and CVD	Human	Vilmi-Kerala et al. [73]	
Osteopontin	Lack of osteopontin prevented from WAT inflammation and macrophage infiltration	Mice	Kiefer et al. [33]	It is positively associated with insulin resistance state
	↑ plasma levels in T2D patients	Human	Barchett [5]	

(continued)

Table 11.1 (continued)

ECM component	Main result	Human/mice	Reference	Main role
	Osteopontin plasma levels correlated with reductions of body weight, BMI and insulin sensitivity improvement	Human	Carbone et al. [12]	
Thrombospondin	↑ expression in WAT of obese mice	Mice	Matsuo et al. [44]	It may implied in the development of obesity and possible crosstalk between the WAT and skeletal muscle
	Treatment with thrombospondin 1 supressed insulin signalling in muscle cells	Mice	Matsugi et al. [43]	
	↑ expression in WAT from obese patients	Human	Matsuo et al. [44]	
Hyaluronan	↑ expression in WAT of obese mice	Mice	Ji et al. [28]	Monocyte adhesion and chemotaxis. It may also be associated with a decrease in adipogenesis
	Hyaluronan inhibitor treatment improved adipose inflammation and insulin resistance	Mice	Ji et al. [28]	
	Hyaluronan decreased adipogenesis in WAT	Human	Wilson et al. [75]	

Symbols (↑) Increase, (↓) decreased levels

Implications of ECM Remodeling of Adipose Tissue in Obesity and Metabolic Disease

Angiogenesis

Angiogenesis is a physiological process characterized by formation of new blood vessels from precursor cells, and it is crutial for normal maintenance of homeostais, remodeling and expansion of the tissue [72]. In particular, angiogenesis happens between the vascular cells such as endothelial cells, smooth muscle cells, pericytes, preadipocytes and adipocytes, stromal vascular cells and rest of immune cells [10]. These cells are able to secrete some pro- and antiangiogenic molecules that control angiogenesis through different mechanisms.

Adipose tissue is highly vascularized, and an wide capillary network nourishes each adipocyte. Blood vessels deliver nutrients, oxygen, hormones, cytokines, and growth factors to the tissue [36]. Besides, the infiltration of immune cells into the tissue are also supplied by blood vessels. Some studies have reported that angiogenesis often precedes adipogenesis, and adipose tissue expansion is associated with angiogenesis. On the contrary, inhibition of the latter can prevent enlargement of the tissue, concluding the existence of an interaction between endothelial cells and adipocytes [22]. During obesity, a decrease in intracellular adipocyte oxygen tension with increased HIF-1α protein expression precedes macrophage accumulation and pro-inflammatory gene expression, contributing to adipocyte dysfunction in obesity. Recent literature reveals that high levels of intracellular saturated free fatty (FFA) causes a higher adenine nucleotide translocase (ANT)-2-dependent in uncoupled mitochondrial respiration [35]. Consequently, this triggers to higher O_2 consumption and a state of adipocyte hypoxia with an up regulation of *HIF-1α* gene expression. Thus, deletion of adipocyte *Ant2* ameliorates inflammation in adipose tissue and insulin resistance, but no changes in body weight or energy expenditure were observed in HFD fed mice. Mechanistically, adipose-specific *Ant2* knockout inhibits the obesity-induced increase in uncoupled respiration in adipocytes, causing an increased in the intracellular O_2 tension and a lower *HIF-1α* gene expression [61].

In obesity, adipose tissue expansion drives the formation of new blood vessels, which also stimulates adipocyte differentiation [23]. Nevertheless, hypertrophy of the adipocytes is not frequently accompanied by a comparable increased of angiogenesis, which causes tissue dysfunction [11]. VEGFs, fibroblast growth factor-2, angiopoietins 1-2, adiponectin, leptin, and plasminogen activator inhibitor-1, among others, are implicated in the angiogenesis [20]. Accordingly, VEGFs and, in particular, VEGF-A, through VEGF receptor-2, participates actively in the angiogenesis [47]. In this line, VEGF-A function has been reported to contribute to angiogenesis in both animal models and human. Therefore, increased VEGF levels in serum is associated with BMI in people with obesity and overweight [39]. In contrast, other studies have failed to reproduce these findings [57]. In this framework, a decrease in *Vegf* gene expression in WAT has been reported in mice [69] and obese humans [52]. This observation can be partially explained by *Vegf* overexpression in mice that showed protection to HFD-induced inflammation and insulin resistance due to the higher VEGF levels observed in obese subjects [17]. A meta-analysis has also showed a strong association between higher VEGFs expressions and metabolic syndrome, though evidence in obesity is not completely clear [76]. In WAT, VEGF levels have been shown to be selectively increased in response to fasting [25]; while long-term intermittent fasting results in an increase in angiogenesis in WAT as well as browning in subcutaneous WAT. Such changes were associated with an improvement in insulin resistance and attenuation of inflammatory processes [25]. Interestingly, the effects of fasting are markedly reduced in liver-specific FGF21 knock out mice, indicating that FGF21 regulates WAT VEGF levels [25].

Several ECM components have been linked to angiogenesis in adipose tissue. A study has described that CD248 affects hypoxia and modulates the vascularization in adipocytes. Moreover, CD248 expression is up regulated in human adipocytes and it is associated with obesity and metabolic disturbance [54]. F13A1, a transglutaminase that is linked to adipogenesis in cells and obesity in humans and mice, is highly increased in WAT with acquired excess weight and associated with pro-inflammatory, cell stress, angiogenesis, and extracellular matrix remodeling [30].

Finally, modulation of angiogenesis in WAT could benefit obesity treatment and metabolic dysfuntion [23]. Indeed, novel subcutaneous WAT implantation of the adipose matrix with adipogenic and angiogenic factors stimulate adipogenesis in mice [34]. Therefore, a complete understanding of WAT components and how it regulates vasculature in obesity might be crutial to develop effective obesity treatments.

ECM Remodeling, Insulin Signaling, and Glucose Homeostasis

One of the most established mechanisms of metabolically dysfunctional adipose tissue is the excessive accumulation of ECM components. In particular, collagen deposition is the main physical barrier limiting adipose tissue expandability. Therefore, a healthy adipocyte expansion during the development of obesity promotes the shunting of lipids into other tissues, being this process known as ectopic lipid accumulation [74]. Collagen deposition can promote insulin resistance and thickening of capillary basement membrane that is a signal of diabetic microangiopathy. Although the specific mechanism is not completely understood, excessive ECM components deposition in adipose tissue prompts changes in gene expression implicating metabolically unfavorable processes, such as angiogenesis inhibition, adipocyte death, and proinflammatory macrophage infiltration, which could cause insulin resistance. In this context, the accumulation of ECM components in adipose tissue decrease the expansion of adipocytes, triggering cell apoptosis and/or necrosis [31]. Consequently, adipose inflammation and insulin resistance are produced due to the ability of necrotic adipocytes to attract immune cells [65]. While adipose tissue inflammation as a primary event that results in insulin resistance remains to be completely understood, a recent study has suggested that insulin resistance causes an inflammatory response whereby there is an accumulation of M1 macrophages; an observation that was supported by the occurrence of insulin resistance in adipocytes in mTORC2 knock out mice [62]. Indeed, monocyte recruitment and differentiation into M1 macrophages in visceral WAT results in inflammation. Thus, taken together, it is evident that it is the insulin resistnce in adipose tissue that leads to inflammation and not inflammation itself that leads to adipose tissue insulin resistance. In line with these findings, Ruiz-Ojeda et al. [59] demonstrated that isolated adipocytes from subcutaneous WAT of adipose-specific knockout of Kindlin-2, an integrin activator, showed local insulin resistance read by reduced Akt phosphorylation and decreased glucose uptake. However, no differences were observed in the proinflammatory markers in the adipose tissue of chow diet-fed mice [59].

Concerning human studies insulin resistance is associated with collagen accumulation. A recent study revealed an association between collagen content in WAT and insulin resistance in both Caucasian and Chinese populations [4]. Similarly, they detected the same results among subjects with obesity where insulin resistance was evaluated through hyperinsulinemic-euglycemic clamp. The amount of fibrosis in WAT was higher in subject with high levels of insulin resistance, concluding that fibrosis was directly liked to decrease in the insulin sensitivity [21]. Another study also reinforced the knowledge that insulin resistance is accompanied by a high rise of collagens (type I and type III) in biopsies from WAT of healthy males [70]. It has also been detected that excessive collagen accumulation in WAT might inhibit the angiogenesis [64]. These studies suggest an important role in collagen accumulation in the insulin resistance in humans. Our research group found an upregulation in genes involved in ECM regulation (*TNMD* and *NQO1*), adipogenesis (*CRYAB* and *AFF1*) and inflammation (*ANXA1*); and a downregulation in *CALCRL* gene expression in VAT of prepubertal children with obesity compared to normal weight, demonstrating a different gene expression profile in ECM-genes in childhood obesity [2]. In conclusion, proper ECM remodeling maintain or restore insulin sensitivity, inflammation, and angiogenesis in adipocytes, as well as in the regulation of whole body energy metabolism. Nevertheless, novel therapeutic approaches are needed for the effective treatment of obesity and metabolic associated diseases in order to maintain or restore insulin sensitivity.

Conclusions

Adipose tissue expandability drives a tone of cellular responses that are dynamically altered and can disturb the adipocyte function. It has been established that appropriate ECM reorganization and remodeling, and angiogenesis are crucial to prevent fibrosis, inflammation, and insulin resistance in adipose tissue. In addition, modulation of ECM remodeling and angiogenesis may promote adipocyte differentiation, preventing the hypertrophic adipocytes in obesity, and thereby the ectopic lipid accumulation in other tissues. Furthermore, adipose ECM remodeling is modulated by the inflammatory responses of different immune cell types and could drive insulin resistance in adipocytes. However, new evidence proposes that insulin resistance in adipose tissue leads to inflammation rather than vice versa. A deeper mechanistic understanding of adipose ECM remodeling might help to develop new therapeutic approaches against obesity-induced insulin resistance and metabolic disease.

References

1. Abdelaziz R et al (2015) Tissue inhibitors of metalloproteinase-1 and 2 and obesity related non-alcoholic fatty liver disease: is there a relationship. Digestion 92(3):130–137
2. Aguilera CM et al (2015) Genome-wide expression in visceral adipose tissue from obese prepubertal children. Int J Mol Sci 16(4):7723–7737
3. Åkra S, Aksnes TA, Flaa A, Eggesbø HB, Opstad TB, Njerve IU, Seljeflot I (2020) Markers of remodeling in subcutaneous adipose tissue are strongly associated with overweight and insulin sensitivity in healthy non-obese men. Sci Rep 10(1):14055. https://doi.org/10.1038/s41598-020-71109-4
4. Alba DL et al (2018) Subcutaneous fat fibrosis links obesity to insulin resistance in Chinese Americans. J Clin Endocrinol Metab 103(9):3194–3204
5. Barchetta I et al (2019) Impaired bone matrix glycoprotein pattern is associated with increased cardio-metabolic risk profile in patients with type 2 diabetes mellitus. J Endocrinol Invest 42(5):513–520
6. Berg G, Barchuk M, Miksztowicz V (2019) Behavior of metalloproteinases in adipose tissue, liver and arterial wall: an update of extracellular matrix remodeling. Cells 8(2):158

7. Bonnans C, Chou J, Werb Z (2014) Remodelling the extracellular matrix in development and disease. Nat Rev Mol Cell Biol 15(12):786–801
8. Bourboulia D, Stetler-Stevenson WG (2010) Matrix metalloproteinases (MMPs) and tissue inhibitors of metalloproteinases (TIMPs): Positive and negative regulators in tumor cell adhesion. Semin Cancer Biol 20(3):161–168
9. Buechler C, Krautbauer S, Eisinger K (2015) Adipose tissue fibrosis. World J Diabetes 6(4):548–553
10. Cao Y (2013) Angiogenesis and vascular functions in modulation of obesity, adipose metabolism, and insulin sensitivity. Cell Metab 18(4):478–489
11. Cao Y (2014) Angiogenesis as a therapeutic target for obesity and metabolic diseases. Chem Immunol Allergy 99:170–179
12. Carbone F, Adami G, Liberale L, Bonaventura A, Bertolotto M, Andraghetti G, Scopinaro N, Camerini GB, Papadia FS, Cordera R et al (2018) Serum levels of osteopontin predict diabetes remission after bariatric surgery. Diabetes Metab. https://doi.org/10.1016/j.diabet.2018.09.007
13. Choi JS, Kim BS, Kim JY, Kim JD, Choi YC, Yang HJ, Park K, Lee HY, Cho YW (2011). Decellularized extracellular matrix derived from human adipose tissue as a potential scaffold for allograft tissue engineering. J Biomed Mater Res Part A 97:292–299. https://doi.org/10.1002/jbm.a.33056
14. Daquinag AC et al. (2020) Glycosaminoglycan modification of Decorin depends on MMP14 activity and regulates collagen assembly. Cells 9(12)
15. DeMarsilis AJ et al. (2014) Elastin insufficiency predisposes mice to impaired glucose metabolism. J Mol Genet Med 8(3)
16. Derosa G et al (2008) Matrix metalloproteinase-2 and -9 levels in obese patients. Endothelium 15(4):219–224
17. Elias I et al (2012) Adipose tissue overexpression of vascular endothelial growth factor protects against diet-induced obesity and insulin resistance. Diabetes 61(7):1801–1813
18. Ezzati-Mobaser S. et al. (2020) The up-regulation of markers of adipose tissue fibrosis by visfatin in pre-adipocytes as well as obese children and adolescents. Cytokine 134:155193
19. Gao H et al. (2019) Lipoatrophy and metabolic disturbance in mice with adipose-specific deletion of kindlin-2. JCI Insight 4(13)
20. Gealekman O et al (2014) Control of adipose tissue expandability in response to high fat diet by the insulin-like growth factor-binding protein-4. J Biol Chem 289(26):18327–18338
21. Guglielmi V et al (2015) Omental adipose tissue fibrosis and insulin resistance in severe obesity. Nutr Diabetes 5(8):e175–e175
22. Gupta RK et al (2012) Zfp423 expression identifies committed preadipocytes and localizes to adipose endothelial and perivascular cells. Cell Metab 15(2):230–239
23. Hammarstedt A et al (2018) Impaired adipogenesis and dysfunctional adipose tissue in human hypertrophic obesity. Physiol Rev 98(4):1911–1941
24. Han CY et al (2007) Adipocyte-derived serum amyloid A3 and hyaluronan play a role in monocyte recruitment and adhesion. Diabetes 56(9):2260–2273
25. Hua L et al. (2021) Dietary Intake Regulates White Adipose Tissues Angiogenesis via Liver Fibroblast Growth Factor 21 in Male Mice. Endocrinology 162(3)
26. Jaiswal A et al (2011) Comparative analysis of human matrix metalloproteinases: emerging therapeutic targets in diseases. Bioinformation 6(1):23–30
27. Jaworski DM et al (2011) Sexually dimorphic diet-induced insulin resistance in obese tissue inhibitor of metalloproteinase-2 (TIMP-2)-deficient mice. Endocrinology 152(4):1300–1313
28. Ji E et al (2014) Inhibition of adipogenesis in 3T3-L1 cells and suppression of abdominal fat accumulation in high-fat diet-feeding C57BL/6J mice after downregulation of hyaluronic acid. Int J Obes (Lond) 38(8):1035–1043
29. Jiang H et al (2007) TIMP-1 transgenic mice recover from diabetes induced by multiple low-dose streptozotocin. Diabetes 56(1):49–56
30. Kaartinen MT et al. (2020) F13A1 transglutaminase expression in human adipose tissue increases in acquired excess weight and associates with inflammatory status of adipocytes. Int J Obes (Lond)

31. Khan T et al (2009) Metabolic dysregulation and adipose tissue fibrosis: role of collagen VI. Mol Cell Biol 29(6):1575–1591
32. Kiefer FW et al (2008) Osteopontin expression in human and murine obesity: extensive local up-regulation in adipose tissue but minimal systemic alterations. Endocrinology 149(3):1350–1357
33. Kiefer FW et al (2010) Neutralization of osteopontin inhibits obesity-induced inflammation and insulin resistance. Diabetes 59(4):935–946
34. Kokai LE et al (2019) Injectable allograft adipose matrix supports adipogenic tissue remodeling in the nude mouse and human. Plast Reconstr Surg 143(2):299e–309e
35. Lee YS et al (2014) Increased adipocyte O2 consumption triggers HIF-1α, causing inflammation and insulin resistance in obesity. Cell 157(6):1339–1352
36. Lemoine AY, Ledoux S, Larger E (2013) Adipose tissue angiogenesis in obesity. Thromb Haemost 110(4):661–668
37. Lin TH Chun, L Kang (2016) Adipose extracellular matrix remodelling in obesity and insulin resistance. Biochem Pharmacol 119:8–16
38. Liu Y et al (2018) Integrated analysis of long noncoding RNA and mRNA expression profile in children with obesity by microarray analysis. Sci Rep 8(1):8750
39. Loebig M et al. (2010) Evidence for a relationship between VEGF and BMI independent of insulin sensitivity by glucose clamp procedure in a homogenous group healthy young men. PLOS One 5(9):e12610
40. Luk CT et al (2017) FAK signalling controls insulin sensitivity through regulation of adipocyte survival. Nat Commun 8:14360
41. Maquoi E et al (2002) Modulation of adipose tissue expression of murine matrix metalloproteinases and their tissue inhibitors with obesity. Diabetes 51(4):1093–1101
42. Mariman ECM, Wang P (2010) Adipocyte extracellular matrix composition, dynamics and role in obesity. Cell Mol Life Sci 67(8):1277–1292
43. Matsugi K et al (2016) Thrombospondin 1 suppresses insulin signaling in C2C12 myotubes. Kobe J Med Sci 62(1):E13–E18
44. Matsuo Y et al. (2015) Thrombospondin 1 as a novel biological marker of obesity and metabolic syndrome. Metabol Clin Experim 64(11):1490–1499
45. McCulloch LJ et al (2015) COL6A3 is regulated by leptin in human adipose tissue and reduced in obesity. Endocrinology 156(1):134–146
46. Menghini R et al (2009) Tissue inhibitor of metalloproteinase 3 deficiency causes hepatic steatosis and adipose tissue inflammation in mice. Gastroenterology 136(2):663–72.e4
47. Moens S et al (2014) The multifaceted activity of VEGF in angiogenesis—Implications for therapy responses. Cytokine Growth Factor Rev 25(4):473–482
48. Mori S et al (2014) Characteristic expression of extracellular matrix in subcutaneous adipose tissue development and adipogenesis; comparison with visceral adipose tissue. Int J Biol Sci 10(8):825–833
49. Murphy G (2011) Tissue inhibitors of metalloproteinases. Genome Biol 12(11):233
50. Nagase H, Visse R, Murphy G (2006) Structure and function of matrix metalloproteinases and TIMPs. Cardiovasc Res 69(3):562–573
51. Oh J et al. (2021) Type VI collagen and its cleavage product, endotrophin, cooperatively regulate the adipogenic and lipolytic capacity of adipocytes. Metabolism 114:154430
52. Pasarica M et al (2009) Reduced adipose tissue oxygenation in human obesity: evidence for rarefaction, macrophage chemotaxis, and inflammation without an angiogenic response. Diabetes 58(3):718–725
53. Petäistö T, Vicente D, Mäkelä KA, Finnilä MA, Miinalainen I, Koivunen J, Izzi V, Aikio M, Karppinen SM, Devarajan R, Thevenot J, Herzig KH, Heljasvaara R, Pihlajaniemi T (2020) Lack of collagen XVIII leads to lipodystrophy and perturbs hepatic glucose and lipid homeostasis. J Physiol 598(16):3373–3393. https://doi.org/10.1113/JP279559
54. Petrus P et al (2018) Adipocyte expression of SLC19A1 links DNA hypermethylation to adipose tissue inflammation and insulin resistance. J Clin Endocrinol Metab 103(2):710–721

55. Poltavets V et al. (2018) The role of the extracellular matrix and its molecular and cellular regulators in cancer cell plasticity. Front Oncol 8(431)
56. Rebuffat SA et al (2018) Adipose tissue derived-factors impaired pancreatic β-cell function in diabetes. Biochim Biophys Acta Mol Basis Dis 1864(10):3378–3387
57. Rehman J et al (2003) Obesity is associated with increased levels of circulating hepatocyte growth factor. J Am Coll Cardiol 41(8):1408–1413
58. Ritter AM et al (2017) Crosstalk between obesity and MMP-9 in cardiac remodelling -a cross-sectional study in apparent treatment-resistant hypertension. Blood Press 26(2):122–129
59. Ruiz-Ojeda FJ, Wang J, Bäcker T, Krueger M, Zamani S et al. (2021) Active integrins regulate white adipose tissue insulin sensitivity and brown fat thermogenesis. Mol Metab 45:101147
60. Sakamuri S et al (2017) Absence of tissue inhibitor of metalloproteinase-4 (TIMP4) ameliorates high fat diet-induced obesity in mice due to defective lipid absorption. Sci Rep 7(1):6210
61. Seo JB et al (2019) Knockdown of Ant2 reduces adipocyte hypoxia and improves insulin resistance in obesity. Nat Metab 1(1):86–97
62. Shimobayashi M et al (2018) Insulin resistance causes inflammation in adipose tissue. J Clin Invest 128(4):1538–1550
63. Signorelli SS et al (2005) Plasma levels and zymographic activities of matrix metalloproteinases 2 and 9 in type II diabetics with peripheral arterial disease. Vasc Med 10(1):1–6
64. Spencer M et al (2011) Adipose tissue extracellular matrix and vascular abnormalities in obesity and insulin resistance. J Clin Endocrinol Metab 96(12):E1990–E1998
65. Strissel KJ et al (2007) Adipocyte death, adipose tissue remodeling, and obesity complications. Diabetes 56(12):2910–2918
66. Sun K et al (2013) Fibrosis and adipose tissue dysfunction. Cell Metab 18(4):470–477
67. Sun Z, Costell M, Fässler R (2019) Integrin activation by talin, kindlin and mechanical forces. Nat Cell Biol 21(1):25–31
68. Sun K, Kusminski CM, Scherer PE (2011) Adipose tissue remodeling and obesity. J Clin Invest 121(6):2094–2101
69. Sung HK et al (2013) Adipose vascular endothelial growth factor regulates metabolic homeostasis through angiogenesis. Cell Metab 17(1):61–72
70. Tam CS et al (2014) Weight gain reveals dramatic increases in skeletal muscle extracellular matrix remodeling. J Clin Endocrinol Metab 99(5):1749–1757
71. Tinahones FJ et al (2012) Obesity-associated insulin resistance is correlated to adipose tissue vascular endothelial growth factors and metalloproteinase levels. BMC Physiol 12:4
72. Viallard C, Larrivée B (2017) Tumor angiogenesis and vascular normalization: alternative therapeutic targets. Angiogenesis 20(4):409–426
73. Vilmi-Kerälä T et al (2017) Subclinical inflammation associated with prolonged TIMP-1 upregulation and arterial stiffness after gestational diabetes mellitus: a hospital-based cohort study. Cardiovasc Diabetol 16(1):49
74. Williams AS, Kang L, Wasserman DH (2015) The extracellular matrix and insulin resistance. Trends Endocrinol Metab 26(7):357–366
75. Wilson N et al (2019) Role of hyaluronan in human adipogenesis: evidence from in-vitro and in-vivo studies. Int J Mol Sci 20(11):2675
76. Zafar MI et al (2018) Association between the expression of vascular endothelial growth factors and metabolic syndrome or its components: a systematic review and meta-analysis. Diabetol Metab Syndr 10:62–62

Part II
Therapeutic Mechanisms of Obesity

Chapter 12
Obesity; Its Prevalence, Consequences and Potential Therapies

Tanya Sharma, Husam Salah, Naga Sai Shravan Turaga, and Jawahar L. Mehta

Abstract Overweight and obesity are exponentially growing health concerns, with prevalence fast approaching 2 billion people affected- a number that has tripled over the last few decades. Obesity is a known risk factor for a gamut of diseases including diabetes, hypertension, coronary artery disease, arrhythmias, and heart failure. The relationship of obesity and cardiovascular disease is now extensively studied. Here we discuss this association in detail and elaborate on the underlying mechanisms including metabolic syndrome, hemodynamic compensations, inflammation, hypoxia, and alterations in microbiome. We also evaluate the importance of screening for and treating obesity in the clinical practice, as well as the currently available modalities in management.

Keywords Obesity · Cardiovascular disease · Metabolic syndrome · Hypoxia · Inflammation · Microbiome · Dietary interventions · Physical activity · Behavioral therapy · Pharmacotherapy · Bariatric surgery

Introduction

The World Health Organization (WHO) defines overweight and obesity as an abnormal or excessive accumulation of fat that may impair health [1]. Under the 2030 Agenda for Sustainable Development, WHO described obesity as a major challenge due to its contribution to non-communicable disease burden. Table 12.1 presents some striking facts from the 2020 WHO factsheet on obesity. Obesity has been rightfully described as a global epidemic. The average body mass index (BMI), which is derived by weight (in kg) per square of height (in meter), has increased by 0.4 kg/m^2 per decade between 1980–2008 [2]. This alarming rise in the prevalence of obesity over the latter half of the last century has attracted attention to the proportional rise in obesity related morbidity and mortality.

T. Sharma · H. Salah · N. S. S. Turaga · J. L. Mehta (✉)
Division of Cardiology, University of Arkansas for Medical Sciences and the Central Arkansas Veterans Healthcare System, Little Rock, AR, USA
e-mail: mehtajl@uams.edu

© Springer Nature Switzerland AG 2021

P. S. Tappia et al. (eds.), *Cellular and Biochemical Mechanisms of Obesity*, Advances in Biochemistry in Health and Disease 23,
https://doi.org/10.1007/978-3-030-84763-0_12

Table 12.1 WHO key facts on Obesity and Overweight*

• Worldwide obesity has nearly tripled between 1975 and 2016
• In 2016, more than 1.9 billion adults were overweight. Of these, over 650 million were obese
• 39% of adults were overweight in 2016
• 13% of the world's adult population (11% of men and 15% of women) were obese in 2016
• Most of the world's population live in countries where overweight and obesity kills more people than underweight

*From WHO fact sheet on Obesity and overweight[1]

United States has the highest BMI prevalence of all high-income countries [2]. In 2007–2008, the age-adjusted prevalence of obesity in the US was 33.8% and that for overweight and obesity combined 68% [3]. While the trends from this report show a slower rise in BMI over the last decade, overall future estimates do not appear to be promising in terms of controlling the obesity-epidemic. It is estimated that by the year 2030 nearly 1 in 2 adults in the United States would be classified as obese (defined as BMI 25 to < 30 kg/m2) and nearly 1 in 4 adults would be severely obese (BMI > 35 kg/m2) [4].

Historically considered a disease of high-income populations, overweight and obesity are sharply increasing in low and middle-income regions of the world as well. Over the last few decades the BMI increased in all but a few countries [2]. In 2008, Oceania countries recorded the highest BMI globally as a group; with Nauru, Cook Islands and Tonga seeing increases or more than 2 kg/m2/decade since the 1980s [2]. Childhood obesity trends in low and middle-income countries portend an ominous trend in the upcoming years. According to WHO data the number of overweight children under the age of 5 has increased by nearly 24% since 2000. As of 2019, almost 50% of all overweight or obese children under the age of 5 years liven in Asia [1].

While there is an upward trend in obesity globally, the extent of the prevalence is not the same across population groups. Gender, for example presents a physiological inequitability when it comes to being overweight and obese. Worldwide, women are 4% more likely to be obese as compared to men [1]. Furthermore, obesity impacts women more adversely than men. In USA, overweight and obese women had a 6.6-times and 1.8 times higher burden of disease, respectively, when compared to overweight and obese men [5]. Obese women suffered 70,000 additional deaths than normal-weight women, whereas the comparative number is 42,000 in men. Across the population, women lost an additional 1.5 million quality adjusted life years to obesity than men.

Obesity has been known to be a factor in the development of cardiovascular disease. It is closely connected with several other diseases which indirectly affect the occurrence of cardiovascular disease as well, such as hypertension and diabetes. In this chapter, we aim to review the mechanistic relationship between obesity and cardiovascular disease and its implication in clinical practice of cardiovascular medicine.

Obesity and Relationship with Cardiovascular Disease

Obesity has been associated with atherosclerotic disease, heart failure, as well as, arrhythmias and sudden cardiac death [6]. It has been debated whether it is an independent risk factor for cardiovascular disease or is associated by proxy since the impact of obesity on co-existing risk modifiers is remarkable. Cross-sectional analysis of the National Health and Nutrition Examination Survey (NHANES) from 2000 showed that the prevalence of hypertension amongst obese (42.5%) and overweight (27.8%) individuals was starkly different from those with BMI < 25 kg/m2(15.3%). In addition, approximately 75% of obese participants had prehypertension, while prevalence in the normal weight group was 47% [7]. Obesity predisposes to impaired glucose tolerance and type 2 diabetes. Even in state of euglycemia, obese people have about 30% less insulin sensitivity and hence need to secrete more insulin, which leads to loss of compensatory insulin hypersecretion in overt diabetes [8]. It has been shown that even modest weight reductions of 5% to 10% of body weight impact blood pressure, glycemic control, total cholesterol, and reduce the severity of obstructive sleep apnea [9].

While the associations of obesity with other known risk factors cannot be ignored, it is more likely to be only one of the mechanisms involved in predisposing to cardiovascular disease. All major epidemiological studies in the field conclude that risk for cardiovascular disease persists after adjustment for other comorbidities, ascertaining obesity as an independent risk factor [6]. In the Framingham Heart Study, multivariate logistic regression adjusted for age, systolic blood pressure, serum cholesterol, cigarettes smoking, glucose intolerance, and left ventricular hypertrophy, yielded a significant association between body weight and cardiovascular disease. Furthermore, amongst the obese, those without any additional risk factors, cardiovascular disease incidence increased with weight in both men and women [10].

In the Framingham cohort, coronary artery disease (CAD) was the most frequent manifestation of cardiovascular disease. The incidence was seen to incrementally increase with weight and the curve was steeper amongst the younger sub-population [10]. In a more recent meta-analysis of over a million participants, obesity was associated with a nearly 60% higher risk of CAD. Additionally, each unit increment in BMI was associated with an increase in risk of CAD by 5% in men and 4% in women, respectively [11].

The risk of clinical heart failure was seen to almost double with the presence of obesity [12]. The risk increased by 5% in men and 7% in women for each 1 kg/m2 increase in BMI. Heart failure mortality revealed a non-linear relationship with a J-shaped curve with a threshold at BMI ~ 23-23 kg/m2 [13]. BMI is a strong predictor for heart failure with preserved ejection fraction [14]. This is in keeping with the observation that a large waist circumference is associated with left ventricular dysfunction [15].

Obesity also adversely affects cardiac structure and repolarization activity, predisposing to arrhythmias and sudden death. In a meta-analysis of over 100,000 participants, obese individuals have almost a 1.5 times the risk of developing atrial fibrillation [16]. Obesity is seen to be significantly associated with prolonged corrected QT interval, and the QT interval seems to decrease with weight loss with diet, exercise or bariatric surgery [17]. In the Framingham Heart Study, obesity was seen to be associated with a 40-fold increase in the risk of sudden cardiac death [18].

The impact of obesity on cardiovascular outcomes is undeniably reflected throughout literature and can be explained by complex interplay of mechanisms. The product of current prevalence of obesity and the degree of impact on incidence of cardiovascular disease makes obesity a prominent target for risk factor modification to improve morbidity and mortality outcomes of cardiovascular disease.

Pathophysiology of Cardiovascular Disease in Obesity

The pathophysiological link between obesity and cardiovascular disease is complex. A number of mechanisms play a role in the development and progression of cardiovascular disease in obesity. These mechanisms likely stem from the altered metabolic profile, tissue oxygenation and neurohormonal profiles, and the different structural and functional cardiac adaptations that occur as a result of adipose tissue accumulation. Figures 12.1 and 12.2 show a schematic representation of these adaptation and structural effects.

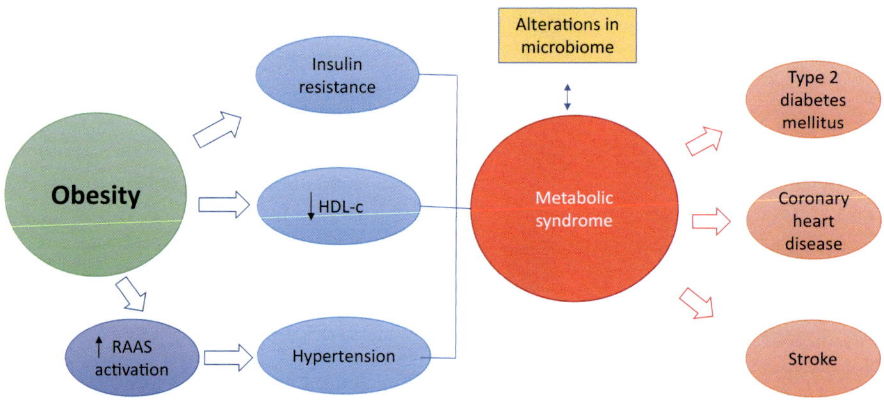

Fig. 12.1 Obesity and its relationship to metabolic syndrome

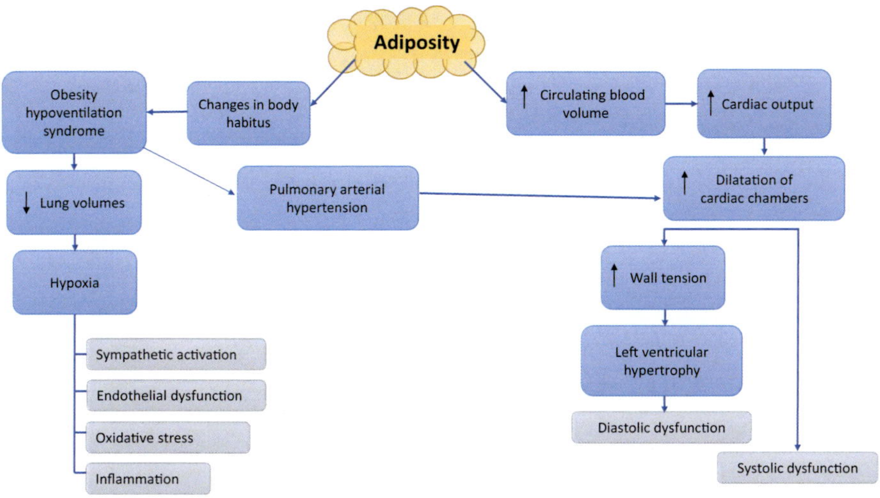

Fig. 12.2 Physiological alterations in cardio-pulmonary mechanisms in obesity. There is disordered cardiac function and vascular homeostasis secondary to increased adipose mass

Metabolic Syndrome

The metabolic syndrome is a cluster of metabolic risk factors that include atherogenic dyslipidemia, hypertension, high glucose level with insulin resistance, prothrombotic state and proinflammatory state, all of which are known risk factors for atherosclerotic cardiovascular disease [19, 20]. Obesity can induce metabolic syndrome by contributing to each of these risk factors, as shown in Fig. 12.3. (1) Obesity decreases

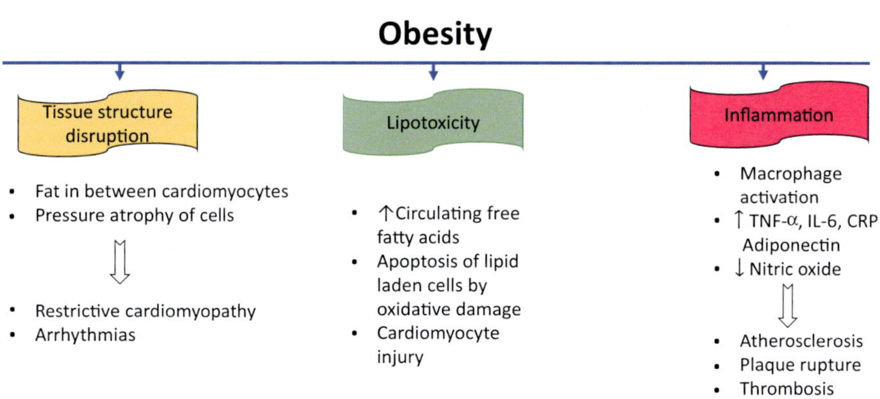

Fig. 12.3 Structural and systemic alterations in obesity. Effect on obesity on cardiovascular system mediated by structural alterations in cardiac tissue, lipotoxicity by free fatty acids and increased inflammation

high-density lipoprotein (HDL) level, a hallmark of atherogenic dyslipidemia [20]. (2) Obesity is a major risk for hypertension; compared to healthy individuals, obese patients have higher level of renin-angiotensin system activation, which is a major driver for hypertension [21]. (3) Obesity is associated with insulin resistance. The exact pathophysiological mechanism is not well understood but is believed to be related to mitochondrial dysfunction, oxidative stress, and lipotoxicity [22]. (4) Obesity drives a chronic inflammatory state and is associated with fibrinolysis impairment, both of which induces a prothrombotic state [23]. The constellation of all of these factors puts individuals with obesity at higher risk for cardiovascular disease.

Hypoxia

Littleton et al. demonstrated a negative correlation between the body mass index and partial pressure of arterial blood oxygen independent of hypoventilation [24]. Obesity can cause hypoxia as a result of decreased lung volumes. Additionally, obesity is associated with an increased risk for obstructive sleep apnea, in which episodic upper airway narrowing/closure occurs during sleep, resulting in intermittent hypoxia [25]. Hypoxia leads to sympathetic activation, endothelial dysfunction, oxidative stress, and a state of systemic inflammation [25]. All of these factors increase the risk for cardiovascular disease in obesity individuals.

Inflammation

An increasing body of evidence links various cardiovascular diseases, such as atherosclerosis, heart failure, acute coronary syndrome, and coronary artery disease to inflammation [26, 27]. Systemic inflammation is a hallmark of obesity and primarily results from macrophage activation [27]. Macrophages act as scavengers of apoptotic adipocytes; therefore, an increase in adipose tissue leads to accumulation and activation of macrophages with subsequent inflammation and release of proinflammatory mediators, such as tumor necrosis factor-alpha (TNF-alpha), interleukin 6 (IL-6), and C-reactive protein (CRP). Additionally, obesity is associated with a significant reduction in adiponectin, [28], a protein derived from adipocytes with potent anti-inflammatory effects. Adiponectin increases the production of nitric oxide, which, in turn, reduces oxidative stress, inhibits plaque formation and thrombosis, and has an important vasoprotective role [29–31]. A possible mechanism of cardiomyopathy in obesity is myocardium lipotoxicity due to free fatty acid resulting in apoptosis of lipid-laden cells including cardiomyocytes [32].

Altered Hemodynamics

Obesity results in an increase in blood volume with subsequent increase in cardiac output by increasing the stroke volume [33, 34]. Consequently, the venous return to the right heart increases, which, in turn, produces dilatation of the heart chambers. This, in turn, results in higher wall tension and left ventricular hypertrophy (LVH), which is typically eccentric, accompanied by decreased diastolic compliance [35, 36]. The end result of these changes is typically left ventricular (LV) enlargement due to increased left ventricular filling pressure. LVH initially adapts to the LV enlargement preserving the systolic function. However, as LV dilatation and wall tension increase, systolic function develops [33, 34].

Myocyte Degeneration

A number of mechanisms underlie cardiomyopathy in patients with obesity. Myocyte degeneration can occur as a result of the accumulation of fat between cords of cells, resulting in adipositas cordis [33, 37–39]. In this syndrome, bands of irregular aggregates of adipose tissue result in separation of myocardial cells with subsequent pressure-induced atrophy in these cells [33]. As fat accumulation in the myocardium progresses, a pattern of restrictive cardiomyopathy may develop [40]. Occasionally, the sinus node, the atrioventricular node, the right bundle branch can be involved, resulting in conduction disorders [38, 39].

Microbiome in Obesity

Gastrointestinal tract is an immunologically active organ system which harbors very diverse microbiota that play a vital role in the host metabolism and energy homeostasis. The gut microbiota composition is altered by a multitude of host and external factors like dietary habits, lifestyle, circadian rhythm, psychological stress, hygiene, vaccinations, antibiotic use and physical activity. Human beings and microbiota have co-evolved and co-adapted, at an individual host level or at a population level over time suggesting the potential of host-microbiome relationship in metabolic syndrome and related disorders [41]. The "old friend and "missing friend" hypothesis suggest that lifestyle changes resulted in loss of specific bacteria resulting in less diverse modern-day microbiota, potentially triggering autoimmune, allergic and inflammatory disorders [42]. Individuals with obesity and metabolic syndrome have a higher *Firmicutes*-to-*Bacteroidetes* ratio and a reduction in the abundance of firmicutes was observed post gastric bypass surgery suggesting a potential association [43]. Similarly, fecal microbiota transplantation (FMT) from a lean donor to a patient with metabolic syndrome resulted in improvement in insulin sensitivity and FMT from an overweight donor was associated with new-onset obesity [44, 45]. Microbiota affect metabolism through various mechanisms like bile acid metabolism which affects glucose metabolism through activation of farnesoid X receptor and G-protein coupled bile acid receptor [43], by-products of choline metabolism which are predictors of cardiovascular mortality [46], short chain fatty acid metabolism which modulate insulin signaling in pancreatic beta cells through G protein-coupled receptor 41 and 43 signaling pathways [47]. Gut microbiota are also a source of pathogen associated molecular patterns (PAMPs) which can activate Toll-like receptors and Nod-like receptors [48]. PAMPs can translocate across intestinal barrier and cause inflammatory response and systemic endotoxemia. Further research into mechanism, potential therapies and lifestyle adjustments affecting the host-microbiota relationship are needed and can open up a great avenue in management of metabolic syndrome and various other immunological diseases [48, 49].

Clinical Aspects

Obesity is a sensitive, complex, and stigmatized disease with many social, behavioral, psychological, economic and health related repercussions. Hence, understanding the conundrum of obesity is necessary to formulate individualized strategies for prevention, treatment and preventing relapse of obesity.

Prevention

Prevention of obesity is a lifelong process which starts in infancy and continues all along adulthood. Multiple medical societies including American Academy of Pediatrics (AAP), Center for Disease Control (CDC), U.S. Preventive Services Task Force (USPSTF), American Heart Association (AHA),the National Heart Lung and Blood Institute (NHLBI), the Institute of Medicine (IOM) have developed evidence based consensus and guidelines focusing on early detection of high risk population, physical activity, healthy diet, counselling, behavioral modifications and sleep.

Primary prevention aims at creating awareness about obesity and encouraging healthy habits at a large scale. It includes public health initiatives like health promotion at events, workshops, TV advertisements or education and behavioral training at schools or communities. Primary prevention helps to improve the overall health and awareness of the general population.

Secondary prevention aims at identifying at-risk individuals by screening and formulating necessary corrective measures. It can be done through pediatricians, primary care physicians and public health officials at an individual and family level.

Tertiary prevention aims at identifying individuals who are already obese or at very high risk, and to form an individualized structured multidisciplinary program to cope up with obesity and obesity related complications [50].

Management

Treatment of obesity is multifaceted and complex. It requires a multidisciplinary team effort with a specifically tailored approach for each individual patient. It is a coordinated effort with the primary care physician as the center of the cogwheel working with dietitians, exercise physiologists, behavioral therapists and specialists like endocrinologists and bariatric surgeons for advanced therapies.

Management of obesity aims at decreasing body fat to improve physical appearance, mental health, functional capacity, quality of life and medical health. Even though physical appearance and functional status improves with procedures like liposuction, it does not have any beneficial effects on cardiovascular metabolic risk factors. The benefit was observed with fat loss induced by negative energy balance with various treatment modalities [51].

Dietary Intervention

Maintaining a negative net calorie balance can be achieved by multiple methods, consuming less than the age appropriate daily calorie requirement, usually 1200 to 1500 kcal/d for women and 1500 to 1800 kcal/d for men; Estimating individual energy requirements and maintaining energy deficit of 500 kcal/d or 750 kcal/d or 30% energy deficit [52–54]; or by Ad libitum approaches by restricting/eliminating or prescribing certain foods. Various successful dietary approaches (eg, Mediterranean-style diet, lower-fat lacto–ovo-vegetarian or vegan-style diet, or lower-fat diet with high dairy/calcium with added fiber and/or low–glycemic-index [low–glycemic-load] foods) with varying proportions of protein, carbohydrates and fats are available [55]. An individualized dietary plan needs to be formulated considering the dietary preferences, socioeconomic and cultural variables. It is estimated that average weight loss is maximal at 6 months(4 kg to 12 kg), with smaller losses for up to 2 years(4 kg to 10 kg at 1 year and 3 kg to 4 kg at 2 years) in overweight and obese adults [55].

Physical Activity

Weight loss can be attained through a negative energy balance by increasing expenditure through physical activity or by decreasing intake by dietary modifications or both. Weight loss by itself is not an effective method for weight loss as It is generally easier to reduce the intake than to increase the expenditure. Physical activity is more beneficial in maintaining weight loss and preventing weight regain. Long Term compliance with physical activity is difficult and needs behavioral strategies like counselling, frequent reminders by telephone or by mail to enhance adoption and maintenance of physical therapy. At least 150 min of moderate-intensity aerobic exercise weekly or 75 min of vigorous-intensity aerobic exercise weekly is recommended for adults [55–57].

Behavioral Therapy

Behavioral therapy aims to identify, analyze and modify potentially self-destructive or unhealthy behaviors that increase body weight. It provides techniques to maintain continued compliance with other weight loss therapies by changing lifestyle habits. Various behavioral therapies have been described like goal setting, self-monitoring, stimulus control, cognitive restructuring, problem solving, relapse prevention, stress management, contingency management and social support [58–60].

Pharmacotherapy

Obesity is a chronic disease, so pharmacotherapy should be initiated in conjunction with other treatment modalities with intent to treat long term, if not lifelong [61]. Anti-obesity medications (AOM) have been shown to attain a weight loss range of 5–10% total body weight. AOMs are long term medications with multiple side effects and cessation of drugs leads to rebound weight gain. Hence, it is important to make an informed shared decision after weighing in the risks and benefits with patients before initiating the therapy [62]. Current guidelines approve AOMs for individuals with BMI $\geq$ 30 kg/m2 or BMI $\geq$ 27 kg/m2 with comorbidities [55]. Commonly used AOMs include Orlistat, phentermine, Liraglutide, naltrexone and bupropion. Their mechanism of action, side effects and contraindications are summarized in Table 12.2 [62–65].

Bariatric Surgery

Bariatric surgery is currently recommended for patients with BMI of 35.0 to 39.9 kg/m2 plus $\geq$ 1 severe obesity-related medical complication such as hypertension, type 2 diabetes mellitus, heart failure, or OSA and persons with a BMI $\geq$ 40 kg/m2 based on the National Institutes of Health 1991 consensus statement [55, 66]. Bariatric surgery has a low perioperative mortality of approximately 1%. Gastric bypass, gastric banding, gastroplasty and Biliopancreatic diversion $\pm$ duodenal switch are different bariatric surgeries which are currently being performed. Depending on the type of procedure weight loss at 2 to 3 years ranges from 20 to 35% of initial weight in patients with presurgical BMI $\geq$ 30 [67–70]. Long term complications of bariatric procedures include intestinal obstruction, marginal ulcer, ventral

Table 12.2 Pharmacotherapy in obesity

Name	Mechanism of action	Side effects	Contraindication
Orlistat	Intestinal lipase inhibitor / Reduces fat absorption by up to 30%	Gastrointestinal side effects like flatulence and greasy stools. Fat-soluble vitamin deficiency. Reduced absorption of lipophilic medications if both drugs are taken simultaneously	Pregnancy, patients with malabsorption disorders, reduced gallbladder function Caution is advised in patients with obstructed bile duct, impaired liver function, or pancreatic disease
Phentermine	Stimulates the release of norepinephrine and dopamine from nerve terminals. Appetite suppressant Appetite suppressant and sympathomimetic agent	Dry mouth, constipation, and insomnia. Dose related Increase in heart rate blood pressure and can prevent weight loss–induced decrease in blood pressure	Contraindicated in pregnancy, patients older than 65 years of age, and patients with a history of drug misuse Relative contraindications include severe hypertension, coronary artery disease, concomitant use of SSRIs, MAO inhibitors, tricyclic antidepressants, and stimulants
Liraglutide	GLP-1 receptor agonist / Decreases appetite, increases fullness, increases satiety	Nausea, vomiting, diarrhea, constipation, abdominal pain, dyspepsia, injection site reactions, headache, cholelithiasis, pancreatitis (rare), hypoglycemia in diabetics on hypoglycemic medications	Pregnancy, Personal or family history of medullary thyroid cancer or multiple endocrine neoplasia syndrome type 2.
Naltrexone / Bupropion	Combination of a dopamine and norepinephrine reuptake inhibitor and mu-opioid receptor antagonist / Decreases appetite and cravings	Nausea, vomiting, diarrhea, constipation, headache, insomnia or sleep disorder, dizziness, anxiety, depression, seizure, increase in blood pressure and heart rate, dry mouth, tremor, open angle glaucoma	Pregnancy. Hypersensitivity to bupropion or naltrexone Caution is advised in patients with a history of seizures, serious psychiatric illness, drug or alcohol misuse, or eating disorders

hernia, gastric wall erosion, metabolic complications like hypoglycemia, mineral and vitamin deficiencies [71].

Future Directions

Obesity is a complex medical and socio-economic problem which has a great potential and necessity for innovation at various levels. There is a great potential of using the technological advances in creating social awareness, education and promoting healthy habits and diet. There has been a surge in the use of personal devices like smartphones, mobile applications and smart watches and has a great potential to impact obesity at an individual level. Government and administrative agencies need to formulate policies to promote awareness, control dietary sources of high calories by taxation, and provide incentives for promoting healthy lifestyles. Over the last two decades, there has been great advancements in medical research in understanding and managing obesity, but there are multiple aspects of obesity which are still not well understood. Multiple studies are currently being conducted to understand biomarkers, microbiomes, mechanisms and molecular genetics of obesity, and to develop therapeutics and new laparoscopic and minimally invasive procedures which require funding from governmental agencies and industrial collaboration. Hence, there is a great potential in the field of obesity and requires efforts and collaboration at various levels.

References

1. Obesity and overweight [Internet]. [cited 2020 Nov 5]. Available from: https://www.who.int/news-room/fact-sheets/detail/obesity-and-overweight
2. Finucane MM, Stevens GA, Cowan MJ et al (2011) National, regional, and global trends in body-mass index since 1980: systematic analysis of health examination surveys and epidemiological studies with 960 country-years and 9·1 million participants. Lancet (London, England). 377:557–567
3. Flegal KM, Carroll MD, Ogden CLCL (2010) Prevalence and trends in obesity among US adults, 1999–2008. JAMA 303:235–241
4. Ward ZJ, Bleich SN, Cradock AL, Barrett JL, Giles CM, Flax C, Long MW GS (2019) Projected U.S. state-level prevalence of adult obesity and severe obesity. N Engl J Med 381:2440–50
5. Muennig P, Lubetkin E, Jia H, Franks P (2006) Gender and the burden of disease attributable to obesity. Am J Public Health 96:1662–1668
6. Koliaki C, Liatis SKA (2019) Obesity and cardiovascular disease: revisiting an old relationship. Metabolism 92:98–107
7. Wang YWQ (2004) The prevalence of prehypertension and hypertension among US adults according to the new joint national committee guidelines: new challenges of the old problem. Arch Intern Med 164:2126–2134
8. Zaccardi F, Webb DR, Yates TDM (2016) Pathophysiology of type 1 and type 2 diabetes mellitus: a 90-year perspective. Postgr Med J. 92:63–69
9. Eckel RHKR (1998) American heart association call to action: obesity as a major risk factor for coronary heart disease AHA Nutrition Committee. Circulation 97:2099–2100

10. Hubert HB, Feinleib M, McNamara PMCW (1983) Obesity as an independent risk factor for cardiovascular disease: a 26-year follow-up of participants in the Framingham Heart Study. Circulation 67:968–977
11. Mongraw-Chaffin ML, Peters SAE, Huxley RRWM (2015) The sex-specific association between BMI and coronary heart disease: a systematic review and meta-analysis of 95 cohorts with 1·2 million participants. Lancet Diabetes Endocrinol 3:437–449
12. Kenchaiah S, Evans JC, Levy D, Wilson PW, Benjamin EJ, Larson MG, Kannel WB, Vasan RS (2002) Obesity and the risk of heart failure. N Engl J Med 347:305–313
13. Aune D, Sen A, Norat T, Janszky I, Romundstad P, Tonstad S, Vatten LJ (2016) Body mass index, abdominal fatness, and heart failure incidence and mortality: a systematic review and dose-response meta-analysis of prospective studies. Circulation 16(133):639–649
14. Ho JE, Lyass A, Lee DS, Vasan RS, Kannel WB, Larson MG, Levy D (2013) Predictors of new-onset heart failure: differences in preserved versus reduced ejection fraction. Circ Heart Fail 6:279–286
15. Canepa M, Strait JB, Abramov D, Milaneschi Y, AlGhatrif M, Moni M, Ramachandran R, Najjar SS, Brunelli C, Abraham TP, Lakatta EG, Ferrucci L (2012) Contribution of central adiposity to left ventricular diastolic function (from the Baltimore Longitudinal Study of Aging). Am J Cardiol 109:1171–1178
16. Wanahita N, Messerli FH, Bangalore S, Gami AS, Somers VK, Steinberg JS (2008) Atrial fibrillation and obesity–results of a meta-analysis. Am Heart J 155:310–315
17. Omran J, Firwana B, Koerber S, Bostick B, Alpert MA (2016) Effect of obesity and weight loss on ventricular repolarization: a systematic review and meta-analysis. Obes Rev 17:520–530
18. Kannel WB, Plehn JF, Cupples LA (1988) Cardiac failure and sudden death in the Framingham study. Am Heart J 115:869–875
19. Isomaa B, Almgren P, Tuomi T, Forsen B, Lahti K, Nissen M, Taskinen MR, Groop L (2001) Cardiovascular morbidity and mortality associated with the metabolic syndrome. Diabetes Care 24:683–689
20. Grundy SM (2004) Obesity, metabolic syndrome, and cardiovascular disease. J Clin Endocrinol Metab 89:2595–2600
21. Jiang S-Z, Lu W, Zong X-F, Ruan H-Y, Liu Y (2016) Obesity and hypertension. Exp Ther Med 12:2395–2399
22. Ye J (2013) Mechanisms of insulin resistance in obesity. Front Med 7:14–24
23. Blokhin IO, Lentz SR (2013) Mechanisms of thrombosis in obesity. Curr Opin Hematol 20:437–444
24. Littleton SW, Tulaimat A (2017) The effects of obesity on lung volumes and oxygenation. Respir Med 124:15–20
25. Turnbull CD (2018) Intermittent hypoxia, cardiovascular disease and obstructive sleep apnoea. J Thorac Dis 10:S33–S39
26. Ruparelia N, Chai JT, Fisher EA, Choudhury RP (2017) Inflammatory processes in cardiovascular disease: a route to targeted therapies. Nat Rev Cardiol 14:314
27. Wang Z, Nakayama T (2010) Inflammation, a link between obesity and cardiovascular disease. Mediators Inflamm 2010:535918
28. Ricci R, Bevilacqua F (2012) The potential role of leptin and adiponectin in obesity: a comparative review. Vet J 191:292–298
29. Szmitko PE, Teoh H, Stewart DJ, Verma S (2007) Adiponectin and cardiovascular disease: state of the art? Am J Physiol Heart Circ Physiol 292:H1655–H1663
30. Matsuda M, Shimomura I, Sata M, Arita Y, Nishida M, Maeda N, Kumada M, Okamoto Y, Nagaretani H, Nishizawa H, Kishida K, Komuro R, Ouchi N, Kihara S, Nagai R, Funahashi T, Matsuzawa Y (2002) Role of adiponectin in preventing vascular stenosis. The missing link of adipo-vascular axis. J Biol Chem 277:37487–91
31. Ellulu MS, Patimah I, Khaza'ai H, Rahmat A, Abed Y (2017) Obesity and inflammation: the linking mechanism and the complications. Arch Med Sci 13:851–863
32. Zhou YT, Grayburn P, Karim A, Shimabukuro M, Higa M, Baetens D, Orci L, Unger RH (2000) Lipotoxic heart disease in obese rats: implications for human obesity. Proc Natl Acad Sci USA 97:1784–1789

33. Poirier P, Giles TD, Bray GA, Hong Y, Stern JS, Pi-Sunyer FX, Eckel RH (2006) American heart A, obesity committee of the council on nutrition PA and metabolism. Obesity and cardiovascular disease: pathophysiology, evaluation, and effect of weight loss: an update of the 1997 American Heart Association Scientific Statement on Obesity and Heart Disease from the Obesity Committee of the Council on Nutrition, Physical Activity, and Metabolism. Circulation 113:898–918
34. Poirier P, Giles TD, Bray GA, Hong Y, Stern JS, Pi-Sunyer FX, Eckel RH (2006) Obesity and cardiovascular disease: pathophysiology, evaluation, and effect of weight loss. Arterioscler Thromb Vasc Biol 26:968–976
35. Messerli FH (1986) Cardiopathy of obesity–a not-so-Victorian disease. N Engl J Med 314:378–380
36. Ku CS, Lin SL, Wang DJ, Chang SK, Lee WJ (1994) Left ventricular filling in young normotensive obese adults. Am J Cardiol 73:613–615
37. Carpenter HM (1962) Myocardial fat infiltration. Am Heart J 63:491–496
38. Balsaver AM, Morales AR, Whitehouse FW (1967) Fat infiltration of myocardium as a cause of cardiac conduction defect. Am J Cardiol 19:261–265
39. Spain DM, Cathcart RT (1946) Heart block caused by fat infiltration of the interventricular septum (cor adiposum). Am Heart J 32:659–664
40. De Scheerder I, Cuvelier C, Verhaaren R, De Buyzere M, De Backer G, Clement D (1987) Restrictive cardiomyopathy caused by adipositas cordis. Eur Heart J 8:661–663
41. Moossavi S, Bishehsari F (2019) Microbes: possible link between modern lifestyle transition and the rise of metabolic syndrome. Obes Rev 20:407e19
42. Rook GA (2010) 99th Dahlem conference on infection, inflammation and chronic inflammatory disorders: darwinian medicine and the 'hygiene' or 'old friends' hypothesis. Clin Exp Immunol 160(1):70–79
43. Tremaroli V, Backhed F (2012) Functional interactions between thegut microbiota and host metabolism. Nature 489(7415):242–249
44. Alang N, Kelly CR (2015) Weight gain after fecal microbiota transplantation. Open Forum Infect Dis 2(1):ofv004
45. Vrieze A, Van Nood E, Holleman F et al (2012) Transfer of intestinal microbiota from lean donors increases insulin sensitivity in individuals with metabolic syndrome. Gastroenterology 143(4):913–6.e7
46. Tang WH, Wang Z, Levison BS et al (2013) Intestinal microbial metabolism of phosphatidylcholine and cardiovascular risk. N Engl J Med 368(17):1575–1584
47. Ichimura A, Hirasawa A, Hara T, Tsujimoto G (2009) Free fatty acid receptors act as nutrient sensors to regulate energy homeostasis.Prostaglandins Other Lipid Mediat 89(3–4):82–88
48. Yiu JH, Dorweiler B, Woo CW (2017) Interaction between gut microbiota and toll-like receptor: from immunity to metabolism. J Mol Med (Berl) 95(1):13–20
49. Thaiss CA, Levy M, Grosheva I et al (2018) Hyperglycemia drives intestinal barrier dysfunction and risk for enteric infection. Science 359(6382):1376–1383
50. Weihrauch-Blüher S, Kromeyer-Hauschild K, Graf C et al (2018) Current guidelines for obesity prevention in childhood and adolescence. Obes Facts 11:263–276
51. Klein S, Fontana L, Young VL, Coggan AR, Kilo C, Patterson BW, Mohammed BS (2004) Absence of an effect of liposuction on insulin action and risk factors for coronary heart disease. N Engl J Med 350:2549–2557
52. Harris JA, Benedict FG (1919) A biometric study of basal metabolism in man. Carnegie Institute of Washington publication 279.Washington, DC: Carnegie Institution of Washington
53. Mifflin MD, St Jeor ST, Hill LA et al (1990) A new predictive equation for resting energy expenditure in healthy individuals. Am J Clin Nutr 51:241–247
54. Joint FAO/WHO/UNU Expert Consultation on Energy and Protein Requirements (1985) Energy and Protein Requirements: report of Joint FAO/WHO/UNU Expert Consultation. Technical Report Series No. 724.Geneva: World Health Organization 1–206
55. Jensen MD, Ryan DH, Apovian CM, Ard JD, Comuzzie AG, Donato KA, Hu FB, Hubbard VS, Jakicic JM, Kushner RF, Loria CM, Millen BE, Nonas CA, Pi-Sunyer FX, Stevens J, Stevens

VJ, Wadden TA, Wolfe BM, Yanovski SZ, Jordan HS, Kendall KA, Lux LJ, Mentor-Marcel R, Morgan LC, Trisolini MG, Wnek J, Anderson JL, Halperin JL, Albert NM, Bozkurt B, Brindis RG, Curtis LH, DeMets D, Hochman JS, Kovacs RJ, Ohman EM, Pressler SJ, Sellke FW, Shen WK, Smith SC Jr, Tomaselli GF (2014) American College of Cardiology/American Heart Association Task Force on Practice Guidelines; Obesity Society. 2013 AHA/ACC/TOS guideline for the management of overweight and obesity in adults: a report of the American College of Cardiology/American Heart Association Task Force on Practice Guidelines and The Obesity Society. Circulation. 129(25 Suppl 2):S102–38. doi: https://doi.org/10.1161/01.cir.0000437739.71477.ee. Epub 2013 Nov 12. Erratum in: Circulation. 2014;129(25 Suppl 2):S139–40
56. Wing RR (1999) Physical activity in the treatment of the adulthood overweight and obesity: current evidence and research issues. Med Sci Sports Exerc 31:S547–S552
57. Saris WH, Blair SN, van Baak MA, Eaton SB, Davies PS, Di Pietro L, Fogelholm M, Rissanen A, Schoeller D, Swinburn B et al (2003) How much physical activity is enough to prevent unhealthy weight gain? Outcome of the IASO 1st Stock Conference and consensus statement. Obes Rev 4:101–114
58. Foreyt JP, Poston WS II (1998) The role of the behavioral counselor in obesity treatment. J Am Diet Assoc 98:S27–S30
59. Wing RR (1998) Behavioral approaches to the treatment of obesity. In: Bray GA, Bouchard C, James WPT (eds) Handbook of obesity. Marcel Dekker, New York, NY, pp 855–877
60. Klein S, Burke LE, Bray GA, Blair S, Allison DB, Pi-Sunyer X, Hong Y, Eckel RH (2004) American Heart association council on nutrition, physical activity, and metabolism. Clinical implications of obesity with specific focus on cardiovascular disease: a statement for professionals from the American heart association council on nutrition, physical activity, and metabolism: endorsed by the American college of cardiology foundation. Circulation. 110:2952–67
61. Wadden TA, Berkowitz RI, Sarwer DB, Prus-Wisniewski R, Steinberg C (2001) Benefits of lifestyle modification in the pharmacologic treatment of obesity: a randomized trial. Arch Intern Med 161:218–227
62. Rössner S, Sjöström L, Noack R, Meinders AE, Noseda G (2000) Weight loss, weight maintenance, and improved cardiovascular risk factors after 2 years treatment with orlistat for obesity. Eur Orlistat Obesity Study Group. Obes Res 8:49–61
63. Kang JG, Park CY, Kang JH, Park YW, Park SW (2010) Randomized controlled trial to investigate the effects of a newly developed formulation of phentermine diffuse-controlled release for obesity. Diabetes Obes Metab 12:876–882
64. Pi-Sunyer X, Astrup A, Fujioka K, Greenway F, Halpern A, Krempf M, Lau DC, le Roux CW, Violante Ortiz R, Jensen CB, Wilding JP (2015) SCALE obesity and prediabetes NN8022–1839 study group. A randomized, controlled trial of 3.0 mg of Liraglutide in weight management. N Engl J Med 373:11–22
65. Greenway FL, Fujioka K, Plodkowski RA, Mudaliar S, Guttadauria M, Erickson J, Kim DD, Dunayevich E (2010) COR-I study group. Effect of naltrexone plus bupropion on weight loss in overweight and obese adults (COR-I): a multicentre, randomised, double-blind, placebo-controlled, phase 3 trial. Lancet. 2010;376:595–605. Erratum in: Lancet. 2010;376:594. Erratum in: Lancet. 2010;376:1392
66. Conference NIH (1991) Gastrointestinal surgery for severe obesity: consensus development conference panel. Ann Intern Med 115:956–961
67. MacLean LD, Rhode BM, Sampalis J, Forse RA (1993) Results of the surgical treatment of obesity. Am J Surg 165:155–162
68. Sugerman HJ, Starkey JV, Birkenhauer R (1987) A randomized prospective trial of gastric bypass versus vertical banded gastroplasty for morbid obesity and their effects on sweets versus non-sweets eaters. Ann Surg 205:613–624

69. Hall JC, Watts JM, O'Brien PE, Dunstan RE, Walsh JF, Slavotinek AH, Elmslie RG (1990) Gastric surgery for morbid obesity. The Adelaide Study. Ann Surg. 211:419–427
70. Howard L, Malone M, Michalek A, Carter J, Alger S, Van Woert J (1995) Gastric bypass and vertical banded gastroplasty—a prospective randomized comparison and 5-year follow-up. Obes Surg 5:55–60
71. Puzziferri N, Roshek TB 3rd, Mayo HG, Gallagher R, Belle SH, Livingston EH (2014) Long-term follow-up after bariatric surgery: a systematic review. JAMA 312:934–942

Chapter 13
Obesity: Molecular Mechanisms, Epidemiology, Complications and Pharmacotherapy

Saeeda Al Jaberi, Athena Cohen, Zulqarnain Saeed, Shreesh Ojha, Jaipaul Singh, and Ernest Adeghate

Abstract Obesity is a common disorder affecting millions of people worldwide. The number of overweight and obese subjects, globally, is currently 2 billion and 800 million, respectively. Projected estimates show that the number of overweight citizens will approach 60% of the world's population by the year 2030. Oxidative stress facilitates the development of obesity by stimulating pre-adipocyte differentiation and eventual adipose accumulation. Large deposits of fat release excessive quantities of adipocytokines, resulting in chronic inflammation. The obesity-induced chronic inflammation paves the way for a large variety of systemic complications including but not limited to diabetes mellitus, hyperlipidemia, atherosclerotic lesions, cardiovascular diseases tissue and malignancy. In addition, other obesity-inducers, such as increased insulin growth factor 1, insulin resistance, and increased tissue level of leptin and low concentration of adiponectin may lead to the development of tissue malignancy. Increased physical activity coupled with a healthy food intake is crucial to the management of obesity. Anti-obesity drugs such as sibutramine, qsymia (a combination of phentermine and topiramate), and orlistat have been used to treat obesity with variable degrees of efficacy. Bariatric surgery becomes a choice in severe cases when physical activity and pharmacotherapy fail. In the obese patient with diabetes mellitus, the choice of hypoglycemic agent is important. Metformin,

S. Al Jaberi · A. Cohen · E. Adeghate (✉)
Department of Anatomy, College of Medicine & Health Sciences, United Arab Emirates University, Al Ain, United Arab Emirates
e-mail: eadeghate@uaeu.ac.ae

A. Cohen
Department of Chemistry, Emory University, Atlanta, GA 30322, USA

Z. Saeed
Department of Psychology and Social Work, Flinders University, Adelaide, Australia

S. Ojha
Department of Pharmacology, College of Medicine & Health Sciences, United Arab Emirates University, Al Ain, United Arab Emirates

J. Singh
School of Natural Sciences, University of Central Lancashire, Preston PR1 2HE, England, UK

© Springer Nature Switzerland AG 2021
P. S. Tappia et al. (eds.), *Cellular and Biochemical Mechanisms of Obesity*,
Advances in Biochemistry in Health and Disease 23,
https://doi.org/10.1007/978-3-030-84763-0_13

and sodium glucose cotransporters 2 (SGLT2) inhibitors, a new set of antidiabetic drugs can significantly reduce body weight and improve cardiorenal function. SGLT2 inhibitors, thus belong to a class of potential drugs that can be used to treat obesity. In conclusion, obesity is a "deadly" condition that can predispose individuals to many life threatening health conditions.

Keywords Obesity · Adipocytokines · Cardiovascular disease · Diabetes mellitus · Chronic inflammation · Cancer · Oxidative stress · Hypoglycemic agents

Introduction

Obesity represents a condition of excess body fat. It is a chronic multifactorial disease, commonly assessed using body mass index (BMI), which is calculated by dividing the body weight in kg by the height in meters squared (kg/m^2). Individuals with BMI measurements of less than 18.5 kg/m^2 are considered underweight, 18.5–24.9 kg/m^2 healthy, 25–29 kg/m^2 overweight, and those with 30 kg/m^2 and above are considered obese [1]. Besides the BMI, other measurements that are helpful in quantifying obesity include waist circumference. Increases in waist circumference is associated with high mortality even in individuals with normal BMI measurements. A review of waist circumference and mortality performed on 650,000 adults showed that increased waist circumference above normal resulted in a higher mortality rate [2]. Using a combination of both methods helps in generating a more accurate and clearer way to determine whether high BMI measurements are related to increased muscle content or elevated adiposity. In addition, more advanced methods but less commonly used for measuring total body fat are magnetic resonance imaging (MRI) and dual energy X-ray absorptiometry. These methods are not used routinely because they are expensive and not easily available.

Obesity is associated with significant risks and implications for chronic diseases such as type 2 diabetes mellitus (DM), high blood pressure, malignant tumors, heart and vascular diseases and cerebral stroke [3–6]. Obesity is also associated with various diseases including gastroesophageal reflux disease, colorectal polyps, malignant disease of the colon and hepatic conditions such as non-alcoholic fatty liver disease, cirrhosis and hepatocellular carcinoma. Furthermore, obesity has been shown to reduce the quality of life, as it may result in overt DM and coronary artery disease (CAD), which contribute to the reduction of the quality of life among those patients [7]. Obesity and type 2 DM go together and have been referred to as the "twin epidemics". It is not surprising therefore, that 61.3% of diabetic adults are overweight and 87.5% are obese [8].

Epidemiology

The prevalence of obesity continues to increase worldwide. For example, an estimated 1.1 billion adults were considered overweight in 2005 [9]. Indeed, the number of people considered to be overweight has doubled in the last 15 years, exceeding 2 billion in 2020 [10]. This shows that approximately one third of the world population is either overweight or obese. It is estimated that by the year 2030, 57.8% of the world population will be obese if the prevalence rate continues rising at its current pace [11]. On a global scale, men with BMI $\geq$ 25 kg/m^2 increased from 25.4% to 38.5% accounting for a 5% percentage increase to 10.1% from the period between 1980 and 2015, and in females it was from 27.8% to 39.4% accounting for a percentage increase from 8.9% to 14.8% during the same time period. Moreover, the prevalence of obesity is more pronounced in women when compared to men. It was projected that by the year 2025, the prevalence of obesity will reach 18% in men and 21% in women [12]. However, there is hope because the obesity rate has flattened out in the past 10 years in many developed countries. For instance, the current prevalence of obesity in the US remained around 30–34% and in the UK around 23–24%, in the period between 2005 and 2015. On the other hand, the global obesity prevalence is still expanding as the trend has picked up in other regions of the world, where the majority of people live [13]. The South East Asian, Western Pacific and African regions had the lowest prevalence of obesity in 1980, which accounted for 0.4% in Vietnam, 0.6% in Bangladesh and China, 1.2% in Ethiopia, 1.4% in Indonesia, 1.5% in Myanmar, 1.7% in Nigeria and India, 1.8% in Japan, and 1.9% in South Korea. However, by the year 2015, only one country (Vietnam, 1.6%) was able to maintain obesity prevalence rate below 2%. Numerous countries have recorded large increases in the prevalence of obesity. The robust rising trends will likely continue in low income countries, where 2 out of 3 of the world's obese individuals currently reside [14, 15].

Molecular Mechanisms Underlying the Development of Obesity

Obesity and Oxidative Stress

Oxidative stress, a condition where endogenous antioxidants is overwhelmed by large production of free radicals, has been shown to play a key role in preadipocyte differentiation and subsequent development of obesity. A multitude of experimental evidence indicates that redox balance is very important in the development, maturation and deposition of adipose tissue [16–19]. For example, murine models lacking the non-selenocysteine-containing phospholipid hydroperoxide glutathione peroxidase (NPGPx), a sensor and transducer of tissue oxidative stress has been shown to have increased level of reactive oxygen species (ROS), increased accumulation of

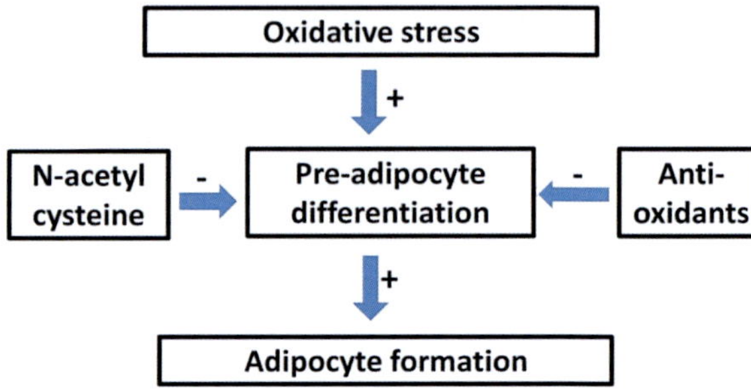

Fig. 13.1 Pathways by which oxidative stress induces adipocyte formation

lipid, and develop adipocyte hypertrophy. This effect was inhibited by treating the animal with N-acetylcysteine (NAC), an antioxidant. Moreover, the single nucleotide polymorphisms of the NPGPx gene in humans, associated with a reduced level of NPGPx expression in fat tissue is strongly associated with obesity [18].

Another study investigated the role of ROS and antioxidant in the development of adipose tissue. The study showed that NAC inhibited ROS-mediated conversion of mesenchymal stem cells to adipose tissue [20]. This confirms the role of ROS in the development of obesity. The administration of RNA interference inhibited the adipocyte differentiation and maturation in 3T3-L1 pre-adipocytes. It also caused a reduction in total body weight, decreased triglyceride level, and reduced, NADPH oxidase 2 and ROS formation [21]. Yan et al., showed that inhibition of mitochondrial ROS production in human mesenchymal stem cells with the antioxidant, rotenone, prevented stem cells from differentiating into mature adipocytes [22] Fig. 13.1.

The Role of Adipocytokines in the Development of Obesity

Adipocytokines have been implicated in the development of several debilitating diseases including obesity, DM and malignant tumors [23]. Large and excessive adipose tissue deposition and adipocyte hypertrophy has been shown to cause chronic inflammation. This obesity-associated chronic inflammation is caused by the high level of pro-inflammatory adipocytokines including interlukin-6 (IL-6), tumor necrosis factor alpha (TNF-α) [24], leptin [25], and C-reactive protein (CRP) released into blood circulation. CRP, an acute phase protein made by hepatocytes is released into the blood circulation in response to a large variety of inflammatory adipocytokines, especially IL-6. It was reported that adolescent, who are physically inactive and overweight have a higher CRP blood level when compared to their lean and physically active mates. This shows that overweight and physically inactive individuals suffer from low-grade inflammation [26]. Several experimental and human studies have confirmed that adipocytokines play important roles in the pathogenesis of obesity and obesity-associated conditions such as DM and malignant tumors. In a study conducted on obese African American female subjects to investigate whether

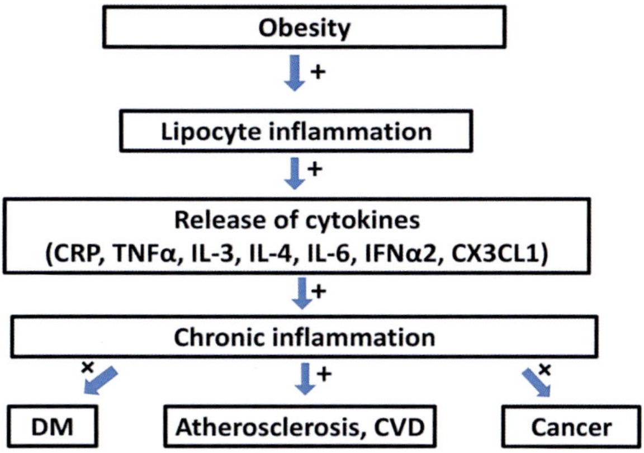

Fig. 13.2 Pathways by which obesity facilitates chronic inflammation, heart and vascular diseases, diabetes mellitus and cancer

adipocytokine and chemokine serum concentration correlate with the serum levels of HbA1c and glucose. It was shown that the serum levels of IL-3, IL-4, IL-7, TNF-α, IFN-α2 and CX3CL1 increased significantly, when compared to DM patients with normal HbA1c. This suggests that IL-3, IL-4, IL-7, TNF-α, IFN-α2 and CX3CL1 may take part in the pathogenesis of type 2 DM in obese people [27]. Another adipocytokine that has been implicated in the development of obesity at the molecular and cellular level is IL-6. Increased levels of IL-6 in adipocytes increase leptin release via enhanced lipid breakdown and fatty acid oxidation through the JAK-, and AMPK-mediated pathways [28]. Indeed, the more severe the obesity, the higher the serum levels of TNF-α, indicating a strong correlation between obesity and TNF-α. In addition, many reports have also shown that obesity and excessive lipid accumulation is a cause of insulin resistance [29–31]. TNF-α also contributes to the development of obesity by altering the metabolism of leptin [32] Fig. 13.2.

Obesity and Hormones Regulating Food Intake and Satiety

Hormones regulating appetite, satiety and food are numerous. The serum levels of many hormones including, glucagon-like peptide 1 (GLP-1), oxyntomodulin (OXM) and peptide YY 36 (PYY) are elevated during high caloric ingestion, signaling to the brain that the stomach is now full. This group of hormones is referred to as suppressors of appetite [33]. They are thus involved in determining whether obesity occurs or not and any dysregulation of these hormones may lead to the development of obesity. In addition, GLP-1 as a member of the incretin family is capable of inducing insulin production with a concomitant inhibition of glucagon release [34–37]. Another hormone that plays a role in the regulation of food intake is ghrelin, a peptide produced by cells of the gastric epithelium and increases the urge to eat [38–40]. In fact, treatment of both obese and non-obese individuals with these hormones stimulates the urge to eat [41]. Excessive serum level of ghrelin caused by an ectopic tumor may result in obesity. Leptin, which is produced by adipocytes, increases with the level of body fat and correlates positively with BMI. Leptin interacts with hypothalamic orexinergic centers to enhance energy expenditure and improve satiety [42, 43]. When the serum level of leptin is high for a long period of time, the hypothalamic neurons do not really sense that information any more. The brain thinks that the person is still hungry, thereby eating more to reach the point of satiety leads to developing obesity [44, 45]. This will make the obesity even more severe. Another hormone that regulates the pathogenesis of obesity is adiponectin. The serum level of this adipocytokine correlates inversely with BMI [46]. Adiponectin protects the body against inflammation [46, 47] and increases insulin sensitivity [48]. A reduced serum level of adiponectin increases the likelihood of insulin resistance with a concomitant elevation in adipocyte differentiation [48]. In fact, a reduction in the expression of adiponectin mRNA in fat cells of obese subjects has been reported [49].

What are the Outcomes of Obesity?

Heart and Vascular Complications

A major outcome of obesity is cardiovascular disease (CVDs). Obesity, coupled with inactive lifestyle significantly increases the risks of developing hypertension, dyslipidemia, atherosclerosis, and DM. Hypertension will put a lot of pressure on the heart while lipidemia will increase the risk of developing atherosclerosis. All of these will also contribute to the stiffening of the coronary arteries, leading to coronary artery disease (CAD) and eventually myocardial infarction (MI) [50]. The pro-inflammatory adipocytokines and other molecules released into blood circulation by excessive fat cells play a role in the pathogenesis of obesity [51]. The low-grade chronic inflammation seen in the obese is deleterious for other organs especially the heart and the vascular bed. For example, IL-6, a biomarker for a chronic inflammation state is increased in CAD, MI, peripheral vascular disease, and other metabolic diseases such as DM [52, 53]. IL-6 does not only trigger and maintain atherosclerosis, it also enhances the formation of C-reactive protein (CRP) from hepatocytes [52–54], an important marker of the inflammation of the cardiovascular system. Many studies have clearly indicated that obesity is not good for the heart and the vascular system. For example, it has been shown that the higher the BMI, the more established, more frequent and more advanced the atherosclerotic vascular lesions are, when compared to subjects with normal BMI [55]. In addition, it has been reported that a mere 10 kg increase in total body weight significantly increases the risk of CAD by a staggering 12% [56]. Data from the Framingham Heart Study also showed that an elevation in BMI by 1 kg/m^2 significantly increases the risk of developing cardiac failure by 5% in male and 7% in female subjects [57]. Individuals with obesity commonly suffer from hypertension. It has been reported that a 5 kg/m^2 increase in BMI results in a 5 mmHg rise in systolic blood pressure [58] due to the activation of the renin–angiotensin–aldosterone system and increased activity of the sympathetic nervous system [59]. Obesity increases both aldosterone levels and the mineralocorticoid receptor expression, which induce interstitial cardiac fibrosis, platelet aggregation, and endothelial dysfunction (Fig. 13.2). In a large data from about 2 million people, Flint et al., showed that persistent hypertension is a major risk factor for CAD, ischemic and hemorrhagic strokes and MI [60].

Malignant Tumors

Globally, benign and malignant tumors (cancer) are the second leading causes of death [61]. Unfortunately, the incidence of cancer is expected to continue rising because of the high prevalence of obesity and DM, as they are amongst the leading risk factors for the development and progression of cancer. Recent data have indicated that there is a complex relationship between obesity, insulin resistance (INS resistance) and adipocytokines in the etiology and pathogenesis of cancer. A large number of pathways that link obesity to the development of malignant tumors have been proposed. The suggested factors and conditions that predispose an individual developing malignant tumors include, hyperinsulinemia, insulin resistance, abnormalities of the insulin growth factor 1 (IGF-1) system, chronic inflammation, overproduction of ROS, impaired immune function, and increased circulatory levels of adipocytokines [23, 62–66]. Although leptin, IL-6, TNF-α, resistin adipocytokines secreted by fat cells have also been implicated in the development of a large variety of malignant tumors, there are good "actors" amongst the cytokines produced by fat cells. The good cytokines, which protect against the development of malignant tumors include adiponectin and omentin-1 [23]. For example, adiponectin has the capacity to inhibit the progression of tumors of the white blood cells, adenocarcinoma of the mammary gland, and fibrosarcoma [67]. In contrast, a reduced blood level of adiponectin facilitate cancer cell formation and progression [62, 68]. In addition, the serum levels of IGF-1, TNF-a, IL-6 is significantly elevated if adiponectin level is low [69]. IGF-1, TNF-a, and IL-6 enhances cell proliferation seen in cancer [69]. Leptin is yet another adipose hormone which has been implicated in the development of many malignant tumors. Leptin has indeed been shown to play a role in the development of malignant tumors of the mammary gland in women [70, 71], and cancer in prostate cell lines [72]. Leptin is also capable of stimulating malignant tumor formation in the gastrointestinal system [46], where leptin has been shown to promote the invasion of nearby tissues and organs by metastatic cells via the Rho/ROCK pathway [73]. The level of leptin correlates with aggressiveness of the malignant tumor of the colorectal segment of the gastrointestinal system [74] Fig. 13.3.

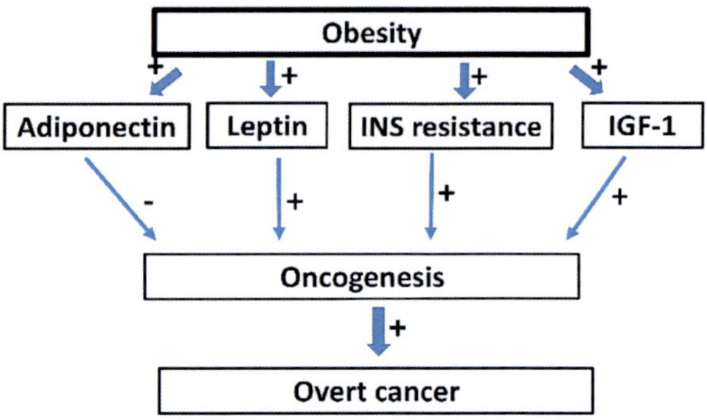

Fig. 13.3 Pathways by which obesity facilitates the development of cancer

Diabetes Mellitus

One of the major health consequence of obesity in overt DM. Obesity is highly associated with the disturbance of glucose metabolism and one of the main risk factors for the development of insulin resistance (INS resistance) and type 2 DM [75]. In obesity, adipocytes and other cells involved in glucose uptake such as hepatocytes and skeletal muscle are dysfunctional, leading to insulin resistance [76]. In fact, some reports have suggested that obesity downregulates the expression of glucose transporters on the cell membrane of glucose sensitive cells such as hepatocytes and skeletal muscle, which facilitate the uptake of glucose by those cells [77]. Lipotoxicity caused by the release of free fatty acids (FFA) from fat deposits may also contribute to insulin resistance in hepatocytes and skeletal muscle cells [78, 79]. The liver releases more glucose via gluconeogenesis in response to hepatic insulin resistance thus increasing the prospect of developing hyperglycemia and DM [79, 80]. In a follow up study, which investigated the effects of BMI and waist circumference on the risk of developing type 2 DM, Wang et al. [81] showed that waist circumference, waist hip ratio and abdominal obesity correlate proportionally with the development of type 2 DM. The higher BMIs, in the 27.2 and 54.2 kg/m^2 range result in an eight fold increased in the risk of developing type 2 DM when compared to individuals with a BMI of less than 22.8 kg/m^2 [82]. Moreover, individuals with extreme waist circumference of 101.6–157.5 cm, showed more than 12-fold increased risk for developing type 2 DM [81].

Management of Diabetic Patients with Obesity

Diet and Lifestyle Modification

The type of food and diet approaches that we take seem to play a critical role in our health. Consumption of foods with low glycemic index also has an important role in reducing the complications of obesity and DM [83]. Physical inactivity and or a sedentary lifestyle are considered one of the main risk factors in developing obesity and insulin resistance. On the other hand, dietary modification and management accompanied by initiation of physical workouts has long been proven to be beneficial in reducing body weight and bettering insulin sensitivity [84]. A meta-analysis on the effects of low-carbohydrate diets on CVD risk factors resulted in marked decreases in total body weight, BMI, abdominal circumference, systolic and diastolic blood pressure, triglycerides, fasting glucose, $HbA1_C$, plasma insulin and CRP, with a concomitant increase in HDL-cholesterol [85]. Moreover, a combination of medications and intense lifestyle modification resulted in weight reduction and decreased levels of HbA1c when compared to normally treated cases, indicating that lifestyle modification are beneficial in the management of obesity and DM [86].

Hypoglycemic Drugs Used in the Treatment of Obesity

Metformin

Metformin, an insulin sensitizer, is one of the most commonly used drug for the management of type2 DM in adults. It is one of the most studied and widely use hypoglycemic drug. Metformin has shown weight reducing capabilities in various clinical studies with a HbA1c reduction of 1%, especially when used with increased physical activities [87–89]. Metformin exhibits weight reducing and hypoglycemic effects by blocking gluconeogenesis in the liver, stimulating the uptake of glucose in skeletal muscle cells and hepatocytes, and dreducing appetite [50]. Although the mechanisms of actions of metformin are not completely understood, the proposed mechanism of action is said to be via the regulation of mitochondrial molecules involved in energy balance and fuel utilization [90–92]. A clinical trial conducted by the Diabetes Prevention Program/Diabetes Prevention Program Outcomes Study showed that metformin treatment can reduce the incidence of type 2 DM by as much as 31% in obese indiviuals [93]. Metformin has also been used to treat polycystic ovary syndrome [94], gestational DM [95], non-alcoholic fatty liver disease [50, 96] and malignant tumors [97].

Inhibitors of Gastrointestinal α-glucosidase

Acarbose, voglibose and miglitol are the major drugs belonging to the group of inhibitors of α-glucosidase. α-glucosidase is an enzyme found in the small intestinal epithelium where it cleaves monosaccharides from polysaccharides such as starch and promote their absorption. For that, α-glucosidase inhibitors are used to prevent carbohydrate absorption in the small intestine. In a new meta-analysis study, α-glucosidase inhibitors significantly reduced HbA1c, body weight and glucose when compared to control. The gastrointestinal side effects such as flatulence and diarrhea are not popular with users [50, 87, 98].

Inhibitors of the Dipeptidyl Peptidase-4 Enzyme

Dipeptidyl Peptidase-4 (DPP-4) inhibitors are endogenous incretin enhancers that act by the inhibition of DPP-4 enzyme that in turn prevents the degradation of GLP-1 and glucose-dependent insulinotropic polypeptide (GIP), and thereby improving the lifespan of the active form of GLP-1 and GIP [36, 99, 100]. These hormones are known to enhance glucose-dependent pancreatic insulin response, suppression of glucagon release, and increase satiety [99, 101]. The use of DPP-4 inhibitors was associated with enhancement of glycemic control, insulin secretion, and β-cell function in murine models [99, 101] and in patients with type 2 DM [99]. In patients with type 2 DM, DPP-4 inhibitors decreased postprandial glucose elevation, fasting plasma glucose and HbA1c levels with low risk of developing hypoglycemia and body weight gain [36]. DPP-4 inhibitors stimulates pancreatic beta cell function with no effect on insulin resistance [101]. In this drug class no weight gain has been detected and average HbA1c decrease of 0.5–1.0% has been reported [102]. DPP-4 inhibitors is thus safe for diabetic patients suffering from obesity.

Sodium Glucose Cotransporter 2 Inhibitors

Sodium glucose cotransporter 2 (SGLT2) inhibitors are a new set of anti-diabetic drugs [103]. They possess a significant ability to reduce HbA1c, body weight and blood pressure [103]. SGLT2 inhibitors are not without side effects because people taking SGLT2 inhibitors may have genitourinary tract infections and increased risk of fracturing their bones [104]. However, the use of SGLT2 inhibitors is associated with a significant reduction in body weight and cardiorenal benefits [50]. This is yet another group of hypoglycemic drug, which can be used to treat obesity and DM since the two conditions commonly co-exist.

Drugs Used in the Treatment of Obesity

Drugs are also available to treat obese individuals who do not have the co-morbidity of DM. Although the health-risks associated with obesity can be mitigated through the adoption of an exercise regime and healthy diet, there are numerous promising drugs that can be used in conjunction with these life-style changes to help treat obesity.

Sibutramine and orlistat have been approved by the U.S. Food and Drug Administration (FDA) for pharmacotherapy of obesity. These drugs are recommended for obese patients with a high BMI [105]. Orlistat prevents the absorption of fat by inhibiting the enzyme lipase secreted by the pancreas and the epithelial cells the small intestine. Orlistat can stimulat weight loss of up to 3 kg on average, in addition to managing cardiovascular risk factors and DM in severely obese individuals [106]. In a double blind, controlled, randomized study conducted on 3,305 obese subjects, orlistat significantly reduced total body weight and the risk of developing type 2 DM [107]. However, orlistat comes with uncomfortable side effects including fatty and frequent stool [105]. Sibutramine prevents the re-uptake of monoamines such as serotonin/norepinephrine, reduces total body weight by inhibiting appetite and increasesing satiety [105]. Sibutramine also reduces HbA1c and blood pressure in DM patients [107, 108].

Qsymia, a combination of phentermine and topiramate, utilizes the anorectic and an antiepileptic capabilities of these agents to suppress appetite. Qsymia has recently been approved by the FDA to help manage the body weight of severely obese patients (>BMI 30 kg/m^2). The mode of action of phentermine is still elusive, but it was suggested that it acts via the release of catecholamines by hypothalamic neurons. These catecholamines (serotonin, adrenaline, noradrenaline) cause a reduction in appetite and food intake, culminating in weight loss. The exact pathway by which topiramate (the second drug in Qsymia) exerts its effect is equally unknown. This antiepileptic drug is said to exert its effects by modifying the action of gamma-aminobutyrate and it receptors. This event then leads to a reduced appetite and enhance satiety in the individual [109].

Bariatric Surgery

Weight loss has key benefits on metabolic parameters, which made people with type 2 DM and or obesity pursue more invasive interventions to cut their excess body weight. The use of bariatric surgery to treat obesity has grown exponentially in the past decade [110]. Bariatric surgery has been shown to induce more than 73% remission of type 2 DM compared to control group [111], leading to normal $HbA1_C$, blood glucose and lipids levels [112]. Reports also showed that bariatric surgery was able to reduce the rate of mortality from DM by 92% and from CAD by 59% when compared with non-diabetic controls [113].

Conclusion

In conclusion, obesity is a common health condition affecting millions of people worldwide. Obesity increases the risks for cardiovascular diseases, diabetes mellitus, and cancer due to increases in the levels of cytokines, lipid, blood glucose and free radicals. The impact of obesity can be reduced by either lifestyle modifications or pharmacotherapy, and can also be cured by bariatric surgery.

References

1. Nuttall FQ (2015) Body mass index: obesity, BMI, and health: a critical review. Nutr Today 50(3):117–128. https://doi.org/10.1097/NT.0000000000000092
2. Cerhan JR, Moore SC, Jacobs EJ, Kitahara CM, Rosenberg PS et al (2014) A pooled analysis of waist circumference and mortality in 650,000 adults. Mayo Clin Proc 89(3):335–345. https://doi.org/10.1016/j.mayocp.2013.11.011
3. Nakamura K, Fuster JJ, Walsh K et al (2014) Adipokines: a link between obesity and cardiovascular disease. J Cardiol 63(4):250–259. https://doi.org/10.1016/j.jjcc.2013.11.006
4. Fantuzzi G (2005) Adipose tissue, adipokines, and inflammation. J Allergy Clin Immunol 115(5):911–919. https://doi.org/10.1016/j.jaci.2005.02.023
5. Preuss HG, Bagchi M, Bagchi D, Kaats GR (2004) Obesity and cancer. Phytopharmac Cancer Chemoprev 197–204. https://doi.org/10.1634/theoncologist.2009-0285
6. Field AE, Coakley EH, Must A, Spadano JL, Laird N, Dietz WH et al (2001) Impact of overweight on the risk of developing common chronic diseases during a 10-year period. Arch Intern Med 161(13):1581–1586. https://doi.org/10.1001/archinte.161.13.1581
7. Hlatky MA, Chung SC, Escobedo J, Hillegass WB, Melsop K, Rogers W, Brooks MM (2010) The effect of obesity on quality of life in patients with diabetes and coronary artery disease. Am Heart J 159(2):292–300. https://doi.org/10.1016/j.ahj.2009.11.004
8. McCafferty BJ, Hill JO, Gunn AJ (2020) Obesity: scope, lifestyle interventions, and medical management. Techn Vascul Intervent Radiol 23(1):100653
9. Haslam DW, James WPT (2005) Obesity. Lancet 366(9492):1197–1209. https://doi.org/10.1016/S0140-6736(05)67483-1
10. So IYH (2020) Obesity and its complications. Adv Biochem Health Dis 19:43–58
11. Kelly T, Yang W, Chen CS, Reynolds K, He J (2008) Global burden of obesity in 2005 and projections to 2030. Int J Obes 32(9):1431–1437. https://doi.org/10.1038/ijo.2008.102
12. Di Cesare M, Bentham J, Stevens GA, Zhou B, Danaei G (2016) Trends in adult body-mass index in 200 countries from 1975 to 2014: a pooled analysis of 1698 population-based measurement studies with 19.2 million participants. The Lancet 387(10026):1377–1396. https://doi.org/10.1016/S0140-6736(16)30054-X
13. He Y, Pan A, Wang Y, Yang Y, Xu J et al. (2017) Prevalence of overweight and obesity in 15.8 million men aged 15–49 years in rural China from 2010 to 2014. Sci Reports 7(1):1–10. https://doi.org/10.1038/s41598-017-04135-4
14. Ng M, Fleming T, Robinson M, Thomson B, Graetz N et al (2014) Global, regional, and national prevalence of overweight and obesity in children and adults during 1980–2013: a systematic analysis for the global burden of disease study 2013. The Lancet 384(9945):766–781. https://doi.org/10.1016/S0140-6736(14)60460-8
15. WHO (2016) https://www.who.int/data/gho/data/indicators/indicator-details/GHO/prevalence-of-obesity-among-adults-bmi-=-30-(crude-estimate)-(-). Accessed January 9, 2021
16. Higuchi M, Dusting GJ, Peshavariya H, Jiang F, Hsiao STF, Chan EC, Liu GS (2013) Differentiation of human adipose-derived stem cells into fat involves reactive oxygen species and forkhead box o1 mediated upregulation of antioxidant enzymes. Stem Cells Devel 22(6):878–888. https://doi.org/10.1089/scd.2012.0306
17. Huh JY, Kim Y, Jeong J, Park J, Kim I, Huh KH, Ha H (2012) Peroxiredoxin 3 is a key molecule regulating adipocyte oxidative stress, mitochondrial biogenesis, and adipokine expression. Antiox Redox Sign 16(3):229–243. https://doi.org/10.1089/ars.2010.3766
18. Chang YC, Yu YH, Shew JY, Lee WJ, Hwang JJ, Chen YH, Lee WH (2013) Deficiency of NPGPx, an oxidative stress sensor, leads to obesity in mice and human. EMBO Mol Med 5(8):1165–1179. https://doi.org/10.1002/emmm.201302679

19. Youn J, Siu KL, Lob HE, Itani H, Harrison DG, Cai H (2014) Role of vascular oxidative stress in obesity and metabolic syndrome. Diabetes 63(7):2344–2355. https://doi.org/10.2337/db13-0719
20. Ristow M, Wolfrum C (2013) A radical opposition in body weight control. EMBO molecular medicine, 5(8), 1147–1148. https://doi.org/10.1002/emmm.201303094
21. Kanda Y, Hinata T, Kang SW, Watanabe Y (2011) Reactive oxygen species mediate adipocyte differentiation in mesenchymal stem cells. Life Sci 89(7–8):250–258. https://doi.org/10.1016/j.lfs.2011.06.007
22. Nam WS, Park KM, Park JW (2012) RNA interference targeting cytosolic NADP+-dependent isocitrate dehydrogenase exerts anti-obesity effect in vitro and in vivo. Biochimica et Biophysica Acta—Molec Basis Dis 1822(8):1181–1188. https://doi.org/10.1016/j.bbadis.2012.04.003
23. Yan H, Aziz E, Shillabeer G, Wong A, Shanghavi D, Kermouni A (2002) Nitric oxide promotes differentiation of rat white preadipocytes in culture. J Lipid Res 43(12):2123–2129. https://doi.org/10.1194/jlr.M200305-JLR200
24. Spyrou N, Avgerinos KI, Mantzoros CS, Dalamaga M (2018) Classic and novel Adipocytokines at the intersection of obesity and cancer: diagnostic and therapeutic strategies. Curr Obes Rep 7(4):260–275. https://doi.org/10.1007/s13679-018-0318-7
25. Roytblat L, Rachinsky M, Fisher A, Greemberg L, Shapira Y, Douvdevani A, Gelman S (2000) Raised interleukin-6 levels in obese patients. Obes Res 8(9):673–675. https://doi.org/10.1038/oby.2000.86
26. Dubey S, Kabra M, Bajpai A, Pandey RM, Hasan M, Gautam RK, Menon PSN (2007) Serum leptin levels in obese Indian children: relation to clinical and biochemical parameters. Indian Pediatr 44(4):257–262
27. Halle M, Korsten-Reck U, Wolfarth B, Berg A (2004) Low-grade systemic inflammation in overweight children: impact of physical fitness. Exerc Immunol Rev 10:66–74
28. Williams A, Greene N, Kimbro K (2020) Increased circulating cytokine levels in african American women with obesity and elevated HbA1c. Cytokine 128(January). https://doi.org/10.1016/j.cyto.2020.154989
29. Li F, Li Y, Duan Y, Hu CAA, Tang Y, Yin Y (2017) Myokines and adipokines: involvement in the crosstalk between skeletal muscle and adipose tissue. Cytokine Growth Factor Rev 33:73–82. https://doi.org/10.1016/j.cytogfr.2016.10.003
30. Kaila B, Raman M (2008) Obesity: a review of pathogenesis and management strategies. Can J Gastroenterol 22(1):61–68. https://doi.org/10.1155/2008/609039
31. Coppack SW (2001) Pro-inflammatory cytokines and adipose tissue. Proceed Nutr Soc 60(3):349–356. https://doi.org/10.1079/pns2001110
32. Nieto-Vazquez I, Fernández-Veledo S, Krämer DK, Vila-Bedmar R, Garcia-Guerra L, Lorenzo M (2008) Insulin resistance associated to obesity: The link TNF-alpha. Arch Physiol Biochem 114(3):183–194. https://doi.org/10.1080/13813450802181047
33. Kirchgessner TG, Uysal KT, Wiesbrock SM, Marino MW, Hotamisligil GS (1997) Tumor necrosis factor-α contributes to obesity-related hyperleptinemia by regulating leptin release from adipocytes. J Clin Investig 100(11):2777–2782. https://doi.org/10.1172/JCI119824
34. Latorre R, Sternini C, De Giorgio R, Greenwood-Van Meerveld B (2016) Enteroendocrine cells: a review of their role in brain-gut communication. Neurogastroenterol Motil 28(5):620–630. https://doi.org/10.1111/nmo.12754
35. Manning S, Pucci A, Batterham RL (2015) GLP-1: a mediator of the beneficial metabolic effects of bariatric surgery? Physiology 30(1):50–62. https://doi.org/10.1152/physiol.00027.2014
36. Lotfy M, Singh J, Rashed H, Tariq S, Zilahi E, Adeghate E (2014) The effect of glucagon-like peptide-1 in the management of diabetes mellitus: cellular and molecular mechanisms. Cell Tissue Res 358(2):343–358. https://doi.org/10.1007/s00441-014-1959-9
37. Lotfy M, Singh J, Kalász H, Tekes K, Adeghate E (2011) Medicinal chemistry and applications of incretins and DPP-4 inhibitors in the treatment of Type 2 diabetes mellitus. The Open Med Chem J 5(Suppl 2):82–92. https://doi.org/10.2174/1874104501105010082

38. Lotfy M, Singh J, Rashed H, Tariq S, Zilahi E, Adeghate E (2014) Mechanism of the beneficial and protective effects of exenatide in diabetic rats. J Endocrinol 220(3):291–304. https://doi.org/10.1530/JOE-13-0426
39. Müller TD, Nogueiras R, Andermann ML, Andrews ZB, Anker SD, Argente J, … Tschöp MH (2015) Ghrelin. Molecular Metabolism 4(6):437–460. https://doi.org/10.1016/j.molmet.2015.03.005
40. Elabadlah H, Hameed R, D'Souza C, Mohsin S, Adeghate EA (2020) Exogenous ghrelin increases plasma insulin level in diabetic rats. Biomolecules 10(4):633. https://doi.org/10.3390/biom10040633
41. Adeghate E, Ponery AS (2002) Ghrelin stimulates insulin secretion from the pancreas of normal and diabetic rats. J Neuroendocrinol 14(7):555–560. https://doi.org/10.1046/j.1365-2826.2002.00811.x
42. Druce MR, Wren AM, Park AJ, Milton JE, Patterson M, Frost G, … Bloom SR (2005) Ghrelin increases food intake in obese as well as lean subjects. Int J Obesity 29(9):1130–1136. https://doi.org/10.1038/sj.ijo.0803001
43. Mizuno TM, Kelley KA, Pasinetti GM, Roberts JL, Mobbs CV (2003) Transgenic neuronal expression of proopiomelanocortin attenuates hyperphagic response to fasting and reverses metabolic impairments in leptin-deficient obese mice. Diabetes 52(11):2675–2683. https://doi.org/10.2337/diabetes.52.11.2675
44. Adeghate E, Lotfy M, D'Souza C, Alseiari SM, Alsaadi AA, Qahtan SA (2020) Hypocretin/orexin modulates body weight and the metabolism of glucose and insulin. Diabetes Metab Res Rev 36(3), e3229. https://doi.org/10.1002/dmrr.3229
45. Perez-Suarez I, Ponce-González JG, De La Calle-Herrero J, Losa-Reyna J, Martin-Rincon M, Morales-Alamo D, … Calbet JAL (2017) Severe energy deficit upregulates leptin receptors, leptin signaling, and PTP1B in human skeletal muscle. J Appl Physiol 123(5):1276–1287. https://doi.org/10.1152/japplphysiol.00454.2017
46. Yang XN, Zhang CY, Bing-Wei W, Zhu SG, Zheng RM (2015) Leptin signalings and leptin resistance. Sheng Li Ke Xue Jin Zhan [Progress in Physiology] 46(5):327–333. https://doi.org/10.1007/s11684-013-0263-5
47. Booth A, Magnuson A, Fouts J, Foster M (2015) Adipose tissue, obesity and adipokines: role in cancer promotion. Horm Mol Biol Clin Invest 21(1):57–74. https://doi.org/10.1515/hmbci-2014-0037
48. Nigro E, Scudiero O, Monaco ML, Palmieri A, Mazzarella G, Costagliola C, … Daniele A (2014) New insight into adiponectin role in obesity and obesity-related diseases. BioMed Res Int. 2014:658913https://doi.org/10.1155/2014/658913
49. Fu L, Isobe K, Zeng Q, Suzukawa K, Takekoshi K, Kawakami Y (2007) B-Adrenoceptor agonists downregulate Adiponectin, but upregulate Adiponectin receptor 2 and Tumor necrosis factor-α expression in adipocytes. Eur J Pharmacol 569(1–2):155–162. https://doi.org/10.1016/j.ejphar.2007.05.005
50. Matsuzawa Y, Funahashi T, Kihara S, Shimomura I (2004) Adiponectin and metabolic syndrome. Arterioscler Thromb Vasc Biol 24(1):29–33. https://doi.org/10.1161/01.ATV.0000099786.99623.EF
51. Adeghate EA, Kalász H, Al Jaberi S, Adeghate J, Tekes K. (2021) Tackling type 2 diabetes-associated cardiovascular and renal comorbidities: a key challenge for drug development. Expert opinion on investigational drugs, 30(2), 85–93. https://doi.org/10.1080/13543784.2021.1865914
52. Maury E, Brichard SM (2010) Adipokine dysregulation, adipose tissue inflammation and metabolic syndrome. Mol Cell Endocrinol 314(1):1–16
53. Shoelson SE, Herrero L, Naaz A (2007) Obesity, inflammation, and insulin resistance. Gastroenterology 132(6):2169–2180. https://doi.org/10.1053/j.gastro.2007.03.059
54. Pradhan AD, Manson JE, Rifai N, Buring JE, Ridker PM (2001) C-reactive protein, interleukin 6, and risk of developing type 2 diabetes mellitus. J Am Med Assoc 286(3):327–334. https://doi.org/10.1001/jama.286.3.327

55. Ziegler L, Lundqvist J, Dreij K, Wallén H, de Faire U, Paulsson-Berne G, Hedin U, Matic L, Gigante B (2020) Expression of Interleukin 6 signaling receptors in carotid atherosclerosis. Vascular medicine (London, England), 1358863X20977662. Advance online publication. https://doi.org/10.1177/1358863X20977662
56. McGill HC, McMahan CA, Herderick EE, Zieske AW, Malcom GT, Tracy RE, Strong JP (2002) Obesity accelerates the progression of coronary atherosclerosis in young men. Circulation 105(23):2712–2718. https://doi.org/10.1161/01.CIR.0000018121.67607.CE
57. Din-Dzietham R, Liu Y, Bielo MV, Shamsa F (2007) High blood pressure trends in children and adolescents in national surveys, 1963 to 2002. Circulation 116(13):1488–1496. https://doi.org/10.1161/CIRCULATIONAHA.106.683243
58. Kenchaiah S, Evans JC, Levy D, Wilson PW, Benjamin EJ, Larson MG, Kannel WB, Vasan RS (2002) Obesity and the risk of heart failure. N Engl J Med 347(5):305–313. https://doi.org/10.1056/NEJMoa020245
59. Schmieder RE, Messerli FH (1993) Does obesity influence early target organ damage in hypertensive patients? Circulation 87(5):1482–1488. https://doi.org/10.1161/01.CIR.87.5.1482
60. Rabbia F, Silke B, Conterno A, Grosso T, De Vito B, Rabbone I. ... Veglio F (2003) Assessment of cardiac autonomic modulation during adolescent obesity. Obesity Res 11(4):541–548. https://doi.org/10.1038/oby.2003.76
61. Flint AC, Conell C, Ren X et al (2019) Effect of systolic and diastolic blood pressure on cardiovascular outcomes. N Engl J Med 381:243
62. Jemal A, Miller KD, Ma J, Siegel RL, Fedewa SA, Islami F, ... Thun MJ (2018) Higher lung cancer incidence in young women than young men in the United States. New England J Med 378(21):1999–2009. https://doi.org/10.1056/NEJMoa1715907
63. Dalamaga M, Diakopoulos KN, Mantzoros CS (2012) The role of adiponectin in cancer: a review of current evidence. Endocr Rev 33(4):547–594. https://doi.org/10.1210/er.2011-1015
64. Dalamaga M (2013) Obesity, insulin resistance, adipocytokines and breast cancer: new biomarkers and attractive therapeutic targets. World J Experim Med 3(3). https://doi.org/10.5493/wjem.v3.i3.34
65. Pickens CA, Sordillo LM, Zhang C, Fenton JI (2017) Obesity is positively associated with arachidonic acid-derived 5- and 11-hydroxyeicosatetraenoic acid (HETE). Metabol Clinic Experim 70:177–191. https://doi.org/10.1016/j.metabol.2017.01.034
66. Lee MK, Kim JY, Kim D Il, Kang, DW, Park J hye, Ahn KY, ... Jeon JY (2017) Effect of home-based exercise intervention on fasting insulin and Adipocytokines in colorectal cancer survivors: a randomized controlled trial. Metabol Clin Experim 76:23–31. https://doi.org/10.1016/j.metabol.2017.07.005
67. Mendonça FM, De Sousa FR, Barbosa AL, Martins SC, Araújo RL, Soares, R, Abreu C (2015) Metabolic syndrome and risk of cancer: which link? Metabol Clin Experim 64(2):182–189. https://doi.org/10.1016/j.metabol.2014.10.008
68. Diedrich J, Gusky HC, Podgorski I (2015) Adipose tissue dysfunction and its effects on tumor metabolism. Horm Mol Biol Clin Invest 21(1):17–41. https://doi.org/10.1515/hmbci-2014-0045
69. Pérez-Hernández AI, Catalán V, Gómez-Ambrosi J, Rodríguez A, Frühbeck G (2014) Mechanisms linking excess adiposity and carcinogenesis promotion. Front Endocrinol 5:1–17. https://doi.org/10.3389/fendo.2014.00065
70. Wang Y, Lam KS, Xu A (2007) Adiponectin as a negative regulator in obesity-related mammary carcinogenesis. Cell Res 17(4):280–282. https://doi.org/10.1038/cr.2007.14
71. Harris HR, Tworoger SS, Hankinson SE, Rosner BA, Michels KB (2011) Plasma leptin levels and risk of breast cancer in premenopausal women. Cancer Prev Res 4(9):1449–1456. https://doi.org/10.1158/1940-6207.CAPR-11-0125
72. Niu J, Jiang L, Guo W, Shao L, Liu Y, Wang L (2013) The association between leptin level and breast cancer: a meta-analysis. PLoS ONE 8(6). https://doi.org/10.1371/journal.pone.0067349

73. Lee CH, Woo YC, Wang Y, Yeung CY, Xu A, Lam KSL (2015) Obesity, adipokines and cancer: an update. Clin Endocrinol 83(2):147–156. https://doi.org/10.1111/cen.12667
74. Dong, Z., Fu, S., Xu, X., Yang, Y., Du, L., Li, W., … Wang, C. (2014). Leptin-mediated regulation of ICAM-1 is Rho/ROCK dependent and enhances gastric cancer cell migration. British Journal of Cancer, 110(7), 1801–1810. https://doi.org/10.1038/bjc.2014.70
75. Koda M, Sulkowska M, Kanczuga-Koda L, Surmacz E, Sulkowski S (2007) Overexpression of the obesity hormone leptin in human colorectal cancer. J Clin Pathol 60(8):902–906. https://doi.org/10.1136/jcp.2006.041004
76. Chobot A, Górowska-Kowolik K, Sokołowska M, Jarosz-Chobot P (2018) Obesity and diabetes—not only a simple link between two epidemics. Diabetes Metab Res Rev 34(7):1–9. https://doi.org/10.1002/dmrr.3042
77. Bastien M, Poirier P, Lemieux I, Després JP (2014) Overview of epidemiology and contribution of obesity to cardiovascular disease. Prog Cardiovasc Dis 56(4):369–381. https://doi.org/10.1016/j.pcad.2013.10.016
78. Abel ED, Peroni O, Kim JK, Kim YB, Boss O, Hadro E, … Kahn BB (2001) Adipose-selective targeting of the GLUT4 gene impairs insulin action in muscle and liver. Nature 409(6821):729–733. https://doi.org/10.1038/35055575
79. Belfort R, Mandarino L, Kashyap S, Wirfel K, Pratipanawatr T, Berria R, ... Cusi K (2005) Dose-response effect of elevated plasma free fatty acid on insulin signaling. Diabetes 54(6):1640−1648
80. Samuel VT, Shulman GI (2016) The pathogenesis of insulin resistance: integrating signaling pathways and substrate flux. J Clin Investig 126(1):12–22. https://doi.org/10.1172/JCI77812
81. Shai I, Jiang R, Manson JAE, Stampfer MJ, Willett WC, Colditz GA, Hu FB (2006) Ethnicity, obesity, and risk of type 2 diabetes in women: a 20-year follow-up study. Diabetes Care 29(7):1585–1590. https://doi.org/10.2337/dc06-0057
82. Wang Y, Rimm EB, Stampfer MJ, Willett WC, Hu FB (2005) Comparison of abdominal adiposity and overall obesity in predicting risk of type 2 diabetes among men. Am J Clin Nutr 81(3):555–563. https://doi.org/10.1093/ajcn/81.3.555
83. Oguma Y, Sesso HD, Paffenbarger RS Jr, Lee IM (2005) Weight change and risk of developing type 2 diabetes. Obes Res 13(5):945–951
84. Brand-Miller J, Hayne S, Petocz P, Colagiuri S (2003) Low-glycemic index diets in the management of diabetes: a meta-analysis of randomized controlled trials. Diabetes Care 26(8):2261–2267. https://doi.org/10.2337/diacare.26.8.2261
85. Horton ES (1988) Role and management of exercise in diabetes mellitus. Diabetes Care 11(2):201–211. https://doi.org/10.2337/diacare.11.2.201
86. Santos FL, Esteves SS, da Costa Pereira A, Yancy WS, Nunes JPL (2012) Systematic review and meta-analysis of clinical trials of the effects of low carbohydrate diets on cardiovascular risk factors. Obes Rev 13(11):1048–1066. https://doi.org/10.1111/j.1467-789X.2012.01021.x
87. Gummesson A, Nyman E, Knutsson M, Karpefors M (2017) Effect of weight reduction on glycated haemoglobin in weight loss trials in patients with type 2 diabetes. Diabetes Obes Metab 19(9):1295–1305
88. Adeghate E, Kalasz H, Veress G, Teke K (2010) Medicinal chemistry of drugs used in diabetic cardiomyopathy. Curr Med Chem 17(6):517–551. https://doi.org/10.2174/092986710790416281
89. Domecq JP, Prutsky G, Leppin A, Sonbol MB, Altayar O, Undavalli C, … Murad MH (2015) Drugs commonly associated with weight change: a systematic review and meta-analysis. J Clin Endocrinol Metabol 100(2):363–370. https://doi.org/10.1210/jc.2014-3421
90. Meneghini LF, Orozco-Beltran D, Khunti K, Caputo S, Damçi T, Liebl A, Ross SA (2011) Weight beneficial treatments for type 2 diabetes. J Clin Endocrinol Metab 96(11):3337–3353. https://doi.org/10.1210/jc.2011-1074
91. Zhou G, Goodyear LJ, Moller DE, Zhou G, Myers R, Li Y, … Moller DE (2001) Role of AMP-activated protein kinase in mechanism of metformin action find the latest version: role of AMP-activated protein kinase in mechanism of metformin action. J Clin Invest 108(8):1167–1174. https://doi.org/10.1172/JCI200113505.Introduction

92. Madiraju AK, Erion DM, Rahimi Y, Zhang XM, Braddock DT, Albright RA, … Shulman GI (2014) Metformin suppresses gluconeogenesis by inhibiting mitochondrial glycerophosphate dehydrogenase. Nature 510(7506):542–546. https://doi.org/10.1038/nature13270
93. Rena G, Pearson ER, Sakamoto K (2013) Molecular mechanism of action of metformin: old or new insights? Diabetologia 56(9):1898–1906. https://doi.org/10.1007/s00125-013-2991-0
94. Aroda VR, Knowler WC, Crandall JP, Perreault L, Edelstein SL, Jeffries SL, … Nathan DM (2017) Metformin for diabetes prevention: insights gained from the diabetes prevention program/diabetes prevention program outcomes study. Diabetologia 60(9):1601–1611. https://doi.org/10.1007/s00125-017-4361-9
95. Naderpoor N, Shorakae S, De Courten B, Misso ML, Moran LJ, Teede HJ (2015) Metformin and lifestyle modification in polycysticovary syndrome: systematic review and meta-analysis. Hum Reprod Update 21(5):560–574. https://doi.org/10.1093/humupd/dmv025
96. Jiang YF, Chen XY, Ding T, Wang XF, Zhu ZN, Su SW (2015) Comparative efficacy and safety of OADs in management of GDM: network meta-analysis of randomized controlled trials. J Clin Endocrinol Metab 100(5):2071–2080. https://doi.org/10.1210/jc.2014-4403
97. Chalasani N, Younossi Z, Lavine JE, Diehl AM, Brunt EM, Cusi K, … Sanyal AJ (2012) The diagnosis and management of non-alcoholic fatty liver disease: practice guideline by the American association for the study of liver diseases, American college of gastroenterology, and the American gastroenterological association. Hepatology 55(6):2005–2023. https://doi.org/10.1002/hep.25762
98. Wu L, Zhu J, Prokop LJ, Hassan Murad M (2015) Pharmacologic therapy of diabetes and overall cancer risk and mortality: a meta-analysis of 265 studies. Sci Rep 5(June):1–10. https://doi.org/10.1038/srep10147
99. Gao X, Cai X, Yang W, Chen Y, Han X, Ji L (2018) Meta-analysis and critical review on the efficacy and safety of alpha-glucosidase inhibitors in Asian and non-Asian populations. J Diabet Invest 9(2):321–331. https://doi.org/10.1111/jdi.12711
100. Herman GA, Bergman A, Stevens C, Kotey P, Yi B, Zhao P, Wagner JA (2006) Effect of single oral doses of sitagliptin, a dipeptidyl peptidase-4 inhibitor, on incretin and plasma glucose levels after an oral glucose tolerance test in patients with type 2 diabetes. J Clin Endocrinol Metab 91(11):4612–4619. https://doi.org/10.1210/jc.2006-1009
101. Mohsin S, Baniyas MM, AlDarmaki RS, Tekes K, Kalász H, Adeghate EA (2019) An update on therapies for the treatment of diabetes-induced osteoporosis. Expert Opin Biol Ther 19(9):937–948. https://doi.org/10.1080/14712598.2019.1618266
102. Ahrén B (2013) Incretin therapy for type 2 diabetes: GLP-1 receptor agonists and DPP-4 inhibitors. Eur Diabet Nurs 10(1):31–36. https://doi.org/10.1002/edn.221
103. Pospisilik JA, Stafford SG, Demuth HU, Brownsey R, Parkhouse W, Finegood DT, … Pederson RA (2002) Long-term treatment with the dipeptidyl peptidase IV inhibitor P32/98 causes sustained improvements in glucose tolerance, insulin sensitivity, hyperinsulinemia, and β-cell glucose responsiveness in VDF (fa/fa) Zucker rats. Diabetes 51(4):943–950. https://doi.org/10.2337/diabetes.51.4.943
104. Lyu X, Zhu X, Zhao B, Du L, Chen D, Wang C, … Ran X (2017) Effects of dipeptidyl peptidase-4 inhibitors on beta-cell function and insulin resistance in type 2 diabetes: meta-analysis of randomized controlled trials. Sci Reports 7(37):1–11. https://doi.org/10.1038/srep44865
105. Pappachan JM (2015) Incretin manipulation in diabetes management. World J Diab 6(6):774. https://doi.org/10.4239/wjd.v6.i6.774
106. Adeghate E, Mohsin S, Adi F, Ahmed F, Yahya A, Kalász H, Tekes K, Adeghate EA (2019) An update of SGLT1 and SGLT2 inhibitors in early phase diabetes-type 2 clinical trials. Expert Opin Investig Drugs 28(9):811–820. https://doi.org/10.1080/13543784.2019.1655539
107. Yang X, Liu Q, Li Y, Ding Y, Zhao Y, Tang Q, Wu T, Chen L, Pu S, Cheng S, Zhang J, Zhang Z, Huang Y, Li R, Zhao Y, Zou M, Shi X, Jiang W, Wang R, He J (2021) SGLT2 inhibition by canagliflozin ameliorates diet-induced obesity by promoting intra-adipose sympathetic innervation. Br J Pharmacol. https://doi.org/10.1111/bph.15381.Advanceonlinepublication. doi:10.1111/bph.15381

108. Padwal RS, Majumdar SR (2007) Drug treatments for obesity: orlistat, sibutramine, and rimonabant. Lancet 369(9555):71–77. https://doi.org/10.1016/S0140-6736(07)60033-6
109. Jarl S, Mark N (2004) XENical in the prevention of diabetes in obese subjects (XENDOS) study. Diabetes Care 27(1):155–161. https://doi.org/10.2337/diacare.27.1.155
110. Sharma AM (2001) Sibutramine in overweight/obese hypertensive patients. Int J Obes 25:S20–S23. https://doi.org/10.1038/sj.ijo.0801934
111. Vettor R, Serra R, Fabris R, Pagano C, Federspil G (2005) Effect of sibutramine on weight management and metabolic control in type 2 diabetes: a meta-analysis of clinical studies. Diabetes Care 28(4):942–949. https://doi.org/10.2337/diacare.28.4.942
112. Shibuya K, Ali KF, Ji X, Milinoivh A, Bauman J, Kattan MW, Pantalone KM, Burguera B (2019) The benefit of short-term weight loss with anti-obesity medications in real-world clinical practice. Endocr Pract 25(10):1022–1028. https://doi.org/10.4158/EP-2019-0081
113. Smith FJ, Holman CDAJ, Moorin RE, Fletcher DR (2008) Incidence of bariatric surgery and postoperative outcomes: a population-based analysis in Western Australia. Med J Aust 189(4):198–202. https://doi.org/10.5694/j.1326-5377.2008.tb01981.x
114. Dixon JB, O'Brien PE, Playfair J, Chapman L, Schachter LM, Skinner S, … Anderson M (2008) Adjustable gastric banding and conventional therapy for type 2 diabetes: a randomized controlled trial. Obstet Gynecol Survey 63(6):372–373. https://doi.org/10.1097/01.ogx.0000314848.71777.69
115. Blackstone R, Bunt JC, Cortés MC, Sugerman HJ (2012) Type 2 diabetes after gastric bypass: remission in five models using HbA1c, fasting blood glucose, and medication status. Surg Obesity Relat Dis 8(5):548–555
116. Adams TD, Gress RE, Smith SC, Halverson RC, Simper SC, Rosamond WD, … Hunt SC (2007) Long-term mortality after gastric bypass surgery. New England J Med 357(8):753–761. https://doi.org/10.1056/NEJMoa066603

Chapter 14
Telomere Shortening and Calorie Restriction in Obesity

Naoki Makino and Toyoki Maeda

Abstract Obesity, which predisposes to persistant metabolic disorders such as diabetes, dyslipidemia, and hypertension, has been reported to correlate with accelerated biological aging with shortened somatic telomere length. In these disease states, production of peroxidative biochemical compounds is systemically promoted, which is believed to be the major cause of telomere depletion. Therefore, obesity is an underlying condition that accelerates biological aging due to telomere depletion. Obesity itself also contributes to mild telomere shortening due to somatic cell peroxidation. In addition, obesity may be associated with subtelomeric demethylation, a biomarker for accelerated telomere shortening. Calorie restriction (CR) is now the most powerful and efficient intervention and involves the administration of a well-balanced and nutrient-dense diet that reduces calorie intake. This study was to investigate the impact of CR on cardiac senescence in an animal model of diabetes and examine the signal transduction mechanisms for changes in cell survival as well as cardiac function. Male 8-week-old Otsuka Long-Evans Tokushima fatty (OLETF) diabetic rats were divided into 2 groups: a group fed ad libitum (AL), and a group fed with CR (30% energy reduction). At 40 weeks of age, the telomere length was not altered in experimental rats with or without CR, however, telomerase activity in both strains of CR rats was significantly elevated. Obesity, as well as accelerated aging, can be considered a premetabolic disease state. Preventing obesity and managing your diet to keep your body mass index within normal limits are important steps in anti-aging.

Keywords Telomere · Leucocytes · Methylation · Type 2 diabetes · Calorie restriction · Autophagy · Diastolic function · Signal transduction

N. Makino (✉) · T. Maeda
Division of Cardiology and Clinical Gerontology, Department of Internal Medicine, Kyushu University Beppu Hospital, 4546 Tsurumihara, Beppu 8740838, Japan

© Springer Nature Switzerland AG 2021

P. S. Tappia et al. (eds.), *Cellular and Biochemical Mechanisms of Obesity*, Advances in Biochemistry in Health and Disease 23,
https://doi.org/10.1007/978-3-030-84763-0_14

Introduction

Data from epidemiological, experimental and clinical studies strongly indicate that maintaining a healthy body weight and preventing the fat accumulation is essential for the prevention of multiple chronic diseases and the promotion of healthy aging [1, 2]. Obesity is characterized by chronic, low-grade inflammation in adipose tissue, and this can directly enhance oxidative stress [3]. Further, obesity has been associated with decreased DNA repair processes that are essential cell responses to DNA damage [4, 5], and shorter telomere length (TL) [6, 7].

Telomere is DNA–protein complexes at the ends of chromosome which maintain chromosomal stability [7]. Telomere shortens as individual ages and therefore telomere attrition is considered a hallmark of ageing [8]. This presentation introduced the analysis of TL distribution and subtelomeric methylation estimated by modified genomic Southern blotting method [9, 10]. For better understanding of telomere biology in obesity, this study examined responses to calorie restriction (CR) with intermittent moderate intervals in Otsuka Long-Evans Tokushima fatty (OLETF) rats, which is a model animal of obesity and diabetes [11–13]. In addition, molecular mechanisms, especially with regard to myocardial cellular survival, senescence, protection against cellular damage, and the activation of cardiac remodeling [14, 15] were evaluated in association with aging-related diastolic malfunction. Our results strongly suggest that CR improves diastolic function of diabetic hearts, ameliorating the decrease in telomerase activity and the degradation of cardiac protein.

Obesity and Reactive Oxygen Species

Obesity usually results from excessive fat accumulation in the body caused by excessive food intake. Excess fat accumulates as visceral fat, which causes fatty liver and fat deposits on the walls of blood vessels, which in turn causes arteriosclerosis [16, 17]. The fat accumulated in such bodies serves as a substrate for the subsequent oxidative reaction, producing oxidants and peroxides [18]. Oxidized fat is a source of reactive oxygen species (ROS). Fatty agents are involved as a source of peroxide compounds in additional biochemical reactions mediated by locally activated local enzymes and cytokines [19]. The ROS produced by these series of reactions degrade the surrounding somatic cells and damage the final DNA structure, nuclear DNA structure including telomeric DNA [18]. Telomeres are complex structures composed of repetitive DNA components and supporting protein molecules located at the extreme ends of eukaryotic chromosomes [20]. ROS causes somatic DNA mutations and also destroys telomere structure. This telomere damage is followed by pathological telomere depletion. Telomeres are physiologically shortened in all cell divisions due to defective DNA replication mechanisms [20]. Telomere shortening is restricted to suppress the survival of very old cells that have undergone excessively repeated cell divisions that carry the risk of carcinogenesis caused by somatic mutations that accumulate during DNA replication throughout the history of cell division. [20]. When the telomere length reaches its limit, the cell stops its cell division, enters a cellular senescent state, and dies. Therefore, normal somatic cells die before being converted into cancer cells. Telomere shortening is considered the cell cycle history of cells and reflects the age of biological cells. Somatic obesity systemically produces excess ROS, followed by accelerated telomere shortening. In fact, many past reports have shown that obese people have shorter somatic telomeres than non-obese people [21]. Obesity predisposes to metabolic disorders that induce DM, dyslipidemia, and hypertension. These conditions have been reported to cause higher ROS production [22] and contribute to accelerated telomere shortening [23]. Type 2 DM patients have been reported to tolerate telomeres with short somatic telomere length [24]. For dyslipidemia, serum HDL levels are negatively associated with somatic telomere length [25]. On the other hand, serum levels of LDL or triglycerides do not affect a patient's telomere length. Hypertension is also associated with shortening telomeres [24, 26]. Therefore, TL wasting is affected by complex biochemical and physiological conditions in metabolic disorders that are primarily derived from obesity.

Obesity and Telomere Shortening

Metabolic disorders cause systemic arteriosclerosis that underlies cardiovascular disease. In addition, diabetics are more susceptible to infections due to immunosuppression. On the other hand, shortening telomeres is prone to heart disease and infectious diseases, which is life-threatening [27]. Second, populations with short TL of somatic cells are reported to have shorter life expectancy than those with long TL [25]. Somatic TL reflects the TL of stem cells. After consuming peripheral cells, stem cell division induces complementation of lost cells. In the case of an infectious disease, the immune response is activated and white blood cells and lymphocytes function and are lost in a series of immune responses. When such an immune response stimulating event is repeated, peripheral immune cells are rapidly consumed, and the shortening of TL of immune stem cells is accelerated. Indeed, serum gamma globulin levels, which are considered to reflect a history of repeated immune response stimulating events, are negatively correlated with peripheral leukocyte TL, which reflects blood stem cell TL [28]. Obesity induces metabolic disorders that lead to life-threatening diseases such as cardiovascular and infectious diseases. Each of these accelerates TL shortening. In addition, shorter telomeres are more susceptible to infections and heart disease and shorter life expectancy [27]. Therefore, when obesity occurs, telomeres are at risk of a vicious cycle of TL shortening. Dietary restrictions and daily exercise are important to stop the acceleration of TL shortening. Furthermore, it is highly desirable to initiate treatment of metabolic disorders early. Escape from obesity is a very effective way to treat DM, dyslipidemia and hypertension in the case of obesity.

Obesity and Subtelomeric Methylation

In treating obesity-related disorders, it is essential to detect not only the TL itself, but also the rate of TL shortening. As a candidate for detecting the telomere shortening rate, we propose to evaluate the distribution of telomere length and the extent of subtelomeric epigenetic changes such as methylation of subtelomeres. TL measurements are made in several ways. PCR-mediated TL measurement is widely used in recent telomere surveys. For analysis of clinical samples, genomic Southern blotting using telomere DNA probe is performed. In contrast to PCR-mediated methods, Southern blots can detect telomere DNA length distribution. Proportional changes in the proportion of long and short telomeres indicate the proportion of young and old cells, respectively. From our observations, the long telomere rate decreases and the short telomere rate increases with physiological aging. On the other hand, the proportion of long telomeres decreases, but the proportion of short telomeres also decreases with pathological aging, such as excessive radiation exposure. Therefore, biological aging can be assessed by the Southern method. Also, in clinical cases, analysis of TL distribution can detect pathological aging. In some neurodegenerative diseases such as Parkinson's disease and Alzheimer's disease, accelerated biological senescence associated with short somatic TL is not always detectable [28–31]. As a result, the average TL reduction in these cases is obscured. Thus, the reduction of short TL fractions interferes with the detection of TL shortening. In addition to telomere length distribution, TL evaluated from genomic Southern blots is modified to detect subtelomeric methylation. We used two types of restriction enzymes. The former and the latter cut the same nucleotide sequence, but the former cannot cut the sequence when methylated. Thus, methylation-sensitive isoschizomers detect the total length of TLs (mTL: TLs containing methylated subtelomeres) and the extent of highly adjacent subtelomeric methylated regions. By comparing mTL with TL, the degree of methylation of subtelomeres can be quantitatively evaluated (Fig. 1a). Obesity is thought to lead to overweight fatigue and knee joint pain, which interferes with daily physical function and lack of exercise. It also triggers a vicious circle for obesity. On the other hand, muscle mass has a positive correlation with TL [32]. Lack of exercise shortens TL. Among patients with chronic illness, those with higher physical performance tend to have longer TLs (Fig. 1b), and higher methylation in the subtelomeric region is a predictor of effective recovery of physical function after hospitalization physical therapy. (Fig. 1c) [9]. Furthermore, subtelomeric hypomethylation has been shown to correlate with accelerated telomere shortening in diabetes [10]. Here we present a new study on BMI and telomere length in patients with stable chronic disease. The telomere parameters and BMI of patients (83 males, 40 females) were further analyzed. The higher BMI group (mean BMI 26.7 $\pm$ 2.1 for men, 27.7 $\pm$ 4.6 for women) had TL 6.1 $\pm$ 0.6 kb in men and subtelomeric methylation range (met TL-TL) 0.9 $\pm$ 0.4 kb and TL 6.7 $\pm$ 0.7 kb and mTL- shows female TL 0.85 $\pm$ 0.4 kb and lower BMI group (mean BMI 20.9 for men, 25.2 $\pm$ 3.2 for women) shows mean TL 6.1 $\pm$ 0.5 and meets TL-TL 1.0 $\pm$ 0.2 for men, TL 6.3 $\pm$ 0.7 kb and mTL-filled. TL 1.1 $\pm$ 0.3 kb in women (p-value; TL 0.691 and mTL-TL 0.840 in men, TL

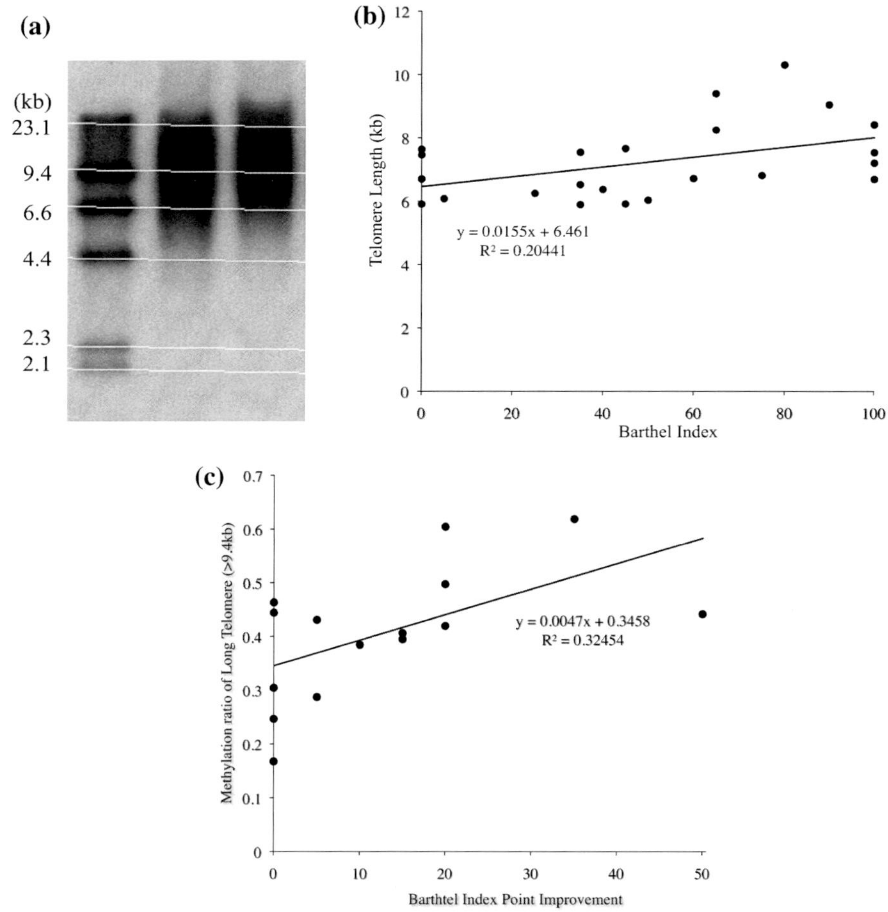

Fig. 14.1 **a** Southern blot results of leukocyte DNA. The left lane shows the length of the DNA fragment (kb). The middle and right lanes are examples of genomic Southern blots using a telomere DNA probe.b. The methylation-insensitive restriction enzyme (Msp I) is used in the middle lane. A methylation insensitive restriction enzyme (HpaII) is used in the right lane. Both enzymes cleave the tetranucleotide sequence CCGG Note that the right lane smear (methylated telomere length) is longer than the middle lane smear (standard telomere length). **b** Regression analysis between white blood cell telomere length and physical performance [32]. Physical performance is assessed using the Barthel Index. Notice that the two variables show a positive correlation. **c** Regression analysis between improvement of physical performance and subtelomeric methylation status of long telomeres [32]. Physical fitness recovery of inpatients with physical therapy was assessed by comparing the Barthel Index assessed at admission and discharge. The degree of subtelomeric methylation of telomeres longer than 9.4 kb is positively associated with improved Barthel index

0.126 and mTL-TL 0.038 in women). Although no significant correlation between telomere length telomeres and BMI was detected, this analysis detected subtelomeric demethylation with high BMI (unpublished data). Several previous reports failed to reveal a significant negative correlation between obesity and telomere length [33–35]. Presumably, background medical conditions can lead to loss of correlation between BMI and telomere parameters. However, higher BMI is associated with subtelomeric demethylation consistent with an accelerated stage of telomere depletion. Therefore, obesity can be considered as a preliminary condition that promotes aging, even before the shortening of TL is apparent. Obesity therefore needs to be corrected by dietary restrictions and proper exercise to prevent further metabolic disorders.

General Characteristic in Animal Study

The OLETF rat model represents some of the phenotypes of type 2 diabetes and the pathological features of human metabolic syndrome including late onset hyperglycemia, mild obesity, increased insulin resistance, hyperlipidemia, and hypertension [11–13]. Mean body weight (BW) at 8 weeks of age was similar to that of each of all other experimental rat model groups (Table 14.1). Average weekly food consumption was lower in the LETO-CR and OLETF-CR groups than in all other groups. The food intake of high-fat calorie diet (HFD) rats [36] was not different from that of OLETF rats fed ad libitum. Caloric restriction with a 30% energy reduction resulted in significantly reduction in BW in the LETO-CR and OLETF-CR groups compared to other groups fed ad libitum. Fasting blood glucose levels at 40 weeks of age were significantly lower in OLETF-CR rats than OLETF fed ad libitum, and were higher in HFD than in LETO and LETO-CR rats. The homeostasis model assessment of insulin resistance (HOMA-IR), a marker of insulin resistance, was determined based on both plasma glucose and serum insulin levels. HOMA-IR was significantly less in LETO or OLETF-CR interventions at 40 weeks of age, compared to each AL-fed group. Serum cholesterol and triglyceride levels were significantly elevated in HFD, OLETF and OLETF-CR rats at 40 weeks of age. However, serum adiponectin levels were higher in each of the two CR groups than in LETO or OLETF without CR rats at the end of the study, respectively. Systolic blood pressure was comparable between the groups under study. And calorie restriction.

Telomere Biology and CR

Telomeric DNA length was assessed by dot-blot analysis of 40-week-old rat cardiac cells (Fig. 2a). Telomere DNA length was significantly shorter in the heart tissue of HFD rats, but was not altered by CR in experimental rats with or without CR [37]. On the other hand, telomerase activity quantified using a TRAP assay was significantly elevated in the hearts of both LETO-CR and OLETF-CR rats (Fig. 2b). However,

Table 14.1 General characteristics in LETO rats divided into groups fed ad libitum, high fatty diet or caloric restriction and OLETF diabetic rats divided into groups fed ad libitum, or caloric restriction at 8 and 40 weeks of age

	LETO						OLETF			
	AL (n = 6)		CR (n = 6)		HFD (n = 6)		AL (n = 6)		CR (n = 5)	
	8 wks	40 wks	8 wks	40 wks	8 wks	40 wks	8 wks	40 wks	8 wks	40 wks
BW.(g)	245 ± 12	445 ± 22[a]	239 ± 11	395 ± 25[a,b]	240 ± 10	716 ± 26[a,b]	246 ± 11	608 ± 31[a]	249 ± 16	559 ± 29[a,b]
Glucose	98 ± 4.8	108 ± 6.8	111 ± 7.1	116 ± 5.8	119 ± 4.6	138 ± 6.8[b]	141 ± 8.2[a]	178 ± 7.3	138 ± 6.9[a]	139 ± 7.1[b]
Serum Insulin	2.2 ± 0.1	2.5 ± 0.1	2.4 ± 0.1	1.8 ± 0.10[b]	2.3 ± 0.15	5.8 ± 0.36[a,b]	7.6 ± 0.32[a]	10.3 ± 0.53[a]	8.1 ± 0.38[a]	4.6 ± 0.4[a,b]
HOMA-IR	0.52 ± 0.03	0.57 ± 0.22	0.67 ± 0.05	0.51 ± 0.04	0.51 ± 0.04	1.97 ± 0.06[a,b]	2.64 ± 0.11[a]	4.53 ± 0.23[a]	2.74 ± 0.18[a]	1.58 ± 0.06[a,b]
Total Cholesterol	108 ± 6	111 ± 7	110 ± 5	106 ± 6	108 ± 5	131 ± 6[a,b]	129 ± 9[a]	148 ± 11	125 ± 7[a]	129 ± 6[a]
Triglyceride	41 ± 2	45 ± 3	42 ± 2	38 ± 3	42 ± 2	163 ± 8[b]	164 ± 10[a]	241 ± 15[a]	153 ± 9[a]	128 ± 5[a,b]
Adiponectin	2.1 ± 0.10	1.9 ± 0.14	2.2 ± 0.16	4.1 ± 0.14[a,b]	2.1 ± 0.13	1.4 ± 0.07[a,b]	1.8 ± 0.08[a]	1.8 ± 0.07	1.9 ± 0.12	3.7 ± 0.2[a,b]

Each value is means ± SE AL; ad libitum, CR; calorie restriction, HFD; high fatty diet BW; body weight, Heart W; heart weight, Heart to BW; g/Kg, Glucose, mg/dl, Total cholesterol and Triglyceride; mg/dl, Insulin; (μU/ml), Adiponectin; μg/ml, a: $p < 0.05$ versus 8 weeks old LETO-AL rats. b; $p < 0.05$ versus 40 weeks old LETO-AL or OLETF –AL rats in the same strain

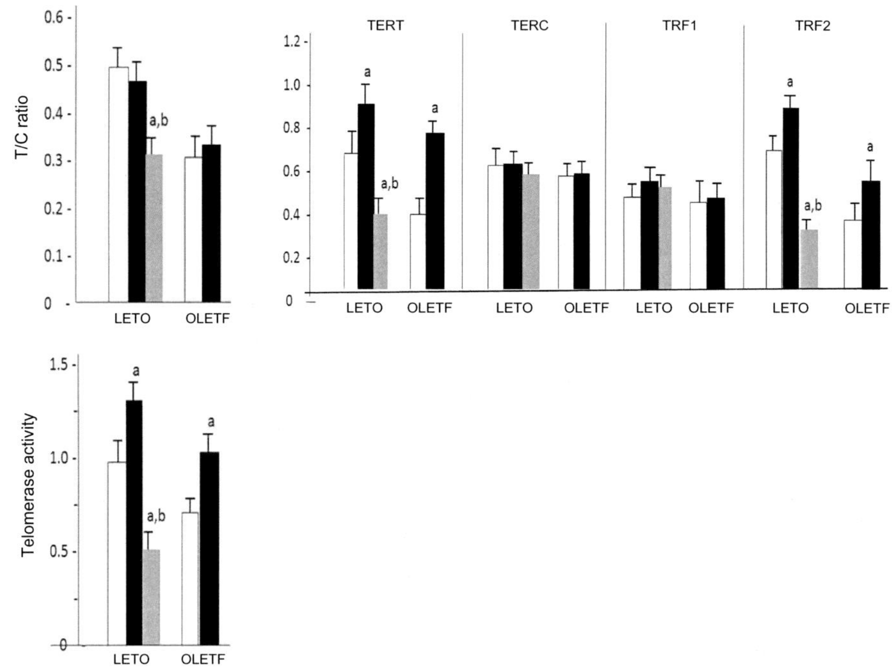

Fig. 14.2 **a** Summarized results as relative the telomeric-to-centromeric DNA content (T/C) ratio. **b** Summarized data of telomerase activity. Each value is presented in comparison with data of LETO rats fed ad libitum. **c** Summarized data are shown protein expression levels of telomere reverse transcriptase (TERT), telomere RNA component (TERC), telomere repeat binding factor (TRF)1 and TRF2 in heart tissue from LETO and OLETF rats with or without CR. Each value is presented as the ratio to GAPDH. The open bar indicates data from LETO or OLETF rats fed ad libitum; the solid bar indicates data from LETO-CR or OLETF-CR rats; the gray bar indicates data from LETO rats fed high calorie food diets (HFD). Each group contained 6 animals. Values are means ±SE. a, $P < 0.05$ versus LETO or OLETF rats; b, $P < 0.05$ versus LETO-CR rats

this activity was depressed in the hearts of HFD rats. Next, protein expression levels of the catalytic subunits TERT and TERC were assessed (Fig. 2c). The expression level of TERT was lower in OLETF rats than in LETO rats, however, this difference was erased by CR diet in OLETF rats. Similar results were obtained for telomere-associated protein TRF2. The expression of TERC and TRF-1 was similar among the groups [37].

Cell Signaling and CR

Since protein expression for the cell signaling of survival factors is associated with CR intervention, protein expression of IGF-1, p53, phospho-Akt, FoxO1 and Sirt1 in the heart was investigated from LETO and OLETF rats of CR (Fig. 3a). Protein

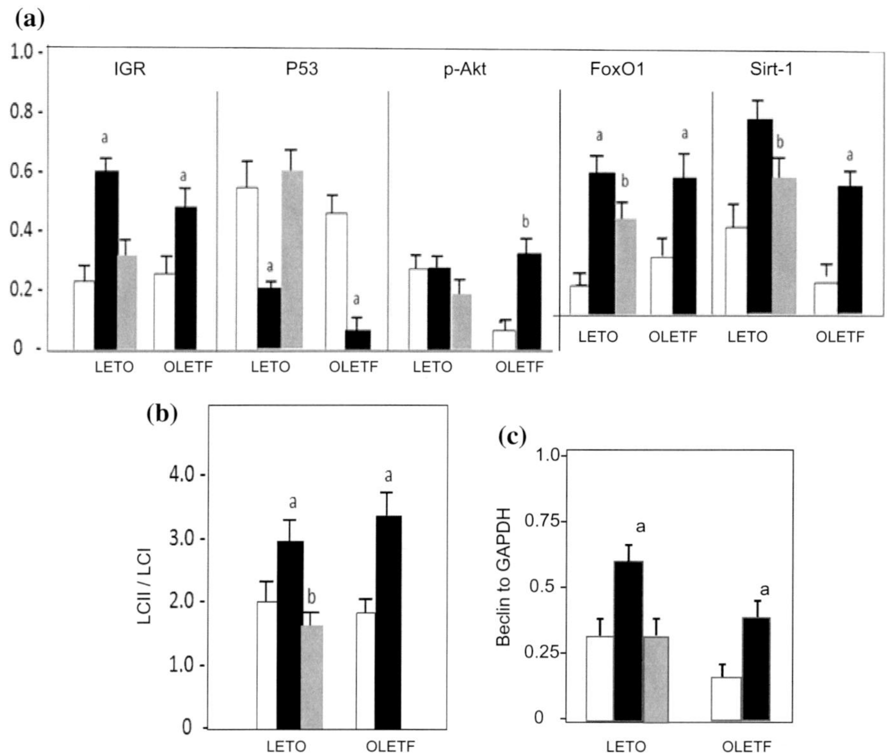

Fig. 14.3 **a** Summarized data of Western immunoblotting are shown protein expressions for insulin-like growth factor (IGF)-1, p53, phospho-Akt (p-Akt), forkhead transcription factor (FoxO1) and Sirt1 in the hearts from LETO and OLETF rats with or without CR. Each value is presented as the ratio to GAPDH. Each bar has the same patterns as Fig. 14.1. Summarized data of Western immunoblots showing the expression of and the LC3-II/LC3-I ratio (**b**) and beclin1/GAPDH (**c**). Each group contained 6 animals. Values are means ±SE. a, $P < 0.05$ versus LETO or OLETF rats; b, $P < 0.05$ versus LETO-CR or OLETF-CR rats

levels for IGF-1, phospho-FoxO1 and Sirt 1 were higher in the hearts of LETO-CR or OLETF-CR rats at 40 weeks old age. However, p53 protein levels were significantly reduced by CR in both strains of rats. Next, we examined the expression of LC3-II in LETO or OLETF rats with CR intervention and assessed the effects of CR on autophagic flux to determine which mechanism is primarily involved in intracellular signaling of cell survival (Fig. 3b). An increase in the ratio of LC3-II to the cytosolic form of LC3 (LC3-I) was observed in CR rats. In addition, the expression levels of beclin1 were also similar observations as seen in results of LC3-II from CR rats (Fig. 3c).

Table 14.2 Results of Doppler Echocardiography in AL, CR or HFD Treated with LETO rats and AL or CR Treated with OLETF rats at 40 wk age

	LETO			OLETF	
	AL (n = 6)	CR (n = 6)	HFD (n = 5)	AL (n = 6)	CR (n = 6)
LVEDD (mm)	8.13 ± 0.36	7.88 ± 0.46	9.36 ± 0.38	8.28 ± 0.49	7.86 ± 0.36
FS (%)	42.6 ± 2.9	42.5 ±0.24	41.8 ± 0.18	4.30 ± 0.21	41.9 ± 2.4
E/A ratio	1.57 ± 0.08	1.62 ± 0.07	1.84 ± 0.07[a]	1.25 ± 0.05[a]	1.42 ± 0.08[b]
DTE	40.0 ± 2.8	39.4 ± 2.1		46.7 ±2.6[a]	47.8 ± 2.8[a]

Values are means ± SE. LVEDD, Fractional shortening (FS) was determined by the following equation: FS = [(LVEDD—LVESD)/LVEDD] × 100 (%). E/A ratio, peak velocity of early transmitral inflow (E)-to-peak velocity of late transmitral inflow (A) ratio; DTE, decelaration time of the E wave. a; $p < 0.05$ versus LETO rats with AL; b; $p < 0.05$ versus OLETF rats with AL

Cardiac Diastolic Function and CR

Echocardiography was performed in different groups of 40-week-old rats (Table 14.2). The E/A ratio was significantly longer in HFD than in LETO rats, and was significantly shorter in OLETF than in OLETF-CR rats. Conversely, the deceleration time of the E wave (DTE) was extended in HFD or OLETF rats, and shortened in OLETF-CR rats, compared to OLETF rats. Left ventricular end-diastolic dimension (LVEDD), left ventricular end-systolic dimension (LVESD) and fractional shortening (all indexes of systolic function) were not significantly different among the different groups.

Conclusions

The present study shows the distribution of telomere length and the extent of subtelomeric epigenetic changes such as subtelomeric methylation in obesity. No significant correlation was detected between telomere length and BMI, however, higher BMI is associated with subtelomeric demethylation consistent with an accelerated stage of telomere depletion. Obesity can be considered as a preliminary condition that promotes aging, even before the shortening of TL is apparent. The harmful effects of obesity on cardiovascular disease have been demonstrated in many studies. This study implies that CR increases telomerase activity and delays cardiac senescence, which may be associated with functional improvement of diastolic dysfunction in the heart of diabetic rats. In addition, these results may be due in part to enhanced autophagy flux, and attenuated muscle protein degradation by CR. Taken together, these data suggest that there may be a benefit of CR in patients with diabetes, and may provide a metabolic switch that translates the dietary changes of CR into a program of health and survival. However, despite the positive anti-aging benefits of

CR reported herein, compliance remains a major impediment because the lifestyle changes associated with CR require considerable patient commitment.

Acknowledgements The OLETF and LETO rats were a generous gift from the Tokushima Research Institute, Otsuka Pharmaceutical (Tokushima, Japan). We thank Ms. Keiko Tsuchida and Yasuko Ueda for excellent technical assistance to this work.

Disclosure The authors report no conflicts of interest in this work.

References

1. Martin-Rodriguez E, Guillen-Grima F, Marti A, Brugos-Larumbe A (2015) Comorbidity associated with obesity in a large population: the APNA study. Obes Res Clin Pract 9:435–447
2. Lauby-Secretan B, Scoccianti C, Loomis D, Grosse Y, Bianchini F, Straif K (2016) International Agency for Research on cancer handbook working G: body fatness and cancer—viewpoint of the IARC Working Group. N Engl J Med 375:794–798
3. Marseglia L, Manti S, D'Angelo G, Nicotera A, Parisi E, Di Rosa G, Gitto E, Arrigo T (2015) Oxidative stress in obesity: a critical component in human diseases. Int J Mol Sci 16:378–400
4. Sampath H, Batra AK, Vartanian V, Carmical JR, Prusak D, King IB, Lowell B, Earley LF, Wood TG, Marks DL, McCullough AK (2011) L RS: Variable penetrance of metabolic phenotypes and development of high-fat diet-induced adiposity in NEIL1-deficient mice. Am J Physiol Endocrinol Metab 300:E724–734
5. Vartanian V, Lowell B, Minko IG, Wood TG, Ceci JD, George S, Ballinger SW, Corless CL, McCullough AK, Lloyd RS (2006) The metabolic syndrome resulting from a knockout of the NEIL1 DNA glycosylase. Proc Natl Acad Sci U S A 103:1864–1869
6. Mason C, Risques RA, Xiao L, Duggan CR, Imayama I, Campbell KL, Kong A, Foster-Schubert KE, Wang CY, Alfano CM, Blackburn GL, Rabinovitch PS, McTiernan A (2013) Independent and combined effects of dietary weight loss and exercise on leukocyte telomere length in postmenopausal women. Obesity (Silver Spring) 21:E549-554
7. Blackburn EH (1991) Structure and function of telomeres. Nature 350:569–573
8. López-Otín C, Blasco MA, Partridge L, Serrano M, Kroemer G (2013) The hallmarks of aging. Cell 153:1194–1217
9. Maeda T, Oyama J, Higuchi Y, Nishiyama Y, Kudo Y, Yamori T, Nakazono T, Arima T, Mimori K, Makino N (2011) The physical ability of Japanese female elderly with cerebrovascular disease correlates with the telomere length and subtelomeric methylation status in their peripheral blood leukocytes. Gerontology. PMID: 20453489
10. Makino N, Maeda T, Abe N (2019) Short telomere subtelomeric hypomethylation is associated with telomere attrition in elderly diabetic patients. Can J Physiol Pharmacol 97:335–339
11. Kawano K, Hirashima T, Mori S, Saitoh Y, Kurosumi M, Natori T (1992) Spontaneous long-term hyperglycemic rat with diabetic complications: Otsuka Long-Evans Tokushima Fatty (OLETF) strain. Diabetes 41:1422–1428
12. Mizushige K [1], Yao L, Noma T, Kiyomoto H, Yu Y, Hosomi Y, Ohmori K, Matsuo K (2000) Alteration in left ventricular diastolic filling and accumulation of myocardial collagen at insulin-resistant prediabetic stage of a type II diabetic rat model. Circulation 101:899–907
13. Makino N, Maeda T, Oyama J, Higuchi Y, Mimori K (2009) Improving insulin sensitivity via activation of PPAR-γ increases telomerase activity in the heart of OLETF rats. Am J Physiol (Heart Circ Physiol) 297:H2188–H2196
14. Picca A, Pesce V, Serena Lezza AM (2017) Does eating less make you live longer and better? An update on calorie restriction. Clin Interv Aging 12:1887–1902
15. Most J, Tosti V, Redman LM, Fontana L (2017) Calorie restriction in humans: an update. Ageing Res Rev 39:36–45

16. Madamanchi NR, Hakim ZS, Runge MS (2005) Oxidative stress in atherogenesis and arterial thrombosis: the disconnect between cellular studies and clinical outcomes. J Thromb Haemost 3:254–267
17. Poch E, Carbonell P, Franco S, Díez-Juan A, Blasco MA, Andrés V (2004) Short telomeres protect from diet-induced atherosclerosis in apolipoprotein E-null mice. FASEB J 18(2):418–20. https://doi.org/10.1096/fj.03-0710fje
18. Kawanishi S, Oikawa S (2004) Mechanism of telomere shortening by oxidative stress. Ann N Y Acad Sci 2004(1019):278–284. https://doi.org/10.1196/annals.1297.047
19. Lastra G, Sowers JR (2013) Obesity and cardiovascular disease: role of adipose tissue, inflammation, and the renin-angiotensin-aldosterone system. Horm Mol Biol Clin Investig 15(2):49–57. https://doi.org/10.1515/hmbci-2013-0025. PMID: 25436732
20. Blackburn EH, Greider CW, Szostak JW (2006) Telomeres and telomerase: the path from maize, Tetrahymena and yeast to human cancer and aging. Nat Med 2006(12):1133–1138. https://doi.org/10.1038/nm1006-1133
21. Mundstock E, Sarria EE, Zatti H, Mattos Louzada F, Kich Grun L, Herbert Jones M, Guma FT, Mazzola In Memoriam J, Epifanio M, Stein RT, Barbé-Tuana FM, Mattiello R (2015) Effect of obesity on telomere length: Systematic review and meta-analysis. Obesity (Silver Spring) 23(11):2165–2174. https://doi.org/10.1002/oby.21183. Epub 2015 Sep 26. PMID: 26407932
22. Rani V, Deep G, Singh RK, Palle K, Yadav UC (2016) Oxidative stress and metabolic disorders: pathogenesis and therapeutic strategies. Life Sci 148:183–93. https://doi.org/10.1016/j.lfs.2016.02.002. Epub 2016 Feb 3. PMID: 26851532
23. Zhao J, Miao K, Wang H, Ding H, Wang DW (2013) Association between telomere length and type 2 diabetes mellitus: a meta-analysis. PLoS ONE 8(e79993):2013
24. Rizvi S, Raza ST, Mahdi F (2014) Telomere length variations in aging and age-related diseases. Curr Aging Sci 7(3):161–167. https://doi.org/10.2174/1874609808666150122153151.=
25. Chen YF, Zhou KW, Yang GZ, Chen C (2019) Association between lipoproteins and telomere length in US adults: data from the NHANES 1999–2002. Lipids Health Dis18(1):80, 1999–2002. https://doi.org/10.1186/s12944-019-1030-7.PMID: 30935416
26. Tellechea ML, Pirola CJ (2017) The impact of hypertension on leukocyte telomere length: a systematic review and meta-analysis of human studies. J Hum Hypertens 31(2):99–105. https://doi.org/10.1038/jhh.2016.45. Epub 2016 Jun 30. PMID: 27357526
27. Cawthon RM, Smith KR, O'Brien E, Sivatchenko A (2003) Kerber RA (2003) Association between telomere length in blood and mortality in people aged 60 years or older. Lancet 361:393–395. https://doi.org/10.1016/S0140-6736(03)12384-7
28. Choudhary B et al (2012) Telomere and telomerase in stem cells: relevance in ageing and disease. Front Biosci (Schol Ed). PMID: 22202040
29. Cai Z, Yan LJ, Ratka A (2013) Telomere shortening and Alzheimer's disease. Neuromolecular Med 15(1):25–48. https://doi.org/10.1007/s12017-012-8207-9
30. Forero DA et al (2016) Telomere length in Parkinson's disease: a meta-analysis. Exp Gerontol. PMID: 26772888
31. Guan JZ et al (2008) A percentage analysis of the telomere length in Parkinson's disease patients. J Gerontol A Biol Sci Med Sci. PMID: 18511749
32. Maeda T, Oyama JI, Higuchi Y, Koyanagi M, Sasaki M, Arima T, Mimori K, Makino N (2011) The correlation between clinical laboratory data and telomeric status of male patients with metabolic disorders and no clinical history of vascular events. Aging Male PMID: 20670100
33. MacEneaney OJ, Kushner EJ, Westby CM et al (2010) Endothelial progenitor cell function, apoptosis, and telomere length in overweight/obese humans. Obesity (Silver Spring) 2010(18):1677–1682
34. Zannolli R, Mohn A, Buoni S et al (2008) Telomere length and obesity. Acta Paediatr 2008(97):952–954
35. Das B, Pawar N, Saini D, Seshadri M (2009) Genetic association study of selected candidate genes (ApoB, LPL, Leptin) and telomere length in obese and hypertensive individuals. BMC Med Genet 10:99

36. Hancock CR, Han D-H, Chen M, Terada S, Yasuda T, Wright DC, Holloszy JO (2008) High-fat diets cause insulin resistance despite an increase in muscle mitochondria. Proc Natl Acad Sci 105:7815–7820
37. Makino N, Maeda T (2020) Calorie restriction delays cardiac senescence and improves cardiac function in obese diabetic rats. Mol Cell Biochem. https://doi.org/10.1007/s11010-020-03899-0. Online ahead of print.PMID: 32918706

Chapter 15
Sympathetic Nervous System and Cardiovascular Alterations Due to Food Restrictions

Anureet K. Shah and Naranjan S. Dhalla

Abstract Excessive food intake over a prolonged period is known to result in obesity, a risk factor for heart disease, associated with derangements of myocardial metabolism. Furthermore, food restriction has been shown to reduce obesity, attenuate metabolic changes in the myocardium and improve hemodynamic performance. This article is therefore intended to discuss the impact of food restriction on cardiovascular performance and sympathetic activity to understand the mechanisms for its beneficial effects. Feeding rats 25% of ad libitum diet (severe food restriction) increased blood pressure and left ventricular function during 2 to 7 days but decreased these parameters as well as heart rate during 10 to 14 days. Both plasma and cardiac norepinephrine levels were increased during 2 to 14 days whereas plasma epinephrine levels were elevated during 2 to 7 days and cardiac epinephrine content was increased during 10 to 14 days of feeding 25% of ad libitum diet. Restriction of food (25% ad libitum) for 14 days also attenuated the epinephrine—induced inotropic effect as well as different types of arrhythmias; this was associated with reduction of β—adrenoceptor density in the myocardium. An increase in sympathetic activity was also seen due to starvation of animals for 4 days. Feeding (50% ad libitum) diet (moderate food restriction) for 14 and 28 days showed depression in different parameters of cardiovascular function as well as increased cardiac norepinephrine and epinephrine content whereas plasma norepinephrine levels were increased at 14 days and decreased at 28 days without any changes in plasma epinephrine levels. These studies indicate that augmentation of sympathetic activity is associated with enhanced cardiovascular performance at initial stages followed by depression in

A. K. Shah (✉)
School of Kinesiology, Nutrition and Food Science, California State University, Los Angeles, CA 90032, USA
e-mail: akaur23@calstatela.edu

N. S. Dhalla
Institute of Cardiovascular Sciences, St. Boniface Hospital Albrechtsen Research Centre, Department of Physiology and Pathophysiology, Max Rady College of Medicine, University of Manitoba, Winnipeg R2H 2A6, Canada

© Springer Nature Switzerland AG 2021

P. S. Tappia et al. (eds.), *Cellular and Biochemical Mechanisms of Obesity*,
Advances in Biochemistry in Health and Disease 23,
https://doi.org/10.1007/978-3-030-84763-0_15

heart function due to severe food restriction. Furthermore, depression in cardiovascular function due to food restriction, in spite of increased cardiac catecholamine content, may be related to reduced effect of sympathetic activity.

Keywords Food restriction · Obesity · Sympathetic activity · Cardiovascular function · Cardiac catecholamines · Plasma catecholamines · Starvation

Introduction

It is now well accepted that obesity is associated with accumulation of fat in the body and is considered to be a major risk factor for the development of heart disease [1–5]. In fact, excessive intake of food is known to be the main cause of obesity and a wide variety of metabolic and functional changes have been shown to occur in the heart during its development. Since food restriction has been demonstrated to suppress obesity and attenuate different metabolic derangements in the heart. [6–10], this article is intended to describe the role of metabolic modifications due to the action of food restriction on cardiovascular function. Furthermore, it is pointed out that the sympathetic nervous system is known to regulate cardiac function as well as several metabolic processes and changes in its activity have been reported to play a significant role in obesity and food restriction [11–15]. Accordingly, this review is focused to highlight the evidence regarding the relationship between changes in the status of sympathetic activity and cardiovascular performance due to food restriction. However, in order to fully appreciate the mechanisms of the beneficial effects of food restriction, it is important first to briefly outline the pathophysiology of cardiac dysfunction in obesity. Although both exercise and fasting are commonly employed for weight reduction and management of obesity [16–20], discussion of these events in this article is centered around the influence of food restriction on some parameters of heart function and sympathetic activity.

Obesity-Induced Cardiovascular Complications

Several cardiovascular abnormalities including hypertension, atherosclerosis, myocardial infarction, heart failure, cardiomyopathy and arrhythmias are seen in obese subjects as a consequence of excessive food consumption for a prolonged period [4, 10, 18, 21–25]. Although obesity is considered as a risk factor for all these cardiovascular complications (Fig. 15.1), the pathophysiologic mechanisms associated with these different diseases are poorly understood. It is noteworthy that all these cardiovascular diseases in obesity occur over a prolonged period and it is not clear whether these abnormalities appear at the same or different times. Nonetheless, it is generally held that elevated levels of circulating lipids in obesity [26–29] may induce defects in endothelium for the occurrence of hypertension and atherosclerosis,

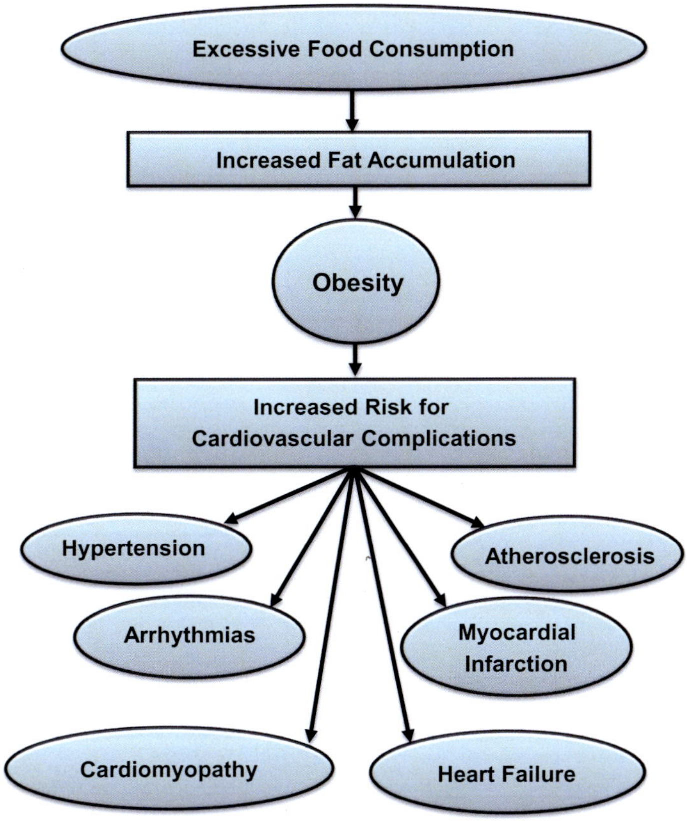

Fig. 15.1 Schematic representation of some cardiovascular complications associated with obesity due to excessive food consumption

elevated levels of free fatty acids may account for arrhythmias [30], as well as deposition of lipids in the heart may explain the development of lipotoxic cardiomyopathy. Furthermore, myocardial infarction and heart failure associated with obesity may be due to the development of coronary disease and cardiac remodeling [21], respectively. It is interesting to point out that such cardiovascular abnormalities in obesity may be a consequence of a generalized defect "metabolic syndrome" because diabetes has also been considered to be a risk factor for the development of similar cardiovascular complications. [31–37]. It should also be noted that obesity associated cardiovascular complications are age-dependent [38] and there are some reports which show paradoxical effects of obesity, as this clinical syndrome has been indicated to delay or attenuate cardiovascular abnormalities [39].

Hyperlipidemia and Hyperglycemia due to excessive food intake for a prolonged period are known to increase body weight, deposition of triglycerides in adipocytes and storage of glycogen in the liver [4, 5, 23, 26, 34, 40]. The elevated levels of circulating free fatty acids and triglycerides in obese subjects are considered to cause a wide variety of metabolic derangements in the heart [41–48]. These alterations include increased uptake and oxidation of free fatty acids, deposition of triglycerides as well as depressed utilization of glucose as a substrate in the myocardium. Excessive β-oxidation of free fatty acids over a time period has been shown to depress mitochondrial oxidative phosphorylation, impair electron transport and promote the generation of oxidative stress in cardiomyocytes. On the other hand, deposition of triglycerides in the myocardium will induce lipotoxicity, development of cardiomyopathy and occurrence of cardiac dysfunction in obesity [4, 23, 49–51]. Expanded adipocytes in obese subjects not only release lipids in the circulation but are also known to secrete different hormones such as leptin, adiponectin and resistin to maintain metabolic homeostasis in the body [40, 44, 52]. In addition, these adipocytes secrete interleukin (1L–6) and tumor necrosis factor (TNF–α) for the development of low grade inflammation [52]. Although decreased insulin sensitivity and glucose utilization by myocardium have been shown to occur in obesity, it is not clear whether these changes are directly due to increased level of free fatty acids or the consequence of increased development of oxidative stress. Both pro-inflammatory cytokines and oxidative stress are known to induce myocardial fibrosis, apoptosis, necrosis and cardiac dysfunction [36, 53, 54]. Some of these mechanisms which may be intimately associated with metabolic derangements in the myocardium due to obesity are shown in Fig. 15.2. There is a good possibility that these mechanisms are associated with metabolic derangements and may predispose the cardiovascular system to different pathological factors for the development of different diseases in obesity.

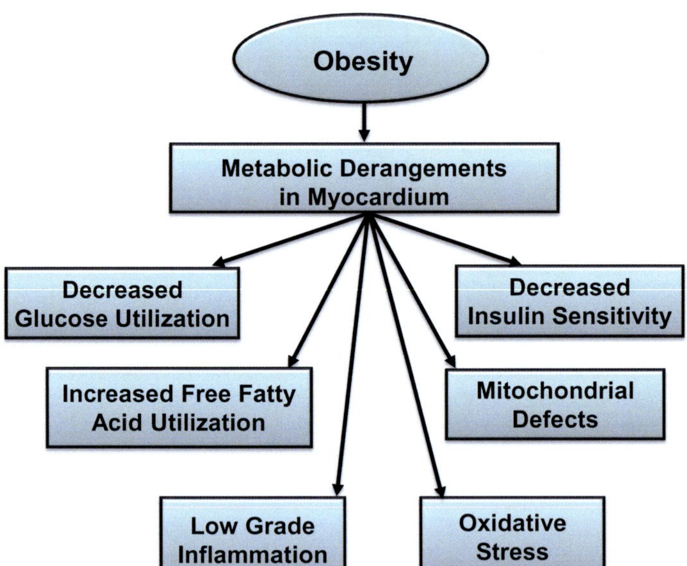

Fig. 15.2 Schematic representation of some metabolic derangements in the heart as a consequence of obesity

Beneficial Effects of Food Restriction in Obesity

Extensive research has been carried out to show that food restriction and caloric reduction exert several positive actions in reducing the cardiovascular complications in obesity [6–11, 55]. Food restriction for a prolonged period not only improves the overall health and metabolic status of obese subjects but also reduces the incidence of cardiovascular abnormalities [8, 14, 17, 56]. Some of the events associated with the loss of body weight and improvement of metabolic status due to prolonged food restriction are represented in Fig. 15.3. It is pointed out that there occurs mobilization and utilization of endogenously stored fuel for the maintenance of metabolic homeostasis upon food restriction. Particularly, glycogen stores and lipid deposits are mobilized first to increase the circulating levels of glucose, free fatty acids and triglycerides for use in the myocardium [6–9, 57]. Thereafter, skeletal muscle proteins are catabolized to amino acids and converted to appropriate fuel for utilization in the heart. Food restriction has also been reported to improve insulin—sensitivity [19,

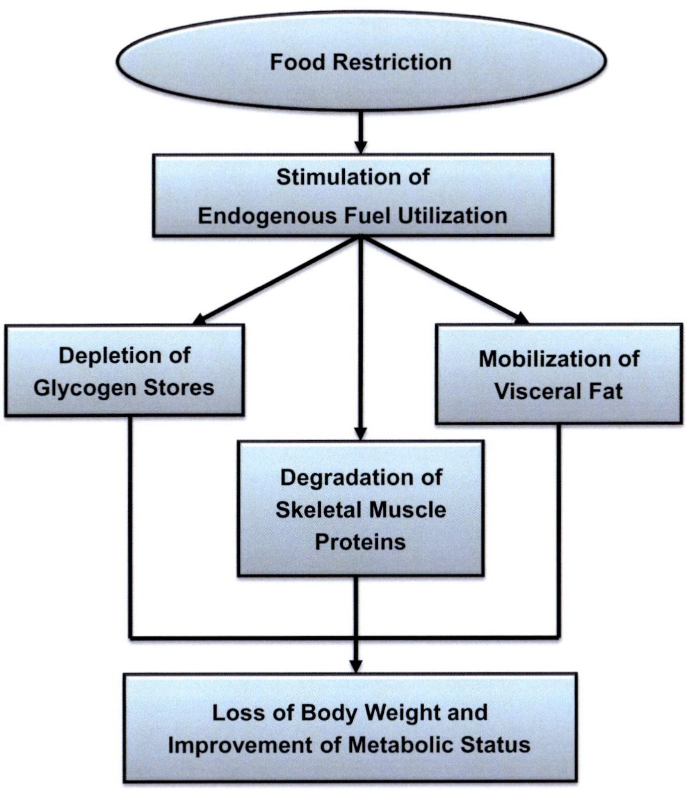

Fig. 15.3 Schematic representation of events depicting the impact of food restriction for the loss of body weight and improvement of metabolic status

58] and this can be seen to promote glycogenolysis and lipolysis as well as the transport of glucose and free fatty acids. Furthermore, food restriction has been shown to attenuate both oxidative stress and inflammation [53, 54, 59, 60] and these effects may also be involved in the improvement of metabolic status and cardiovascular function.

Although heart dysfunction has been demonstrated to be attenuated upon food restriction [6, 7, 61, 62], the status of Ca^{2+}-handling by subcellular organelles remains to be determined in the myocardium during the development of obesity as well as upon the institution of food restriction. In view of the observations that food restriction improves glucose transport and establishes a balance between the utilization of glucose and free fatty acids [8, 9, 44], it is evident that the improvement of cardiac function due to food restriction may be a consequence of attenuation of metabolic defects in the myocardium. Since endothelial function is improved by low-caloric diet [63], such an effect of food restriction can be seen to partly explain for the attenuation of coronary disease [18], atherosclerosis [24] and hypertension [64–66]. Suppression of heart failure in obesity due to food restriction [2, 22, 67] is attributed to the reduction of myocardial fibrosis and apoptosis as a consequence of attenuated inflammations and oxidative stress. Nonetheless, caloric restriction has been shown to impart cardio protection and induce tolerance towards ischemic injury [68–70]. These observations support the view that food restriction reduces incidence for the development of heart disease by improving the metabolic status of myocardium in obesity.

Impact of Severe Food Restriction on Cardiovascular Function

In view of the fact that food restriction has been shown to result in progressive weight loss and reduction in the development of cardiovascular complications, some investigators have attempted to examine its effect on cardiovascular function [25, 71, 72]. When rats were fed 25% ad libitum diet (severe food restriction) for 14 days [71] there was a gradual decrease in body and heart weights during 3 to 14 days whereas the aortic systolic pressure and mean blood pressure were increased during 2 to 7 days but thereafter started decreasing during 10 to 14 days (Table 15.1). The aortic diastolic pressure was increased during 5 to 7 days but was decreased at 14 days of food restriction [71]. Furthermore, the left ventricular systolic and diastolic pressures as well as the left ventricular $\pm$ dP/dt (rates of contraction and relaxation) were increased at 2 and 7 days and decreased at 14 days (Table 15.2). It can also be seen that the heart rate did not change during 2 to 7 days but was decreased during 10 to 14 days of food restriction [71]. These observations suggest that the cardiovascular function is augmented at initial stages (within 2 to 7 days) but is depressed thereafter (within 10 to 14 days) due to severe food restriction in normal animals. However,

Table 15.1 Influence of severe food restriction for different days on growth parameters and blood pressure in normal rats

	Body Wt. (g)	Heart Wt. (g)	Aortic systolic pressure (mm Hg)	Arterial mean pressure (mm Hg)
Control	419 ± 9	1.31 ± 0.04	102 ± 3.6	87 ± 3.0
2 days	378 ± 20	1.18 ± 0.06	123 ± 2.9*	108 ± 5.4
7 days	332 ± 10*	1.00 ± 0.03*	133 ± 6.1*	118 ± 4.8*
10 days	284 ± 10*	0.99 ± 0.03*	107 ± 11.6	92 ± 11.8
14 days	254 ± 11*	0.83 ± 0.04*	70 ± 7.5*	57 ± 7.3*

Control rats were fed ad libitum whereas experimental animals were given 8 g/day regular rat chow (25% of ad libitum food intake) for 14 days. The aortic diastolic pressure was increased at 5 and 7 days and decreased at 14 days. The data are taken from our paper—McKnight et al. 1999 [71]. *—P < 0.05 versus respective control value

Table 15.2 Influence of severe food restriction for different days on heart rate, LV systolic and diastolic pressures, and LV dP/dt in normal rats

	Heart rate (beats/min)	LV systolic pressure (mm Hg)	LV diastolic pressure (mm Hg)	LV + dP/dt (mm Hg/sec)
Control	360 ± 15	135 ± 5	3.1 ± 0.4	2230 ± 80
2 days	374 ± 21	165 ± 4*	6.2 ± 0.7*	2540 ± 95*
7 days	358 ± 17	154 ± 3*	8.4 ± 0.5*	2836 ± 100*
10 days	290 ± 11*	120 ± 6	4.2 ± 0.6	2428 ± 102
14 days	230 ± 16*	96 ± 3*	2.6 ± 0.3	1620 ± 75*

Control rats were fed ad libitum whereas experimental animals were given 8 g/day (25% of ad libitum food intake) for 14 days. LV—dP/dt was increased at 2 and 7 days and decreased at 14 days. The data are taken from our paper—McKnight et al. 1999 [71]. LV—left ventricle; dP/dt—rate of pressure change; *—P < 0.05 versus respective control value

no information regarding the time-course changes in cardiovascular function due to food restriction in obese animals is available in the literature.

Impact of Severe Food Restriction on Sympathetic Activity

In order to understand the mechanism of biphasic changes in cardiovascular function due to severe food restriction, the status of sympathetic nervous system was examined by measuring the levels of plasma and cardiac catecholamines (both norepinephrine and epinephrine). Feeding rats 25% of the ad libitum diet for 14 days [71] was found to increase plasma norepinephrine and epinephrine levels at 1 to 7 days; thereafter plasma norepinephrine levels started to decline but were still elevated significantly at 10 and 14 days, unlike plasma epinephrine levels which declined towards normal values (Table 15.3). There was a gradual increase in cardiac norepinephrine content

Table 15.3 Influence of severe food restriction for different days on plasma and cardiac catecholamine levels in normal rats

	Plasma catecholamines (pg/ml)		Cardiac catecholamines (ng/g)	
	Norepinephrine	Epinephrine	Norepinephrine	Epinephrine
Control	526 ± 31	194 ± 22	614 ± 34	36 ± 5
2 days	2000 ± 140*	1080 ± 156*	876 ± 54	46 ± 5
7 days	2650 ± 228*	1240 ± 210*	940 ± 38*	44 ± 4
10 days	1765 ± 124*	164 ± 56	1250 ± 56*	55 ± 5*
14 days	1140 ± 138*	158 ± 48	1494 ± 72*	61 ± 7*

Control rats were fed ad libitum whereas experimental animals were given 8 g/day regular rat chow (25% ad libitum food intake) for 14 days. The data are taken from our paper—McKnight et al. 1999 [71]. *–$P < 0.05$ versus respective control value

between 2 to 14 days whereas cardiac epinephrine content was significantly increased at 10 and 14 days upon food restriction (Table 15.3). Although the pattern of changes in the levels of plasma and cardiac content of catecholamines is differential in nature, such a difference may be due to the fact that norepinephrine in synthesized in the sympathetic nerve endings in the heart whereas epinephrine is mainly synthesized in the adrenal medulla and is taken up by the heart from circulation. These observations suggest that the sympathetic nervous system is activated [71] and support the view that the increased sympathetic activity may be intimately involved in stimulating the cardiovascular system upon starting at initial stages of severe food restriction. It should be noted that the sympathetic nervous system has been reported to be controlled by nutritional status and is markedly affected by food restriction [12–14, 74].

Influence of Fasting and Starvation on Sympathetic Activity

Fasting is commonly used for the reduction of obesity as well as induction of weight loss, and has been reported to affect heart function, myocardial metabolism and sympathetic nervous system [75–78]. Furthermore, self-induced starvation in anorexia has been shown to affect the heart function and induce marked changes in the sympathetic activity [79–82]. Starvation has also been demonstrated to alter cardiac function, oxidative metabolism and sympathetic activity as well as carbohydrate and lipid metabolism [15, 83–85]. Starvation of rats for 5 days was found to increase norepinephrine content in the heart during 3 to 5 days [86]. The data in Table 15.4 for 4 days of starvation show that increased cardiac content of norepinephrine as accompanied by an increase in the plasma level of norepinephrine. In addition, starvation was observed to decrease the presence of norepinephrine in the granular fraction and increase in the soluble fraction of the heart [Table 15.4]. Such observations provide evidence that increased sympathetic activity due to starvation

Table 15.4 Effect of starvation on cardiac and plasma norepinephrine levels as well as subcellular distribution in normal rats

	Control rats	4 Days starved rats
Cardiac norepinephrine (μg/g heart)	0.702 ± 0.044	1.181 ± 0.167*
Plasma norepinephrine (ng/100 ml)	438 ± 22	586 ± 23*
Subcellular distribution (% Norepinephrine)		
Granular fraction	70.87 ± 0.50	46.73 ± 1.35*
Soluble fraction	29.13 ± 0.50	53. 27 ± 1.35*

Control rats were fed ad libitum whereas experimental animals were starved for 4 days. The data are taken from our paper—Balasubramanian and Dhalla, 1972 [86]. *–$P < 0.05$ versus respective control value

is associated with increased synthesis of norepinephrine [86]. Thus, it is evident that increased levels of plasma and cardiac catecholamines at early stages of severe food restriction are the consequence of sympathetic nervous system activation.

Alterations in Cardiovascular and Sympathetic Activities Due to Prolonged Food Restriction

Since feeding animals 25% of ad libitum diet (severe food restriction) was found to depress left ventricular LV systolic pressure and LV rate of contraction and relaxation at 14 days [72], it was considered of critical importance to investigate if alterations in plasma and cardiac catecholamines at this time were associated with electrocardiographic (EKG) changes [25]. The data in the Table 15.5 indicate that increased plasma norepinephrine, unlike plasma epinephrine, as well as increased cardiac norepinephrine and epinephrine at 14 days of food restriction were associated by bradycardia (depressed heart rate), decreased β–adrenoceptor density, increased QRS duration and increased QT interval without any changes in PR interval [25]. Because increased QT interval is a well known risk marker for the development arrhythmias (which may be attributed to the increased sympathetic activity), some caution should be exercised while using severe food restriction for the management of obesity. Nonetheless, it is pointed out that the incidence of various types of arrhythmias due to epinephrine in animals for 14 days of food restriction was attenuated (Table 15.5). Furthermore, no fibrillation or mortality occurred in animals on 14 days of food restriction (Table 15.5) [25]. Thus, in spite of some adverse cardiac effects of severe food restriction, this intervention can be seen to exert several beneficial actions for the management of obesity.

In another series of experiment, the effects of a moderate degree of food restriction (50% of ad libitum diet) were examined for a longer period (at 14 days and 28 days of food restriction) on cardiovascular and sympathetic activities [72]. The data in Table 15.6 indicate that the loss of body weight and heart weight due to moderate food

Table 15.5 Influence of severe food restriction for 14 days on epinephrine—induced changes in EKG and sympathetic activities in normal rats

	Control	Food Restriction
A. EKG and sympathetic parameters:		
Heart rate (beats/min)	350 ± 19	220 ± 34*
PR interval (msec)	4.46 ± 0.15	5.55 ± 0.76
QRS duration (msec)	1.46 ± 0.08	1.80 ± 0.12*
QT interval (msec)	4.70 ± 0.15	7.50 ± 0.98*
Plasma norepinephrine (pg/ml)	548 ± 69	1180 ± 31*
Plasma epinephrine (pg/ml)	154 ± 77	140 ± 11
Cardiac norepinephrine (ng/g)	753 ± 20	1450 ± 136*
Cardiac epinephrine (ng/g)	30 ± 6	60 ± 8*
β-adrenergic receptor Density (fmol / mg protein)	69 ± 3.3	50 ± 4.4*
B. **Epinephrine—induced Arrhythmias:**		
Incidence (%)	100	66
Fibrillation (%)	33	None

Control rats were fed ad libitum whereas experimental animals were given 25% of food for 14 days. Epinephrine was administered as cumulative intravenous bolus injections (6 μg/kg) over a 30 min period. The data are taken from our paper—McKnight et al. 1996 [25] *–P < 0.05 versus respective control value

restriction for 14 and 28 days was associated with reduced heart rate. Furthermore, mean arterial blood pressure and left ventricular systolic pressure were depressed without any significant changes in left ventricular diastolic pressure at 14 days and 28 days of moderate food restriction [72]. It should also be noted from data in Table 15.6 that cardiac norepinephrine and epinephrine content were increased at both 14 and 28 days of food restriction. On the other hand, plasma norepinephrine levels were increased at 14 days and decreased at 28 days without any changes in plasma epinephrine levels due to food restriction (Table 15.6). These observations indicate that although sympathetic system was activated by moderate food restriction for a prolonged period, but the cardiovascular function was depressed. Such results can be explained on the basis that cardiovascular effects of the increased sympathetic activity due to prolonged food restriction may be attenuated as a consequence of reduced β–adrenoceptor density as indicated in Table 15.5. However, further work regarding the effects of moderate food restriction in obese animals is needed for marking a meaningful conclusion.

Table 15.6 Influence of moderate food restriction (50% ad libitum) for 14 days and 28 days on cardiovascular and sympathetic activity parameters in normal rats

	Control	Food restriction	
		14 days	28 days
Body Wt. (g)	490 ± 12	332 ± 10*	315 ± 17*
Heart Wt. (g)	1.31 ± 0.6	1.13 ± 0.05	0.92 ± 04*
Heart rate (beats/min)	370 ± 16	305 ± 12*	250 ± 10*
Mean arterial pressure (mm Hg)	126 ± 6	83 ± 10*	90 ± 4*
LV systolic pressure (mm Hg)	155 ± 10	116 ± 8*	100 ± 15*
Plasma norepinephrine (pg/ml)	230 ± 17	458 ± 21*	165 ± 9*
Plasma epinephrine (pg/ml)	180 ± 15	194 ± 13	204 ± 16
Cardiac norepinephrine (ng/g heart)	210 ± 18	510 ± 45*	630 ± 54*
Cardiac epinephrine (ng/g heart)	26 ± 4	45 ± 6*	65 ± 8*

Control rats were fed ad libitum whereas experimental animals were given 50% ad libitum rat chow for 14 days and 28 days. The LV diastolic pressure values (mm Hg) for control as well as 14 days and 28 days food restriction were 120 ± 15, 112 ± 10 and 150 ± 21, respectively. The data were taken from our paper—Hilderman et al. 1996 [72]. LV—left ventricle. *–$P < 0.05$ versus respective control value

Conclusions

From the foregoing discussion, it is evident that excessive food intake for a prolonged period results in obesity associated with metabolic derangements and several cardiovascular complications. Although food restriction has been demonstrated to exert beneficial effects by promoting weight loss and reducing metabolic abnormalities, the exact mechanism for improvements in cardiovascular function are not fully understood. In view of the marked actions of the sympathetic nervous system on myocardial metabolism and cardiac function, this article has focused on examining the effects of food restriction on the status of sympathetic activity and cardiovascular function. Severe food restriction in normal rats was associated with increase in the levels of plasma and cardiac catecholamines as well as augmentation of cardiovascular function during 2 to 7 days. However, the intensity of sympathetic activation due to severe food restriction started to decline and in fact was associated with reduced heart rate and cardiac function during 10 to 14 days. It was noteworthy that the increase sympathetic activity at 14 days of severe food restriction was associated with depressed cardiovascular function as well as reduced β–adrenoceptor density and attenuation of catecholamine–induced arrhythmias. Furthermore, moderate degree of food restriction for 14 days and 28 days suppressed cardiovascular function, increased cardiac catecholamines whereas plasma norepinephrine was increased at 14 days and decreased at 28 days without any changes in plasma epinephrine. Thus, it appears that food restriction may activate the sympathetic nervous system as a consequence of hunger stress; however, over a prolonged period the impact of sympathetic activity is attenuated possibly due to reduction in β–adrenoceptor density as well as some other mechanisms and results in improved cardiovascular function. A schematic representation of some of the events associated with improvement of cardiovascular function due to food restriction is shown in Fig. 15.4. Nonetheless, it should be emphasized that an extensive amount of research work needs to be carried out for investigating the role of sympathetic nervous system in affecting the cardiovascular function and metabolism due to food restriction by employing obese subjects.

Acknowledgements The infrastructural support for this work was provided by the St. Boniface Hospital Albrechtsen Research Centre. Thanks, are also due to Ms. Andrea Opsima for typing this manuscript.

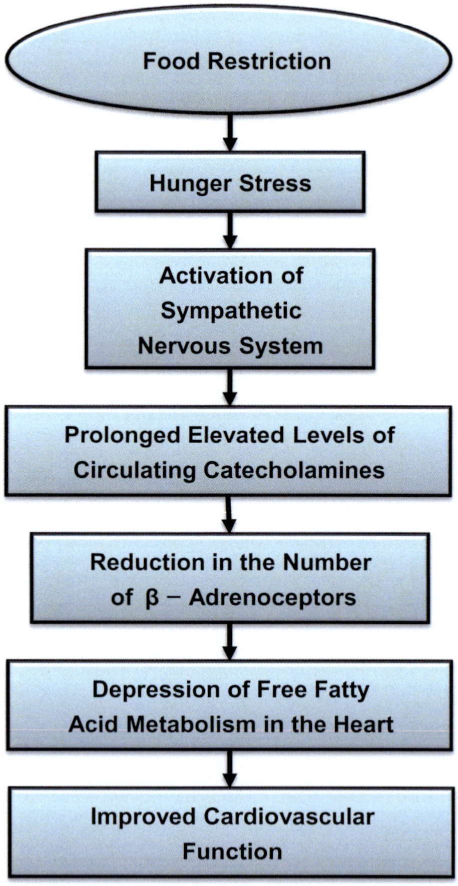

Fig. 15.4 Proposed mechanism of improved cardiovascular function involving reduction in the number of adrenoceptors due to prolonged food restriction

Conflict of Interest The authors do not have any conflict of interest.

References

1. Alpert MA (2001) Obesity cardiomyopathy: pathophysiology and evolution of the clinical syndrome. Am J Med Sci 321:225–236
2. Chess DJ, Stanley WC (2008) Role of diet and fuel overabundance in the development and progression of heart failure. Cardiovasc Res 79:269–278
3. Hubert HB, Feinleib M, McNammara PM, Castelli WP (1983) Obesity as an independent risk factor for cardiovascular disease: a 26-year-follow-up of participants in the Framingham heart study. Circulation 67:968–977
4. Zhou YT, Grayburn P, Karim A et al (2000) Lipotoxic heart disease in obese rats: implications for human obesity. Proc Natl Acad Sci USA 97:1784–1789
5. Wormser D, Kaptoge S, Di Angelantonio E et al (2011) Separate and combined associations of body-mass index and abdominal adiposity with cardiovascular disease: collaborative analysis of 58 prospective studies. Lancet 377:1085–1095
6. Hammer S, Snel M, Lamb HJ et al (2008) Prolonged caloric restriction in obese patients with type 2 diabetes mellitus decreases myocardial triglyceride content and improves myocardial function. J Am Coll Cardiol 52:1006–1012
7. Hammer S, van der Meer RW, Lamb HJ et al (2008) Progressive caloric restriction induces dose dependent changes in myocardial triglyceride content and diastolic function healthy men. J Clin Endocrinol Metab 295:E714–E718
8. Fontana L (2008) Calorie restriction and cardiometabolic health. Eur J Cardiovascu Preven Rehabil 15:3–9
9. Speakman JR, Mitchell SE (2011) Caloric restriction. Mol Aspects Med 32:159–221
10. Masoro EJ (2005) Overview of caloric restriction and ageing. Mech Ageing Dev 126:913–922
11. De Jonge L, AM Moreira E, Martin CK et al. (2010) Impact of six-month caloric restriction on autonomic nervous system activity in healthy, overweight, individuals. Obesity (Silver Spring) 18:414–416
12. Landsberg L, Young JB (1982) Effects of nutritional status on autonomic nervous system function. J Clin Nut 35:1234–1240
13. Davis-Street J, Johnston J (1989) Nutritional effects on cardiac norepinephrine turnover: a critical review. J Appl Cardiol 4:11–17
14. Andersson BM, Wallin BG, Bjorntorp P, Andersson OK (1991) Effect of energy-restricted diet on sympathetic muscle nerve-activity in obese women. Hypertension 18:783–789
15. Sakaguchi TK, Arase JS, Fisler JS, Bray GA (1988) Effect of starvation and food intake on sympathetic activity. Am J Physiol 255 (Regulatory Integrative Comp Physiol 24) R284–R288
16. Weiss EP, Albert SG, Reeds DN (2016) Effects of matched weight loss from calorie restriction, exercise, or both on cardiovascular disease risk factors: a randomized intervention trial. Am J Clin Nutr 104:576–586
17. Goodpaster BH, Delany JP, Otto AD et al (2010) Effects of diet and physical activity interventions on weight loss and cardiometabolic risk factors in severely obese adults: a randomized trial. JAMA 304:1795–1802
18. Fontana L, Villareal DT, Weiss EP et al (2007) Calorie restriction or exercise: effects on coronary heart disease risk factors. A Randomized, controlled trial. Am J Physiol Endocrinol Metab 293:E197–E202
19. Yassine HN, Marchetti CM, Krishnan RK et al (2009) Effects of exercise and caloric restriction on insulin resistance and cardiometabolic risk factors in older obese adults—a randomized clinical trial. J Gerontol A Biol Sci Med Sci 64:90–95

20. Blumenthal JA, Babyak MA, Sherwood A et al (2010) The effects of the dash diet alone and in combination with exercise and caloric restriction on insulin sensitivity and lipids. Hypertension 55:1199–1205
21. Abel ED, Liwin SE, Sweeny G (2008) Cardiac remodeling in obesity. Physiol Rev 88:389–419
22. Kenchaiah S, Evans JC, Levy D, Wilson PWF et al (2002) Obesity and the risk of heart failure. N Engl J Med 347:305–313
23. Wong C, Marwick TH (2007) Obesity cardiomyopathy: pathogenesis and pathophysiology. Nat Clin Pract Cardiovasc Med 4:436–443
24. Fontana L, Meyer TE, Klein S, Holloszy JO (2004) Long-term calorie restriction is highly effective in reducing the risk for atherosclerosis in humans. Proc Natl Acad Sci USA 101:6659–6663
25. McKnight KA, Rupp H, Beamish RE, Dhalla NS (1996) Modification of catecholamine-induced changes in heart function by food restriction in rats. Cardiovasc Drugs Ther 10:239–246
26. Boden G (2008) Obesity and free fatty acids. Endocrinol Metab Clin North An 37:635–646
27. Luiken JJ, Arumugam Y, Dyck DJ et al (2001) Increased rates of fatty acid uptake and plasmalemmal fatty acid transporters in obese Zucker rats. J Biol Chem 276:40567–40573
28. Van der Meer RW, Hammer S, Smit JW et al (2007) Short-term caloric restriction induces accumulation of myocardial triglycerides and decreases left ventricular diastolic function in healthy subjects. Diabetes 56:2849–2853
29. Me Y, Guthrie PH, Razeghi P et al (2002) Impaired long-chain fatty acid oxidation and contractile dysfunction in the obese Zucker rat heart. Diabetes 51:2587–2595
30. Kurien VA, Oliver MF (1971) Free fatty acids during acute myocardial infarction. Prog Cardivasc Dis 13:361–373
31. Aasum E, Belke DD, Severson DL et al (2002) Cardiac function and metabolism in Type 2 diabetic mice after treatment with BM 17.0744, a novel PPAR-alpha activator. Am J Physiol Heart Circ Physiol 283:H949–H957
32. Boudina S, Sena S, Theobald H et al (2007) Mitochondrial energetics in the heart in obesity-related diabetes: direct evidence for increased uncoupled respiration and activation of uncoupling proteins. Diabetes 56:2457–2466
33. Coort SL, Bonen A, van der Vusse GJ et al (2007) Cardiac substrate uptake and metabolism in obesity and type-2 diabetes: role of sarcolemmal substrate transporters. Mol Cell Biochem 299:5–18
34. Maggio CA, Pi-Sunyer FX (2003) Obesity and type 2 diabetes. Endocrinol Metab Clin North Am 32:805–822
35. Harder H, Dinesen B, Astrup A (2004) The effect of a rapid weight loss on lipid profile and glycemic control in obese type 2 diabetic patients. Int J Obes Relat Metab Disord 28:180–182
36. Dhalla NS, Shah AK, Tappia PS (2020) Role of oxidative stress in metabolic and subcellular abnormalities in diabetic cardiomyopathy. Int J Mol Sci 21:2413. Doi: https://doi.org/10.3390/ijms21072413
37. Dhalla NS, Ganguly P, Bhullar SK, Tappia PS (2019) Role of catecholamines in the pathogenesis of diabetic cardiomyopathy. Can J Physiol Pharmacol 97:815–819
38. Sheng Y, Lv S, Huang M et al. (2017) Opposing effects on cardiac function by calorie restriction in different-aged mice. Aging Cell 16:1155–1167
39. Lavie CJ, Milani RV (2003) Obesity and cardiovascular disease: the Hippocrates paradox? J Am Coll Cardiol 42:677–679
40. Gray SL, Vidal-Puig AJ (2007) Adipose tissue expandability in the maintenance of metabolic homeostasis. Nutr Rev 65:S7–S12
41. Lopaschuk GD, Folmes CD, Stanley WC (2007) Cardiac energy metabolism in obesity. Circ Res 101:335–347
42. Peterson LR, Herrero P, Schechtman KB et al (2004) Effect of obesity and insulin resistance on myocardial substrate metabolism and efficiency in young women. Circulation 109:2191–2196
43. Apfelbaum M (1978) Adaptation to changes in caloric intake. Prog Food Nutr Sci 2:543–559

44. Lopaschuk GD, Ussher JR, Folmes DL et al (2010) Myocardial fatty acid metabolism in health and disease. Physiol Rev 90:207–258
45. Boivin A, Deshaies Y (2000) Contribution of hyperinsulinemia to modulation of lipoprotein lipase activity in the obese Zucker rat. Metabolism 49:134–140
46. Boudina S, Sena S, O'Neill BT et al (2005) Reduced mitochondrial oxidative capacity and increased mitochondrial uncoupling impair myocardial energetics in obesity. Circulation 112:2686–2695
47. Bryson JM, Cooney GJ, Wensley VR et al (1996) The effects of the inhibition of fatty acid oxidation on pyruvate dehydrogenase complex activity in tissues of lean and obese mice. Int J Obes Relat Metab Disord 20:738–744
48. Buchanan J, Mazumder PK, Hu P et al (2005) Reduced cardiac efficiency and altered substrate metabolism precedes the onset of hyperglycemia and contractile dysfunction in two mouse models of insulin resistance and obesity. Endocrinology 146:5341–5349
49. Rosen P, Herberg L, Reinauer H (1986) Different types of postinsulin receptor defects contribute to insulin resistance in heart of obese Zucker rats. Endocrinology 119:1285–1291
50. Steppan CM, Lazar MA (2002) Resistin and obesity-associated insulin resistance. Trends Endocrinol Metab 13:18–23
51. Alden PB, Madoff RD, Stahl TJ et al. (1987) Left ventricular function in malnutrition. Am J Physiol (Heart Circ Physiol 22):H380–H387
52. Waki H, Tontonoz P (2007) Endocrine functions of adipose tissue. Annu Rev Pathol 2:31–56
53. Sohal RS, Weindruch R (1996) Oxidative stress, caloric restriction, and aging. Science 273:59–63
54. Csiszar A, Labinskyy N, Jiminez R et al (2009) Anti-oxidative and anti-inflammatory vaso-protective effects of caloric restriction in aging: role of circulating factors and SIRT1. Mech Ageing Dev 130:518–527
55. Lockwood DH, Amatruda JM (1984) Very low calorie diets in the management of obesity. Annu Rev Med 35:373–381
56. Davis-Street JE, Johnston JL (1990) Effects of energy restriction on norepinephrine turnover and serum glucose and fatty acid in lean mice. Pharmacol Behav 35:677–683
57. Jedeiken LA (1966) Cardiac glycogen metabolism *in vivo* in fed and starved rats. Biochem J 99:6-7P
58. Zhu M, Minura J, Lu LX et al (2004) Circulating adiponectin levels increase in rats on caloric restrictions: the potential for insulin sensitization. Exp Gerontol 39:1049–1059
59. Chandrasekar B, Nelson JF, Colston JT et al (2001) Calorie restriction attenuates inflammatory responses to myocardial ischemia reperfusion injury. Am J Physiol Heart Circ Physiol 280:H2094–H2102
60. Shinmura K (2013) Effects of caloric restriction on cardiac oxidative stress and mitochondrial bioenergetics: potential role of cardiac sirtuins. Oxidat Med Cell Long Art ID 528935:1–7. https://doi.org/10.1155/2013/528935
61. Meyer TE, Kovacs SJ, Ehsani AA et al (2006) Long-term caloric restriction ameliorates the decline in diastolic function in humans. J Am Coll Cardiol 47:398–402
62. Han X, Ren J (2010) Caloric restriction and heart function: is there a sensible link? Acta Pharmacol Sin 31:1111–1117
63. Sasaki S, Higashi Y, Nakagawa K et al (2002) A low-calorie diet improves endothelium-dependent vasodilation in obese patients with essential hypertension. Am J Hypertens 15:302–309
64. Kushiro T, Kobayashi F, Osada H et al (1991) Role of sympathetic activity in blood pressure reduction with low calorie regimen. Hypertension 17:965–968
65. VanNess JM, DeMaria JE, Overton JM (1999) Increased NPY activity in the PVN contributes to food-restriction induced reductions in blood pressure in aortic coarctation hypertensive rats. Brain Res 821:263–269
66. Nicoll R, Henerin MY (2018) Caloric restriction and its effect on blood pressure, heart rate variability and arterial stiffness and dilation: a review of the evidence. Int J Mol Sci 19:1–18

67. Haxhe JJ (1967) Experimental undernutrition: Its effects on cardiac output. Metabolism 16:1086–1091
68. Shinmura K, Tamaki K, Saito K et al (2007) Cardioprotective effects of short-term caloric restriction are mediated by adiponectin via activation of AMP-activated protein kinase. Circulation 116:2809–2817
69. Marzetti E, Wohlgemuth SE, Anton SD et al (2009) Cellular mechanisms of cardio protection by calorie restriction: state of the science and future perspectives. Clin Geriatr Med 25:715–732
70. Shinmura K, Tamaki K, Bolli R (2005) Short-term caloric restriction improves ischemic tolerance independent of opening of ATP-sensitive K^+ channels in both young and aged hearts. J Mol Cell Cardiol 39:285–296
71. McKnight KA, Rupp H, Dhalla KS et al (1999) Biphasic changes in heart performance with food restriction in rats. Appl Physiol 87:1909–1913
72. Hilderman T, McKnight K, Dhalla KS et al (1996) Effects long-term dietary restriction on cardiovascular function and plasma catecholamines in the rat. Cardiovsac Drugs Ther 10:247–250
73. Rappaport EB, Young JB, Landsberg L (1982) Initiation, duration and dissipation of diet-induced changes in sympathetic nervous system activity in the rat. Metabolism 31:143–146
74. Overton JM, VanNess JM, Casto RM (1997) Food restriction reduces sympathetic support of blood pressure in spontaneously hypertensive rats. J Nutr 127:655–660
75. Opie LH, Evans JR, Shipp JC (1963) Effect of fasting on glucose and palmitate metabolism of perfused rat heart. Am J Physiol 205:1203–1208
76. Goodale WT, Olson RE, Hackel DB (1959) The effects of fasting and diabetes mellitus on myocardial metabolism in man. Am J Med 27:212–220
77. De Boer SF, Koopmans SJ, Slangen JL, Van der GugTen J (1989) Effects of fasting on plasma catecholamine, corticosterone and glucose concentrations under basal stress conditions in individual rats. Physiol Behav 45:989–994
78. Landsberg L, Young JB (1978) Fasting, feeding and regulation of the sympathetic nervous system. N Engl J Med 298:1295–1301
79. Gotldiener JS, Gross HA, Henry WL et al (1978) Effect of self-induced starvation on cardiac size and function in anorexia nervosa. Circulation 53:425–433
80. Moodie DS (1987) Anorexia and the heart. Results of studies to assess effects. Postgrad Med 81:46–61
81. Halrmi KA, Dekirmengian H, Davis JM et al (1978) Catecholamine metabolism in anorexia nervosa. Arch Gen Psych 35:458–466
82. Gross HA, Lake CR, Ebert MH et al (1979) Catecholamine metabolism in primary anorexia nervosa. J Clin Endocrinol Met 49:805–809
83. Mager M, Iampietro PF (1966) The effect of prolonged cold and starvation and subsequent refeeding on plasma lipids and glucose of normal men. Metabolism 15:9–16
84. Mayes PA, Felts JM (1967) Comparison of oxidative metabolism in starved, fat-fed and carbohydrate-fed rats. Biochem J 103:400–406
85. Fisler JS (1992) Cardiac effects of starvation and semistarvation diests: safety and mechanisms of action. Am J Clin Nutr 56:230S-234S
86. Balasubramanian V, Dhalla NS (1972) Biochemical basis of heart function. V. effect of starvation on storage, transport, and synthesis of cardiac norepinephrine in rats. Can J Physiol Pharmacol 50:238–243

Chapter 16
The Relevance of Metabotrophic Factors in Pathobiology and Therapy of Obesity and Related Diseases

George N. Chaldakov, Luigi Aloe, Gorana Rancic, Rouzha Z. Pancheva, Marcia Hiriart, Marco Fiore, and Stanislav Yanev

Abstract Currently, the most widespread global ailment is not COVID-19 or any other such devastating infectious diseases. In fact, obesity has been recognized as a prime risk in the development of cardiometabolic diseases (CMD), neurodegenerative diseases (NDD) and cancer and their morbidity and mortality signature. The pathobiology and therapy of obesity and related diseases are immensely complex at the cellular and molecular levels. This scenario raises the question of how such a complexity may be grappled in a more tangible manner. Since 2003, we have been thinking "what nobody has yet thought about that which everybody sees", namely, metabotrophic factors (MTF or metabotrophins, metabokines). They include mainly (i) the neurotrophins nerve growth factor (NGF) and brain-derived

---

Author contributions: All the authors contributed to this chapter equally.

G. N. Chaldakov (✉)
Department of Anatomy and Cell Biology, Medical University, 9002 Varna, Bulgaria

Institute for Advanced Study, Varna, Bulgaria

L. Aloe
Fondazione Iret Tecnopolo R. Levi-Montalcini, Rome, Italy

G. Rancic
Department of Histology, Faculty of Medicine, Nis University, Niš, Serbia

R. Z. Pancheva
Department of Hygiene and Epidemiology, Faculty of Public Health, Medical University, Varna, Bulgaria

M. Hiriart
Department of Physiology and Cognitive Neuroscience, Autonomous National University of Mexico (UNAM), Mexico City, Mexico

M. Fiore
Institute of Biochemistry and Cell Biology, Section of Neurobiology, National Research Council (CNR), Rome, Italy

S. Yanev
Department of Drug Toxicology, Institute of Neurobiology, Bulgarian Academy of Sciences, Sofia, Bulgaria

© Springer Nature Switzerland AG 2021
P. S. Tappia et al. (eds.), *Cellular and Biochemical Mechanisms of Obesity*, Advances in Biochemistry in Health and Disease 23,
https://doi.org/10.1007/978-3-030-84763-0_16

neurotrophic factor (BDNF), and (ii) the adipomyokines adiponectin, irisin, BDNF, fibroblast growth factor-21 alike as adipose- and skeletal muscle-derived signaling proteins. Herein, we argue that obesity and related CMD and NDD, particularly Alzheimer's disease, may be viewed as MTF-deficient diseases. Further studies on MTF signatures and ramifications in these diseases are required. This may open up an intriguing line of scientific enquiry that will ally adipobiologists with neurobiologists and myobiologists in the fight against obesity. These would provide greater insights on how we can make MTF work for the improvement of physiological and psychological quality of human life.

Keywords Obesity · Cardiometabolic diseases · Metabotrophic factors · Adipomyokines · NGF · BDNF · Adiponectin · Irisin · Alzheimer's disease

Introduction

> Thus, the task is not so much to see what no one has yet seen, but to think what nobody has yet thought about that which everybody sees.
>
> Arthur Schopenhauer

Life at both the local and systemic levels requires nutritional, immune, neurotrophic and metabotrophic support. Any dysfunction or deficit in this support may lead to illness, such as obesity and related cardiometabolic diseases (CMD) [e.g., atherosclerosis, hypertension, type 2 diabetes mellitus (T2DM) and metabolic syndrome]. At its core, obesity may be classified as dysmetabolic disorder featured by: (i) accumulation, hypoxia and inflammation of white adipose tissue [1–4], and (ii) dysfunction of brown adipose tissue [5–9]. Then, the adipose-derived proinflammatory and dysmetabolic signals are disseminated to many organs of the body. This leads to the development of CMD and neurodegenerative diseases (NDD), particularly Alzheimer's disease (AD), which we shall discuss in the present Chapter. The list of obesity-associated disorders also include: non-alcoholic steatohepatitis, non-alcoholic fatty liver disease, polycystic ovarian syndrome, obstructive sleep apnea, inflammatory bowel disease, thyroid-associated ophthalmopathy, and cancers.

Neurotrophins

In the 1950s at Washington University Medical School, St Louis, MO, Rita Levi-Montalcini and Stanley Cohen discovered a protein with nerve growth-stimulating effect, and they named it nerve growth factor (NGF). This *Eureka* provided a conceptual framework for the formulation of the neurotrophic hypothesis: particular neuronal types require specific trophic factors for their differentiation, function and survival [10–13].

Table 16.1 A selected list of adipose-derived neurotrophic factors (ADNF)[a]

NGF, BDNF, Glial cell line-derived neurotrophic factor
Ciliary neurotrophic factor, Vascular endothelial growth factor
Leptin, Adiponectin, Meteorin-like (Metrnl, also known as Cometin, Subfatin), Neprilysin (β-amyloid peptide-degrading enzyme)
Fibroblast growth factor-21, Metallothionein-I, -II, Angiopoietin-1

[a]References are indicated in the text

Today, NGF and brain-derived neurotrophic factor (BDNF) and their relatives are collectively designated neurotrophins. The latter include: NGF, BDNF, neurotrophin-3 (NT-3), NT-4/5, and NT-6, also pro-NGF and pro-BDNF which are as active as their respective mature forms. Neurotrophins, particularly NGF and BDNF, were recognized as mediators of multiple biological processes, ranging from the neurotrophic [10] through immunotrophic [14] to metabotrophic effects over glucose, lipid, energy and cardiovascular homeostasis [2, 13, 15–20]. Consequently, NGF and BDNF were implicated in the pathobiology of a large spectrum of neuronal and non-neuronal diseases, ranging from AD and other NDD to obesity and CMD.

Adipose cells also secrete various neurotrophic factors (Table 16.1).

Adipokines, Myokines, and Adipomyokines

Clear-cut demarcations do not exist between white adipose tissue (WAT) and brown adipose tissue (BAT). The infiltration of white adipocytes can be found in BAT (whitening of BAT, a pathogenic phenomenon) and of brown adipocytes in WAT (browning of WAT, a sanogenic phenomenon) [5–9, 21].

The first endocrine protein secreted from adipose cells was discovered in 1987 and named adipsin (complement factor D) [22]. It is a serine protease of the alternative pathway of the complement, which along with triggering of the natural defense against infections, is also involved in the pathogenesis of CMD [23].

However, it was since the discovery of leptin in 1994 by Jeffrey Friedman and colleagues [24] and of adiponectin in 1995 by Yuji Matsuzawa et al. [25] that adipose tissue was recognized as a major human endocrine and paracrine organ. In effect, this was conceptualized into a novel research field, adipobiology [9, 26, 27], or adiposcience [28]. Accordingly, adipose-derived signaling proteins were collectively named adipocytokines [25] or adipokines [17, 26, 29], which became more acceptable term. About 30% of genes in WAT cells (adipocytes, stromal vascular cells, and associated immune cells) encode for adipokines. Recent transcriptomic and proteomic analyses revealed that more than 500 adipokines are secreted by adipose cells [30], e.g. leptin, adiponectin, resistin, tumor necrosis factor-α, interleukins, visfatin/SIRT-2, vascular endothelial growth factor, also NGF and BDNF [31; reviewed in 2, 17, 32]. The adipokines provide communication between adipose tissue and the rest of

Table 16.2 A selected list of myokines[a]

Irisin, Brain-derived neurotrophic factor
Adiponectin, Interleukins (IL-6, IL-15)
Angiopoietin-like protein-4
Fibroblast growth factor 21
Monocyte chemoattractant protein-1 (MCP-1/CCL2)
Leukemia inhibitory factor
Myonectin, Myostatin (GDF-8—growth differentiation factor 8)

[a]References are indicated in the text

the body including the brain. Moreover, brain also produces various adipokines with neuroprotective action, such as leptin and adiponectin [33].

Skeletal muscles, accounting for about 30 – 40% of the body, were also reported to be an endocrine and paracrine organ, secreting signaling proteins dubbed myokines [34–36]. Many myokines are induced by exercise (muscle-derived exerkines), most of them circulate throughout the body by means of the extracellular vesicles exosomes [37]. In this regard, Table 16.2 presents a list of myokines.

Adipokines and myokines secreted from both adipose tissue and skeletal muscles were correspondingly termed adipomyokines [34, 35, 38–41]. In mice, 119 myokines, 79 adipokines and 22 adipomyokines were identified [42]. All these are multifunctional proteins involved in the regulation of a wide range of biological processes including lipid, glucose and energy metabolism.

Neurotrophins and Adipomyokines "Became" Metabotrophic Factors and "Make" Obesity a Metabotrophic Factors-Deficient Disease

In 2003 NGF-and-BDNF's physiological profile was enriched with one more extraneuronal activity, namely, the improvement of metabolism of glucose and lipids, also of pancreatic beta cell and cardiovascular homeostasis. These neurotrophins were also named *metabotrophic factors* (MTF) or *metabotrophins* (Greek *metabole*, and *trophe*, nutrition, means "nutritious for metabolism") [2, 15, 17] also *metabokines* [16, 31].

The proof-of-concept was based on results demonstrating that the circulating and/or local NGF and BDNF levels are decreased in (i) human coronary atherosclerosis and in patients with *advanced stage* of metabolic syndrome [43], (ii) T2DM [44, 45], and (iii) AD which is considered recently T3DM [46–48; for adipose AD, see 49]. In contrast, the circulating levels of NGF and BDNF were significantly elevated in patients with *early stage* of metabolic syndrome [50]. It remains to be elucidated whether the metabolically protective reserve of the organism is limited with the progression of (dys)metabolic syndrome.

Table 16.3 Metabotrophic effects of NGF, BDNF, and adiponectin (APN)[a]

NGF shares homology with proinsulin
NGF and BDNF are produced by pancreatic beta cells and exert insulinotropic effect
NGF and BDNF are trophic factors for pancreatic beta cells
APN is anti-obesity, anti-diabetogenic, anti-atherogenic adipokine
BDNF- and APN-deficient mice develop abnormalities similar to metabolic syndrome
BDNF and APN improve cognitive processes
NGF up-regulates expression of LDL receptor-related protein
NGF up-regulates expression of PPAR-gamma
NGF inhibits glucose-induced down-regulation of caveolin-1
NGF improves skin and corneal wound healing
NGF and APN improve vascular (atheroma) wound healing
NGF rescues silent myocardial ischemia in diabetes mellitus
NGF improves diabetic erectile dysfunction
Healthy lifestyle increases brain and/or circulating levels of NGF, BDNF, APN
Atherogenic diet decreases brain BDNF levels
BDNF-deficient mice develop abnormalities similar to the metabolic syndrome
BDNF improves cognitive processes

[a]For references, see [60–63]

Furthermore, circulating levels of NGF and BDNF in patients with acute coronary syndromes were measured, and they were found to be reduced significantly [51, 52]. It was reported that in response to experimental stress or diabetes, NGF and BDNF levels were altered, both in WAT and BAT [53]. Further, it was demonstrated that pancreatic beta cells secrete NGF and express its tyrosine kinase receptor ($TrkA^{NGF}$), findings implicated in the pathogenesis of T2DM and metabolic syndrome [43, 44, 51, 54–57]. Synergistically with leptin, BDNF reduces food intake [58]. Accordingly, mutations of *Bdnf* gene in mice or *Ntr2k2* (encoding $TrkB^{BDNF}$ receptor) in patients are associated with hyperphagia and severe obesity [see 12, 59]. Table 16.3 presents list of metabolically protective effects induced by MTF.

In the same context, some adipomyokines (e.g., irisin, adiponectin, FGF-21) have been implicated in the control of glucose and lipid metabolism, also considered anorexigenic signals in the central control of food intake [17, 45, 64–67]. Thus, several adipomyokines, also other metabolically protective molecules, were also incorporated in the list of MTF (Table 16.4).

Internal (Hidden) Obesity: An Inside *Versus* Outside View

Human adipose tissue is partitioned into two large depots (subcutaneous and visceral), and many small depots associated with internal organs, e.g. heart, blood vessels, major lymph nodes, pancreas, prostate gland and ovaries (Table 16.5).

Table 16.4 A selected list of endogenous metabotrophic factors (metabotrophins)[a]

Nerve growth factor, Brain-derived neurotrophic factor
Ciliary neurotrophic factor, Vascular endothelial growth factor
Leptin, Adiponectin, Irisin, Fibroblast growth factor-21
Sirtuins (Visfatin/SIRT-2, SIRT-1), Klotho, Humanin, Omentin, Chemerin, Apelin, Otopetrin-1, Interleukin-10, Metalothionein-I,-II,
Incretins (Glucagon-like peptide-1, Glucose-dependent insulinotropic polypeptide)
Kisspeptin-1, Progranulin, Kallistatin, Adipsin, Aquaporin-7, Angiopoietin-like protein 4

[a]For references, see [23, 60–63, 68–80]

Table 16.5 Adipotopography: localization of adipose tissue in the human body—variants[a]

TOFI**	Thin Outside, Fat Inside
TOTI*****	Thin Outside, Thin Inside
FOFI*	Fat Outside, Fat Inside
FOTI***	Fat Outside, Thin Inside

[a]The number of asterisks signifies quality of health. Higher the number of asterisks, better the quality

Metabotrophic Factors as Therapeutic Targets in Drug Discovery

NGF and BDNF

As discussed, obesity and related CMD are featured by reduced circulating levels of NGF and BDNF. Most probably, hypometabotrophinemia is metabolically harmful, thus staying in the heart of a complex network of factors orchestrated in the pathobiology of these dysmetabolic diseases. If so, drugs facilitating (boosting) the intracellular secretory pathways [81, 82] of NGF, BDNF, adiponectin, irisin as well as other MTF may represent a novel pharmacotherapeutic approach in these diseases. However, our knowledge of the secretory pathways (synthesis, translocation, folding, targeting, sorting, storage, and exocytosis) of MTF remain limited.

Neurotrophins ligated two different types of receptors on the surface of neurones and other target cells: (i) high-affinity neurotrophin receptors belong to the Trk (pronounced "track") family of tyrosine kinase receptors (TrkA, TrkB and TrkC) which bind to specific neurotrophins (e.g., $TrkA^{NGF}$, $TrkB^{BDNF}$), and (ii) low-affinity panneurotrophin receptor, $p75^{NTR}$, which lacks a tyrosine kinase endodomain. Hence, an other approach for the discovery of novel therapeutics and nutraceuticals for obesity and its diseased relatives may indeed lie in exploring Trk^{NGF} and $TrkB^{BDNF}$ receptor agonists [for trackins, see 83, for olive oil polyphenols, see 84].

As reviewed [13, 60], increasing number of reports demonstrated that damage to some tissues can be cured by the administration of NGF. For instance, (i) wounded diabetic skin, characterized by increased levels of NGF, will benefit by local treatment with NGF, (ii) elevated local NGF levels in experimentally-induced cardiac ischemia is improved by exogenous administration of NGF, and (iii) NGF administration in diabetic rodents promotes repair of pancreatic islets injured.

Last but not least, pro-NGF and pro-BDNF should be studied for their eventual metabotrophic or dysmetabolic potential.

Ciliary Neurotrophic Factor

Recent observations indicated that ciliary neurotrophic factor (CNTF), which is also produced by adipose cells and its receptor expressed on adipocytes, reduce adipose accumulation in the body, even in leptin-resistant obesity [85]. Accordingly, Axokine, a synthetic CNTF analogue, was under development for the potential therapy of obesity and T2DM [86].

Irisin

Irisin (named after the Greek mythology goddess Iris, a messenger of the gods) is a newly identified adipomyokine. It is a cleavage protein of fibronectin type III domain 5 (FNDC5), the latter converted to irisin after exercise. Decreased levels of irisin were found to be independently associated with endothelial dysfunction in nonhypertensive, nondiabetic obese subjects [87]. Today, obesity and related CMD might, at least in part, be viewed as irisin-deficient disorders [87–99].

Adiponectin

Adiponectin is an adipocyte-secreted protein sharing significant similarity with collagens type VIII and type X and complement protein C1q. It is a multipotent adipomyokine that exerts anti-inflammatory, anti-atherogenic, anti-diabetic, anti-obesity, anti-fibrotic, and anti-cancer effects, and therefore, a sanogenic factor indeed [36].

How can adiponectin be targeted therapeutically? Possible answers might be via: (i) boosting intracellular secretory pathways thus increasing its circulating and/or local levels, and (ii) stimulating its receptors. Some new approaches indicated that sustained weight reductions through lifestyle modifications can enhance adiponectin levels [100]. In the same vein, the thiazolidinediones (TZD) pioglitazone and rosiglitazone increase adiponectin expression through the activation of the transcription factor peroxisome proliferator-activated receptor gamma (PPARγ) [101]. A beneficial function of activators of PPARγ, such as rosiglitazone and lycopene, is to stimulate the browning of WAT [102]. To improve the safety usage of these compounds, several approaches were developed to increase their target-directed actions with new drug forms as nanoparticles [103] or liposomes [104]. The adiponectin receptors agonists adiporon and osmotin [105] might also be considered.

Sirtuins

These proteins encoded by the gene *sir* (silent information regulator) belong to sirtuin (SIRT) family of nicotinamide adenine dinucleotide (NAD^+)-dependent deacylases which consists of seven members in mammals (SIRT 1–7) [106, 107]. Mice treated with sirtuin-activating compounds, such as resveratrol (in human clinical trials) or with NAD^+ precursors, improved their cardiometabolic health *via* weight loss and improving insulin sensitivity [108–111]. Further, nicotinamide phosphoribosyl-transferase is NAD biosynthetic enzyme, including in adipose tissue [112].

Klotho

There are alpha-, beta- and gamma-Klotho, the most studied being alha-Klotho. They are transmembrane proteins (the enzyme beta-glucuronidase) and in a secreted, blood circulating form by ectodomain shedding of the membrane-bound form. Soluble Klotho exerts a variety of metabotrophic effects in CMD, also aging [113–116].

Angiopoietin-Like Protein 4

Angiopoietin-like (ANGPTL) proteins are principal regulators of plasma lipid metabolism by functioning as potent inhibitors of lipoprotein lipase [117]. New therapeutic approaches include monoclonal antibodies and antisense oligonucleotides targeted ANGPTL proteins as efficient therapeutic strategy for hypertriglyceridemia and cardiovascular risk reduction [118].

Aquaporin 7

Aquaporin 7 (AQP7) is a water–glycerol transporter that is present in the plasma membrane of adipocytes [74, 119]. It may regulate insulin sensitivity in obese people [120, 121]. Experiments in mice indicate that ginsenoside Rb1, the most active component of *Panax ginseng*, a traditional herbal medicine, can promote lipid transport and ameliorate obesity by upregulating AQP7 through the PPARγ pathway [122].

Incretins

It is known that glucagon-like peptide-1 (GLP-1) is secreted from small intestine enterocytes when food enters the duodenum. It is also synthesized by brain neuronal cells. GLP-1 stimulates insulin secretion, reduces appetite, and protects β cells from apoptosis. Taken together, GLP-1 is featured by an excellent metabotrophic profile. Accordingly, there are already successful clinical trials with GLP-1 receptors agonists (liraglutide, exenatide, etc.) in patients with T2DM and obesity, which also provide "a hope for therapy of AD" [75, 76]. Emerging approaches for optimization of clinical usage of GLP-1 agonists are in the direction of prolongation of action by microspheres or nanocapsules [77]. A promising therapy approach for obesity combines GLP-1 receptor agonists with GIPR (glucose-dependent insulinotropic polypeptide) agonists (e.g. tirzepatide) [78–80].

Neprilysin

Amyloid precursor protein (APP), amyloid-beta (Abeta) peptide and *tau* hyperphosphorylation are the classical molecular signatures of AD. Noteworthy, it was reported that APP expression is increased in adipocytes in obesity [123]. Among several proteases involved in the proteolysis of Abeta peptide, neprilysin (neutral endopeptidase, NEP), a membrane-associated enzyme, appears to be the most important Abeta-degrading enzyme in the brain. It was reported that human adipose tissue-derived stem cells (ADSC) secrete exosomes carrying enzymatically active NEP. When ADSC-derived exosomes were transferred into cultured nerve cells, it resulted in a decrease in both secreted and intracellular Abeta levels [124]. These observations indicate the therapeutic relevance of such extracellular vesicles for AD. In sum, (i) both brain and adipose tissue have elevated APP levels in obesity, (ii) there is extra-neuronal production of both APP and Abeta peptide, including in the adipose tissue, and (iii) the administration of streptozotocin, a well-known experimental model for type 1 diabetes, induces brain insulin resistance and cognitive alterations resembling the status of AD patients [125]. Hence, an intriguing question emerges [49]: can these AD-associated molecules spread from adipose tissue to the brain?

Conclusion

Many basic and clinical studies have demonstrated that circulating and/or tissue levels of MTF (metabotrophins, metabokines) are reduced in individuals with obesity and related CMD. The scheme within the box below illustrates the possible involvement of MTF in the pathobiology and therapy of obesity viewed as a MTF-deficient disease.

MTF deficit → OBESITY ← MTF therapy

In this connection, it is known that insulin resistance and hypercholesterolemia, which are essential pathobiological signatures of obesity and T2DM, have negative impact on cognitive functions [46–49, 63, 125]. Figure 16.1 illustrates our concept of the potential significance of MTF for the pathogenesis of obesity-related CMD and NDD which, particularly AD, might be considered neurometabolic diseases.

Since 2003, we have been "thinking what nobody has yet thought about that which everybody sees" with respect to the concept of MTF and its relevance for the cellular and molecular mechanisms of obesity and related CMD and NDD.

Yet, we have to keep in mind Robert Frost's poem *The Secret Sits*:

We dance round in a ring and suppose,
But the Secret sits in the middle and knows.

Future *dance round* on MTF signature in obesity may therefore cultivate a more relevant thinking about how we can make these talented biomolecules working for the improvement of physical and mental qualities of *Homo sapiens'* life.

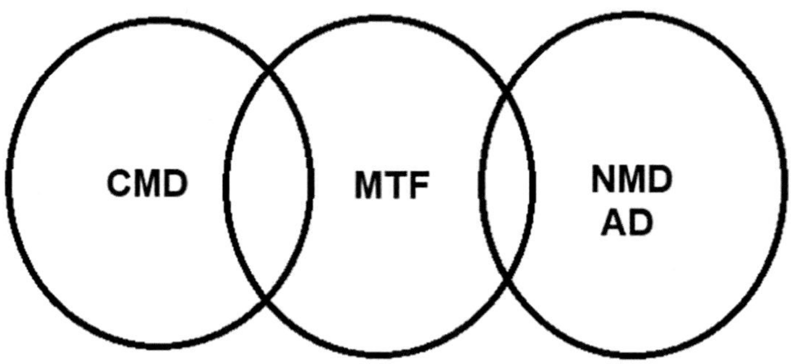

Fig. 16.1 Metabotrophic factors (MTF) on the cross-road of cardiometabolic diseases (CMD) and neurometabolic diseases (NMD), particularly Alzheimer's disease (AD). Credit for Nikifor N. Chaldakov

Acknowledgements This Chapter expresses our tribute to Rita Levi-Montalcini, who with Stanley Cohen, won the Nobel Prize in Physiology or Medicine 1986 for the discovery of NGF. We appreciate the critical and creative reading, also English improvement, of the manuscript by Dr Kelath Murali Manoj (The Science & Ethics Foundation, Kerala, India). We apologize to the authors of many relevant articles that were not quoted here for reasons of brevity.

Conflict of Interest Statement The authors declare that no conflicts of interest exist.

References

1. Trayhurn P (2013) Hypoxia and adipose tissue function and dysfunction in obesity. Physiol Rev 93(1):1–21
2. Chaldakov GN, Fiore M, Hristova MG, Aloe L (2003) Metabotrophic potential of neurotrophins: implication in obesity and related diseases? Med Sci Monit 9(10):HY19–HY21
3. Chaldakov GN, Aloe L, Tonchev AB, Fiore M (2014) From Homo obesus to Homo diabesus: neuroadipology insight. In: Nóbrega C, Rodriguez-López R (eds) Molecular mechanisms underpinning the development of obesity. Springer International Publishing, Switzerland, pp 167–178
4. Behl S, Singh J (2020) Adipocytes under environmental assault: targets for obesity? In: Tappia PS et al (eds) Advances in biochemistry in health and disease, vol 19. Springer Nature, Switzerland, pp 23–41
5. Frühbeck G, Becerril S, Sáinz N et al (2009) BAT: a new target for human obesity? Trends Pharmacol Sci 30(8):387–396
6. Iacobellis G, Di Gioia C, Petramala L et al (2013) Brown fat expresses adiponectin in humans. Int J Endocrinol 2013:126751–126751
7. Blirando K (2016) Epigenetic regulation of adipocytes phenotype: implication for perivascular adipose tissue contribution to cardiometabolic diseases. Adipobiology 8:21–36
8. Sacks H, Symonds ME (2013) Anatomical locations of human brown adipose tissue: functional relevance and implications in obesity and type 2 diabetes. Diabetes 62(6):1783–1790
9. Villarroya F, Cereijo R, Villarroya J et al (2018) Toward an understanding of how immune cells control brown and beige adipobiology. Cell Metab 27(5):954–961
10. Levi-Montalcini R (1987) The nerve growth factor 35 years later. Science 237(4819):1154–1162
11. Fiore M, Chaldakov GN, Aloe L (2009) Nerve growth factor as a signaling molecule for nerve cells and also for the neuroendocrine-immune systems. Rev Neurosci 20(2):133–145
12. Aloe L, Tonchev A, Fiore M, Chaldakov G (2012) Homo diabesus: involvement of metabotrophic factors. Adipobiology 5:45–49
13. Rocco ML, Soligo M, Manni L, Aloe L (2018) Nerve growth factor: early studies and recent clinical trials. Curr Neuropharmacol 16(10):1455–1465
14. Aloe L, Levi-Montalcini R (1977) Mast cells increase in tissues of neonatal rats injected with the nerve growth factor. Brain Res 133(2):358–366
15. Chaldakov GN, Fiore M, Tonchev AB et al (2007) Homo obesus: a metabotrophin-deficient species. Pharmacology and nutrition insight. Curr Pharm Des 13(21):2176–2179
16. Chaldakov G (2011) The metabotrophic NGF and BDNF: an emerging concept. Arch Ital Biol 149(2):257–263
17. Töre F, Tonchev A, Fiore M et al (2007) From adipose tissue protein secretion to adipopharmacology of disease. Immunol Endocr Metab Agents Med Chem 7:149–155
18. Yanev S, Aloe L, Fiore M, Chaldakov GN (2013) Neurotrophic and metabotrophic potential of nerve growth factor and brain-derived neurotrophic factor: linking cardiometabolic and neuropsychiatric diseases. World J Pharmacol 2(4):92–99

19. Rosenbaum T, Vidaltamayo R, Sánchez-Soto MC et al (1998) Pancreatic beta cells synthesize and secrete nerve growth factor. Proc Nat Acad Sci USA 95(13):7784–7788
20. Vidaltamayo R, Mery C, Angeles-Angeles A et al (2003) Expression of nerve growth factor in human pancreatic beta cells. Growth Factors (Switzerland) 21:103–107
21. Rancic G, Fiore M, Tuncel N et al (2016) PVAT and atherogenesis: a crossroad of white and brown adipobiology 8:35–38
22. Cook KS, Min HY, Johnson D et al (1987) Adipsin: a circulating serine protease homolog secreted by adipose tissue and sciatic nerve. Science 237(4813):402–405
23. Lo James C, Ljubicic S, Leibiger B et al (2014) Adipsin is an adipokine that improves β cell function in diabetes. Cell 158(1):41–53
24. Zhang Y, Proenca R, Maffei M et al (1994) Positional cloning of the mouse obese gene and its human homologue. Nature 372(6505):425–432
25. Matsuzawa Y (1997) Pathophysiology and molecular mechanisms of visceral fat syndrome: the Japanese experience. Diabetes Metab Rev 13(1):3–13
26. Chaldakov GN, Stankulov IS, Hristova M, Ghenev PI (2003) Adipobiology of disease: adipokines and adipokine-targeted pharmacology. Curr Pharm Des 9(12):1023–1031
27. Maria MM, Alberto D-R, Rocio G-R et al (2013) Adipobiology for novel therapeutic approaches in metabolic syndrome. Curr Vasc Pharmacol 11(6):954–967
28. Matsuzawa Y (2010) Adiponectin: a key player in obesity related disorders. Curr Pharm Des 16(17):1896–1901
29. Chaldakov G, Fiore M, Ghenev P et al (2000) Atherosclerotic lesions: possible interactive involvement of intima, adventitia and associated adipose tissue. Adipobiology 7:43–49
30. Renes J, Mariman E (2013) Application of proteomics technology in adipocyte biology. Mol Biosyst 9(6):1076–1091
31. Sornelli F, Fiore M, Chaldakov G, Aloe L (2007) Brain-derived neurotrophic factor: a new adipokine. Biomed Rev 18:85–88
32. Trayhurn P, Wood IS (2004) Adipokines: inflammation and the pleiotropic role of white adipose tissue. Br J Nutr 92(3):347–355
33. Tang BL (2008) Leptin as a neuroprotective agent. Biochem Biophys Res Commun 368(2):181–185
34. Trayhurn P, Drevon CA, Eckel J (2011) Secreted proteins from adipose tissue and skeletal muscle—adipokines, myokines and adipose/muscle cross-talk. Arch Physiol Biochem 117(2):47–56
35. Li F, Li Y, Duan Y et al (2017) Myokines and adipokines: involvement in the crosstalk between skeletal muscle and adipose tissue. Cytokine Growth Factor Rev 33:73–82
36. Martinez-Huenchullan SF, Tam CS, Ban LA et al (2020) Skeletal muscle adiponectin induction in obesity and exercise. Metabolism 102:154008
37. Piccirillo R (2019) Exercise-induced myokines with therapeutic potential for muscle wasting. Front Physiol 10:287
38. Raschke S, Eckel J (2013) Adipo-myokines: two sides of the same coin—mediators of inflammation and mediators of exercise. Mediat Inflamm 2013:Article ID 320724
39. Görgens SW, Eckardt K, Jensen J et al (2015) Chapter Thirteen—Exercise and regulation of adipokine and myokine production. In: Bouchard C (ed) Progress in molecular biology and translational science, vol 135. Academic Press, pp 313–336
40. Oh K-J, Lee DS, Kim WK et al (2016) Metabolic adaptation in obesity and type II diabetes: myokines, adipokines and hepatokines. Int J Mol Sci 18(8):1–31
41. Chung HS, Choi KM (2018) Adipokines and myokines: a pivotal role in metabolic and cardiovascular disorders. Curr Med Chem 25(20):2401–2415
42. Schering L, Hoene M, Kanzleiter T et al (2015) Identification of novel putative adipomyokines by a cross-species annotation of secretomes and expression profiles. Arch Physiol Biochem 121(5):194–205
43. Chaldakov GN, Fiore M, Stankulov IS et al (2004) Neurotrophin presence in human coronary atherosclerosis and metabolic syndrome: a role for NGF and BDNF in cardiovascular disease? Prog Brain Res 146:279–289

44. Krabbe KS, Nielsen AR, Krogh-Madsen R et al (2007) Brain-derived neurotrophic factor (BDNF) and type 2 diabetes. Diabetologia 50(2):431–438
45. Yamanaka M, Itakura Y, Ono-Kishino M et al (2008) Intermittent administration of brain-derived neurotrophic factor (BDNF) ameliorates glucose metabolism and prevents pancreatic exhaustion in diabetic mice. J Biosci Bioeng 105(4):395–402
46. de la Monte SM, Wands JR (2008) Alzheimer's disease is type 3 diabetes—evidence reviewed. J Diabetes Sci Technol 2(6):1101–1113
47. Dar TA, Sheikh IA, Ganie SA et al (2014) Molecular linkages between diabetes and Alzheimer's disease: current scenario and future prospects. CNS Neurol Disord Drug Targets 13(2):290–298
48. Sridhar GR, Thota H, Allam AR et al (2006) Alzheimer's disease and type 2 diabetes mellitus: the cholinesterase connection? Lipids Health Dis 5:28
49. Aloe L, Tonchev AB, Maucher A et al (2015) Adipobiology of the brain: from brain diabetes to adipose Alzheimer's disease. Adipobiology 7:37–42
50. Hristova M (2002) NGF and BDNF in patients with metabolic syndrome. PhD thesis (in Bulgarian), Medical University Varna, Bulgaria
51. Manni L, Nikolova V, Vyagova D et al (2005) Reduced plasma levels of NGF and BDNF in patients with acute coronary syndromes. Int J Cardiol 102(1):169–171
52. Ejiri J, Inoue N, Kobayashi S et al (2005) Possible role of brain-derived neurotrophic factor in the pathogenesis of coronary artery disease. Circulation 112(14):2114–2120
53. Sornelli F, Fiore M, Chaldakov GN, Aloe L (2009) Adipose tissue-derived nerve growth factor and brain-derived neurotrophic factor: results from experimental stress and diabetes. Gen Physiol Biophys 28:179–183
54. Larrieta M, Vital P, Mendoza-Rodríguez A et al (2006) Nerve growth factor increases in pancreatic beta cells after streptozotocin-induced damage in rats. Exp Biol Med (Maywood, N.J.) 231:396–402
55. Geroldi D, Minoretti P, Emanuele E (2006) Brain-derived neurotrophic factor and the metabolic syndrome: more than just a hypothesis. Med Hypotheses 67(1):195–196
56. Schulte-Herbrüggen O, Braun A, Wronski S et al (2007) Neurotrophic factors—a tool for therapeutic strategies in neurological, neuropsychiatric and neuroimmunological diseases? Curr Med Chem 14:2318–2329
57. Pedersen BK, Pedersen M, Krabbe KS et al (2009) Role of exercise-induced brain-derived neurotrophic factor production in the regulation of energy homeostasis in mammals. Exp Physiol 94(12):1153–1160
58. Bariohay B, Lebrun B, Moyse E, Jean A (2005) Brain-derived neurotrophic factor plays a role as an anorexigenic factor in the dorsal vagal complex. Endocrinology 146(12):5612–5620
59. Gomez-Pinilla F, Vaynman S, Ying Z (2008) Brain-derived neurotrophic factor functions as a metabotrophin to mediate the effects of exercise on cognition. Eur J Neurosci 28(11):2278–2287
60. Aloe L, Tirassa P, Lambiase A (2008) The topical application of nerve growth factor as a pharmacological tool for human corneal and skin ulcers. Pharmacol Res 57(4):253–258
61. Karatzas A, Katsanos K, Lilis I et al (2013) NGF promotes hemodynamic recovery in a rabbit hindlimb ischemic model through TrkA and VEGFR2-dependent pathways. J Cardiovasc Pharmacol 62(3):270–277
62. Meek TH, Wisse BE, Thaler JP et al (2013) BDNF action in the brain attenuates diabetic hyperglycemia via insulin-independent inhibition of hepatic glucose production. Diabetes 62(5):1512–1518
63. Rao AA (2013) Views and opinion on BDNF as a target for diabetic cognitive dysfunction. Bioinformation 9(11):551–554
64. Lebrun B, Bariohay B, Moyse E, Jean A (2006) Brain-derived neurotrophic factor (BDNF) and food intake regulation: a minireview. Auton Neurosci 126–127:30–38
65. Nicholson JR, Peter J-C, Lecourt A-C et al (2007) Melanocortin-4 receptor activation stimulates hypothalamic brain-derived neurotrophic factor release to regulate food intake, body temperature and cardiovascular function. J Neuroendocrinol 19(12):974–982

66. Fujinami A, Ohta K, Obayashi H et al (2008) Serum brain-derived neurotrophic factor in patients with type 2 diabetes mellitus: relationship to glucose metabolism and biomarkers of insulin resistance. Clin Biochem 41(10):812–817
67. Yamanaka M, Itakura Y, Tsuchida A et al (2007) Comparison of the antidiabetic effects of brain-derived neurotrophic factor and thiazolidinediones in obese diabetic mice. Diabet Obesity Metab 9(6):879–888
68. Iacobellis G, Sharma A (2007) Epicardial adipose tissue as new cardio-metabolic risk marker and potential therapeutic target in the metabolic syndrome. Curr Pharm Des 13:2180–2184
69. Tan BK, Adya R, Randeva HS (2010) Omentin: a novel link between inflammation, diabesity, and cardiovascular disease. Trends Cardiovasc Med 20(5):143–148
70. Castan-Laurell I, Dray C, Attané C et al (2011) Apelin, diabetes, and obesity. Endocrine 40(1):1–9
71. Wang GX, Cho KW, Uhm M et al (2014) Otopetrin 1 protects mice from obesity-associated metabolic dysfunction through attenuating adipose tissue inflammation. Diabetes 63(4):1340–1352
72. Abella V, Pino J, Scotece M et al (2017) Progranulin as a biomarker and potential therapeutic agent. Drug Disc Today 22(10):1557–1564
73. Frühbeck G, Gómez-Ambrosi J, Rodríguez A et al (2018) Novel protective role of kallistatin in obesity by limiting adipose tissue low grade inflammation and oxidative stress. Metabolism 87:123–135
74. Benga G (2012) On the definition, nomenclature and classification of water channel proteins (aquaporins and relatives). Mol Asp Med 33(5):514–517
75. Ahrén B (2011) The future of incretin-based therapy: novel avenues—novel targets. Diabetes Obes Metab 13(1):158–166
76. Hira T, Pinyo J, Hara H (2020) What Is GLP-1 really doing in obesity? Trends Endocrinol Metab 31(2):71–80
77. Xu Y, Van Hul M, Suriano F et al (2020) Novel strategy for oral peptide delivery in incretin-based diabetes treatment. Gut 69(5):911–919
78. Bailey CJ (2020) GIP analogues and the treatment of obesity-diabetes. Peptides 125:170202
79. Holst JJ, Rosenkilde MM (2020) GIP as a therapeutic target in diabetes and obesity: insight from incretin co-agonists. J Clin Endocrinol Metab 105(8):e2710–e2716
80. Svendsen B, Capozzi ME, Nui J et al (2020) Pharmacological antagonism of the incretin system protects against diet-induced obesity. Mol Metab 32:44–55
81. Chaldakov GN, Vankov VN (1986) Morphological aspects of secretion in the arterial smooth muscle cell, with special reference to the golgi complex and microtubular cytoskeleton. Atherosclerosis 61(3):175–192
82. Chaldakov G (2016) Human body as a multicrine system, with special reference to cell protein secretion: from vascular smooth muscles to adipose tissue. Biomed Rev 27:VIII–XIX
83. Yanev S, Fiore M, Hinev A et al (2017) From antitubulins to trackins. Biomed Rev 27:45–53
84. Carito V, Venditti A, Bianco A et al (2014) Effects of olive leaf polyphenols on male mouse brain NGF, BDNF and their receptors TrkA, TrkB and p75. Nat Prod Res 28(22):1970–1984
85. Xu B, Xie X (2016) Neurotrophic factor control of satiety and body weight. Nat Rev Neurosci 17(5):282–292
86. Preti A (2003) Axokine (Regeneron). IDrugs 6(7):696–701
87. Xiang L, Xiang G, Yue L et al (2014) Circulating irisin levels are positively associated with endothelium-dependent vasodilation in newly diagnosed type 2 diabetic patients without clinical angiopathy. Atherosclerosis 235(2):328–333
88. Hofmann T, Elbelt U, Stengel A (2014) Irisin as a muscle-derived hormone stimulating thermogenesis—a critical update. Peptides 54:89–100
89. Kurdiova T, Balaz M, Vician M et al (2014) Effects of obesity, diabetes and exercise on Fndc5 gene expression and irisin release in human skeletal muscle and adipose tissue: in vivo and in vitro studies. J Physiol 592(5):1091–1107
90. Perakakis N, Triantafyllou GA, Fernández-Real JM et al (2017) Physiology and role of irisin in glucose homeostasis. Nat Rev Endocrinol 13(6):324–337

91. Moreno-Navarrete JM, Ortega F, Serrano M et al (2013) Irisin is expressed and produced by human muscle and adipose tissue in association with obesity and insulin resistance. J Clin Endocrinol Metab 98(4):E769-778
92. Ozturk G, Demirel O, Tekatas A et al (2019) Circulating irisin levels in newly diagnosed obstructive sleep apnea patients. Scr Sci Med 51(1):16–20
93. Aydin S, Aydin S, Kobat MA et al (2014) Decreased saliva/serum irisin concentrations in the acute myocardial infarction promising for being a new candidate biomarker for diagnosis of this pathology. Peptides 56:141–145
94. Zhu D, Wang H, Zhang J et al (2015) Irisin improves endothelial function in type 2 diabetes through reducing oxidative/nitrative stresses. J Mol Cell Cardiol 87:138–147
95. Hou N, Han F, Sun X (2014) The relationship between circulating irisin levels and endothelial function in lean and obese subjects. Clin Endocrinol 83
96. Park KH, Zaichenko L, Brinkoetter M et al (2013) Circulating irisin in relation to insulin resistance and the metabolic syndrome. J Clin Endocrinol Metab 98(12):4899–4907
97. Sesti G, Andreozzi F, Fiorentino TV et al (2014) High circulating irisin levels are associated with insulin resistance and vascular atherosclerosis in a cohort of nondiabetic adult subjects. Acta Diabetol 51(5):705–713
98. More CE, Papp C, Harsanyi S et al (2019) Altered irisin/BDNF axis parallels excessive daytime sleepiness in obstructive sleep apnea patients. Respir Res 20(1):67–67
99. Gamas L, Matafome P, Seiça R (2015) Irisin and myonectin regulation in the insulin resistant muscle: implications to adipose tissue-muscle crosstalk. J Diab Res 2015:1–8
100. Liu Y, Vu V, Sweeney G (2019) Examining the potential of developing and implementing use of adiponectin-targeted therapeutics for metabolic and cardiovascular diseases. Front Endocrinol 10(842):1–16
101. Wu H, Li X, Shen C (2020) Peroxisome proliferator-activated receptor gamma in white and brown adipocyte regulation and differentiation. Physiol Res 69:759–773
102. Zhu R, Wei J, Liu H et al (2020) Lycopene attenuates body weight gain through induction of browning via regulation of peroxisome proliferator-activated receptor γ in high-fat diet-induced obese mice. J Nutr Biochem 78:108335
103. Hiradate R, Khalil IA, Matsuda A et al (2020) A novel dual-targeted rosiglitazone-loaded nanoparticle for the prevention of diet-induced obesity via the browning of white adipose tissue. J Control Release
104. Osinski V, Bauknight DK, Dasa SSK et al (2020) In vivo liposomal delivery of PPARα/γ dual agonist tesaglitazar in a model of obesity enriches macrophage targeting and limits liver and kidney drug effects. Theranostics 10(2):585–601
105. Otvos L (2019) Potential adiponectin receptor response modifier therapeutics. Front Endocrinol 10:539
106. Dali-Youcef N, Lagouge M, Froelich S et al (2007) Sirtuins: the "magnificent seven", function, metabolism and longevity. Ann Med 39:335–345
107. Moniot S, Weyand M, Steegborn C (2012) Structures, substrates, and regulators of mammalian sirtuins—opportunities and challenges for drug development. Front Pharmacol 3:16
108. Bonkowski MS, Sinclair DA (2016) Slowing ageing by design: the rise of NAD^+ and sirtuin-activating compounds. Nat Rev Mol Cell Biol 17(11):679–690
109. Dai H, Sinclair DA, Ellis JL, Steegborn C (2018) Sirtuin activators and inhibitors: promises, achievements, and challenges. Pharmacol Ther 188:140–154
110. Mautone N, Zwergel C, Mai A, Rotili D (2020) Sirtuin modulators: where are we now? A review of patents from 2015 to 2019. Expert Opin Ther Pat 30(6):389–407
111. Nijhawan P, Behl T (2020) Role of sirtuins in obesity. Obes Med 17:100156
112. Yoon MJ, Yoshida M, Johnson S et al (2015) SIRT1-mediated eNAMPT secretion from adipose tissue regulates hypothalamic NAD^+ and function in mice. Cell Metab 21(5):706–717
113. Samms RJ, Cheng CC, Kharitonenkov A et al (2016) Overexpression of β-Klotho in adipose tissue sensitizes male mice to endogenous FGF21 and provides protection from diet-induced obesity. Endocrinology 157(4):1467–1480

114. Vo HT, Laszczyk AM, King GD (2018) Klotho, the key to healthy brain aging? Brain Plast 3:183–194
115. Kuro-o M (2011) Klotho and the aging process. Korean J Intern Med 26(2):113–122
116. Dermaku-Sopjani M, Kolgeci S, Abazi S, Sopjani M (2013) Significance of the anti-aging protein Klotho. Mol Membr Biol 30(8):369–385
117. Barchetta I, Chiappetta C, Ceccarelli V et al (2020) Angiopoietin-like protein 4 overexpression in visceral adipose tissue from obese subjects with impaired glucose metabolism and relationship with lipoprotein lipase. Int J Mol Sci 21(19):1–14
118. Morelli MB, Chavez C, Santulli G (2020) Angiopoietin-like proteins as therapeutic targets for cardiovascular disease: focus on lipid disorders. Expert Opin Ther Targets 24(1):79–88
119. Tardelli M, Stulnig TM (2020) Chapter Four—Aquaporin regulation in metabolic organs. In: Litwack G (ed) Vitamins and hormones, vol 112. Academic Press, pp 71–93
120. Frühbeck G, Catalán V, Gómez-Ambrosi J, Rodríguez A (2006) Aquaporin-7 and glycerol permeability as novel obesity drug-target pathways. Trends Pharmacol Sci 27(7):345–347
121. Rodríguez A, Catalán V, Gómez-Ambrosi J, Frühbeck G (2006) Role of aquaporin-7 in the pathophysiological control of fat accumulation in mice. FEBS Lett 580(20):4771–4776
122. Guo R, Wang L, Zeng X et al (2020) Aquaporin 7 involved in GINSENOSIDE-RB1-mediated anti-obesity via peroxisome proliferator-activated receptor gamma pathway. Nutr Metab 17(69):1–17
123. Lee Y-H, Tharp WG, Maple RL et al (2008) Amyloid precursor protein expression is upregulated in adipocytes in obesity. Obesity 16(7):1493–1500
124. Katsuda T, Tsuchiya R, Kosaka N et al (2013) Human adipose tissue-derived mesenchymal stem cells secrete functional neprilysin-bound exosomes. Sci Rep 3:1197
125. Lester-Coll N, Rivera EJ, Soscia SJ et al (2006) Intracerebral streptozotocin model of type 3 diabetes: relevance to sporadic Alzheimer's disease. J Alzheimer's Dis 9:13–33

Chapter 17
New Therapeutic Agents in Obesity-Related Cardiovascular Disorders: Molecular and Cellular Insights

Belma Turan and Deniz Billur

Abstract Several studies emphasized the impact of obesity in a healthy lifespan, association with important health risks for humans, although some studies mentioned a decrease in mortality among not obese but overweight patients in intensive care. Both clinical observations and experimental studies performed in systemic and cellular levels, suggest a complex relationship between human obesity and chronic diseases, such as type 2 diabetes (T2DM), hypertension, and cardiovascular diseases (CVD). The harmful metabolic alterations, at most, associated with visceral adiposity are main contributors to those of above diseases. Of note, it has been also shown that the most of obesity-related dysfunctions are due to the from dyshomeostasis in the central nervous, besides adipose tissue accumulation in mammals. Despite specific altered signaling mechanisms, the accumulation and deposition of fat are resulting in a chronic energy imbalance, and thereby a high circulating level of free fatty acids in tissues, whifh further leads to increase in body weight and several metabolic disturbances in organs. In general aspects, type 2 diabetes (T2DM), atherosclerosis, and metabolic syndrome characterized by insulin resistance besides others are marked syndromes associated with alterations in metabolic and endocrine systems in obesity, as well. New therapeutic agents in clinical approaches to overcome the CVD in obese individuals are including dipeptidyl peptidase-4 inhibitors, glucagon-like peptide 1 receptor agonists, sodium-glucose cotransporter-2 inhibitors, β_3-adrenergic receptor agonists, and agents associated with weight loss. It is well-accepted that obesity is an important syndrome leading to increases the risk of complications in several organ systems. Therefore, there are several recent approaches to use combined therapies to maintain the organ functions in obesity-related disorders.

B. Turan (✉)
Faculty of Medicine, Department of Biophysics, Lokman Hekim University, Ankara, Turkey
e-mail: belma.turan@lokmanhekim.edu.tr; belma.turan@medicine.ankara.edu.tr

D. Billur
Faculty of Medicine, Department of Histology and Embryology, Ankara University, Ankara, Turkey

© Springer Nature Switzerland AG 2021
P. S. Tappia et al. (eds.), *Cellular and Biochemical Mechanisms of Obesity*,
Advances in Biochemistry in Health and Disease 23,
https://doi.org/10.1007/978-3-030-84763-0_17

Introduction

Obesity is one of the biggest health problems of the current century and becoming a worldwide epidemic syndrome. Even before the two-decade, the increase of obesity among young-generation in developed countries can anticipate their next generation and will become more obese than ever numbers [1–3]. Although the exact reasons for the increasing prevalence of obesity are not known yet, experimental and clinical studies documented the leading combination factors such as genetic predisposition, social factors, and feeding habits. Particularly, feedings with high carbohydrate and/or high fat diet are leading to disturbance of energy balance in the body which is further inducing over accumulation of adipose tissue. Over deposited adipose tissue is basically underling the induction of obesity in humans. Not only obesity but also even overweigthness are major risk factors for many chronic diseases, including type 2 diabetes mellitus (T2DM), immunodeficiency, cardiometabolic disorders, and other types of cardiovascular diseases [1, 3–12]. The findings of these studies have documented that obesity-related CVDs are main complications of human obesity, which are originating from comorbidity of several alterations such as coronary artery diseases together with or alone hypertension. Of note, it is not clear whether weight-loss can be able to reverse those syndromes in those individiuals, yet.

Additionally, the novel evidence is pointed out the associated between obesity and COVID-19, due to more severe symptoms, at least, through attenuation in the immune system activity besides chronic inflammation. Therefore, in the light of those data, the authors reported that obesity is a risk factor for not only increased prevalence of COVID-19 but also closely related with lethality among humans [13]. Almost the most of already know documents strongly confirm that obesity is a medical condition with complex pathophysiology, including various mechanisms, thereby underlying several serious chronic organ dysfunctions, such as CVDs.

Several studies emphasized the impact of obesity in a healthy lifespan for humans together with many insights to consider in the drug treatment of obese patients. There are many insights to consider in the drug treatment of obese patients. For long times, although there are great efforts for the development of antiobesity drugs and/or chemical agents with the hope of its prevention, current clinical and experimental findings strongly point out the controversies related to the benefit-risk balance of them in that effort [14–17]. More interestingly, the progression of severe obesity and its associated comorbidities urgently requires the development of novel prevention and/or therapies for obese individuals. However, the cellular, biochemical and molecular mechanisms underlying and/or controlling obesity are not exactly known yet. In these regards, recent studies are focused on anti-obesity drugs/chemical agents in that concept, which will be able to directly or indirectly enhance signaling pathways associated with induction of obesity [18–20]. In this chapter, we aimed to emphasize and describe the current state of the obesity-associated organ dysfunction as well as new drugs with anti-obesity properties and their possible side effects, particularly, CVDs in obese/metabolic syndrome/diabetic patients with insulin resistance.

Obesity-Related Dysfunctions: Discussions at Systemic and Cellular Levels

Obesity, with a simple definition, means too much body fat based on your body mass index. The complexity of obesity in healthy life arises due to events that are a combination of causes and individual factors such as behavior (includes dietary habits, physical inactivity, education, and skills) and genetics. Obesity is presently a serious syndrome because it is also associated with weak mental health outcomes and reduced quality of life. Obesity is also associated with one of the leading causes of death worldwide, mainly including T2DM, cardiometabolic diseases, other types of heart diseases, pulmonary dysfunctions, hypercoagulability, fat-liver diseases, and osteoarthritis besides cancer (Fig. 17.1). However, worldwide documents are demonstrating why obesity causes and/or exacerbates over 200 medical-disorders further leading to worsening disease morbidity and mortality [9, 21]. On the other hand, the pathology of obesity is not related with only one mechanism, on the contrary, there are various reasons underlying the cause of obesity as mentioned in above paragraphs. Notably, it should be added that not every obese individual will likely face to these problems. However, the risk for obesity is highas long as one has a family history related with those conditions.

As summarized in Fig. 17.1, among health risks of obesity and obesity-related pathological conditions, there are several conditions such as T2DM, abnormalities in metabolic, endocrine, and circulatory systems, which are further leading to more

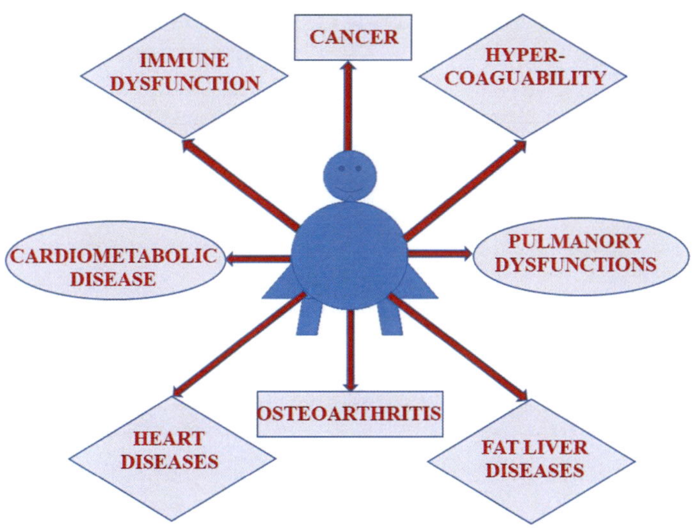

Fig. 17.1 A summarized documents show that, among health risks of obesity and obesity-related pathological conditions, there are T2DM, abnormalities in metabolic, endocrine, and circulatory systems are leading causes to further serious chronic diseases. Secondary diseases include **osteoarthritis, sleep apnea, cancer as well as** psychological and social problems

serious chronic diseases. More over, it should be considered the secondary diseases associated with obesity such as **osteoarthritis, sleep apnea as well as** psychological and social problems [4, 6, 11, 22]. Obesity is the second cause of preventable death in the world after smoking [23]. The common sources of those symptoms in those diseases are originating from dyshomeostasis in several individual and/or combined signaling pathways [24]. For instance, obesity is characterized by dyslipidemia that is associated with high levels of very-low-density lipoprotein (VLDL), cholesterol, triacylglycerols, and total cholesterol, as well as associated with an increase in small dense LDL particles, and lower high-density lipoprotein (HDL) cholesterol levels [25]. As already known, all these changes can have important impacts on CVDs. Despite specific altered signaling mechanisms, the main reasons, leading to obesity-related organ dysfunctions, include accumulation and thereby deposition of fat, thereby, resulting in a chronic positive energy-imbalance in the human body. One of the underlying mechanisms of why fat deposition is further leading to symptoms is due to an increase in adipose tissue that further results in higher circulating levels of free fatty acids [26, 27]. With another definition, the adipose tissue has a < aregard, it has been mentioned about an adipose tissue remodeling in obesity including the changes in the number and size of mature adipocytes, increases in the amount of white adipose tissue, which are closely associated with metabolic complications and insulin resistance, T2DM, CVD) [28]. The adipose tissue plays important role in the cross-correlation between nutrition, metabolism, and immunity, while any inflammation of adipose tissue has as a critical role in the association between obesity and related metabolic complications. Adipose tissue, having active metabolic states, is responsible to release of a large number of cytokines and bioactive mediators (i.e. leptin, adiponectin, interleukin-6). Those can influence not only the control of body-weight but also insulin resistance, T2DM, inflammation, and atherosclerosis [29–31]. In addition, there can be induced many morphological adaptations in cardiac both structure and function of obese individuals [32]. Of note, those cytokines further lead to the activation of signal transduction cascades, inhibiting insulin action in several organs including heart. Supporting this statement, a stimulator role of IL-6 in the production of acute-phase proteins, such as C-reactive protein and fibrinogen has been demonstrated in the liver, which are also important risk factors for CVDs [33–35].

Besides, the alterations in morphology of fat and muscle from the obese mammals are closely correlated with the insulin resistance and other metabolic abnormalities. In these regards, recent studies have demonstrated that muscle and adipose tissue morphology are dramatically changed in obese patients. Those changes are basically hypertrophy in adipocytes, degeneration and necrosis in cells, macrophage infiltration, and also muscle fat accumulation [36]. Under experimental condition, we performed a morphological analysis of some tissues (heart, liver, kidney, and brain) from a high-fat diet-induced obese rats (with a normal blood glucose level and marked insulin-resistance). There were marked changes in those tissues such as serious necrosis and increased collagen accumulation, which were also very densed in the heart tissue (Fig. 17.2). We, particularly, observed macrovesicular fat droplets in hepatocytes, sinüsoidal enlargements and necroinflammatory foci in

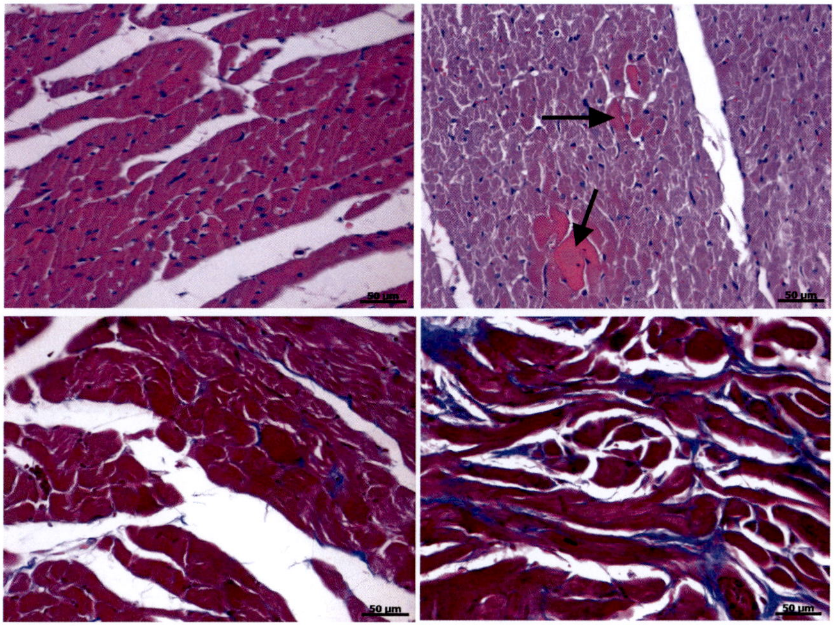

Fig. 17.2 The light microscopy analysis of the heart from obese rats. Heart structure of the control rat showed normal histological features (left in A and B), while in obese rat heart, there was observed necrosis (thin arrow, right in A), there was an increase in collagen fibers observed (with blue stained fibers, right in B). H-E × 400, scale bars are given on the photomicrographs

the liver (Fig. 17.3a) and kidney (Fig. 17.3b, c). Furthermore, in the same group of animals, there were stratum granulosum containing many cells with dark stained pyknotic nuclei scattered as well as stratum granulosa layer and subgranular zone in the hippocampus part of the brain (Fig. 17.4a), while there shrunken Purkinje cells with dark cytoplasm and loss of detail of nucleus (Fig. 17.4b).

It is noteworthy about the complex and conflicting clinal outcomes on how adipose tissue is an active participant in the regulation of physiologic and pathologic processes, including immunity and inflammation. As a response to those regards, cross-talks between immune cells and adipocytes can be one way to reply to those considerations via consideration of the dysregulation of both innate and adaptive immunity [6, 37, 38]. The summarized document gives great explanation how adipose tissue produces several hormones (adipokines, leptin), inflammatory mediators, and non-esterified fatty acids, which are further responsible from induction of chronic inflammation in obese patients [38, 39]. Of note, a long term inflammation can lead to the development of insulin resistance in obese patients. There are also acute effects of obesity on inflammation as well as affected T-cells, which further lead to increased susceptibility to infection in obese patients, as well. Overall, it appears that obesity is associated with chronic inflammation and several impairments in immunity, which further lead to organ dysfunction including CVDs.

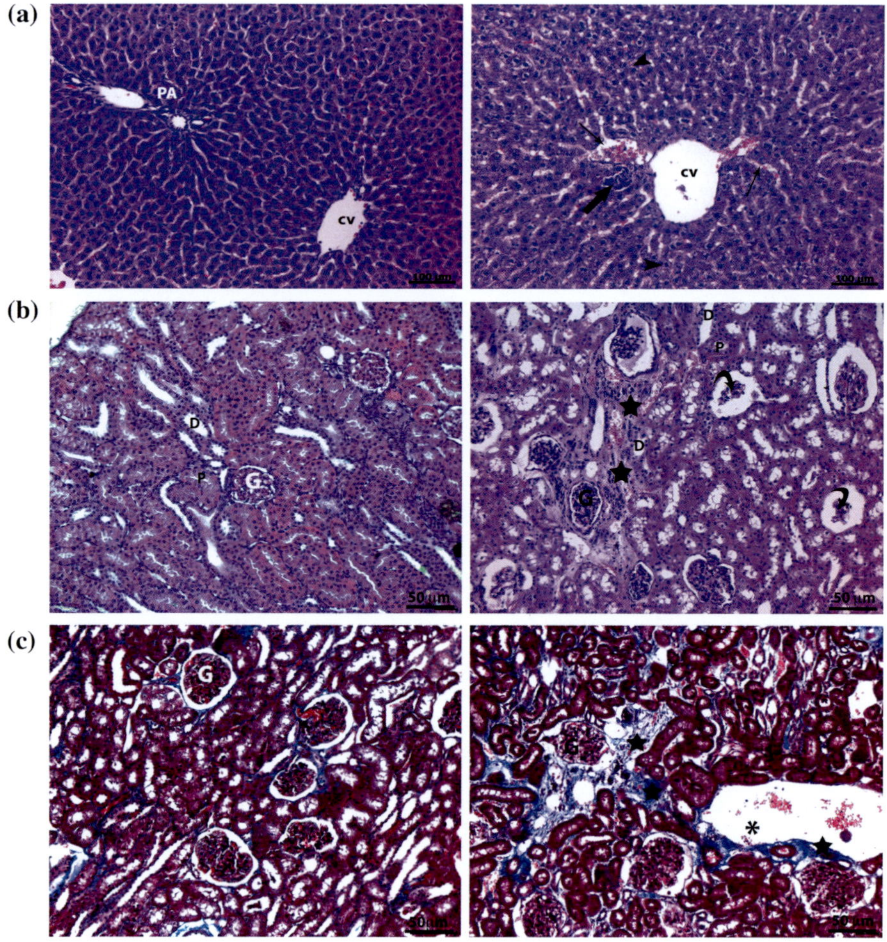

Fig. 17.3 Representative photomicrographs were taken from the histological preparations of the liver and kidney of obese rats compared to those of controls. (**a**) Hematoxylin–Eosin staining of liver sections from control rat (left), showing normal liver histology. Central vein (cv) and portal area were seen in normal appearance. In Obese rat liver sections, macrovesicular fat droplets (arrowhead) in hepatocytes, sinüsoidal enlargements, and necroinflammatory foci were seen, clearly (right). Kidney sections stained Hematoxylin–Eosin (**b**) and Mallory Azan (**c**). Control rats exhibited normal histological appearance (left). In the cortex of this group, glomerulus (G), proximal convoluted tubules, (P), distal convoluted tubules (**d**) were observed. In obese rats, capillary dilatation (Asterix), glomerular degeneration (convoluted arrow), and prominent interstitial fibrosis (star) markedly appeared (left)

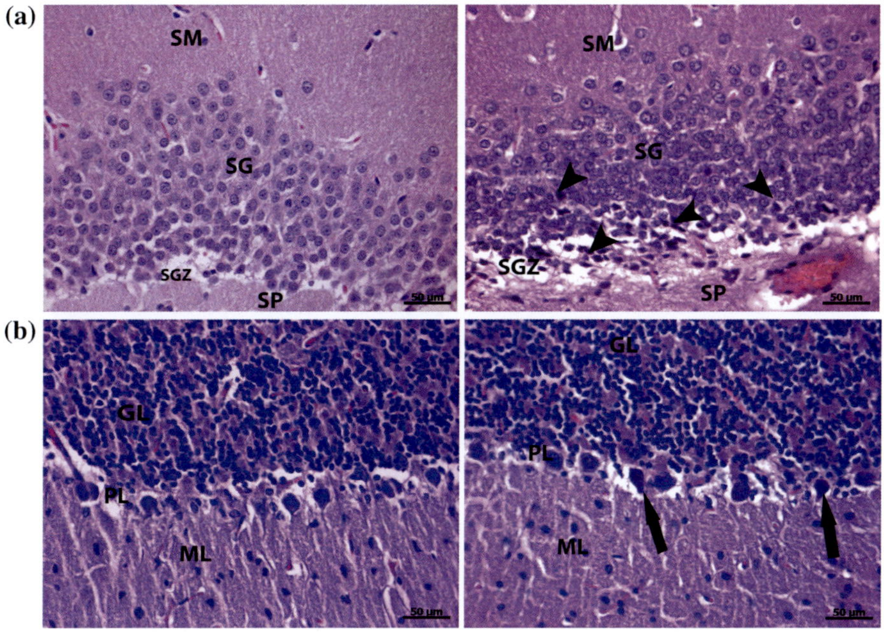

Fig. 17.4 Representative photomicrographs were obtained from the hippocampus and cerebellum from obese rats. (**a**) Each branch of the DG in the hippocampus consists of three layers; Stratum molecular (SM), Stratum granulosum (SG), and Stratum Pleomorph (SP). SG in normal histology, the small, closely-packed neurons of the SG have a limited amount of cytoplasm, and typical euchromatic nucleus obtained in a control rat (left). In obese rats, SG containing many cells with dark stained pyknotic nuclei, scattered as both the SG layer and subgranular zone (right). (**b**) The cortex is layered into the outer molecular cell layer (ML), middle Purkinje cell layer (PL), and inner granular cell layer (GL). In obese rat Purkinje cells (arrow) appear shrunken with dark cytoplasm and loss of nuclear details (left). Control group finding is given in the right of the photomicrographs. H-E $\times$ 400

In addition to the above documents, there are indications to consider obesity with a high-impact event associated with genome stability. Besides inflammation, oxidative stress is commonly occurring in obesity and can induce DNA damage as well as inhibit DNA repair [40–42]. Interestingly, some studies have shown the impact of reversibility of those changes and improvement of dietary habits on the weight loss [43].

The Impact of Obesity on the Cardiovascular System

As discussed in the previous sections, excess fat accumulation in the body, at most, through an imbalance between energy intake and consumption causes obesity in mammals. The increasing prevalence of obesity is recognized to be associated and a major risk for a variety of alterations, mainly in metabolic and endocrine systems [44, 45]. Excess body weight leads to a variety of metabolic changes together with increases in the risk for prevelance of CVDs in mammals. In general, T2DM with insulin-resistance, atherosclerosis characterized with high total cholesterol and/or high levels of triglycerides, and metabolic syndrome (the combination of disorders such as high blood glucose, blood pressure, and serum cholesterol and triglyceride levels) are also marked markers in obesity-related syndromes [6, 46–50].

Noteworthy, the prevalence of CVDs is increasing worldwide with rises of obese populations. Indeed, CVDs are the main cause of mortality while its rate continues to rise with current lifestyle [51]. Considering the current available clinical data, one can emphasize the occurence of obesity-related cardiomyopathy, characterized basically with increased blood-volume and cardiac-output parallel to ventricular hypertrophy and diastolic dysfunction [6, 46, 49, 52]. In addition to those alterations, there are also cardiovascular complications associated with abnormal over fat deposition with its direct effects via immune and endocrine systems and/or its indirect action associated with metabolic syndrome [53]. Although some studies pointed out the limitations related with the benefits of lipid-lowering strategies in obesity [54], there are documents demonstrating the benefits of usage statins to the reduction of lipids and thereby prevention of coronary plaque development in obese individuals [25].

It is noteworthy to take into consideration the coexisting disorders associated with the most serious alterations in obese individuals, of which further leading to organ dysfunction including CVDs. Yet, the reservation of the obesity associated cardiac dysfunction with weight loss is under investigation although several clinical reports mentioned its impact [55]. To clarify those disorders in induction of CVDs, cellular level studies have been developed to show whether or not obesity is a direct consequence of CVDs. In an early study [56], authors used obese Zucker-Diabetic-Fatty rats, and examined the underlying mechanism of cardiac dilatation with reduced contractility via analyzing the myocardial triacylglycerol level. They determined high triacylglycerol level in the heart homogenates of those animals, being associated with less expression level of fatty acid oxidative enzymes and their transcription factors as well as lipoapoptosis followed with markedly development of cardiac dysfunction. In a later study with obese Zucker rats, the cardiac output was significantly depressed, emphasizing the association between the inability of the heart to increase fatty acid oxidation relative to the increased fatty acid availability and lipid accumulation [57].

An important process related with the development of obesity in the body is the correlation between the content of the diet and the body intake of energy. In that content, Fernandes and co-workers [58] studied the effect of a high-fat diet on heart and aorta function in rats. They observed an increased oxidative damage in the aorta, together with increased cyclooxygenase 2 expression and the increased numbers of macrophages. Those changes were accompanied with increases in collagen fibers with decreases in the number of positive α-actin cells and expression of matrix metalloproteinase-2. Furthermore, they observed a marked decline in the antioxidant capacity of hepatic cells and increase in the lipid peroxidation activity. In another study, authors observed a depression in energy expenditure, β-oxidation, and adipogenesis together with adipose tissue accumulation, insulin resistance, high blood pressure as well as high serum cholesterol and triglycerides levels of diet-induced obese rats [59]. Being in line with those results, our histological analysis of the heart tissue from obese rats have shown some degenerations such as multifocal areas of mononuclear cell infiltration and important vacuolation of cardiomyocytes in obese rats (Fig. 17.5a). Moreover, there were also marked alterations in the aortas of those animals such as increased thickness of tunica media, multifocal degeneration of smooth muscle cells, and disorientation of elastic lamellae (Fig. 17.5b, c).

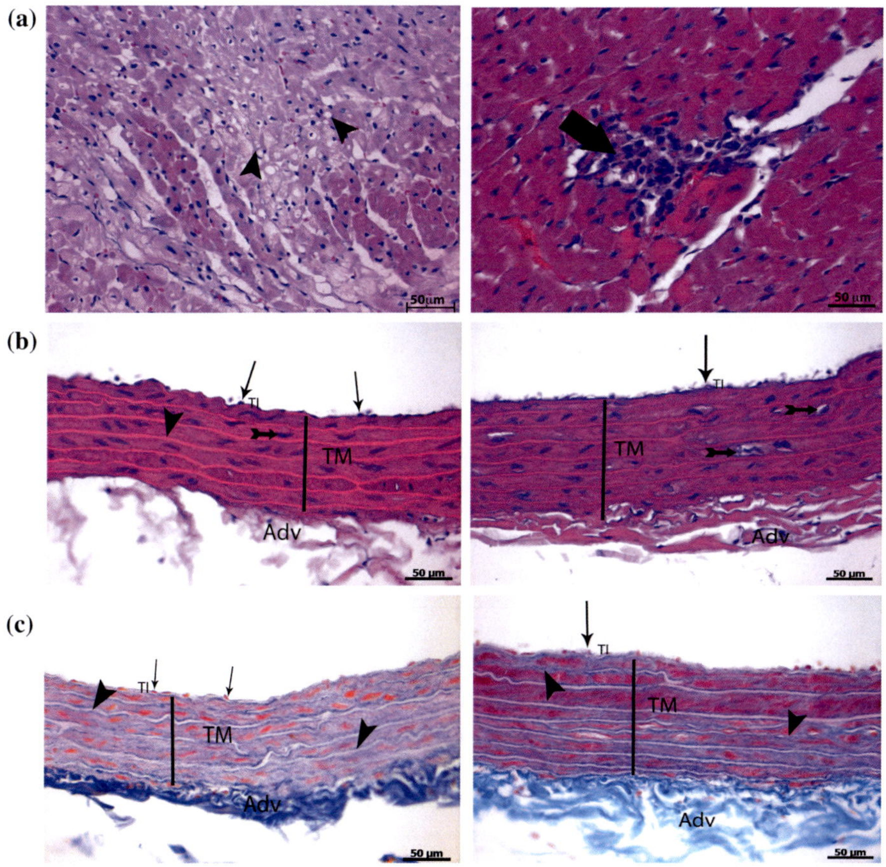

Fig. 17.5 Representative photomicrographs were obtained from the heart and aorta from obese rats. (**a**) Multifocal areas of mononuclear cell infiltration (thick arrow) and vacuolation of cardiomyocytes (arrowhead) in the heart of the obese rat (left). Histological appearance of thoracic aorta stained with Hematoxylin–Eosin (**b**) and Mallory Azan (**c**). Endothelial cells of control rats were lining on the tunica intima (TI) (arrow). Most of the medial smooth muscle cells (tailed arrow) of the tunica media (TM) oriented horizontal to the lumen. Additionally, regularly arranged elastic lamellae (arrowhead) were shown. Normal connective tissue contents were seen in the adventitia (Adv) (left). In obese rats, increasing thickness of tunica media, multifocal degeneration of smooth muscle cells (tailed arrow), and disorientation of elastic lamellae (arrowhead) were seen, clearly (right)

Overall, either animal models or samples from clinical outcomes clearly show a close association between cardiovascular complications and obesity in mammals. Taken into consideration the limiting and divergent data related with obesity-related CDVs, it should be reconsidered the important contributions of alterations in myocardial ultrastructure besides changes in molecular levels.

Cardiac Consequences of the Electrophysiological and Hemodynamic Changes Triggered by Obesity

The worldwide scientific consensus confirmed the obesity-related increases in the risk factors for CVDs, further leading to heart failure. Furthermore, documents longer than the two decades provided wide documents on epidemic proportions of obesity worldwide, while its worldwide prevalence rate has increased rapidly since 1980 [60]. Among cardiovascular complications, hypertension, coronary heart diseases, and stroke are main syndromes associated with obesity-induced disorders [6, 46–50]. Clinical outcomes emphasized a complex combination of complications in obesity-related CVDs, which are basically including ventricular hypertrophy, diastolic dysfunction, and ventricular dilation [6, 46, 49, 52]. Among them, atrial fibrillation and pulmonary hypertension are also frequently observed alterations in obese humans. Likely, elevated left atrial pressure, hypoxia and hypoventilation syndrome, and/or chronic thromboembolism are observed complications in obese individuals as secondary complications. Associated with the above findings, an early and interesting study by Gustafsson and colleagues [50] was performed a retrospective analysis of about 5,000 hospitalized congestive heart failure (CHF) patients with confirmed obesity. They analyzed the interaction between the left ventricular systolic function (LVSF) and obesity on prognosis in CHF. Their data showed that the risk of death in the total population decreased steadily with increasing body-mass-index from the underweight to the obese individuals while they mentioned an association between increasing body-mass-index and a lower mortality in CHF. In addition to those findings, since the induction of several cytokines and inflammatory markers are increased in obese humans, their contribution to the cardiovascular outcome should be one of the mechanisms into in body-mass-index associated CVDs [25].

There are also several clinical outcomes related to the alterations in the myocardium of obese humans [61]. In those obese individuals, there is a marked decrease in ventricular sufficiency as the primary pathophysiological abnormality [62]. Besides, our analysis presented an important data, demonstrating markedly increased collagen deposition and fibrosis in obese rat heart (Figs. 17.2 and 17.4). Regardless of the underlying mechanism of the heart failure in obese humans, known data point out the microvascular fibrosis in the heart tissue as the main contributor [63, 64], which, in turn, lead to abnormalities in proper cardiac capacity [65]. It should be reconsidered that documents can strongly mark the cross-correlation between serious CVDs and obesity, particularly obese children with high risk for obese adults [66,

67]. Of note, obesity can increase cardiovascular morbidity and mortality, further leading to not only arrhythmia and heart failure but also sudden cardiac death [68]. From those points, it is getting more attention to avoid obese overloaded populations worldwide.

There is also an association between abnormal electrocardiogram (ECG) and obesity [69]. Adverse effects of left ventricular (LV) function are characterized by both diastolic and systolic dysfunction as well as shortened ejection fraction, in which there is an impact of LV remodeling in obesity-associated cardiac dysfunction. There are also several experimental studies supporting the human findings to demonstrate the contribution of obesity to cardiac electrical dysfunction. The electrical changes at cellular levels were demonstrated in the isolated myocardial cells from obese samples by using electrophysiological techniques, as well [70, 71]. Since the common types of arrhythmias in obese patients are characterized with long-QT syndrome and/or atrial fibrillation (at most associated with a decrease in the repolarization currents and/or an increase in the depolarization currents). Also, clinical analysis has been documented that there are increases in the outward K^+-current and/or reduces in the inward Ca^{2+}-current which are underling the atrial fibrillation through acceleration of atrial repolarization. Those changes are further promoting to induce arrhythmia in obese humans. Under the light of already published findings, with the consideration the roles of co-metabolic disorders on the electrophysiology of the myocardium in obese mammals, one can first point out the main contribution of changes in several types of ionic-currents of cells to the systemic cardiac dysfunction in those mammals [55, 72, 73]. However, there are a significant amount of inconsistent and contradictory results on this topic. Therefore, it seems it is needed further studies to understand the exact role of obesity on myocardial electrophysiology [74].

In the literature, there are many different animal model studies for generalized obesity, which are aimed to clarify the associated mechanisms between cardiovascular complications and obesity in mammals. Some of them are including diet-induced obesity by feeding animals with either a high-fat or high-carbohydrate diet whereas others are due to genetic mutations resulted in obesity as well as some transgenic animals with targeted mutations in cardiomyocytes. As an example of those studies, Leopoldo and co-workers [75] investigated the cardiac remodeling in hypercaloric diet-induced obese rats by observing their echocardiographic indexes. Those obese rats had severe heart dysfunction including mainly increases in left ventricular mass and depressed systolic function, being parallel to an increase in myocardial collagen content.

As mentioned in above paragraph, there is marked pathological cardiac remodeling characterized with significant systolic dysfunction and increased myocardial stiffness in obese mammals. These pathological changes further induce severe cardiac dysfunction, at most, ending with heart failure. More importantly, the severity of those pathological remodeling in the heart is also gender dependent. Supporting this statement, in an experimental study [76], authors demonstrated the marked differences between male and female rats in terms of pathological remodeling in the heart. Even in an early animal model study, Ricci et al. [77] examined how electrophysiological characteristics of four ionic currents such as the L-type Ca^{2+}-current, the transient outward K^+-current, the delayed rectifier K^+-current, and the Na^+/Ca^{2+} exchange current) were remodeled in isolated left ventricular myocytes from obese rats.

In order to demonstrate the contribution of pathological remodeling of ionic-currents in cardiac dysfunction under insulin resistance, we fed male rats with a high-carbohydrate diet and then to validate the induction of metabolic syndrome (MetS), we determined their body weights, serum leptin and triglyceride levels, and oxidative stress and antioxidant defense status. Those animals were overweight (but not obese) and there were marked increases in both systolic and diastolic pressures, and depressed contractile activity together with marked alterations in the ultrastructure of their left ventricle [78]. We also determined marked pathological remodeling in those animals at cellular level (in cardiomyocytes) such as Ca^{2+}–overload, at most due to leaky cardiac ryanodine receptors (RyR2) through alterations in the phosphorylation of Ca^{2+}-handling proteins [79]. There were also marked decrease in the basal cAMP level, increased protein expression levels of phosphodiesterases (i.e. PDE3, and PDE4) in those cardiomyocytes [80]. Interestingly, there was significantly shorten QT-interval (at least through a less cAMP-release) in the ECGs of 16-week MetS rats (mimicing an early syndrome), being correlated with a depressed cardiac output and an increased heart rate [81]. Furthermore, there was significantly prolonged QT-interval and depressed cardiac contractile function with a clear arrhythmogenic activity in 28–30 weeks MetS rats [82]. These two different types cardiac dysfunction point out the impact of body weight and the types of daily diet into mammalian health. These findings are also indicating the contribution of co-factors in the development of obesity as well as the importance of controlling obesity-related comorbidities into our healthy life. Supporting these statement, considering the close association with altered Ca^{2+}-homeostasis and cardiac arrhythmias, a recent article addressed the molecular changes of some proteins associated with both Ca^{2+}-handling and Ca^{2+}-signaling together with their potential as novel therapeutic targets for obesity [83].

The Role of Oxidative Stress in Obesity-Related Cardiovascular Dysfunction

Oxidative stress has occurred when a critical balance between the generation of ROS and the endogenous antioxidant capacity is disrupted as either systemic or cellular level in biological samples. In obesity and/or diabetes, there are many and different ROS sources, while several factors are affecting the excessive ROS production and thereby the development of organ/tissue/cellular level complications including heart. With a short and brief statement, excess ROS affects heart inducing dysfunction through several ways such as pathological remodeling in myocardial Ca^{2+}-handling. That remodeling mainly in turns can underline underlies the occurrence of arrhythmia and cardiac dysfunction, alone or together with remodeling in vasculature [84].

Excess ROS also regulates multiple vascular cell functions. For instance, it affects endothelial and smooth muscle cell growth, proliferation, and migration, angiogenesis, apoptosis, vascular tone, and genomic stability, further promoting vascular diseases [85–89]. Increased ROS production associated redox-imbalance in cells including cardiomyocytes occurs under many pathological conditions, including diabetes, obesity and MetS. These disorders have cross-correlations and intersections with a positive feedback loop associated with oxidative stress, leading to the development of end-organ damage such as heart failure (Fig. 17.6). Mechanisms of redox-induced regulation of cardiac function in obesity include several signaling pathways besides increased oxidative stress as well as others correlated with increased oxidative stress in organelles. In that regard, a generation of mitochondrial ROS (mitoROS) is closely associated with a depressed ATP level as well as an opening of mitochondrial ATP-sensitive K^+-channels via an increase in cytosolic Ca^{2+}, which further leads to cellular dysfunction, at most, associated with increased oxidative stress in cells [90], including also cardiomyocytes [91]. Therefore, discovery of the novel oxidative stress signaling pathways will have impacts on new therapeutic approaches with novel chemical agents for the prevention and/or diagnosis of ROS-related diseases in humans, particularly in obese patients.

Although there are several biomarkers for obesity-related oxidative stress diagnosis, the serum levels of malondialdehyde (MDA) and thiobarbituric acid reactive substances (TBARS) are well-known two parameters in obese individuals [92], while they are mostly related to excess ROS production from accumulated adipose tissue in the body [93]. Of note, studies emphasized an adipocyte mitochondrial dysfunction as one important way to increase ROS production in obesity [93]. There are also several dysregulated metabolic parameters and complications in obesity and

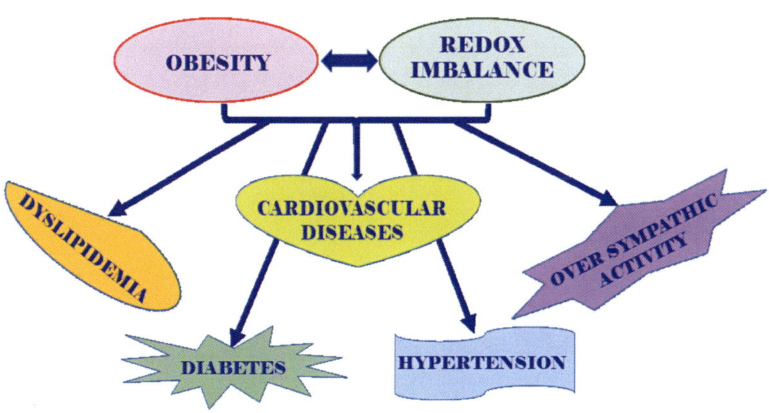

Fig. 17.6 Association between redox-imbalance and obesity in mammalians. Excessive production of ROS (i.e. increased oxidative stress) plays an important role not only in the development of cardiovascular disorders but also development of several pathological conditions which further also underlie the important risks for cardiovascular diseases, including hyperglycemia, hypertension, dyslipidemia, and over sympathetic activity in mammalians

they highly contribute to the amplification of oxidative stress. For note, taking into consideration the leptin level alteration in obesity, for instance, an increased level of leptin can promote inflammation and lipid peroxidation in biological cells. Consequently, dysregulation of all these metabolic parameters can contribute to excess ROS production leading to increases in oxidative stress in obesity [92].

New Therapeutic Agents in Obesity-Related Cardiovascular Disorders

Dysregulation of both ROS and RNS levels and/or productions has an important impact in the pathogenesis of most diseases, including cardiac dysfunction. Those pathological stimuli trigger further a cascade of pathological events in cells although some studies proposed those stimuli as the group of cardioprotective signaling molecules, mainly in pre- and post-conditioning processes. Interestingly, some early and recent studies, particularly obesity studies in animals, highlighted an important anti-obesity and anti-diabetic effect of some selective beta-3 adrenergic receptor (β_3-AR) agonists [94–98]. Indeed, some studies summarized the beneficial effects of some β_3-AR agonists for the treatment of diabetes and obesity as increase lipolysis, fat oxidation, energy expenditure, and insulin action. These observations further lead to intensive research efforts directed at developing novel and more selective β_3-AR agonists especially for the treatment of T2DM and obesity in humans. However, in the line of above results, authors treated either obese or non-insulin-dependent diabetic rats with a β_3-AR agonist and demonstrated its benefits on stimulation of brown fat thermogenesis [95, 99]. These agonists particularly are in usage to overcome the activated lipid metabolism as well as to increase thermogenesis and metabolic rate in the brown adipose tissue, and thereby to reduce the body weight in obese samples. Unfortunately, there were not enough and successful outcomes from human studies to date.

To clarify the roles of β_3-AR activation in myocardial contractility, we performed some cell level studies in isolated rat ventricular cardiomyocytes. Both direct activation of β_3-AR with its agonist and overexpression of β_3-AR have shown emphasized the important deleterious effect of β_3-AR activation in cardiac remodeling under hyperglycemia [100]. In addition, our data demonstrated the association between β_3-AR activation and NO-signaling via increases in cytosolic free Zn^{2+} level while providing significant recovery in ER-stress. Therefore, how long and to which level the β_3-AR agonism would be friend or become foe remains to be a mystery.

Interestingly, studies pointed out the unexpected effect of anti-hyperglycemic drugs such as inducing weight gain and accompany to increase risks for heart failure [101]. This important finding seems to be related to the certain action of glucose-lowering drugs to promote the accumulation of epicardial fat. In clinical trial studies, authors have shown that thiazolidinediones reduced epicardial adipose tissue volume and inflammation and the secretion of proinflammatory adipocytokines [102], and thereby contributing to reduce the risks for myocardial infarction [103]. However, some clinical trials demonstrated their anti-natriuretic action which leads to increases in the risks for heart failure. Therefore, to treat the obese patients with those drugs should have some caution. In those regards, our animal experiments on the treatment of MetS rats with thiazolidinediones (such as pioglitazone) provided slight but significant reduction in high blood glucose level and body weight as well as marked cardioprotection through inhibiting the increased Na^+-influx in the ventricular cardiomyocytes [104]. Interestingly, dipeptidyl peptidase-4 inhibitors also reduce the accumulation of epicardial fat, and provided cardioprotection through leading decline in cardiac fibrosis [55, 105, 106]. Similar to those studies, we treated MetS rats with dipeptidyl peptidase-4 inhibitor sitagliptin and demonstrated its significant cardioprotective effect due to recovery in vascular dysfunction through its role on epigenetic regulation [107]. Another newest drug, glucagon-like peptide 1 (GLP-1) receptor antagonist have reduced the accumulation of epicardial adipose tissue in diabetic and obese patients [108], however, they could not be able to ameliorate its anti-inflammatory properties [109]. Accordingly, these drugs have not be able specifically to reduce the incidence of serious heart failure in clinical trials [110]. In later studies, authors mentioned that the cardiovascular safety and efficacy of GLP-1 agonists in diabetic patients with high cardiovascular risk and those antagonists also provided benefits on the improvement of cardiovascular outcomes of these patients [111, 112]. Furthermore, Sassoon et al. [113] examined the cardiovascular effects of GLP-1 agonists in animals with obesity or MetS as well as humans. In another study, authors also have shown the beneficial effects of GLP-1 activation such as augmentation in cardiac output and improvement in cardiac efficiency after myocardial infarction [114].

Even early studies have shown that there is a close association between obesity, hemodynamic overload, and atrial and ventricular remodeling, while those have a high risk for cardiovascular diseases besides other complications [115, 116]. Interestingly, epidemiological studies reported that there was an association between higher mortality rates and uncontrolled glycemic control, primarily through the use of insulin, in patients with both T2DM and heart failure [116, 117]. Nevertheless, few studies reported the beneficial effects of a non-diabetic drug sodium-glucose cotransporter-2, SGLT2, inhibitor in diabetics with chronic heart failure. The SGLT2-selective inhibitors currently under clinical investigations for use as an anti-diabetic agent and also confirmed by studies with diabetic animals [118]. Human study with an SGLT2 inhibitor was performed by Zinman and co-workers, who demonstrated its beneficial effects on cardiovascular morbidity and mortality in diabetic patients with high cardiovascular risk (EMPA-REG OUTCOME Trial; 119]. However, the mechanisms of that SGLT2 inhibition improves cardiovascular outcomes are not fully understood. Accordingly, we examined the molecular mechanism of the beneficial effect of a SGLT2 inhibitor, dapagliflozin, on cardiac dysfunction in insulin-resistant and overweight MetS rats. Our data focused on its effect in mitochondrial function in the hearts of those animals, and also demonstrated its marked cardioprotective action through augmentation of cellular oxidative stress and in a manner of insulin-independent ways [82].

Since more than eighty percent of even highly motivated patients are unable to achieve weight loss with dietary and lifestyle modifications alone, they prefer to use some drugs and/or other agents. Sibutramine is a norepinephrine and serotonin reuptake inhibitor and approved for weight management in patients. Sibutramine induces an increase in energy expenditure with an impact on satiety [120, 121]. In addition in some patients, sibutramine increases blood pressure, via its sympathomimetic effects [122]. In line with those experimental studies, a trial named as The Sibutramine Cardiovascular Outcomes (SCOUT) evaluated the long-term effects of sibutramine treatment combined with diet and exercise on the rates of cardiovascular events with high cardiovascular risk [123]. In another study, its was documented the cardiovascular risk–benefit profile of sibutramine through the clinical outcomes [124]. Therefore, there is yet a contraindication of the use of this antiobesity agent in patients. To understand whether sibutramine has benefits or risks for the heart, we performed some in vitro experiments in isolated ventricular cardiomyocytes from overweight and insulin-resistant MetS rats. Our unpublished observations, performed in isolated ventricular cardiomyocytes, have revealed that acute sibutramine treatment provided significant prolongations in action potential duration, indicating a pro-arrhythmic action in a concentration-dependent manner, at most, due to its reduction action on voltage-dependent K^+-channel currents. In conclusion, the safety profile of sibutramine seems to be not clear yet particularly in cardiovascular outcomes, and the drug should not be prescribed for overweight/obese patients with a high cardiovascular risk profile.

Concluding Remarks

As discussed in previous sections, the numbers of people having overweightness and obesity are increasing worldwide, and already knowns (from both experimental data and clinical observations) documented that those conditions are mostly depending on excess or abnormal fat accumulation into the body, further leading to increases in risk factors for disturbances in our health. The amount of body fat is generally estimated with the body mass index (BMI). Obesity is strongly linked with several chronic diseases such as diabetes, hypertension, cardiovascular diseases, as well as, mood disorders and physical disabilities.

Recently, there is a wide discussion on the topic of the obesity paradox (an inverse correlation between the risk of obesity-related disease and increased survival in obese patients) although obesity is associated with illness status and appears to be associated with lower mortality but increases the risk of complications in several organ systems [6, 125]. However, more recent data in hospitalized patients or patients with chronic illnesses showed a relationship between BMI and mortality, with overweight and moderate obesity being associated with lower mortality compared with a normal BMI or more severe obesity. However, it is yet under investigation whether the obesity survival paradox represents a real protective effect of adipose tissue [126]. Indeed, in some obese individuals, the high BMI may be related to an increased muscle mass or they may have a more advantageous fat distribution that is not associated with metabolic co-morbidities. Also, adipose tissue may function as a fuel source and provide energy and lipid-soluble nutrients during highly catabolic states [127, 128]. In summary, clinical observations suggest that a metabolically benign adipose tissue phenotype exists and may explain the paradoxically low risk of certain diseases in some but not all obese individuals.

Conflicts of Interest:

The authors declare no conflict of interest.

Acknowledgements The data presented here is supported through a grant by TUBITAK SBAG-216S979 & SBAG-119S661.

References

1. Mokdad AH, Bowman BA, Ford ES, Vinicor F, Marks JS, Koplan JP (2001) The continuing epidemics of obesity and diabetes in the United States. JAMA 286(10):1195–1200
2. Daniels SR (2006) The consequences of childhood overweight and obesity. Future Child 16(1):47–67
3. González-Muniesa P, Mártinez-González MA, Hu FB, Després JP, Matsuzawa Y, Loos RJF et al (2017) Obesity. Nat Rev Dis Primers 3:17034
4. Longo M, Zatterale F, Naderi J, Parrillo L, Formisano P, Raciti GA, et al. (2019) Adipose tissue dysfunction as determinant of obesity-associated metabolic complications. Int J Mol Sci 20(9)
5. Riveros-McKay F, Mistry V, Bounds R, Hendricks A, Keogh JM, Thomas H, et al. (2019) Genetic architecture of human thinness compared to severe obesity. PLoS Genetics. 15(1):e1007603
6. Schetz M, De Jong A, Deane AM, Druml W, Hemelaar P, Pelosi P, et al. (2019) Obesity in the critically ill: a narrative review. Intensive Care Med 2019:1–13
7. Da Costa LA, Arora P, García-Bailo B, Karmali M, El-Sohemy A, Badawi A (2012) The association between obesity, cardiometabolic disease biomarkers, and innate immunity-related inflammation in Canadian adults. Diabetes Metab Syndr Obes. 5:347–355
8. Martyn JA, Kaneki M, Yasuhara S (2008) Obesity-induced insulin resistance and hyperglycemia: etiologic factors and molecular mechanisms. Anesthesiology 109(1):137–148
9. Carroll JF, Zenebe WJ, Strange TB (2006) Cardiovascular function in a rat model of diet-induced obesity. Hypertension 48(1):65–72
10. James PT, Rigby N, Leach R (2004) The obesity epidemic, metabolic syndrome and future prevention strategies. Eur J Cardiovasc Prev Rehabil 11(1):3–8
11. Haffner S, Taegtmeyer H (2003) Epidemic obesity and the metabolic syndrome. Circulation 108(13):1541–1545
12. Modan M, Halkin H, Almog S, Lusky A, Eshkol A, Shefi M, et al. (1985) Hyperinsulinemia. A link between hypertension obesity and glucose intolerance. J Clin Invest 75(3):809–817
13. Petrakis D, Margină D, Tsarouhas K, Tekos F, Stan M, Nikitovic D et al (2020) Obesity—a risk factor for increased COVID-19 prevalence, severity and lethality (Review). Mol Med Rep 22(1):9–19

14. Vallianou NG, Geladari E, Kazazis CE (2017) SGLT-2 inhibitors: their pleiotropic properties. Diabetes Metab Syndr 11(4):311–315
15. Pucci A, Finer N (2015) New medications for treatment of obesity: metabolic and cardiovascular effects. Can J Cardiol 31(2):142–152
16. Heal DJ, Gosden J, Smith SL (2009) Regulatory challenges for new drugs to treat obesity and comorbid metabolic disorders. Br J Clin Pharmacol 68(6):861–874
17. May M, Schindler C, Engeli S (2020) Modern pharmacological treatment of obese patients. Ther Adv Endocrinol Metab. 11:2042018819897527
18. Srivastava G, Apovian CM (2018) Current pharmacotherapy for obesity. Nat Rev Endocrinol 14(1):12–24
19. Hausenloy DJ, Garcia-Dorado D, Bøtker HE, Davidson SM, Downey J, Engel FB et al (2017) Novel targets and future strategies for acute cardioprotection: position paper of the European society of cardiology working group on cellular biology of the heart. Cardiovasc Res 113(6):564–585
20. Mancini MC, Halpern A (2006) Investigational therapies in the treatment of obesity. Expert Opin Investig Drugs 15(8):897–915
21. Gregg EW, Shaw JE (2017) Global health effects of overweight and obesity. N Engl J Med 377(1):80–81
22. Fruhwürth S, Vogel H, Schürmann A, Williams KJ (2018) Novel insights into how overnutrition disrupts the hypothalamic actions of Leptin. Front Endocrinol (Lausanne). 9:89
23. Stewart ST, Cutler DM, Rosen AB (2009) Forecasting the effects of obesity and smoking on US life expectancy. N Engl J Med 361(23):2252–2260
24. Guh DP, Zhang W, Bansback N, Amarsi Z, Birmingham CL, Anis AH (2009) The incidence of co-morbidities related to obesity and overweight: a systematic review and meta-analysis. BMC Public Health 9:88
25. Howard BV, Ruotolo G, Robbins DC (2003) Obesity and dyslipidemia. Endocrinol Metab Clin North Am 32(4):855–867
26. Bray GA, Clearfield MB, Fintel DJ, Nelinson DS (2009) Overweight and obesity: the pathogenesis of cardiometabolic risk. Clin Cornerstone 9(4):30–40; discussion 1–2
27. Czech MP (2020) Mechanisms of insulin resistance related to white, beige, and brown adipocytes. Mol Metab 34:27–42
28. Pellegrinelli V, Carobbio S, Vidal-Puig A (2016) Adipose tissue plasticity: how fat depots respond differently to pathophysiological cues. Diabetologia 59(6):1075–1088
29. Van Gaal LF, Mertens IL, De Block CE (2006) Mechanisms linking obesity with cardiovascular disease. Nature 444(7121):875–880
30. Hotamisligil GS (2006) Inflammation and metabolic disorders. Nature 444(7121):860–867
31. Lau DC, Dhillon B, Yan H, Szmitko PE, Verma S (2005) Adipokines: molecular links between obesity and atheroslcerosis. Am J Physiol Heart Circ Physiol 288(5):H2031–H2041
32. Poirier P, Giles TD, Bray GA, Hong Y, Stern JS, Pi-Sunyer FX et al (2006) Obesity and cardiovascular disease: pathophysiology, evaluation, and effect of weight loss: an update of the 1997 American heart association scientific statement on obesity and heart disease from the obesity committee of the council on nutrition, physical activity, and metabolism. Circulation 113(6):898–918
33. Danesh J, Collins R, Appleby P, Peto R (1998) Association of fibrinogen, C-reactive protein, albumin, or leukocyte count with coronary heart disease: meta-analyses of prospective studies. JAMA 279(18):1477–1482
34. Ridker PM (2007) C-reactive protein and the prediction of cardiovascular events among those at intermediate risk: moving an inflammatory hypothesis toward consensus. J Am Coll Cardiol 49(21):2129–2138
35. Csige I, Ujvárosy D, Szabó Z, Lőrincz I, Paragh G, Harangi M et al (2018) The impact of obesity on the cardiovascular system. J Diabetes Res 2018:3407306
36. Camastra S, Vitali A, Anselmino M, Gastaldelli A, Bellini R, Berta R et al (2017) Muscle and adipose tissue morphology, insulin sensitivity and beta-cell function in diabetic and nondiabetic obese patients: effects of bariatric surgery. Sci Rep 7(1):9007

37. Mraz M, Haluzik M (2014) The role of adipose tissue immune cells in obesity and low-grade inflammation. J Endocrinol 222(3):R113–R127
38. Fantuzzi G (2005) Adipose tissue, adipokines, and inflammation. J Allergy Clin Immunol 115(5):911–919; quiz 20
39. McGuire TR, Brusnahan SK, Bilek LD, Jackson JD, Kessinger MA, Berger AM et al (2011) Inflammation associated with obesity: relationship with blood and bone marrow endothelial cells. Obesity (Silver Spring) 19(11):2130–2136
40. Singh RK, Kumar P, Mahalingam K (2017) Molecular genetics of human obesity: a comprehensive review. C R Biol 340(2):87–108
41. Włodarczyk M, Jabłonowska-Lietz B, Olejarz W, Nowicka G (2018) Anthropometric and dietary factors as predictors of DNA damage in obese women. Nutrients 10(5)
42. Cerdá C, Sánchez C, Climent B, Vázquez A, Iradi A, El Amrani F et al (2014) Oxidative stress and DNA damage in obesity-related tumorigenesis. Adv Exp Med Biol 824:5–17
43. Włodarczyk M, Nowicka G (2019) Obesity, DNA damage, and development of obesity-related diseases. Int J Mol Sci 20(5)
44. Zierath JR (2019) Major advances and discoveries in diabetes—2019 in review. Curr Diab Rep 19(11):118
45. Chaix A, Lin T, Le HD, Chang MW, Panda S (2019) Time-restricted feeding prevents obesity and metabolic syndrome in mice lacking a circadian clock. Cell Metab 29(2):303–19.e4
46. Parsanathan R, Jain SK (2020) Novel invasive and noninvasive cardiac-specific biomarkers in obesity and cardiovascular diseases. Metab Syndr Relat Disord 18(1):10–30
47. Packer M (2018) Epicardial adipose tissue may mediate deleterious effects of obesity and inflammation on the myocardium. J Am Coll Cardiol 71(20):2360–2372
48. Hirayama A, Goto T, Shimada YJ, Faridi MK, Camargo CA, Jr., Hasegawa K (2018) Association of obesity with severity of heart failure exacerbation: a population-based study. J Am Heart Assoc 7(6)
49. Oga EA, Eseyin OR (2016) The obesity paradox and heart failure: a systematic review of a decade of evidence. J Obes 2016:9040248
50. Gustafsson F, Kragelund CB, Torp-Pedersen C, Seibaek M, Burchardt H, Akkan D et al (2005) Effect of obesity and being overweight on long-term mortality in congestive heart failure: influence of left ventricular systolic function. Eur Heart J 26(1):58–64
51. Global, regional, and national incidence, prevalence, and years lived with disability for 328 diseases and injuries for 195 countries, 1990–2016: a systematic analysis for the Global Burden of Disease Study 2016 (2017). Lancet 390(10100):1211–1259
52. Srikanthan K, Shapiro JI, Sodhi K (2016) The role of Na/K-ATPase signaling in oxidative stress related to obesity and cardiovascular disease. Molecules 21(9)
53. Piché ME, Poirier P, Lemieux I, Després JP (2018) Overview of epidemiology and contribution of obesity and body fat distribution to cardiovascular disease: an update. Prog Cardiovasc Dis 61(2):103–113
54. Kwiterovich PO Jr (2002) Clinical relevance of the biochemical, metabolic, and genetic factors that influence low-density lipoprotein heterogeneity. Am J Cardiol 90(8a):30i–47i
55. Alpert MA, Terry BE, Mulekar M, Cohen MV, Massey CV, Fan TM et al (1997) Cardiac morphology and left ventricular function in normotensive morbidly obese patients with and without congestive heart failure, and effect of weight loss. Am J Cardiol 80(6):736–740
56. Zhou YT, Grayburn P, Karim A, Shimabukuro M, Higa M, Baetens D et al (2000) Lipotoxic heart disease in obese rats: implications for human obesity. Proc Natl Acad Sci USA 97(4):1784–1789
57. Young ME, Guthrie PH, Razeghi P, Leighton B, Abbasi S, Patil S et al (2002) Impaired long-chain fatty acid oxidation and contractile dysfunction in the obese Zucker rat heart. Diabetes 51(8):2587–2595
58. Fernandes CR, Kannen V, Mata KM, Frajacomo FT, Jordão Junior AA, Gasparotto B et al (2017) High-fat and fat-enriched diets impair the benefits of moderate physical training in the aorta and the heart in rats. Front Nutr 4:21

59. Grasa-López A, Miliar-García Á, Quevedo-Corona L, Paniagua-Castro N, Escalona-Cardoso G, Reyes-Maldonado E et al (2016) Undaria pinnatifida and fucoxanthin ameliorate lipogenesis and markers of both inflammation and cardiovascular dysfunction in an animal model of diet-induced obesity. Mar Drugs 14(8):148
60. Popkin BM, Doak CM (1998) The obesity epidemic is a worldwide phenomenon. Nutr Rev 56(4 Pt 1):106–114
61. Kitzman DW, Shah SJ (2016) The HFpEF obesity phenotype: the elephant in the room. J Am Coll Cardiol 68(2):200–203
62. Obokata M, Reddy YNV, Pislaru SV, Melenovsky V, Borlaug BA (2017) Evidence supporting the existence of a distinct obese phenotype of heart failure with preserved ejection fraction. Circulation 136(1):6–19
63. Cavalera M, Wang J, Frangogiannis NG (2014) Obesity, metabolic dysfunction, and cardiac fibrosis: pathophysiological pathways, molecular mechanisms, and therapeutic opportunities. Transl Res 164(4):323–335
64. Mohammed SF, Hussain S, Mirzoyev SA, Edwards WD, Maleszewski JJ, Redfield MM (2015) Coronary microvascular rarefaction and myocardial fibrosis in heart failure with preserved ejection fraction. Circulation 131(6):550–559
65. Su M-YM, Lin L-Y, Tseng Y-HE, Chang C-C, Wu C-K, Lin J-L, et al. (2014) CMR-verified diffuse myocardial fibrosis is associated with diastolic dysfunction in HFpEF. JACC: Cardiovascular Imaging 7(10):991–997
66. Jensen MD, Ryan DH, Apovian CM, Ard JD, Comuzzie AG, Donato KA, et al. (2014) 2013 AHA/ACC/TOS guideline for the management of overweight and obesity in adults: a report of the American college of cardiology/American heart association task force on practice guidelines and the obesity society. J Am Coll Cardiol 63(25 Pt B):2985–3023
67. May AL, Kuklina EV, Yoon PW (2012) Prevalence of cardiovascular disease risk factors among US adolescents, 1999–2008. Pediatrics 129(6):1035–1041
68. Kannel WB, Plehn JF, Cupples LA (1988) Cardiac failure and sudden death in the Framingham study. Am Heart J 115(4):869–875
69. Mutiso SK, Rono DK, Bukachi F (2014) Relationship between anthropometric measures and early electrocardiographic changes in obese rats. BMC Res Notes 7(1):931
70. Birse RT, Bodmer R (2011) Lipotoxicity and cardiac dysfunction in mammals and Drosophila. Crit Rev Biochem Mol Biol 46(5):376–385
71. Galinier M, Pathak A, Roncalli J, Massabuau P (2005) Obesity and cardiac failure. Arch Mal Coeur Vaiss 98(1):39
72. Yılmaz M, Altın C, Tekin A, Erol T, Arer İ, Nursal TZ et al (2018) Assessment of atrial fibrillation and ventricular arrhythmia risk after bariatric surgery by P wave/QT interval dispersion. Obes Surg 28(4):932–938
73. Tian M, Dong MQ, Chiu SW, Lau CP, Li GR (2006) Effects of the antifungal antibiotic clotrimazole on human cardiac repolarization potassium currents. Br J Pharmacol 147(3):289–297
74. Aromolaran AS, Boutjdir M (2017) Cardiac ion channel regulation in obesity and the metabolic syndrome: relevance to long QT syndrome and atrial fibrillation. Front Physiol 8:431
75. Leopoldo AS, Sugizaki MM, Lima-Leopoldo AP, do Nascimento AF, Luvizotto RdAM, de Campos DHS, et al. (2010) Cardiac remodeling in a rat model of diet-induced obesity. Canadian J Cardiol 26(8):423–429
76. Nguyen IT, Brandt MM, van de Wouw J, van Drie RW, Wesseling M, Cramer MJ, et al. (2020) Both male and female obese ZSF1 rats develop cardiac dysfunction in obesity-induced heart failure with preserved ejection fraction. Plos One 15(5):e0232399
77. Ricci E, Smallwood S, Chouabe C, Mertani HC, Raccurt M, Morel G et al (2006) Electrophysiological characterization of left ventricular myocytes from obese sprague-dawley rat. Obesity 14(5):778–786
78. Okatan EN, Tuncay E, Hafez G, Turan B (2015) Profiling of cardiac β-adrenoceptor subtypes in the cardiac left ventricle of rats with metabolic syndrome: comparison with streptozotocin-induced diabetic rats. Can J Physiol Pharmacol 93(7):517–525

79. Okatan EN, Durak AT, Turan B (2016) Electrophysiological basis of metabolic-syndrome-induced cardiac dysfunction. Can J Physiol Pharmacol 94(10):1064–1073
80. Okatan EN, Turan B (2019) The contribution of phosphodiesterases to cardiac dysfunction in rats with metabolic syndrome induced by a high-carbohydrate diet. Can J Physiol Pharmacol 97(11):1064–1072
81. Durak A, Olgar Y, Tuncay E, Karaomerlioglu I, Kayki Mutlu G, Arioglu Inan E et al (2017) Onset of decreased heart work is correlated with increased heart rate and shortened QT interval in high-carbohydrate fed overweight rats. Can J Physiol Pharmacol 95(11):1335–1342
82. Durak A, Olgar Y, Degirmenci S, Akkus E, Tuncay E, Turan B (2018) A SGLT2 inhibitor dapagliflozin suppresses prolonged ventricular-repolarization through augmentation of mitochondrial function in insulin-resistant metabolic syndrome rats. Cardiovasc Diabetol 17(1):1–17
83. Njegic A, Wilson C, Cartwright EJ (2020) Targeting Ca2+ handling proteins for the treatment of heart failure and arrhythmias. Front Physiol 11
84. Münzel T, Camici GG, Maack C, Bonetti NR, Fuster V, Kovacic JC (2017) Impact of oxidative stress on the heart and vasculature: part 2 of a 3-part series. J Am Coll Cardiol 70(2):212–229
85. Wagner S, Rokita AG, Anderson ME, Maier LS (2013) Redox regulation of sodium and calcium handling. Antioxid Redox Signal 18(9):1063–1077
86. Burgoyne JR, Mongue-Din H, Eaton P, Shah AM (2012) Redox signaling in cardiac physiology and pathology. Circ Res 111(8):1091–1106
87. Münzel T, Gori T, Bruno RM, Taddei S (2010) Is oxidative stress a therapeutic target in cardiovascular disease? Eur Heart J 31(22):2741–2748
88. Mollnau H, Oelze M, August M, Wendt M, Daiber A, Schulz E et al (2005) Mechanisms of increased vascular superoxide production in an experimental model of idiopathic dilated cardiomyopathy. Arterioscler Thromb Vasc Biol 25(12):2554–2559
89. Belch JJ, Bridges AB, Scott N, Chopra M (1991) Oxygen free radicals and congestive heart failure. Br Heart J 65(5):245–248
90. Liu J, Kennedy DJ, Yan Y, Shapiro JI (2012) Reactive oxygen species modulation of Na/K-ATPase regulates fibrosis and renal proximal tubular sodium handling. Int J Nephrol 2012:381320
91. Weigand KM, Swarts HG, Fedosova NU, Russel FG, Koenderink JB (2012) Na, K-ATPase activity modulates Src activation: a role for ATP/ADP ratio. Biochim Biophys Acta 1818(5):1269–1273
92. Vincent HK, Taylor AG (2006) Biomarkers and potential mechanisms of obesity-induced oxidant stress in humans. Int J Obes (Lond) 30(3):400–418
93. Le Lay S, Simard G, Martinez MC, Andriantsitohaina R (2014) Oxidative stress and metabolic pathologies: from an adipocentric point of view. Oxid Med Cell Longev 2014:908539
94. Chen Y, Wang X, Shen Z, Fan P, Liu R, Liu Y et al (2015) Effect of the beta-3 adrenergic receptor Trp64Arg and uncoupling protein 1–3826 A>G genotypes on lipid and apolipoprotein levels in overweight/obese and non-obese Chinese subjects. Lipids Health Dis 14:34
95. Cypess AM, Weiner LS, Roberts-Toler C, Elía EF, Kessler SH, Kahn PA et al (2015) Activation of human brown adipose tissue by a β3-adrenergic receptor agonist. Cell Metab 21(1):33–38
96. de Souza CJ, Burkey BF (2001) Beta3-adrenoceptor agonists as anti-diabetic and anti-obesity drugs in humans. Curr Pharm Des 7(14):1433–1449
97. Sabri A, Hughie HH, Lucchesi PA (2003) Regulation of hypertrophic and apoptotic signaling pathways by reactive oxygen species in cardiac myocytes. Antioxid Redox Signal 5(6):731–740
98. Yoshida T, Umekawa T, Sakane N, Yoshimoto K, Kondo M (1996) Effect of CL316, 243, a highly specific beta3-adrenoceptor agonist, on sympathetic nervous system activity in mice. Metabolism 45(6):787–791
99. Lowell BB, Flier JS (1997) Brown adipose tissue, beta 3-adrenergic receptors, and obesity. Annu Rev Med 48:307–316
100. Tuncay E, Olgar Y, Durak A, Degirmenci S, Bitirim CV, Turan B (2019) β3-adrenergic receptor activation plays an important role in the depressed myocardial contractility via both elevated levels of cellular free Zn2+ and reactive nitrogen species. J Cell Physiol 234(8):13370–13386

101. Udell JA, Cavender MA, Bhatt DL, Chatterjee S, Farkouh ME, Scirica BM (2015) Glucose-lowering drugs or strategies and cardiovascular outcomes in patients with or at risk for type 2 diabetes: a meta-analysis of randomised controlled trials. Lancet Diabetes Endocrinol 3(5):356–366
102. Grosso AF, de Oliveira SF, Higuchi Mde L, Favarato D, Dallan LA, da Luz PL (2014) Synergistic anti-inflammatory effect: simvastatin and pioglitazone reduce inflammatory markers of plasma and epicardial adipose tissue of coronary patients with metabolic syndrome. Diabetol Metab Syndr 6(1):47
103. Kernan WN, Viscoli CM, Furie KL, Young LH, Inzucchi SE, Gorman M et al (2016) Pioglitazone after ischemic stroke or transient ischemic attack. N Engl J Med 374(14):1321–1331
104. Bilginoglu A, Selcuk MFT, Nakkas H, Turan B (2018) Pioglitazone provides beneficial effect in metabolic syndrome rats via affecting intracellular Na(+) Dyshomeostasis. J Bioenerg Biomembr 50(6):437–445
105. Jackson EK, Zhang Y, Gillespie DD, Zhu X, Cheng D, Jackson TC (2017) SDF-1α (Stromal Cell-Derived Factor 1α) Induces cardiac fibroblasts, renal microvascular smooth muscle cells, and glomerular mesangial cells to proliferate, cause hypertrophy, and produce collagen. J Am Heart Assoc 6(11)
106. Lima-Martínez MM, Paoli M, Rodney M, Balladares N, Contreras M, D'Marco L et al (2016) Effect of sitagliptin on epicardial fat thickness in subjects with type 2 diabetes and obesity: a pilot study. Endocrine 51(3):448–455
107. Cicek FA, Tokcaer-Keskin Z, Ozcinar E, Bozkus Y, Akcali KC, Turan B (2014) Di-peptidyl peptidase-4 inhibitor sitagliptin protects vascular function in metabolic syndrome: possible role of epigenetic regulation. Mol Biol Rep 41(8):4853–4863
108. Dutour A, Abdesselam I, Ancel P, Kober F, Mrad G, Darmon P et al (2016) Exenatide decreases liver fat content and epicardial adipose tissue in patients with obesity and type 2 diabetes: a prospective randomized clinical trial using magnetic resonance imaging and spectroscopy. Diabetes Obes Metab 18(9):882–891
109. Pastel E, McCulloch LJ, Ward R, Joshi S, Gooding KM, Shore AC et al (2017) GLP-1 analogue-induced weight loss does not improve obesity-induced AT dysfunction. Clin Sci (Lond) 131(5):343–353
110. Marso SP, Daniels GH, Brown-Frandsen K, Kristensen P, Mann JF, Nauck MA et al (2016) Liraglutide and cardiovascular outcomes in type 2 diabetes. N Engl J Med 375(4):311–322
111. Bahtiyar G, Pujals-Kury J, Sacerdote A (2018) Cardiovascular effects of different GLP-1 receptor agonists in patients with type 2 diabetes. Curr Diab Rep 18(10):92
112. Ang R, Mastitskaya S, Hosford PS, Basalay M, Specterman M, Aziz Q, et al. (2018) Modulation of cardiac ventricular excitability by GLP-1 (Glucagon-Like Peptide-1). Circ Arrhythm Electrophysiol 11(10):e006740
113. Sassoon DJ, Tune JD, Mather KJ, Noblet JN, Eagleson MA, Conteh AM et al (2017) Glucagon-like peptide 1 receptor activation augments cardiac output and improves cardiac efficiency in obese swine after myocardial infarction. Diabetes 66(8):2230–2240
114. Jones B, Buenaventura T, Kanda N, Chabosseau P, Owen BM, Scott R et al (2018) Targeting GLP-1 receptor trafficking to improve agonist efficacy. Nat Commun 9(1):1602
115. Gottdiener JS, Reda DJ, Massie BM, Materson BJ, Williams DW, Anderson RJ (1997) Effect of single-drug therapy on reduction of left ventricular mass in mild to moderate hypertension: comparison of six antihypertensive agents. The Department of Veterans Affairs Cooperative Study Group on Antihypertensive Agents. Circulation 95(8):2007–2014
116. Lauer MS, Anderson KM, Levy D (1992) Separate and joint influences of obesity and mild hypertension on left ventricular mass and geometry: the Framingham heart study. J Am Coll Cardiol 19(1):130–134
117. Khalaf KI, Taegtmeyer H (2012) After avandia: the use of antidiabetic drugs in patients with heart failure. Tex Heart Inst J 39(2):174–178
118. Han S, Hagan DL, Taylor JR, Xin L, Meng W, Biller SA et al (2008) Dapagliflozin, a selective SGLT2 inhibitor, improves glucose homeostasis in normal and diabetic rats. Diabetes 57(6):1723–1729

119. Zinman B, Wanner C, Lachin JM, Fitchett D, Bluhmki E, Hantel S et al (2015) Empagliflozin, cardiovascular outcomes, and mortality in type 2 diabetes. N Engl J Med 373(22):2117–2128
120. Lean M, Finer N (2006) ABC of obesity. Management: part II--drugs. Bmj 333(7572):794–797
121. Halford JC, Boyland EJ, Cooper SJ, Dovey TM, Huda MS, Dourish CT et al (2010) The effects of sibutramine on the microstructure of eating behaviour and energy expenditure in obese women. J Psychopharmacol 24(1):99–109
122. Florentin M, Liberopoulos EN, Elisaf MS (2008) Sibutramine-associated adverse effects: a practical guide for its safe use. Obes Rev 9(4):378–387
123. James WP, Caterson ID, Coutinho W, Finer N, Van Gaal LF, Maggioni AP et al (2010) Effect of sibutramine on cardiovascular outcomes in overweight and obese subjects. N Engl J Med 363(10):905–917
124. Scheen AJ (2010) Cardiovascular risk-benefit profile of sibutramine. Am J Cardiovasc Drugs 10(5):321–334
125. Antonopoulos AS, Tousoulis D (2017) The molecular mechanisms of obesity paradox. Cardiovasc Res 113(9):1074–1086
126. Hogue CW Jr, Stearns JD, Colantuoni E, Robinson KA, Stierer T, Mitter N et al (2009) The impact of obesity on outcomes after critical illness: a meta-analysis. Intensive Care Med 35(7):1152–1170
127. Marques MB, Langouche L (2013) Endocrine, metabolic, and morphologic alterations of adipose tissue during critical illness. Crit Care Med 41(1):317–325
128. Alipoor E, Mohammad Hosseinzadeh F, Hosseinzadeh-Attar MJ (2018) Adipokines in critical illness: a review of the evidence and knowledge gaps. Biomed Pharmacother 108:1739–1750

Chapter 18
Role of the Synchronization of Circadian Clock by Meal-Timing in Obesity and Type 2 Diabetes

Daniela Jakubowicz, Shani Tsameret, Zohar Landau, and Julio Wainstein

Abstract Obesity and diabetes are increasing worldwide in epidemic proportion. Most alarmingly, is that in the last three decades, no country has successfully succeeded in reducing obesity and diabetes rates, therefore "unless the strategies for combating this epidemic are changed," both obesity and diabetes will increase exponentially in the years to come. Most of the metabolic processes involved in glucose and energy metabolism, i.e., β-cell secretory function, insulin sensitivity, muscular glucose uptake, and hepatic glucose production, display daily oscillation and are controlled by the circadian clock, to anticipate the recurring feeding-fasting cycles and to optimize metabolic efficiency in the appropriate temporal sequence. Growing evidence shows that meal timing not aligned with the light/dark cycle, like skipping breakfast, overeating at night, or snacking all day, including at hours assigned to sleep, lead to asynchrony and disruption of circadian clock gene expression and metabolic and appetite disturbances. It has been suggested, that the circadian misalignment and mistimed meals, typical of the modern society exposed to a 7/24 activity schedule, is the underlying cause of the vertiginous rise of obesity and T2D.

This review will focus on the recent studies reporting that meal timing aligned with the circadian clock, by shifting most calories and carbohydrates to the early hours of the day, through resetting the synchrony of the circadian clock gene expression, may improve glucose and energy metabolism and appetite regulation, resulting in more efficient weight loss, better glycemic control and reduced appetite, thereby preventing obesity and hyperglycemic relapse.

Keywords Circadian rhythms · Clock genes · Weight loss · Glycemic oscillations · Food-craving

---

Daniela Jakubowicz, Shani Tsameret and Julio Wainstein—equal contribution

---

D. Jakubowicz (✉) · Z. Landau · J. Wainstein
Diabetes Unit, Wolfson Medical Center, Sackler Faculty of Medicine, Tel Aviv University, Holon, Israel

S. Tsameret
Institute of Biochemistry, Food Science and Nutrition, The Robert H. Smith Faculty of Agriculture, Food and Environment, The Hebrew University of Jerusalem, Rehovot, Israel

© Springer Nature Switzerland AG 2021
P. S. Tappia et al. (eds.), *Cellular and Biochemical Mechanisms of Obesity*, Advances in Biochemistry in Health and Disease 23,
https://doi.org/10.1007/978-3-030-84763-0_18

Introduction

Obesity is increasing worldwide in epidemic proportion [1]. As it is was predictable, the obesity epidemic resulted in diabetes epidemic, with 9.3% of adults, almost 430 million diabetics worldwide [2]. Diabetes epidemic has further severe health implications, many life-threatening conditions, and increased morbidity and mortality [2]. Most alarmingly, in the last three decades, no country has successfully succeeded in reducing obesity and diabetes rates [1, 2]. Therefore, "unless the strategies for combating this epidemic are changed," both obesity and diabetes will increase exponentially in the years to come.

Meal Timing not Aligned with the Circadian Clock, Increase the Risk of Obesity and Type 2 Diabetes (T2D)

The prevalence of irregular eating patterns, "out of phase" with the external light/dark cycle, like eating at abnormal night-time hours among shift workers, skipping breakfast or snacking all day, have increased over the past decades in the Western society, in parallel with obesity and T2D epidemics [3–6]. Circadian misalignment in particular skipping breakfast and overeating in the evening are directly associated with weight gain, insulin resistance, increased risk for developing metabolic syndrome, obesity, T2D [3–6], and increased cardiometabolic risk [7]. Moreover, in T2D, the omission of breakfast is associated with significant increase in HbA_{1C} and all-day postprandial hyperglycemia, even without overeating in the evening, [6]. Further, it has been suggested that the disruption of the circadian clock, is essential in the pathophysiology of obesity and T2D [8].

Circadian Regulation of Metabolism

The circadian clock has approximately 24 h periodicity and plays a critical role in generating circadian rhythms. Like in the plants which obtain energy through photosynthesis during the daytime, and nocturnal animals forage for food at night, many living organisms, including humans, have developed an intrinsic circadian clock that enables the expression of appropriate metabolic processes, (i.e., β-cell secretion, muscular glucose uptake, hepatic glucose production) at the appropriate time of the day [8, 9]. The circadian clock anticipates the recurring feeding-fasting cycle and optimizes the metabolic efficiency in an appropriate temporal sequence [10–13]. The meal-timing pattern has a critical influence on the clock gene oscillatory expression. Meal timing disruption or misalignment within the circadian rhythms, like skipping breakfast or snacking at night, promotes asynchrony of clock gene expression and metabolic disturbances, i.e., weight gain, hyperglycemia [13]. In contrast, greater intake in the morning than in the evening has a resetting effect on clock gene oscillation [14–16], and is associated with beneficial effects on weight loss, glycemia, and appetite control, independent of total energy intake [15–26].

Central and Peripheral Clocks

The circadian rhythms are driven by the central or master clock, consisting of a group of clock genes localized in the suprachiasmatic nuclei (SCN) of the hypothalamus, and peripheral clock genes disseminated throughout the body [8–10, 27, 28]. The central clock in SCN is synchronized to the light inputs, and coordinate the peripheral clocks to the light/dark cycle, through neuronal pathways, behaviors such as sleep–wake cycle, feeding/fasting cycle, core body temperature, and hormonal rhythms, i.e., melatonin, ACTH, and cortisol [8–10] (Fig. 18.1).

Although the central clock coordinates the clock genes in the peripheral tissues (i.e., adipose tissue, liver, β-cells, a-cells, liver, gut, muscle) [29], they are mostly entrained by food cues, i.e., the daily time of food intake or food availability, allowing to anticipate the secretion of metabolic hormones and enzymes before food intake, at a specific time of the day [10, 11, 30–35]. Noteworthy is that the first meal of the

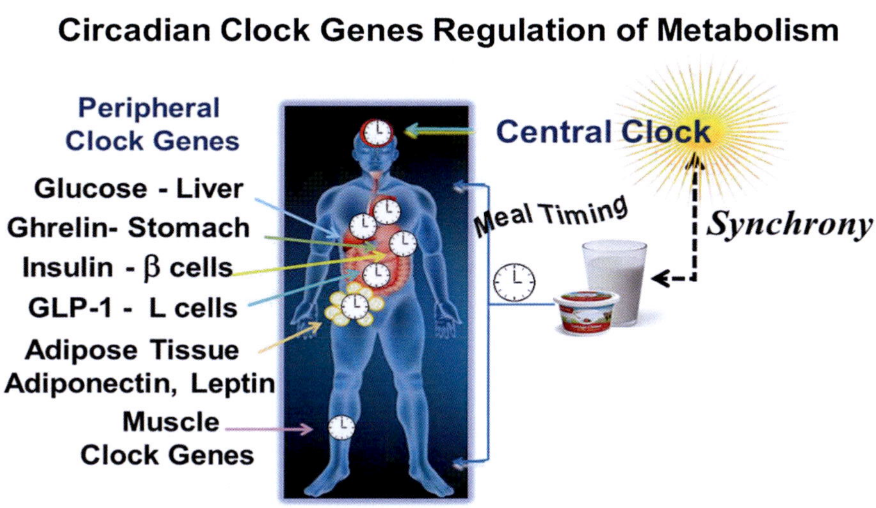

Fig. 18.1 Central and Peripheral Clock Genes. In this illustration is shown a central clock in the SCN, and some examples of peripheral clock disseminated throughout the body i.e., in the liver regulating the hepatic glucose output, in the stomach controlling the ghrelin secretion, and insulin secretion in β-cells, GLP-1 in the intestinal L- cells, leptin and adiponectin in the adipose tissue and the glucose uptake in the muscular cells. We may observe that the central clock is activated by the light (light/dark cycle), while the peripheral clocks are entrained by the time of food intake (eating/fasting cycle). Both, stimuli, the food intake, and the light should occur in synchrony

day, i.e., breakfast, exerts more powerful resetting effect on the clock network than other meals, underscoring the damage caused by the absence or delayed breakfast on the clock regulation of metabolism [15, 36–38].

Molecular Mechanism of the Circadian Clock-Driven Metabolism

The circadian clock molecular mechanism is identical in central and peripheral clocks—consist of self-sustained transcriptional feedback loops. The transcriptional activators, locomotor output cycles protein kaput (*CLOCK*), and the brain and muscle Arnt-like protein 1 (*BMAL1*), act as positive elements [28–30]. The CLOCK:*BMAL1* heterodimer, associated with *SIRT1* deacetylase, drives the transcription of five repressor encoding genes, three period genes (*PER1, PER2, PER3*), and two cryptochrome genes (*CRY1 and CRY2*). The resulting PER and CRY proteins interact to form PER:CRY dimers in the cytoplasm. These dimers are translocated back to the nucleus after ~24 h, to stop and repress their own *CLOCK:BMAL1* induced transcription, thus generating a cycle of transcription that recur every 24 h [9, 28, 30].

In a secondary regulatory loop, *CLOCK:BMAL1 heterodimer* mediates the transcription of the repressor *REV-ERBα* nuclear receptor and one promoter gene, the retinoic acid receptor-related orphan receptor (*RORα*), maintaining further the circadian (~24 h) oscillatory function of the clock, and many other clock-controlled genes, like PPARγ coactivator 1α (PGC-1α), AMPK and SIRT1 [9, 28].

In the peripheral tissues, the *CLOCK: BMAL1* driven transcription of *PERs, CRYs, REV-ERBs*, and *RORs* clock genes, along with *PGC-1α* and *SIRT1* and other transcriptional elements, promote downstream the expression of several proteins encoded by tissue-specific clock-controlled output genes. This relays the clock information to the cellular processes and circadian secretion and activity of most enzymes, hormones, and transport systems involved in glucose and lipid homeostasis. *BMAL-1, RORα*, and *SIRT1* positively regulate the circadian β-cells insulin secretion [31, 33–35], insulin sensitivity [39], muscular GLUT4 activity, and glucose uptake [32]. BMAL1 activity has been also associated with β-cell replicative capacity and survival [8].

BMAL-1 and RORα integrity is also necessary for circadian secretion of glucagon-like peptide-1 (GLP-1) in the intestinal L-cells [40], while, *BMAL-1, RORα*, and *PGC-1α* expressed in the adipose tissue, regulate the nocturnal lipolysis [9, 41]. The expression of *REVERBα* and *RORα, SIRT1* and *PGC-1α,* in the liver, modulates the circadian expression of the gluconeogenic enzyme phosphoenolpyruvate carboxykinase (PEPCK) [42], and the rhythms of hepatic glucose 6-phosphatase (HG6-P) in the glycolysis pathway [42, 43]. In addition, *BMAL1, CRY2, CRY1* and *PER2*, through posttranslational regulation of cAMP signaling, suppress the glucagon-stimulated hepatic glucose production [44], and coordinating the nocturnal oscillation of hepatic glucose output, namely glycogenolysis in the first part of night and gluconeogenesis in the second part, before waking up [44, 45].

Clock Controlled Metabolic Oscillation

The food is assimilated and processed differently when is ingested in the morning versus in the evening. The thermic effect of food, β cell responsiveness, insulin sensitivity, and muscular glucose uptake, all are enhanced in the morning hours compared to afternoon or evening [10, 11, 34, 46–50]. Therefore, human metabolism is optimized for food intake in the morning (i.e., breakfast) while the evening and nighttime are optimal for fasting and sleep [3, 11, 24, 41, 48]. As a result, postprandial glycemia displays a clear circadian pattern showing maximal glucose elevation after identical meal consumed in the afternoon and evening compared to the morning, in healthy [14, 22, 34, 42, 47, 48] and in T2D individuals [16, 24]. Moreover, low glycaemic index foods are of less value in glycaemic control, if are consumed in the evening versus in the morning [49], suggesting that food consuming late in the day has a detrimental metabolic impact irrespective of glycaemic index. Even the consumption of sweets, i.e., baked cake, in the breakfast or after lunch (15:30), has no detrimental effects on glycemic control, in contrast, the consumption of sweets in the evening at 19:30 (post-dinner), not only triggers significantly higher glycemic response but also worsens the glycemic response to the following day's breakfast [50].

Synchrony Between Central and Peripheral Clocks

For the functionality of the circadian system, the individual clocks must be correctly synchronized one to another and to the external environment [14]. This coordination between the central pacemaker and peripheral clocks is achieved when the feeding/fasting cycle is aligned with the light/dark cycle [11–14]. Therefore, both stimuli, "light", and "food" should occur simultaneously "in synchrony". Since breakfast, has a powerful resetting effect on the clock network, the temporal synchronization of breakfast and the light in the morning, is critical, for the achievement of metabolic homeostasis [15, 36–38].

In Fig. 18.2, is shown how the synchrony between breakfast and light, "turns-on" the clock gene machinery in the early morning. This, further regulate the clock-controlled output genes relaying the clock information to the tissue-specific downstream proteins and the circadian function of the cellular processes, such as insulin secretion, muscular glucose uptake, hepatic glucose production, etc. [10, 28, 30].

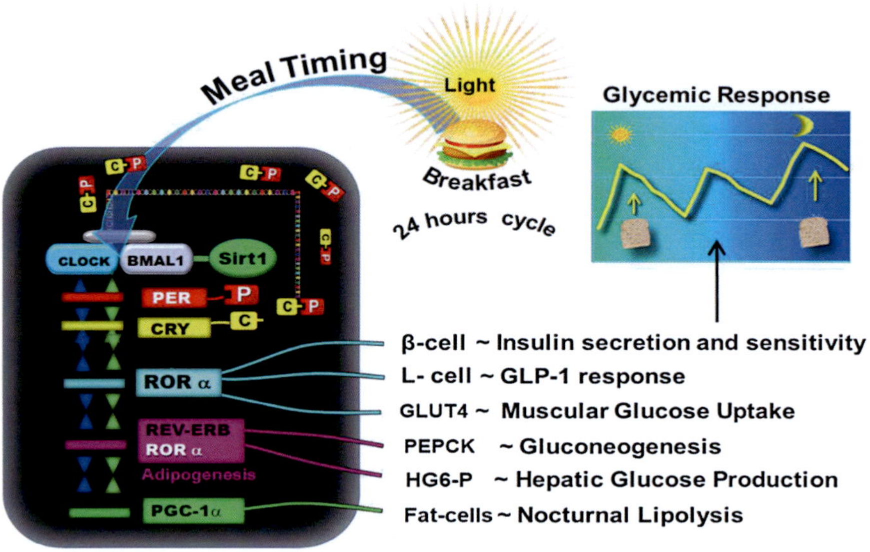

Fig. 18.2 Synchronization between central and peripheral clock genes. In the above illustration we can observe that the breakfast in sync with morning light "turns on" the clock, activating *CLOCK: BMAL1* complex, which promote the transcription of *PERs* and *CRYs* clock genes. The resulting proteins PER (P) and CRY (C) form PER: CRY (C-P) dimers in the cytoplasm, and then return to the nucleus after ~ 24h to repress *CLOCK: BMAL1*, thus maintaining the ~ 24h cycle of the clock. At the same time, *CLOCK: BMAL1* driven transcription of *PERs, CRYs, REV-ERBα*, and *RORα* genes, along with *PGC-1α* and *SIRT1*, promote the transcription of tissue-specific "clock-controlled genes." These tissue-specific genes regulate downstream targets' expression, relaying the clock information to cellular processes such as circadian β- cell insulin secretion and sensitivity, the circadian GLP-1 response, muscular glucose uptake, and hepatic glucose production. As a result, the glucose response after identical meals is significantly higher in the evening versus in the morning

Effects of Asynchrony Between Clocks

Eating and sleeping out of synchrony, by delaying the first meal of the day or by increasing meal frequency, with macronutrients evenly distributed across the day, including at hours assigned to sleep [11–16, 36–38, 51–53], promote the uncoupling or desynchronization of the peripheral clocks from the central pacemaker and disrupted circadian clocks regulation of the metabolic processes. This may result in altered thermogenesis, weight gain, increased lipids, fatty liver, and hyperglycemia [11–16, 36–38, 51–53]. In Fig. 18.3, we observe how misalignment between the hours of eating and sleeping, lead to disrupted clock gene regulation leading to deficient β cell secretion and hyperglycemia, deficient GLUT 4 activity, muscular glucose uptake, increased hepatic glucose output, enhanced adipogenesis, and reduced lipolysis, alteration of GLP-1 secretion and of glucose and lipid intestinal absorption [11].

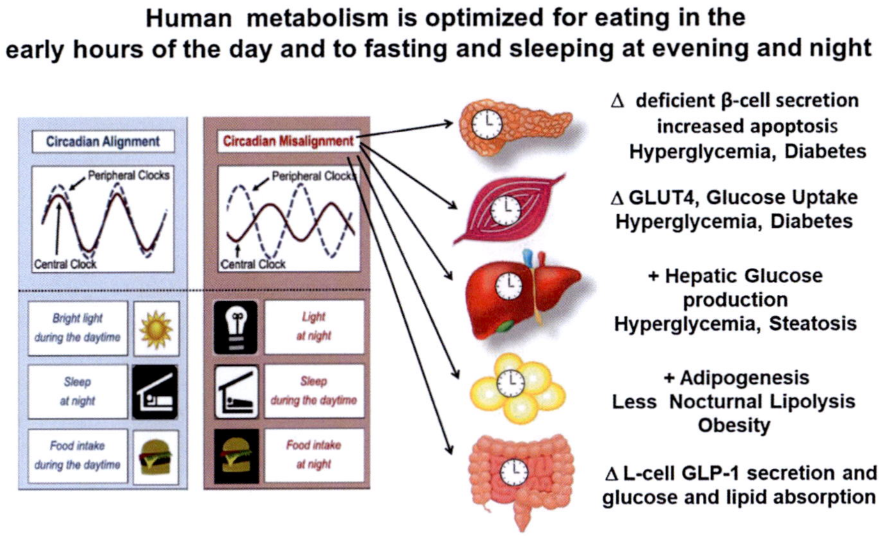

Fig. 18.3 Shown above is a schematic representation of circadian alignment between central and peripheral clocks (left panel) versus misalignment (right panel). Bright light exposure during the daytime, food intake during the daytime, and sleeping during the biological night promote circadian alignment between the central and peripheral clocks. Conversely, light exposure or food intake in the evening or at night, or sleeping during the daytime, misaligns the two clock systems and leads to metabolic dysfunction. The disruption of clock gene regulation lead to deficient β cell secretion and hyperglycemia, deficient muscular glucose uptake, increased hepatic glucose production, increased adipogenesis, decreased lipolysis, alteration of GLP-1 secretion and alteration of glucose and lipid intestinal absorption. "Reproduced and adapted with permission" [11]

In fact, we recently showed that the omission of breakfast led to alteration of the metabolic clock gene expression in both healthy and individuals with T2D [15]. The absence of breakfast, down-regulated the mRNA expression of AMPK and of the pivotal metabolic clocks genes (i.e., *BMAL1, PER1, RORα*), resulting in higher glycemic response and deficient insulin and GLP-1 postprandial secretion after subsequent meals [15]. This shows, that just one day omission of breakfast adversely affects clock and clock-controlled gene expression, and was correlated with an increase of the postprandial glycemic response in both healthy and T2D [15].

Clinical and epidemiological studies have indicated as well, that late meals are associated with obesity and T2D [6, 21, 22]. Likewise, a diet intervention not aligned with the circadian clock by shifting calories and CH to later hours of the day, is associated with less weight loss and higher overall glycemia among obese [21, 22] and inT2D [18, 23, 24].

Beneficial Effects of the Synchronization Between Clocks

In contrast, to circadian misalignment, eating in synchrony with the circadian clock by shifting more calories and especially CH to earlier hours of the day (i.e., high energy and CH breakfast), and reducing energy and CH consumption in the evening hours, facilitate weight loss, and improve glycemic excursions, and also reduces hunger and craving in obese and in T2D, compared to the inverse pattern, i.e., high in energy and CH dinner and reduced breakfast [16, 18, 21–24].

In an acute study in T2D, it was found, that high energy and CH breakfast versus omission of breakfast, led to significant improvement of postprandial glucose, insulin, and GLP-1 response after subsequent isocaloric lunch and dinner [25]. We also reported in healthy and in T2D, that high energy and CH breakfast versus omission of breakfast, led to up-regulation of clock genes (*BMAL1, PER1, PER2, CRY1, RORα, AMPK* and *SIRT1*) mRNA expression only in the day when the breakfast was consumed. This resetting effect on clock genes in "YES" breakfast day, was associated with the reduction of overall glycemia after following meal [15], suggesting that enhanced clock gene expression, driven by high energy and CH breakfast, might be the underlying mechanism of the improvement of overall glycemia when most of calories and CH are shifted to the early hours of the day [24–26].

More recently in T2D patients treated with insulin, we explored during three months, the effects of either one of two isocaloric diet intervention (DI) with different meal timing and distribution. As it is shown in Fig. 18.4, one of the two DI was aligned to the circadian clock, with 3 meals a day consisted of high energy and CH breakfast and low in energy and CH dinner (3Mdiet), the other DI was the traditional diet with 6 small meals with energy and CH evenly distributed along the day without any temporal alignment to the rhythms imposed by the circadian clock [16]. Compared to the traditional diet (6Mdiet), the 3Mdiet led to a significant resetting effect in the oscillatory expression of the clock genes involved in insulin secretion, glucose uptake, and hepatic glucose production, namely *BMAL1, CRY1, PER2, RORα* and to the increase of daily SIRT1 levels [16]. This upregulation of clock gene expression in the 3Mdiet was associated with a significantly more efficient reduction of HbA1c, weight loss fasting glucose and glycemic excursion assessed by continuous glucose monitoring (CGM). The results are shown in Fig. 18.5. Notably, the reduced overall glycemic excursions in 3Mdiet, were also significantly reduced during the nocturnal segment (00:00 to 06:00), suggesting a reduced nocturnal hepatic glucose production and improved hepatic insulin sensitivity in the DI aligned to the circadian clock [45]. The time spent in the normal glucose range was also significantly increased in 3Mdiet compared to 6Mdiet, while the percentage of time spent in hyperglycemia was significantly reduced [16].

Importantly the titration of total daily insulin dose resulted in a significant decrease in insulin requirement by 27.5 units only in the 3Mdiet (Fig. 18.5). In addition, we found that the appetite and craving scores, for all kinds of foods, but especially for sweets, were all significantly reduced only in the 3Mdiet. This improves significantly the adherence to the diet intervention as we already shown in a previous study [21].

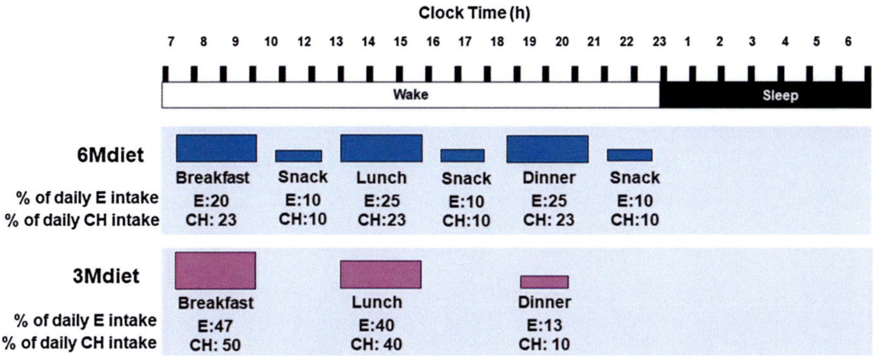

Fig. 18.4 Mealtime and distribution of the 3Mdiet and 6Mdiet. CH, carbohydrates of daily carbohydrate intake; E, energy of the daily caloric intake; ■: 3Mdiet; ■: 6Mdiet. The (3Mdiet)consisted on high energy and CH breakfast, medium sized lunch and low energy and low CH dinner with total of ~ 150 gr of daily CH content in which 50% of daily CH was consumed at breakfast (before 9:30), 40% at lunch (before 15:00) and 10% of CH at dinner not later than 20:00. This diet was compared with isocaloric six meal diet (6Mdiet), with 3 meals and 3 snacks,as model of meal timing not aligned to the circadian clock, with same calories and CH content,but uniformly distributed across the day, with 23% of daily CH intake at breakfast, 23% at lunch and 23% in the dinner and 3 snacks each one with 10% of daily CH intake, including one snack at 22:00. The participants of 6Mdiet, were instructed to consume breakfast: between 8:00-10:00,lunch: 13:00-15:00, dinner:18:00-20:00, and three snacks at 11:00, 17:00 and at 22:00."Reproduced and adapted with permission" [16]

Based on these results, we can assume that meal timing aligned to the circadian clock by shifting most calories and CH to the early hours of the day, upregulated the oscillatory mRNA expression of the pivotal clock genes, associated with improved β cell insulin secretion and sensitivity, less β cell apoptosis, enhanced GLUT4 and muscular glucose uptake, suppression of hepatic glucose production, less adipogenesis and enhanced lipolysis, with improved control of appetite and craving impulses [16]. The upregulation of clock gene expression might be the underlying mechanism of the beneficial effect on weight loss, glycemic control, and appetite, achieved with a DI aligned to the circadian clock.

Reduction of Appetite and "Food-Craving" Should Be a Major Target of Diet Intervention

Obesity is associated with impaired inhibitory control over food intake [54]. Although dietary restriction often results in initial weight loss, the majority of obese and obese T2D patients, cannot maintain for the long term the weight-loss strategies, and fail to maintain their reduced weight, over time [54–57]. Most weight-loss diets result in persistent compensatory metabolic changes, including reduced energy expenditure,

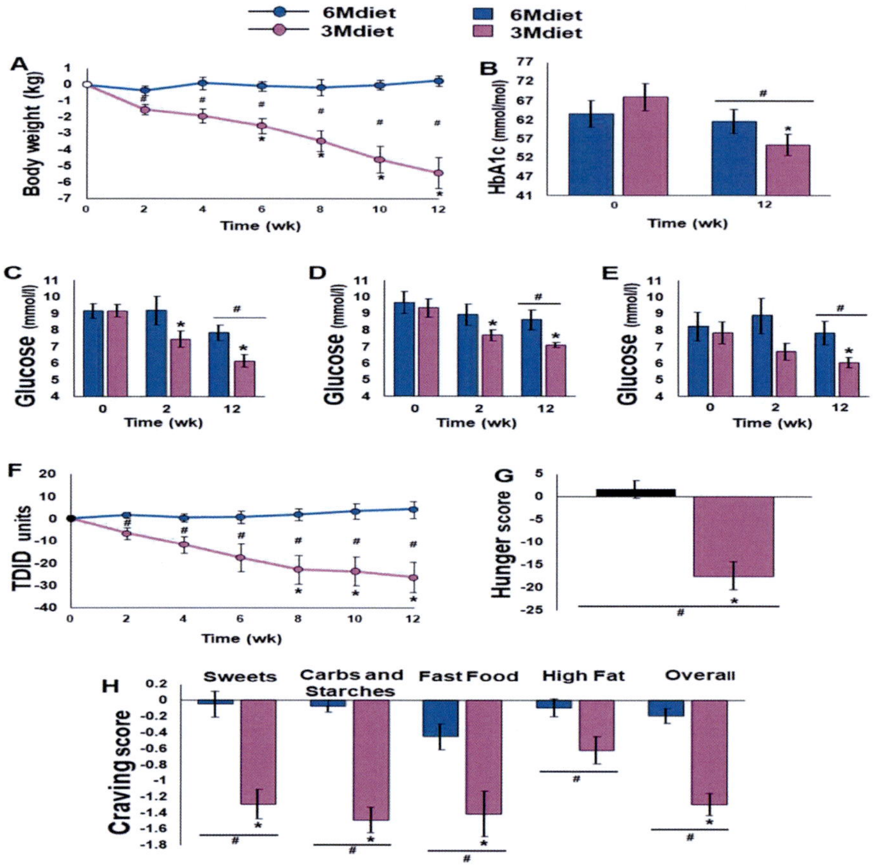

Fig. 18.5 Body weight, HbA1c, glucose levels, TDID, hunger, and cravings at baseline, 2 weeks (wk), and 12 weeks of 3Mdiet and 6Mdiet. A: Weight loss. B: HbA1c. C: Fasting glucose. D: Twenty-four–hour mean glucose. E: Nocturnal (0000–0600 h) mean glucose. F:TDID. G: Hunger scores. H: Mean daily craving scores. Values are mean ± SE. *Significant difference within groups compared with baseline, $P < 0.05$; #significant difference between groups, $P < 0.05$. ■: 3Mdiet ■: 6Mdiet "Reproduced and adapted with permission" [16]

increased appetite, food craving, and decreased postprandial ghrelin suppression [54–59]. Proposed predictors of weight regain after weight loss include the increase of these subjective appetite scores, especially increased food cravings [55–59]. Therefore, long-term strategies to counteract these adverse effects on appetite are needed to facilitate the maintenance of weight loss and diabetes control over time. These strategies may include the change of meal timing and macronutrient distribution.

Conflicting Circadian Rhythms of Hunger and Craving

Many recent studies show the advantage, of meal timing schedule aligned with the circadian clock by shifting more energy and CH to early hours of the day, and reducing energy and CH, in the afternoon and evening, to achieve weight loss and glycemic control [7, 16, 17, 22]. However, and paradoxically, the diurnal variation of appetite and craving scores, show the inverse oscillation, with higher hunger and food reward impulses, in the evening hours, which appears to be in conflict with the recommended meal timing [17, 60]. Hunger and craving scores are at the lowest in the morning hours (8–10 a.m.) and the highest scores are found in the afternoon (4 p.m. to 6 p.m.) [17, 38, 60]. Craving for sweets is slightly dissociated from hunger, increasing throughout the day and peaking in the late afternoon and early evening, around 7 p.m. [16, 56, 59]. Indeed, obese and diabetics are typically not hungry, not ravenous, and have minimal hedonic impulses upon waking up, consequently, breakfast is typically the smallest meal of the day, while in the evening when the high energy and CH consumption is associated with weight gain and higher glycemic responses, is the time when those obese and obese with T2D, perceive more hunger and hedonic impulses, especially for sweets [16, 21, 60]. Further, the omission of breakfast appears to have additive effects, further increasing the appetite and reward responses (i.e., carb-craving) in the evening [17].

The Effect of High Energy and CH Breakfast on Hunger and Food-Craving

As the reduction of hunger and especially carb-craving is critical for the achievement of adequate weight loss and glycemic control over time, we reported both, in obese and in obese with T2D, that compared with low energy and low CH breakfast. a DI consisting of high energy and CH breakfast including small sweet i.e., chocolate, is more effective for weight loss [16, 18, 21–23], for reduction of overall glycemic excursions [16, 22, 23], and for the prevention of weight regain [21]. The high energy and CH breakfast, also led to significant ghrelin suppression and increased postprandial GLP-1 responses, which was associated with increased satiety and reduced mean daily and postprandial hunger VAS scores after breakfast lunch, and dinner, suggesting a day-long effect of high energy and CH breakfast [21–26]. Further, the high energy and CH breakfast diet, significantly ameliorated the desire for sweets and carb-craving in the afternoon and evening [16, 21–26]. Notably, the reduction of the craving for sweets was more significant between 4 p.m. and 7 p.m. [16], when the craving for sweets is at its highest level [58–60] (Fig. 18.5). As proposed predictors of poor adherence to the diet, and of weight regain include increased subjective appetite scores, especially increased hunger and craving for sweets [58, 59], the significant reduction of hunger, cravings in the afternoon, places the high energy and

CH breakfast diet as adequate strategy for the achievement of sustained weight loss and diabetes control [16, 21].

The effectiveness of the high energy and CH breakfast diet, which also has high protein content, is in line with previous studies showing that dietary proteins are the most satiating of the macronutrients in conditions of both energy restriction and energy balance [61]. Meal timing also appears to influence its satiating properties. Specifically, protein consumed at breakfast (compared to lunch or dinner) lead to greater initial and sustained feelings of fullness, increased satiety, and reduced levels of the appetite-regulating hormones such as ghrelin [21, 62].

Moreover, the daily addition of a carbohydrate-rich snack (i.e., sweet) to the breakfast in the high energy and CH breakfast diet, has been shown to reduce the snack's reward value decreasing cravings for sweets, bread, snacks, and fast food in the evening [63, 64]. Furthermore, it was shown that the addition of chocolate in the breakfast has a powerful resetting effect on the clock genes expression, which might be an additional underlying mechanism by which the high energy and CH breakfast diet, significantly reduces the craving for sweets [64].

Conclusions

Circadian clock regulation of glucose and energy metabolism requires coordination and alignment between the central clock in the SCN, which is entrained to light signals, with the peripheral clock genes, i.e., β-cells, muscle, liver, etc., mostly entrained to the hour of food intake and food availability. This synchronization between the central pacemaker and peripheral clocks is achieved when the feeding/fasting cycle is aligned with the light/dark cycle. Therefore, both stimuli, "light", and "food" should occur simultaneously "in synchrony". Since breakfast, has the most powerful resetting effect on the clock network, the alignment between breakfast and the light, in the morning is critical, for the achievement of metabolic homeostasis. Growing evidence shows that a meal timing aligned with the circadian clock, with most of the daily calories and CH assigned to the early hours of the day i.e., high energy and CH breakfast and reduced in energy and CH consumption at dinner and in the late hours of the day, leads to upregulation of clock genes expression and synchrony within the clocks, and is associated with more efficient weight loss, reduced overall glycemic excursions along with reduced appetite and craving scores. Since high energy and CH breakfast, upregulate and synchronize the mRNA expression of pivotal clock genes, involved in β-cell insulin secretion, muscular glucose uptake, hepatic glucose production, nocturnal lipolysis, and appetite regulation, it can be assumed, that upregulation of clock genes expression might be the underlying mechanism of the beneficial effect on weight loss, glycemic control, and appetite when the meal timing is aligned to the circadian clock.

References

1. Bhupathiraju SN, Hu FB (2016) Epidemiology of obesity and diabetes and their cardiovascular complications. Circ Res 118:1723–1735
2. Saeedi P, Petersohn I, Salpea P et al (2019) Global and regional diabetes prevalence estimates for 2019 and projections for 2030 and 2045: results from the international diabetes federation diabetes Atlas, 9th edition. Diabetes Res Clin Pract 157:107843–107854
3. Scheer FA, Hilton MF, Mantzoros CS, Shea SA (2009) Adverse metabolic and cardiovascular consequences of circadian misalignment. Proc Natl Acad Sci U S A 106:4453–4448
4. Mekary RA, Giovannucci E, Willett WC et al (2012) Eating patterns and type 2 diabetes risk in men: breakfast omission, eating frequency, and snacking. Am J Clin Nutr 95:1182–1189
5. Nimitphong H, Siwasaranond N, Saetung S et al (2018) The relationship among breakfast time, morningness-eveningness preference and body mass index in type 2 diabetes. Diabetes Med 35:964–971
6. Reutrakul S, Hood MM, Crowley SJ et al (2013) Chronotype is independently associated with glycemic control in type 2 diabetes. Diabetes Care 36:2523–2529
7. St-Onge MP, Ard J, Baskin ML et al (2017) American Heart Association Obesity Committee of the Council on Lifestyle and Cardiometabolic Health; Council on Cardiovascular Disease in the Young; Council on Clinical Cardiology; and Stroke Council. Meal timing and frequency: implications for cardiovascular disease prevention: a scientific statement from the American Heart Association. Circulation 28:e96–e121
8. Javeed N, Matveyenko AV (2018) Circadian etiology of type 2 diabetes mellitus. Physiology (Bethesda) 33:138–150
9. Froy O, Garaulet M (2018) The circadian clock in white and brown adipose tissue: mechanistic, endocrine, and clinical aspects. Endocr Rev 39:261–273
10. Oike H, Oishi K, Kobori M (2014) Nutrients, clock genes, and chrononutrition. Curr Nutr Rep 3:204–212
11. Poggiogalle E, Jamshed H, Peterson CM (2018) Circadian regulation of glucose, lipid, and energy metabolism in humans. Metabolism 84:11–27
12. Panda S (2016) Circadian physiology of metabolism. Science 354:1008–1015
13. Oosterman JE, Kalsbeek A, la Fleur SE, Belsham DD (2015) Impact of nutrients on circadian rhythmicity. Am J Physiol Regul Integr Comp Physiol 308:R337–R350
14. Wehrens SMT, Christou S, Isherwood C et al (2017) Meal timing regulates the human circadian system. Curr Biol 27:1768–1775.e3
15. Jakubowicz D, Wainstein J, Landau Z et al (2017) Influences of breakfast on clock gene expression and postprandial glycemia in healthy individuals and individuals with diabetes: a randomized clinical trial. Diabetes Care 40:1573–1579
16. Jakubowicz D, Landau Z, Tsameret S et al (2019) Reduction in glycated hemoglobin and daily insulin dose alongside circadian clock upregulation in patients with type 2 diabetes consuming a three-meal diet: a randomized clinical trial. Diabetes Care 42:2171–2180
17. Beaulieu K, Oustric P, Alkahtani S et al (2020) Impact of meal timing and chronotype on food reward and appetite control in young adults. Nutrients 12:1506–15011
18. Kahleova H, Belinova L, Malinska H et al (2014) Eating two larger meals a day (breakfast and lunch) is more effective than six smaller meals in a reduced-energy regimen for patients with type 2 diabetes: a randomised crossover study. Diabetologia 57:1552–1560
19. Allison KC, Hopkins CM, Ruggieri M et al (2020) Prolonged, controlled daytime versus delayed eating impacts weight and metabolism. Curr Biol 20:S0960-9822(20)31669-9
20. Garaulet M, Gómez-Abellán P, Alburquerque-Béjar JJ et al (2013) Timing of food intake predicts weight loss effectiveness. Int J Obes (Lond) 37:604–611
21. Jakubowicz D, Froy O, Wainstein J, Boaz M (2012) Meal timing and composition influence ghrelin levels, appetite scores and weight loss maintenance in overweight and obese adults. Steroids 77:323–331

22. Jakubowicz D, Barnea M, Wainstein J, Froy O (2013) High caloric intake at breakfast vs. dinner differentially influences weight loss of overweight and obese women. Obesity (Silver Spring) 21:2504–2512
23. Rabinovitz HR, Boaz M, Ganz T, Jakubowicz D et al (2014) Big breakfast rich in protein and fat improves glycemic control in type 2 diabetics. Obesity (Silver Spring) 22:E46–E54
24. Jakubowicz D, Wainstein J, Ahren B et al (2015) High-energy breakfast with low-energy dinner decreases overall daily hyperglycemia in type 2 diabetic patients: a randomized clinical trial. Diabetologia 58:912–919
25. Jakubowicz D, Wainstein J, Ahren B et al (2015) Fasting until noon triggers increased postprandial hyperglycemia and impaired insulin response after lunch and dinner in individuals with type 2 diabetes: a randomized clinical trial. Diabetes Care 38:1820–1826
26. Jakubowicz D, Wainstein J, Landau Z et al (2017) High-energy breakfast based on whey protein reduces body weight, postprandial glycemia and HbA_{1C} in Type 2 diabetes. J Nutr Biochem 49:1–7
27. Reinke H, Asher G (2019) Crosstalk between metabolism and circadian clocks. Nat Rev Mol Cell Biol 20:227–241
28. Kim YH, Lazar MA (2020) Transcriptional control of circadian rhythms and metabolism: a matter of time and space. Endocr Rev 1:707–732
29. Pilorz V, Astiz M, Heinen KO, Rawashdeh O, Oster H (2020) The concept of coupling in the mammalian circadian clock network. J Mol Biol 432:3618–3638
30. Vieira E, Burris TP, Quesada I (2014) Clock genes, pancreatic function, and diabetes. Trends Mol Med 20:685–693
31. Gil-Lozano M, Mingomataj EL, Wu WK et al (2014) Circadian secretion of the intestinal hormone GLP-1 by the rodent L cell. Diabetes 63:3674–3685
32. Dyar KA, Ciciliot S, Wright LE, Bienso RS et al (2014) Muscle insulin sensitivity and glucose metabolism are controlled by the intrinsic muscle clock. Mol Metab 3:29–41
33. Sadacca LA, Lamia KA, deLemos AS et al (2011) An intrinsic circadian clock of the pancreas is required for normal insulin release and glucose homeostasis in mice. Diabetologia 54:120–124
34. Saad A, Dalla Man C, Nandy DK et al (2012) Diurnal pattern to insulin secretion and insulin action in healthy individuals. Diabetes 61:2691–2700
35. Kuang J, Hou X, Zhang J et al (2014) Identification of insulin as a novel retinoic acid receptor-related orphan receptor alpha target gene. FEBS Lett 588:1071–1079
36. Shimizu H, Hanzawa F, Kim D et al (2018) Delayed first active-phase meal, a breakfast-skipping model, led to increased body weight and shifted the circadian oscillation of the hepatic clock and lipid metabolism-related genes in rats fed a high-fat diet. PLoS One 13:e02066–e02069
37. Wu T, Sun L, ZhuGe F et al (2011) Differential roles of breakfast and supper in rats of a daily three-meal schedule upon circadian regulation and physiology. Chronobiol Int 28:890–903
38. Sherman H, Genzer Y, Cohen R et al (2012) Timed high-fat diet resets circadian metabolism and prevents obesity. FASEB J 26:3493–3502
39. Sun C, Zhang F, Ge X et al (2007) SIRT1 improves insulin sensitivity under insulin-resistant conditions by repressing PTP1B. Cell Metab 6:307–319
40. Biancolin AD, Martchenko A, Mitova E et al (2020) The core clock gene, Bmal1, and its downstream target, the SNARE regulatory protein secretagogin, are necessary for circadian secretion of glucagon-like peptide-1. Mol Metab 31:124–137
41. Stenvers DJ, Jongejan A, Atiqi S et al (2019) Diurnal rhythms in the white adipose tissue transcriptome are disturbed in obese individuals with type 2 diabetes compared with lean control individuals. Diabetologia 62:704–716
42. Taira A, Arita E, Matsumoto E et al (2019) Systemic oscillator-driven and nutrient-responsive hormonal regulation of daily expression rhythms for gluconeogenic enzyme genes in the mouse liver. Chronobiol Int 36:591–615
43. Pérez-Mendoza M, Rivera-Zavala JB, Díaz-Muñoz M (2014) Daytime restricted feeding modifies the daily variations of liver gluconeogenesis: adaptations in biochemical and endocrine regulators. Chronobiol Int 31:815–828

44. Zhang EE, Liu Y, Dentin R et al (2010) Cryptochrome mediates circadian regulation of cAMP signaling and hepatic gluconeogenesis. Nat Med 16:1152–1156
45. Basu A, Joshi N, Miles J et al (2018) Paradigm shifts in nocturnal glucose control in type 2 diabetes. J Clin Endocrinol Metab 103:3801–3809
46. Bo S, Fadda M, Castiglione A et al (2015) Is the timing of caloric intake associated with variation in diet-induced thermogenesis and in the metabolic pattern? A randomized cross-over study. Int J Obes (Lond) 39:1689–1695
47. Morgan LM, Shi JW, Hampton SM, Frost G (2012) Effect of meal timing and glycaemic index on glucose control and insulin secretion in healthy volunteers. Br J Nutr 108:1286–1291
48. Morris CJ, Yang JN, Garcia JI et al (2015) Endogenous circadian system and circadian misalignment impact glucose tolerance via separate mechanisms in humans. Proc Natl Acad Sci U S A 112:E2225–E2234
49. Gibbs M, Harrington D, Starkey S et al (2014) Diurnal postprandial responses to low and high glycaemic index mixed meals. Clin Nutr 33:889–894
50. Nitta A, Imai S, Kajiyama S et al (2019) Impact of different timing of consuming sweet snack on postprandial glucose excursions in healthy women. Diabetes Metab 45:369–374
51. Arble DM, Bass J, Laposky AD et al (2009) Circadian timing of food intake contributes to weight gain. Obesity (Silver Spring) 17:2100–2102
52. Hatori M, Vollmers C, Zarrinpar A et al (2012) Time-restricted feeding without reducing caloric intake prevents metabolic diseases in mice fed a high-fat diet. Cell Metab 15:848–860
53. Fuse Y, Hirao A, Kuroda H et al (2012) Differential roles of breakfast only (one meal per day) and a bigger breakfast with a small dinner (two meals per day) in mice fed a high-fat diet with regard to induced obesity and lipid metabolism. J Circadian Rhythms 10:4–12
54. Wang GJ, Shokri Kojori E, Yuan K et al (2020) Inhibition of food craving is a metabolically active process in the brain in obese men. Int J Obes (Lond) 44:590–600
55. Sumithran P, Prendergast LA, Delbridge E et al (2011) Long-term persistence of hormonal adaptations to weight loss. N Engl J Med 365:1597–604
56. Phelan S, Hassenstab J, McCaffery JM et al (2011) Cognitive interference from food cues in weight loss maintainers, normal weight, and obese individuals. Obesity (Silver Spring) 19:69–73
57. Epstein LH, Carr KA, Lin H, Fletcher KD (2011) Food reinforcement, energy intake, and macronutrient choice. Am J Clin Nutr 94:12–18
58. Reichenberger J, Richard A, Smyth JM et al (2018) It's craving time: time of day effects on momentary hunger and food craving in daily life. Nutrition 55–56:15–20
59. Bechtold DA, Loudon AS (2013) Hypothalamic clocks and rhythms in feeding behaviour. Trends Neurosci 36:74–82
60. Scheer FA, Morris CJ, Shea SA (2013) The internal circadian clock increases hunger and appetite in the evening independent of food intake and other behaviors. Obesity (Silver Spring) 21:421–423
61. Westerterp-Plantenga MS, Nieuwenhuizen A, Tomé D et al (2009) Dietary protein, weight loss, and weight maintenance. Annu Rev Nutr 29:21–41
62. Leidy HJ, Racki EM (2010) The addition of a protein-rich breakfast and its effects on acute appetite control and food intake in 'breakfast-skipping' adolescents. Int J Obes (Lond) 34:1125–1133
63. Temple JL, Chappel A, Shalik J et al (2008) Daily consumption of individual snack foods decreases their reinforcing value. Eat Behav 9:267–276
64. Escobar C, Espitia-Bautista E, Guzmán-Ruiz MA et al (2020) Chocolate for breakfast prevents circadian desynchrony in experimental models of jet-lag and shift-work. Sci Rep 10:6243–6250

Chapter 19
Methylglyoxal and Its Role in Obesity-Associated Heart Failure with Preserved Ejection Fraction

Fadhel A. Alomar, Caronda J. Moore, Salah Abohelaika, Fahad Al-Muhanna, Mohammed A. Alshabeed, Frederick Hamel, Cyrus DeSouza, and Keshore R. Bidasee

Abstract Heart failure (HF), including early-onset HF with preserved ejection fraction (HFpEF) is a common cause of morbidity and frequent hospitalizations in obese individuals. Diagnosis of HFpEF remains challenging due to its heterogeneous phenotype. To date, the molecular causes for the early-onset decrease in myocardial energetics, increase in global longitudinal strain, dysregulation of the endothelial cells in the coronary vasculature, myocardial fibrosis and inflammation

---

F. A. Alomar
Department of Pharmacology and Toxicology, College of Clinical Pharmacy, Imam Abdulrahman Bin Faisal University, Dammam, Saudi Arabia

C. J. Moore
Kansas State University Innovation Partners Manhattan, Manhattan, KS 66502, US

S. Abohelaika
Clinical Pharmacology Department, Qatif Central Hospital, Ministry of Health, Qatif, Saudi Arabia

F. Al-Muhanna
Department of Internal Medicine, King Fahd Hospital of the University, College of Medicine, Imam Abdulrahman Bin Faisal University, Dammam, Saudi Arabia

M. A. Alshabeed
Department of Development Medicine, King Adbullah International Medical Research Center, Riyadh, Saudi Arabia

F. Hamel
Research Service, Omaha Veterans Affairs Medical Center, Omaha, US

C. DeSouza
Department of Internal Medicine, Division of Diabetes Endocrinology & Metabolism, University of Nebraska Medical Center, Omaha, NE, US

K. R. Bidasee (✉)
Department of Pharmacology and Experimental Neuroscience, and Environment and Occupational Health, University of Nebraska Medical Center, Omaha, NE 68198, US
e-mail: kbidasee@unmc.edu

Nebraska Redox Biology Center, Lincoln NE, US

© Springer Nature Switzerland AG 2021

P. S. Tappia et al. (eds.), *Cellular and Biochemical Mechanisms of Obesity*, Advances in Biochemistry in Health and Disease 23,
https://doi.org/10.1007/978-3-030-84763-0_19

remain poorly understood. This paucity of information is the primary reason for a lack of specific pharmacologic agents to treat HFpEF. There is now considerable evidence indicate that accumulation of the reactive α-dicabonyl species, methylglyoxal (MG) is the underlying cause for a diverse array of cardiac defects including myocyte and endothelial cell dysfunction, coronary microvascular leakage, inflammation, micro-ischemia and fibrosis. Here we review how MG is synthesized and degraded, the consequence MG accumulation, and present new data showing elevated MG levels in plasma of obese patients with and without HF.

Keywords Obesity · Heart failure · Glycolysis · Methylglyoxal · Endothelial cells · Microvascular leakage · Inflammation · Fibrosis · Inflammation

Introduction

Obesity, defined as body-mass-index of $\geq$30 kg/m^2, has risen exponentially during the past few decades and is now a major cause of poor health in many countries [1, 2]. More than 13% of the world's 7.8 billion people are obese, with rates amongst women significantly higher than that in men (15% of women and 11% of men) [3]. In North American and Middle Eastern populations, more than 40% of individuals are obese [4]. Socioeconomic factors, race and ethnic disparities contribute to higher obesity rates amongst Hispanics/non-Hispanic blacks (47%) in the USA compared to non-Hispanic whites (38%) and Asians (13%) [5]. Chronic obesity is the underlying cause for an expanding array of diseases, including type 2 diabetes mellitus (T2DM), chronic kidney disease, non-alcoholic fatty liver disease, hypertension, several different types of cancers and heart failure (HF) [1]. An apparent "beneficial effect" of obesity has also been reported. Studies have found that individuals with BMI between 30 kg/m^2 and 35 kg/m^2 (class I obese) with coronary heart disease that underwent percutaneous coronary intervention were more likely to have favorable outcomes compared to normal or underweight individuals [6]. This favorable prognosis was not seen in class II (BMI = 35 to < 39 kg/m^2) and class III (>40 kg/m^2) obese individuals. However, Hainer and Aldhoon-Hainerová indicated that this "obesity paradox" is due to a lack of the discriminatory power of BMI to differentiate between lean body mass and fat mass [7]. The direct and indirect medical and non-medical costs of obesity worldwide is estimated to be more than $2 trillion USD [1, 8].

Shortness of breath (dyspnea), fatigue, chest discomfort and edema of the lower extremities are common characteristics of heart failure (HF) in obese individuals with Type 2 diabetes mellitus (T2DM) [9–12]. There is also a higher incidence of HF in postmenopausal women [13]. These symptoms arise from structural and functional impairments in the filling and ejection of blood from the left ventricle of the heart [14]. HF can be subdivided into early-onset HF with preserved ejection fraction (HFpEF) and later-stage HF with reduced ejection fraction (HFrEF). In HFpEF, the fraction of blood ejected remains unchanged (>50%), but the filling rate and extensibility of the left ventricle is compromised, i.e., impaired relaxation of the left ventricle during diastole. In HFrEF, the volume of blood expelled from the left ventricle per contraction is reduced, i.e., ejection fraction is reduced, < 50%. Recently, the European Society of Cardiology (ESC) also introduced HF with mid-range ejection fraction (HFmrEF) to define those patients with ejection fraction between 40–50% [15]. However, the phenotype of patients with ejection fractions between 40–50% is not yet fully defined, as some studies have found they exhibit clinical profiles similar to HFrEF including a higher risk of sudden cardiac death, while others have reported a closer resemblance to HFpEF, including impaired diastolic relaxation [16].

HFpEF is a major cause of early-onset morbidity, frequent hospitalizations, and healthcare costs in individuals with chronic obesity [14, 17]. This condition is challenging to diagnose due to heterogeneity in its phenotype. Metabolic syndrome, type 2 diabetes mellitus (T2DM), pulmonary hypertension and renal insufficiency are usually present in patients with HFpEF [14]. However, it is not clear whether pulmonary hypertension and renal insufficiency are primary causes of HFpEF or occur secondary to the impaired diastolic function [18]. Clinical studies targeting the RAAS which is effective in patients with HFrEF, including I-PRESERVE (Irbesartan in Heart Failure with Preserved Systolic Function), CHARM-Preserved (Candesartan in Heart Failure: Assessment of Reduction in Mortality and Morbidity), TOPCAT (spironolactone), PARAGON-HF, (sacubitril-valsartan and valsartan), and DIG-PEF (Digitalis Investigation Group, preserved ejection fraction) have shown only modest ability to decrease hospitalization rates and slow HFpEF in obese individuals [19–25]. This is likely because of the multitude of contributing pathobiologic mechanisms within cardiomyocytes, the coronary vasculature, and the surrounding tissues. Clinical trials to investigate whether sodium-glucose co-transporter 2 (SGLT2) inhibitors are effective to treat HFpEF is ongoing [26]. Emerging data from several laboratories including ours indicate that accumulation of the cytotoxic α-oxoaldehyde species methylglyoxal (MG) is a contributing cause for the decreased myocardial energetics, endothelial dysfunction microvasculature leakage, inflammation and fibrosis seen in HFpEF [27–30]. Here we discuss generation and degradation of MG in mammalian cells, the consequence MG accumulation, and provide original data showing elevated MG in plasma of obese patients with and without HFpEF.

Methylglyoxal

Synthesis

Methylglyoxal (MG, Fig. 19.1, also known as 2-oxopropanal and pyruvaldehyde) is the most reactive of the endogenous α-oxoaldehydes generated in mammals [31–34]. This reactive dicarbonyl species is generated principally from the spontaneous breakdown (non-enzymatic) of the glycolytic triose intermediates, glyceraldehyde 3-phosphate (G3P) and dihydroxyacetone phosphate (DHAP) during the biochemical

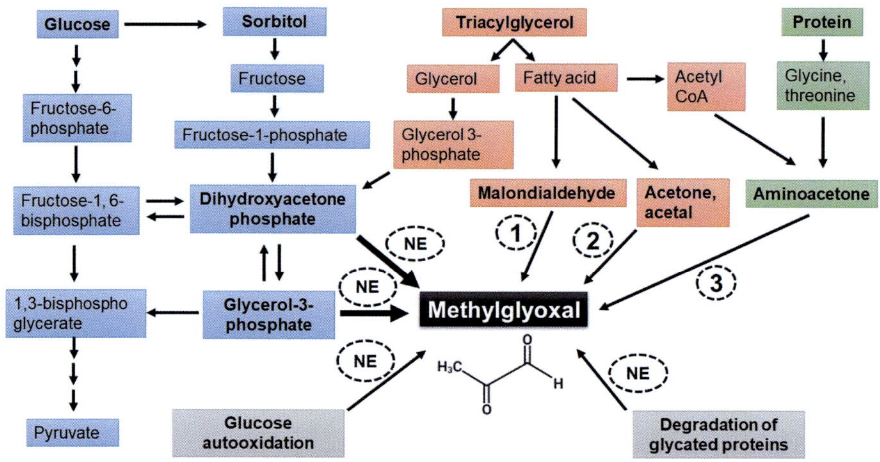

Fig. 19.1 Schematic representation of the main metabolic pathways involved MG production. MG is formed mainly by the spontaneous breakdown of the glycolytic intermediates, glyceraldehyde-3-phosphate (G3P) and dihydroxyacetone phosphate (DHAP), but smaller amounts are also formed also from the metabolism of lipids and proteins. The last enzymes in the enzymatic pathway are numbered 1: phosphoglucose isomerase, 2: acetone monooxygenase, 3: vascular adhesion protein-1, and non-enzymatic (spontaneous) pathways are labeled NE

processes that converts glucose into pyruvate (glycolysis), Fig. 19.1 [34]. About 0.1% of the glucotriose flux is converted into MG [35]. Smaller amounts of MG are also produced from the spontaneous break down of glucose, metabolism of glucose via the sorbitol pathway (polyol pathway), enzymatic degradation of triacylglycerol, proteins (amino acids) and fatty acids (enzymatic breakdown of malondialdehyde in the cytoplasm by phosphoglucoisomerase), and the degradation of glycated proteins [31, 36, 37]. In healthy individuals, MG concentrations in plasma typically ranges between 60 – 250 nM and 1- 5 μM in tissues [38–40].

The majority of MG generated inside healthy cells is reversibly bound to the arginine, lysine and cysteine residues of proteins, with about 1% existing as the free unhydrated, monohydrated, or dihydrated form [40, 41]. Free and reversibly bound forms of MG are also in constant equilibrium with each other. Free MG also readily exchange between cellular and extracellular compartments including juxtaposed cells with a half-life cell membrane permeability of < 15 min [31]. Physiological levels of MG play important roles in regulating cell proliferation, differentiation [41]. Additionally, MG plays an important roles in regulating non-REM sleep and anxiety through its actions as an agonist at $GABA_A$ receptors [42–47]. Studies have suggested that MG may also serve a hormesis role; a favorable biological response to low-dose exposure of a stressor compound [48, 49].

Degradation

Because supraphysiologic level of MG is cytotoxic, cellular accumulation of MG is tightly regulated. Inside cells, free MG rapidly forms a hemithioacetal with reduced glutathione [50]. This MG-GSH hemithioacetal is then rapidly degraded by the dual-enzyme glyoxalase system present in the cytoplasm of every cell to produce D-lactic acid and regenerate the reduced glutathione in two sequential reactions (Fig. 19.2, middle). In the first step, glyoxalase-I, (*GLOI,* EC4.4.1.5, Glo1, also known as lactoylglutathione lyase) converts the MG-GSH hemithioacetal into S, D-lactoylglutathione. In the second step, the S, D-lactoylglutathione is degraded by glyoxalase-II (*GLOII,* EC3.1.2.6, Glo2) in the presence of H_2O to D-lactic acid and reduced glutathione [31, 32]. Glo1 is the rate-limiting step for the detoxification of MG and the rate of MG degradation depends on the amount of free GSH present inside cells. This degradation mechanism continuously reduces the amount free and by extension the amount of MG reversibly bound to amino acid residues inside cells. D-lactate which is metabolized in mammalian tissues by mitochondrial D-lactate dehydrogenase, can be also used as a surrogate indicator of MG flux [51]. Other enzymes, including aldehyde dehydrogenases family (ALDHs) and the

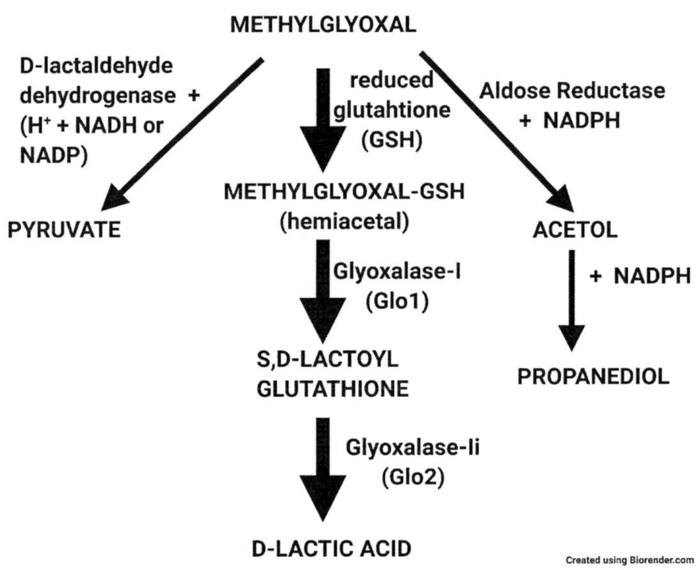

Fig. 19.2 Pathways involved in the degradation of methylglyoxal in mammalian cells. MG is degraded principally via the glyoxalase system which consists of two enzymes glyoxalase-1 (Glo1) and glyoxalase-2 (Glo2). In the first step, the MG-GSH hemithioacetal formed is used as a substrate by Glo1 to form S, D-lactoylglutathione. Glo2 in the presence of water then catalyzes the transformation of S, D-lactoylglutathione into D-lactate and GSH. MG is also degraded by aldose reductase and D-lactaldehyde dehydrogenase with the enzymes using NADPH and NADH as co-factors and with higher Kms'

aldoketo reductases (AKR) family, particularly aldose reductase isoform AKR1B3 in mice, AKR1B4 in rats, and AKR1B1 in humans also degrade MG, albeit with K_{ms}' several orders of magnitude higher than that of Glo-1 and with other cofactors including NADPH and NADP (Fig. 19.3) [31, 52]. However, the relative contribution of ALDHs and AKRs to MG detoxification depends on their cellular expression. Therefore, ALDHs, and AKR are of relevance for the detoxification of MG when the highly efficient glyoxalase system becomes compromised.

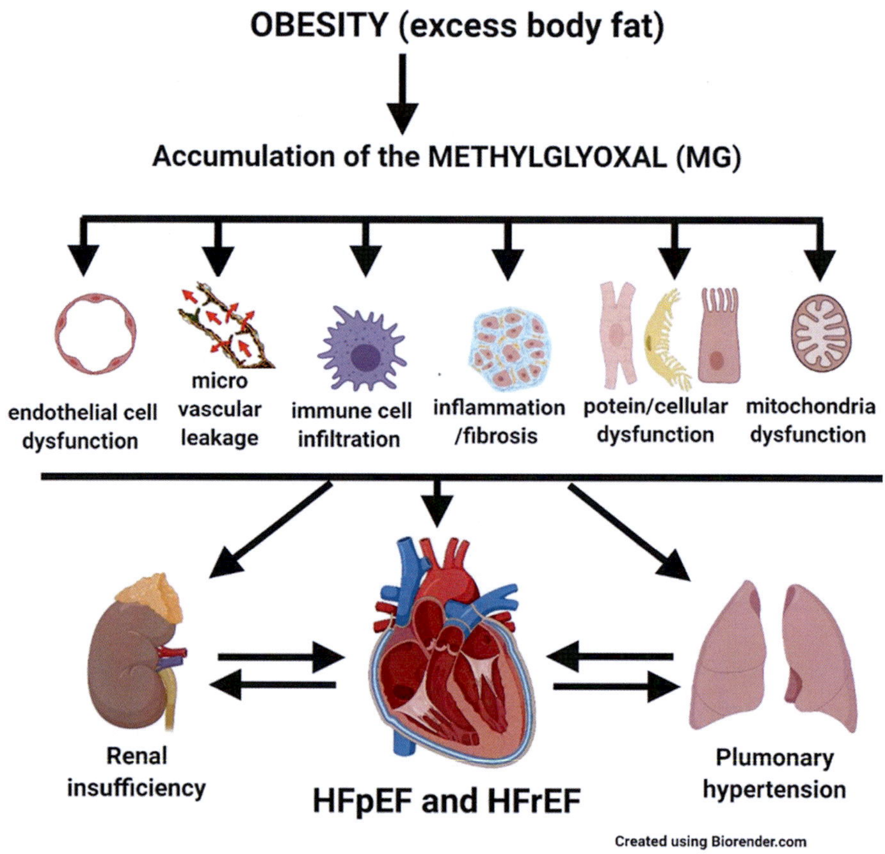

Fig. 19.3 Pathways by which elevated MG contribute to heart failure (HF) with preserved and reduced ejection fractions. Elevated MG can dysregulate microvascular endothelial cells leading to increased microvascular permeability (arising from decreased expression of tight junction proteins), immune cell infiltration, inflammation and fibrosis. These changes can simultaneously occur in the microvasculature of heart, lungs and kidneys

Glyoxalase-1

The human glyoxalase 1 gene (*GLO1*) is located in chromosome 6 (6p21.3-p21.1) between HLA and the centromere [53]. This region is a hotspot for functional copy number variation (CNV), also known as copy number polymorphisms, and can give rise to as much as four-fold higher expression of Glo1 [54, 55]. Thornalley and colleagues reported that the prevalence of *GLO1* copy number increase in the human population is approximately 2% [56]. Gene cloning showed that human *GLO1*-coding regions consists of 12 kb with introns separating five exons. There are two different genetic variants of GLO1 at position 111 that results in glutamate to alanine substitution and three allozymes of Glo1 [31]. Mouse *Glo1* is located in chromosome 17 at locus 17a3.3, approximately 3 cM from the Ss locus of the H-2 histocompatibility region [57, 58].

The activity of Glo1 is regulated both at the transcription and post translational levels [31, 57, 59–61]. The promoter region of human *GLO1* has a functionally operative regulatory insulin-response element (IRE), a metal-response element (MRE) and an antioxidant response element (ARE) [31, 61, 62]. Binding of the anti-oxidant transcription factor nuclear factor erythroid 2-related factor 2 (Nrf2) to the ARE region of *GLO1* induces expression of Glo1 [31, 62, 63]. Thus, activators of Nrf2 will induce expression of Glo1 [63, 64]. Under hypoxic conditions binding of hypoxia inducible factor 1α (HIF-1α) to ARE suppresses Glo1 expression, hence a reason why hypoxia induces MG stress. Ranganathan et al., also reported a two-fold reproducible increase in reporter activity of Glo1 with insulin and ZnCl2 treatments, indicating a functionally operative insulin response element (IRE) and metal response element (MRE) [61].

Glo1 activity is also regulated by post-translational modifications. N-terminal acetylation at alanine 2, a vicinal disulfide bridge between cysteine residues 19 and 20 (mouse and human), mixed disulfide with glutathione on cysteine 139 (mouse and human), phosphorylation at Thr 107 (mouse and human) and NO modifications at Cys-139 have been identified and shown to modulate the activity of Glo1 [31, 65]. Morgenstern and colleagues showed that phosphorylation at Thr 107 by Ca^{2+}/calmodulin-dependent kinase II delta (CaMKIIδ) enhances the catalytic efficiency of Glo1 and blunt formation MG adducts [66]. However, TNF-α induced phosphorylation of Glo1 at Thr 106(7) by protein kinase A (PKA) decreases the ability of Glo1 to degrade MG. de Hemptinne and colleagues reported that NO modification and phosphorylation of Glo1 at Thr 106(7) also suppresses TNFα-induced NF-kB-dependent reporter gene expression [67]. NO-modification and glutathionylation at Cys139 inhibit the ability of Glo1 to metabolize MH-GSH hemithioacetal. However, N-acetylation and the oxidation of Cys19 and 20, do not impact Glo1 activity.

GSH which is crucial for the formation of MG-GSH hemithioacetal is synthesized by two ATP-dependent sequential reactions; ligation of L-glutamate and L-cysteine by γ-glutamylcysteine ligase (GCL; EC 6.3.2.2), and the addition of glycine to γ-glutamylcysteine by glutathione synthetase (GSS; EC 6.3.2.3) [68]. Glutathione reductase (EC 1.8.1.7) also known as glutathione-disulfide reductase (GSR) catalyzes the reduction of oxidized glutathione (GSSG) to the reduced glutathione (GSH), needed for maintaining the reducing environment inside cells. Like Glo1, GCL, GSS and GSR expression are under regulatory control by the anti-oxidant transcription factor, Nrf2. We earlier reported reduced GSH levels in diabetic myocytes [68, 69]. A reduction in the GSH pool will lead to reduced rate of formation of the MG-GSH hemithioacetal. The inflammatory condition induced by increased adiposity also reduces expression of Glo1. The combination of a reduction in GSH and Glo1 will result in accumulation of MG.

Elevated MG is also a potent inducer of mitochondria reactive oxygen species (ROS), by decreasing the activities of complexes I, II, and V of the electron transport chain. MG-induced increase in ROS will further deplete the GSH pool inside cells. We showed that increasing expressing Glo1 in cardiac cells using the promoter of the inflammation-induced protein endothelin-1, after induction of diabetes but before the onset of HF prevented down regulation of GSH [30, 70]. Several mechanisms are likely responsible for the preservation of GSH. First, by having sufficient amount of Glo1 inside cells, there is no buildup of MG-GSH hemithioacetal that traps GSH. Instead, the MG-GSH hemithioacetal is converted by Glo1 to the S-D-lactoylglutathione and the latter is degraded by Glo2, recycling the GSH. It should be noted that Glo2 expression is regulated by a separate set of transcription factors including p63 and p73, steroid hormones, androgen receptor and phosphatase and tensin homologue (PTEN)/phosphoinositide 3-kinase (PI3K)/protein kinase B (AKT)/mammalian target of rapamycin (mTOR) signaling [71–74]. Second, lowering MG will attenuate mitochondria ROS production, thereby "freeing up" GSH. Third, lowering MG will upregulate the ARE-Nrf2 pathway and expression of γ-glutamylcysteine ligase and glutathione reductase, leading to an increase in synthesis of new GSH [68].

Effects of MG Accumulation

Studies have reported increased MG levels in plasma of patients and tissues and plasma of animals with inflammatory conditions including obesity, diabetes, end-stage renal disease, neuro degenerative disorders, infections, and cancers [31, 62, 75–80]. This increase in MG is arising from both increases in MG synthesis and reductions in MG degradation. As shown in Fig. 19.1, an increase in MG production can arises from multiple sources including glycolysis, polyol pathway, and degradation of triacylglycerol, fatty acid and amino acid metabolism. It can also arise from reductions in expression of GSH and Glo1. In addition to reversibly reacting with arginine, lysine and cysteine residues on proteins, at supraphysiologic levels, MG also irreversibly react with nucleophilic sites of proteins, phospholipids, and DNA to form adducts. Studies have identified the MG adducts, hydroimidazolones (MG-H1, MG-H2 and MG-H3), argpyrimidine, tetrahydropyrimidine (THP), N_{ε}-(1-carboxyethyl)lysine (CEL) and 1,3-di(N_{ε}-lysino)-4-methyl-imidazolium on arginine and lysine residues of proteins. Carboxymethyl/carboxyethyl adducts have also been identified on the nitrogen atoms on the head groups of select basic phospholipids. The cross-link adduct 2-ammonio-6-((2-[(4-ammonio-5-oxido-5-oxopentyl)amino]-4-methyl-4,5-dihydro-1H-imidazol-5-ylidene)amino)hexanoate (MODIC) formed between arginine and lysine residues with the same or different proteins have also been reported. Imidazopurinones, 6,7-dihydro-6,7-dihydroxy-7-methylimidazo-[2,3-b]purine-9(8)one and 6,7-dihydro-6,7-dihydroxy-6-methylimidazo-[2,3-b]purine-9(8)one (MGdG) are the primary products formed when MG reacts with DNA [31, 81–83]. Cleaved peptides containing MG adduct, in particular MG-H1 and N_ε-(1-carboxyethyl)lysine can serve as agonist for the receptor of AGEs (RAGE) to activate a number of cellular process including pro-inflammatory pathways to exacerbate inflammation [31]. We earlier showed that not all adducts formed on long-lived proteins negatively impact their functions [70]. Some adducts may have no functional consequence, suggesting that some proteins may serve to scavenge reactive carbonyl species. Cell have two primary degradation mechanisms; lysosomes and proteasomes. Lysosomes degrades extracellular damaged proteins via endocytosis, phagocytosis and autophagy using an array of proteases. Proteins that are extensively modified by MG are resistant and have slower degradation rates in lysosomes. The proteasome generally degraded intracellular misfolded or damaged proteins that adversely affect cellular functions. They undergo poly-ubiquitination and degradation via the 20S core proteasome that contains the proteolytic activities of the 26S proteasome. However, studies have shown that AGE-modified proteins, including MG-modified proteins are resistant against the 20S proteasome degradation [31, 84–86].

To date, no enzymes have been identified in mammalian cells that can degrade MG adducts after they are formed on proteins. The Parkinson disease protein 7 also known as DJ-1 was initially thought to have deglycase activity by hydrolyzing the hemithioacetals and hemiaminals formed when MG reacted with the thiol and amino groups of proteins. However, in a recent report, Andreeva et al., reported that the hemithioacetal formed between MG and reduced glutathione apparent spontaneously decomposes with a half-life of 12 s and this was mistakenly viewed as deglycase activity of DJ-1 [87].

Effects of Elevated MG on the Function of Vascular Endothelial Cells

Coronary atherosclerosis (CA) is a well-known cardiovascular disease (CVD) in obese individuals, especially in those with T2DM [88]. This macrovascular pathology has been extensively studied, and arises from oxidation and accumulation of lipids, foam cell formation, and inflammation that lead to atheromatous plaques in the intima of the arterial walls [89]. Migration and proliferation of smooth muscle cell from the media into the intima and the formation of a fibrous caps also occurs. The impacts of obesity on the coronary microvasculature of the heart are less well studied.

The coronary microvasculature is the system of small blood vessels that transport nutrients to and remove waste from the heart. They also regulate vascular tone and perfusion pressures in the heart. The smallest of these blood vessels are called capillaries [90, 91]. Arterioles and meta-arterioles which contain two to four layers of circumferentially arranged smooth muscle cells (SMCs) that transport nutrients to capillaries. Post capillary venules that transport waste from tissues usually have larger internal diameters than arterioles and contain 1–2 layers of SMCs [90, 91]. The endothelium is a highly specialized, single layer of cells lining the lumen of these microvessels. Endothelial cells (ECs) within the microvasculature regulates vascular tone and perfusion pressures by synthesizing and releasing vasodilatating substances including nitric oxide, prostacyclin, carbon monoxide, and vasoconstricting substances (endothelins, thromboxane A and the endothelium-derived constrictor factor) that regulate the contractile status of vascular SMCs [92–94]. ECs of capillaries are surrounded by pericytes and allow for the diffusion of solutes from the blood into the tissues and vice versa take place [90, 91]. The ECs in post capillary venules also aid in transvascular transport of plasma proteins and immune cells from the blood into tissues by synthesizing adhesion molecules (e.g., platelet endothelial cell adhesion molecule-1 PECAM-1, vascular cell adhesion protein 1, VCAM, and vascular adhesion protein-1, VAP-1). ECs in arterioles and downstream of post capillary venules express large amounts of tight junction (TJ) proteins to prevent para-and trans-cellular passage of substances and immune cells from the blood into myocardium [95].

MG is a potent endothelial cell toxin. ECs have a relatively lower Glo1 content compared to juxtaposed SMCs [96]. Earlier we showed that bathing arterioles of anesthetized control rats with MG for 30 min, attenuated the ability of the endothelial nitric oxide synthase (eNOS)-activating ligand adenosine diphosphate to vasodilate these arterioles, akin to that seen in diabetes [96]. Interestingly, bathing arterioles from control rats with MG did not have any effect on the ability of the SMCs relaxing agent nitroglycerin to vasodilate these vessels. Others have suggested that this may be due in part to MG decreasing phosphorylation of eNOS at Ser^{1177}, which causes eNOS to switch from generation of NO to the generation of superoxide anion [97]. We also show that chronic exposure of ECs to MG decreases expression of tight junction proteins [96]. We also showed that increasing expressing Glo1 in ECs using the promoter of the inflammation-induced protein endothelin-1, shortly after induction of diabetes, prevented the loss of vasodilation of arterioles to ADP challenge [96]. Whether MG generated by ECs are diffusing into SMCs and becoming degraded or MG production in SMCs is being upregulated and this MG diffusing out and dysregulating ECs remains to be delineated.

Elevated Levels of Methylglyoxal in Obese Patients with and Without Type 2 Diabetes Mellitus and Heart Failure

The human heart contains several different cell types; cardiac myocytes (one third), smooth muscle and endothelial cells of the coronary vasculature, fibroblasts and other connective tissue cells, mast and immune system-related cells and cardiac stem cells [98] and dysregulation of any of these cells could impair myocardial function. In the non-disease of the heart (healthy), myocytes obtain 85% of their ATP from β-oxidation of fatty acids and the remaining 15% from aerobic glycolysis. ECs and SMCs obtain ~75–80% of their ATP from anaerobic glycolysis and the remaining 25–20% from aerobic glycolysis [99–102]. Since there are ~2 X more ECs and SMCs in the heart than cardiomyocytes, and anaerobic glycolysis affords ~15 X less ATP molecules per molecule of glucose, the glucose demand in ECs and SMCs is greater than that of cardiac myocytes [98, 102, 103].

During insulin resistance and/or diabetes, conditions typically present in individuals with chronic obesity, G6PDH-mediated entry of glucose into the pentose phosphate pathway is attenuated, lowering production of the primary intracellular reductant, NADPH [104, 105]. Accumulation of ROS including superoxide anions will result in DNA strand breaks and activation of polyADP-ribose polymerase (PARP1). Activated PARP1 ribosylates and reduces the activity of glyceraldehyde-3-phosphate dehydrogenase (GAPDH), the enzyme that converts glyceraldehyde-3-phosphate (G3P) to 1,3-bisphosphoglycerate (1,3BPG) resulting in the build-up of G3P [106, 107]. Increased degradation of G3P will increase MG production. The inflammation associated with insulin resistance and diabetes and increased MG production will downregulate Nrf2 and Glo1. Excess glucose also induces arginase activity, which consumes the NO-precursor arginine and uncouples the NO generating from eNOS activation [104]. Impairment in the function of ECs will lead to dysfunction of the coronary microvasculature of the heart.

Studies have reported increased MG and D-lactate in sera of obese adults. [108]. A low calorie diet and Roux-en-Y gastric bypass also reduced plasma MG in obese women with T2DM, indicating that elevated MG is associated with obesity [109]. Recently, Schalkwijk and colleagues found that elevated plasma MG level during fasting and post-oral glucose tolerance test were associated with [110]. Thornalley and colleagues earlier showed that a co-formulation consisting of the Nrf2 activators *trans*-resveratrol and hesperetin improved glycemic control, and vascular inflammation and function in healthy overweight and obese individuals by inducing expression of Glo1 [111], emphasizing that accumulation of MG is contributing to vascular inflammation. Chronic elevation in MG will impair the function ECs in the coronary microvasculature of the heart resulting in microvascular leakage, micro-ischemia, immune cell infiltration into the myocardium, inflammation and fibrosis [30], (Fig. 19.3). The function of ECs in the macrovasculature of the heart may also be negatively impacted, leading to the development of arteriosclerosis [89]. Accumulation of MG will also result in irreversible post-translational modification and dysregulate cellular proteins including those in the mitochondria leading to deceased ATP production and cardiac dysfunction [70, 112]. The ECs of the vasculature of the kidneys and lungs will also be negatively impacted, which could help explain the deficits in pulmonary and renal functions in individuals with HFpEF (Fig. 19.3).

Here we add to the body of literature showing elevated levels of MG in plasma of male obese individuals with without T2DM and HFpEF relative to non-obese male controls. These de-identified samples (twenty five per group) were collected from the Veteran's Administration Hospital, Omaha, NE and from the Diabetes Clinic, University of Nebraska Medical Center. Age, weight, body mass indices (BMI), HbA1c, ejection fraction, HOMA-IR, C-reactive protein and glomerular filtration rate were obtained. Plasma MG levels were determined using the derivatization procedure recently optimized in our laboratory [30, 113]. A calibration curve was generated using commercially available 2-methylquinoxaline, the product formed between MG and o-phenylenediamine for quantitation.

In this cohort, the mean age of obese individuals was not significantly different from control (47.10 ± 2.70 years and 47.64 ± 2.26, $p > 0.05$). Mean age of obese individuals with T2DM, and T2DM and HFpEF were older (59.00 ± 1.51 and 61.40 ± 1.80 years, $p < 0.05$), Fig. 19.5a. Weight, BMI, and HbA1c of obese individuals with and without T2DM and T2DM HF are also shown in Figs. 19.4b, c, d). Nearly all obese patients in this study were of class II ($<35\ kg/m^2$ BMI $< 40\ kg/m^2$). HOMA-IR (insulin resistance), and plasma levels of C-reactive proteins were higher in obese individuals with T2DM and HFpEF (Figs. 19.5a, b). Glomerular filtration rate was reduced in obese individuals with T2DM and HFpEF (Fig. 19.5c). Percent ejection fraction in all patient groups were >50%, although it was lower in obese individuals with T2DM and HF (Fig. 19.5d). MG levels were 4X higher in obese individuals compared with non-obese individuals of similar ages and persisted in obese individuals with and without T2DM and HFpEF (Fig. 19.5e). These data suggest that plasma MG levels increase with obesity and persist thereafter. A persistent increase in plasma MG will impair the function of ECs of the coronary micro-vasculature (and macro), negatively impacting vascular permeability, perfusion pressures and the development of arteriosclerosis.

Conclusion

This review is intended to introduce readers to the most potent and reactive α-oxoaldehyde species generated in mammals (including humans), namely methylglyoxal (MG) and its role in the development of heart failure with preserved ejection fraction (HFpEF). Data is also provided showing MG is elevated in plasma of obese patients with HFpEF. The primary source of MG in mammals is glycolysis, although smaller amounts are generated from other sources included lipid oxidation. In healthy cells, this MG reversibly binds to exposed arginine, lysine and cysteine residues on select proteins inside cells. During inflammatory conditions including obesity (adipose tissue is an inflammatory organ) and diabetes, MG levels increase

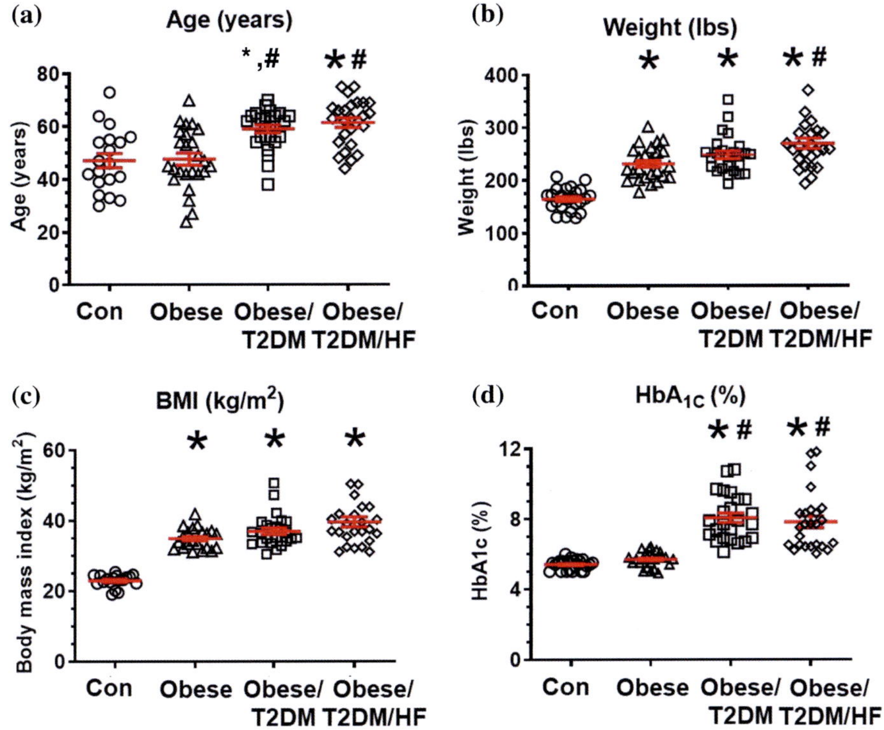

Fig. 19.4 General characteristics of patients used in the study. Panels A-D show physical characteristics including age, weight and body-mass indices and HbA1c, in non-obese controls, class II obese, class II obese with type 2 diabetes mellitus (T2DM) and class II obese with T2DM and heart failure Data shown are mean ± SEM for n = 25 patients per group. *denotes significantly different control, # denotes significantly different from obese

in plasma (including humans shown in Fig. 19.5) in part from increased synthesis and reduced degradation. Accumulation of MG results in irreversible binding of MG to basic residues on proteins, lipid and DNA which can impair cellular function. MG produced by one cell can diffuse and negatively impact the function of juxtaposed cells. Thus, it stands to reasoning that increasing expression of Glo1 and/or the enzymes that synthesizes GSH, including GCL and GSS should blunt accumulation of MG and attenuate dysregulation of ECs seen during inflammation conditions. However, it should be pointed out that expression of Glo1, GCL and GSS are regulated by the anti-oxidant transcription factor Nrf2. The Nrf2 expression is downregulated under inflammation conditions. Thus, stimulators of Nrf2 may have only transient benefits if the underlying inflammatory conditions is not suppressed. Earlier, we used the promoter of the inflammation-induced protein endothelin-1 in an adeno-associated viral (AAV) vector to increase expression of Glo1 in SMCs and myocytes in diabetic rats (that have underlying inflammation) and showed that this strategy attenuated ECs dysregulation and HFpEF. Interestingly, increasing Glo1

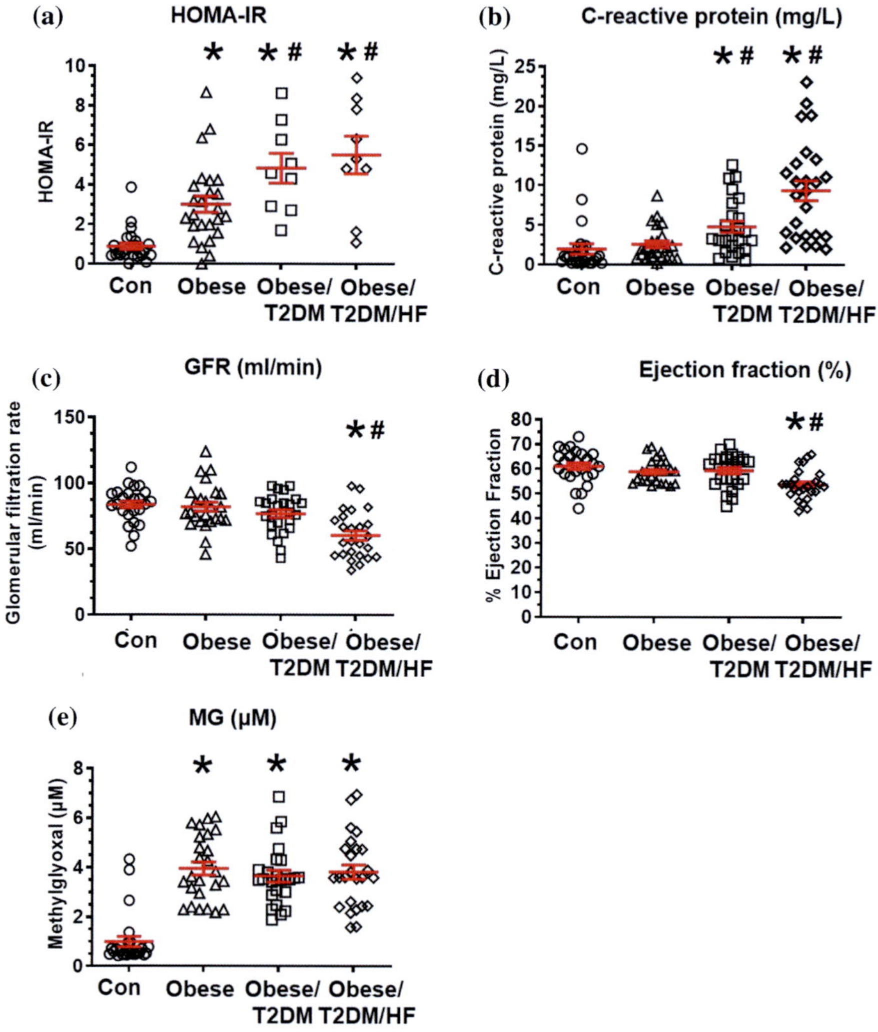

Fig. 19.5 Parameters measured in patients of this study. Panels A-E show HOMA-IR, C-reactive protein, glomerular filtration rate, ejection fraction and plasma MG levels, respectively, in non-obese controls, class II obese, class II obese with type 2 diabetes mellitus (T2DM) and class II obese with T2DM and heart failure. Data shown are mean ± SEM for n = 25 patients per group. *denotes significantly different control, # denotes significantly different from obese

only modestly lowered blood glucose levels. However, it should be pointed out that although AAV may become integrated into the DNA of cells, a specific helper virus is needed for long term replication. Additionally, AAV effects may be rapidly diluted in rapidly dividing or replenishing cells, including ECs.

Acknowledgements This work was funded in part by a grant from the National Institutes of Health (HL151602 - 01A1-R56, and P30 MH062261) to KRB

References

1. Kushner RF, Kahan S (2018) Introduction: the state of obesity in 2017. Med Clin North Am 102:1–11
2. Swinburn BA, Kraak VI, Allender S et al (2019) The global Syndemic of obesity, undernutrition, and climate change: the lancet commission report. Lancet 393:791–846
3. World Health Organization Fact Sheet, Obesity 2020. https://www.who.int/news-room/fact-sheets/detail/obesity-and-overweight. Accessed August 18th 2021
4. World Obesity. Global Obesity Observatory. https://data.worldobesity.org/%23MX%7C1%7CA%7CF. Accessed August 18th 2021
5. Byrd AS, Toth AT, Stanford FC (2018) Racial disparities in obesity treatment. Curr Obes Rep 7(2):130–138
6. Donataccio MP, Vanzo A, Bosello O (2020) Obesity paradox and heart failure. Eat Weight Disord. August, 1–11 (Online ahead of print)
7. Hainer V, Aldhoon-Hainerova I (2013) Obesity paradox does exist. Diabetes Care 36(Suppl 2):S276–S281
8. Anekwe CV, Jarrell AR, Townsend MJ et al (2020) Socioeconomics of Obesity. Curr Obes Rep 9(3):272–279
9. Yoon S, Eom GH (2019) Heart failure with preserved ejection fraction: present status and future directions. Exp Mol Med 51:1–9
10. Iyngkaran P, Thomas MC, Neil C et al (2020) The heart failure with preserved ejection fraction conundrum-redefining the problem and finding common ground? Curr Heart Fail Rep 17:34–42
11. Chrysant SG, Chrysant GS (1995) Obesity-related heart failure with preserved ejection fraction: new treatment strategies. Hosp Pract 47(67–72):2019
12. Iwakura K (2019) Heart failure in patients with type 2 diabetes mellitus: assessment with echocardiography and effects of antihyperglycemic treatments. J Echocardiogr 17:177–186
13. Savarese G, D'Amario D (2018) Sex differences in heart failure. Adv Exp Med Biol 1065:529–544
14. Sharma K, Kass DA (2014) Heart failure with preserved ejection fraction: mechanisms, clinical features, and therapies. Circ Res 115:79–96
15. Oeing CU, Tschope C, Pieske B (2016) The new ESC guidelines for acute and chronic heart failure 2016. Herz 41:655–663
16. Nadar SK, Tariq O (2018) What is heart failure with mid-range ejection fraction? a new subgroup of patients with heart failure. Card Fail Rev 4:6–8
17. Carson PE, Anand IS, Win S et al (2015) The hospitalization burden and post-hospitalization mortality risk in heart failure with preserved ejection fraction: results from the I-PRESERVE Trial (Irbesartan in Heart Failure and Preserved Ejection Fraction). JACC Heart Fail 3:429–441
18. Sharma K, Hill T, Grams M et al (2015) Outcomes and worsening renal function in patients hospitalized with heart failure with preserved ejection fraction. Am J Cardiol 116:1534–1540
19. Abdul-Rahim AH, Perez AC, MacIsaac RL et al (2017) Candesartan in heart failure assessment of reduction in M, Morbidity P and the Irbesartan in heart failure with preserved systolic function steering C, risk of stroke in chronic heart failure patients with preserved ejection fraction, but without atrial fibrillation: analysis of the CHARM-Preserved and I-Preserve trials. Eur Heart J 38:742–750
20. McMurray JJ, Carson PE, Komajda M et al (2008) Heart failure with preserved ejection fraction: clinical characteristics of 4133 patients enrolled in the I-PRESERVE trial. Eur J Heart Fail 10:149–156

21. Campbell RT, Jhund PS, Castagno D et al (2012) What have we learned about patients with heart failure and preserved ejection fraction from DIG-PEF, CHARM-preserved, and I-PRESERVE? J Am Coll Cardiol 60:2349–2356
22. Widimsky J Sr (2015) Effect of spironolactone in patients with heart failure and preserved left ventricular function—TOPCAT study. Vnitr Lek 61:376–380
23. Desai AS, Jhund PS. After TOPCAT (2016). What to do now in heart failure with preserved ejection fraction. Eur Heart J 37(41):3135–3140
24. Solomon SD, Rizkala AR, Gong J et al (2017) Angiotensin receptor neprilysin inhibition in heart failure with preserved ejection fraction: rationale and design of the PARAGON-HF Trial. JACC Heart Fail 5:471–482
25. Kuno T, Ueyama H, Fujisaki T et al (2020) Meta-analysis evaluating the effects of renin-angiotensin-aldosterone system blockade on outcomes of heart failure with preserved ejection fraction. Am J Cardiol 125:1187–1193
26. Galli M, D'Amario D, Sofia C et al (2018) Clinical potential relevance of metabolic properties of SGLT2 inhibitors in patients with heart failure. Expert Opin Drug Metab Toxicol 14:1273–1285
27. Shinohara M, Thornalley PJ, Giardino I et al (1998) Overexpression of glyoxalase-I in bovine endothelial cells inhibits intracellular advanced glycation endproduct formation and prevents hyperglycemia-induced increases in macromolecular endocytosis. J Clin Invest 101(1):142–147
28. Brouwers O, Niessen PM, Miyata T et al (2014) Glyoxalase-1 overexpression reduces endothelial dysfunction and attenuates early renal impairment in a rat model of diabetes. Diabetologia 57:224–235
29. Vulesevic B, McNeill B, Giacco F et al (2016) Methylglyoxal-induced endothelial cell loss and inflammation contribute to the development of diabetic cardiomyopathy. Diabetes 65:1699–1713
30. Alomar FA, Al-Rubaish A, Al-Muhanna F, Al-Amein, AK, Singh J, and Bidasee KR (2020) Adeno-associated viral transfer of glyoxalase-1 blunts carbonyl and oxidative stresses in hearts of type 1 diabetic rats. Antioxidants, 9(7):592–614
31. Schalkwijk CG, Stehouwer CDA (2020) Methylglyoxal, a highly reactive dicarbonyl compound, in diabetes, its vascular complications, and other age-related diseases. Physiol Rev 100:407–461
32. Thornalley PJ (1990) The glyoxalase system: new developments towards functional characterization of a metabolic pathway fundamental to biological life. Biochem J 269:1–11
33. Richard JP (1993) Mechanism for the formation of methylglyoxal from triosephosphates. Biochem Soc Trans 21:549–553
34. Kold-Christensen R, Johannsen M (2020) Methylglyoxal metabolism and aging-related disease: moving from correlation toward causation. Trends Endocrinol Metab 31:81–92
35. Thornalley PJ (1988) Modification of the glyoxalase system in human red blood cells by glucose in vitro. Biochem J 254:751–755
36. Kalapos MP (1994) Methylglyoxal toxicity in mammals. Toxicol Lett 73:3–24
37. Nigro C, Leone A, Raciti GA, Longo M, Mirr, P, Formisano P, Beguinot F and Miele C (2017) Methylglyoxal-Glyoxalase 1 balance: the root of vascular damage. Int J Mol Sci 18(1):188–202
38. Henning C, Liehr K, Girndt M et al (2014) Extending the spectrum of a-dicarbonyl compounds in vivo. J Biol Chem 289:28676–28688
39. Rabbani N, Thornalley PJ (2014) Measurement of methylglyoxal by stable isotopic dilution analysis LC-MS/MS with corroborative prediction in physiological samples. Nat Protoc 9:1969–1979
40. Scheijen JL, Schalkwijk CG (2014) Quantification of glyoxal, methylglyoxal and 3-deoxyglucosone in blood and plasma by ultra performance liquid chromatography tandem mass spectrometry: evaluation of blood specimen. Clin Chem Lab Med 52:85–91

41. Lo TW, Westwood ME, McLellan, Selwood, T and Thornalley PJ. (1994) Binding and modification of proteins by methylglyoxal under physiological conditions. A kinetic and mechanistic study with N alpha-acetylarginine, N alphaacetylcysteine, and N alphaacetyllysine, and bovine serum albumin. J Biol Chem 269 (51):32299–32305
42. Jia X, Chang T, Wilson TW, and Wu L (2012) Methylglyoxal mediates adipocyte proliferation by increasing phosphorylation of Akt1. PLoS One 7(5):e36610
43. Auburger G, Kurz A (2011) The role of glyoxalases for sugar stress and aging, with relevance for dyskinesia, anxiety, dementia and Parkinson's disease. Aging (Albany NY) 3:5–9
44. Hovatta I, Tennant RS, Helton R et al (2005) Glyoxalase 1 and glutathione reductase 1 regulate anxiety in mice. Nature 438:662–666
45. McMurray KM, Du X, Brownlee M, Palmer AA (2016) Neuronal overexpression of Glo1 or amygdalar microinjection of methylglyoxal is sufficient to regulate anxiety-like behavior in mice. Behav Brain Res 301:119–123
46. Distler MG, Plant LD, Sokoloff G et al (2012) Glyoxalase 1 increases anxiety by reducing GABAA receptor agonist methylglyoxal. J Clin Invest 122:2306–2315
47. Jakubcakova V, Curzi ML, Flachskamm C et al (2013) The glycolytic metabolite methylglyoxal induces changes in vigilance by generating low-amplitude non-REM sleep. J Psychopharmacol 27:1070–1075
48. Nokin MJ, Durieux F, Bellier J et al (2017) Hormetic potential of methylglyoxal, a side-product of glycolysis, in switching tumours from growth to death. Sci Rep 7:11722
49. Zemva J, Fink CA, Fleming TH et al (2017) Hormesis enables cells to handle accumulating toxic metabolites during increased energy flux. Redox Biol 13:674–686
50. Rabbani N, Thornalley PJ (2019) Glyoxalase 1 modulation in obesity and diabetes. Antioxid Redox Signal 30:354–374
51. Gugliucci A, Caccavello R (2020) Optimized sensitive and inexpensive method to measure D-lactate as a surrogate marker of methylglyoxal fluxes in metabolically relevant contexts. Methods S1046–2023(20):30101–30108
52. Vander Jagt DL (2003) Hunsaker LA Methylglyoxal metabolism and diabetic complications: roles of aldose reductase, glyoxalase-I, betaine aldehyde dehydrogenase and 2-oxoaldehyde dehydrogenase. Chem Biol Interact 143–144:341–351
53. Tripodis N, Mason R, Humphray SJ, Herberg JA, Trowsdale J, Nizetic D, Senger G, and Ragoussis J (1998). Physical map of human 6p21.2–6p21.3: region flanking the centromeric end of the major histocompatibility complex. Genome Res 8:631–643
54. Shafie A, Xue M, Thornalley PJ, Rabbani N (2014) Copy number variation of glyoxalase I. Biochem Soc Trans 42:500–503
55. Wong KK, deLeeuw RJ, Dosanjh NS et al (2007) A comprehensive analysis of common copynumber variations in the human genome. Am J Hum Genet 80:91–104
56. Rabbani N, Xue M, Thornalley PJ (2014) Activity, regulation, copy number and function in the glyoxalase system. Biochem Soc Trans 42:419–424
57. Thornalley PJ (2003) Glyoxalase I–structure, function and a critical role in the enzymatic defence against glycation. Biochem Soc Trans 31:1343–1348
58. Shafie A, Xue M, Barker G et al (2016) Reappraisal of putative glyoxalase 1-deficient mouse and dicarbonyl stress on embryonic stem cells in vitro. Biochem J 473:4255–4270
59. Bellahcène A, Nokin MJ, Castronovo V, Schalkwijk C (2017) Methylglyoxal-derived stress: an emerging biological factor involved in the onset and progression of cancer. Semin Cancer Biol 49:64–74
60. Antognelli C, Palumbo I, Aristei C, Talesa VN (2014) Glyoxalase I inhibition induces apoptosis in irradiated MCF-7 cells via a novel mechanism involving Hsp27, p 53 and NF-kB. Br J Cancer 111:395–406
61. Ranganathan, Ciaccio PJ, Walsh ES, Tew KD (1999) Genomic sequence of human glyoxalase-I: analysis of promoter activity and its regulation. Gene 240:149–155
62. Thornalley PJ (2003) Glyoxalase I–structure, function and a critical role in the enzymatic defence against glycation. Biochem Soc Trans 31:1343–1348

63. Thornalley PJ (2003) Glyoxalase I–structure, function and a critical role in the enzymatic defence against glycation. Biochem Soc Trans 31:1343–1348
64. He F, Ru X, Wen T (2020) NRF2, a transcription factor for stress response and beyond. Int J Mol Sci 21(13):4777
65. Xue M, Rabbani N, Momiji H, Imbasi P, Anwar MM, Kitteringham N, Park KB, Souma T, Moriguchi T, Yamamoto M, and Thornalley, PJ (2012) Transcriptional control of glyoxalase 1 by Nrf2 provides a stress-responsive defence against dicarbonyl glycation. Biochem J 443(1):213–222
66. Mitsumoto A, Kim KR, Oshima G et al (1999) Glyoxalase I is a novel nitric-oxide-responsive protein. Biochem J 344:837–844
67. Morgenstern J, Katz S, Krebs-Haupenthal J, Chen J, Saadatmand A, Cortizo FG, Moraru A, Zemva J, Campos MC, Teleman A, Backs J, Nawroth P, and Fleming T (2020) Phosphorylation of T107 by CamKIIdelta regulates the detoxification efficiency and proteomic integrity of glyoxalase 1. Cell Rep 32(1):108160
68. de Hemptinne V, Rondas D, Toepoel M, Vancompernolle K (2009) Phosphorylation on Thr-106 and NO-modification of glyoxalase I suppress the TNF-induced transcriptional activity of NF-kappaB. Mol Cell Biochem 325:169–178
69. Forman HJ, Zhang H, Rinna A (2009) Glutathione: overview of its protective roles, measurement, and biosynthesis. Mol Aspects Med 30:1–12
70. Venugopal R, Jaiswal AK (1998) Nrf2 and Nrf1 in association with Jun proteins regulate antioxidant response element-mediated expression and coordinated induction of genes encoding detoxifying enzymes. Oncogene 17(24):3145–3156
71. Shao CH, Tian C, Ouyang S et al (2012) Carbonylation induces heterogeneity in cardiac ryanodine receptor function in diabetes mellitus. Mol Pharmacol 82:383–399
72. Xu Y, Chen X (2006) Glyoxalase II, a detoxifying enzyme of glycolysis byproduct methylglyoxal and a target of p63 and p73, is a pro-survival factor of the p53 family. J Biol Chem 281:26702–26713
73. Antognelli C, Del Buono C, Baldracchini F et al (2007) Alteration of glyoxalase genes expression in response to testosterone in LNCaP and PC3 human prostate cancer cells. Cancer Biol Ther 6:1880–1888
74. Antognelli C, Ferri I, Bellezza G et al (2017) Glyoxalase 2 drives tumorigenesis in human prostate cells in a mechanism involving androgen receptor and p53–p21 axis. Mol Carcinog 56:2112–2126
75. Talesa VN, Ferri I, Bellezza G et al (2017) Glyoxalase 2 is involved in human prostate cancer progression as part of a mechanism driven bPTEN/PI3K/AKT/mTOR Signaling with involvement of PKM2 and ER_. Prostate 77:196–210
76. Gugliucci A (2017) Formation of fructose-mediated advanced glycation end products and their roles in metabolic and inflammatory diseases. Adv Nutr 8:54–62
77. Yacoub R, Nugent M, Cai C, Nadkarni GN, Chaves, LD Abyad S, Honan AM, Thomas, SA , Zheng W, Valiyaparambil SA, Bryniarski MA, Sun Y, Buck M, Genco RJ, Quigg RJ , He, JC, and Uribarri, J (2017) Advanced glycation end products dietary restriction effects on bacterial gut microbiota in peritoneal dialysis patients; a randomized open label controlled trial. PLoS One 12:e0184789
78. Lloret A, Calzone R, Dunster C et al (2008) Different patterns of in vivo pro-oxidant states in a set of cancer- or aging-related genetic diseases. Free Radic Biol Med 44:495–503
79. Kuhla B, Luth HJ, Haferburg D et al (2005) Methylglyoxal, glyoxal, and their detoxification in Alzheimer's disease. Ann N Y Acad Sci 1043:211–216
80. Itokawa M, Miyashita M, Arai M, Dan T, Takahashi K, Tokunaga T, Ishimoto K, Toriumi K, Ichikawa T, Horiuchi Y, Kobori A, Usami S, Yoshikawa T, Amano N, Washizuka S, Okazaki Y, and Miyata T (2018). Pyridoxamine: A novel treatment for schizophrenia with enhanced carbonyl stress. Psychiatry Clin Neurosci 72(1):35-44
81. Lv H, Wei GY, Guo CS, D et al. (2020) 20S proteasome and glyoxalase 1 activities decrease in erythrocytes derived from Alzheimer's disease patients. Neural Regen Res 15:178–183

82. Shuck SC, Wuenschell GE, Termini JS (2018) Product studies and mechanistic analysis of the reaction of methylglyoxal with deoxyguanosine. Chem Res Toxicol 31:105–115
83. Rabbani N, Thornalley PJ (2018) Advanced glycation end products in the pathogenesis of chronic kidney disease. Kidney Int 93:803–813
84. Takahashi K (1977) The reactions of phenylglyoxal and related reagents with amino acids. J Biochem 81:395–402
85. Bulteau AL, Verbeke P, Petropoulos I et al (2001) Proteasome inhibition in glyoxal-treated fibroblasts and resistance of glycated glucose-6-phosphate dehydrogenase to 20S proteasome degradation in vitro. J Biol Chem 276:45662–45668
86. Jaisson S, Gillery P (2014) Impaired proteostasis: role in the pathogenesis of diabetes mellitus. Diabetologia 57:1517–1527
87. Thornalley PJ, Battah S, Ahmed N et al (2003) Quantitative screening of advanced glycation endproducts in cellular and extracellular proteins by tandem mass spectrometry. Biochem J 375:581–592
88. Andreeva A, Bekkhozhin Z, Omertassova N, Baizhumanov T, Yeltay G, Akhmetali M, Toibazar D, and Utepbergenov D (2019). The apparent deglycase activity of DJ-1 results from the conversion of free methylglyoxal present in fast equilibrium with hemithioacetals and hemiaminals. J Biol Chem 294(49):18863-18872
89. Martin-Timon I, Sevillano-Collantes C, Segura-Galindo A, Del Canizo-Gomez FJ (2014) Type 2 diabetes and cardiovascular disease: have all risk factors the same strength? World J Diab 5:444–470
90. Libby P, Buring JE, Badimon L et al (2019) Atherosclerosis. Nat Rev Dis Primers 5:56–63
91. Somani A, Steiner ME, Hebbel RP (2010) The dynamic regulation of microcirculatory conduit function: features relevant to transfusion medicine. Transf Apheresis Sci Off J World Apheresis Assoc Off J Euro Soc Haemaph 43:61–68
92. Granger DN, Senchenkova E (2010), In Inflammation and the microcirculation. San Rafael (CA). Morgan & Claypool Life Sciences; 2010. San Rafael (CA): Morgan & Claypool Life Sciences. Integrated Systems Physiology—From Cell to Function
93. Palade GE, Simionescu M, Simionescu N (1979) Structural aspects of the permeability of the microvascular endothelium. . Acta Physiol Scand Suppl 463:11–32
94. Hirase T, Node K (2012) Endothelial dysfunction as a cellular mechanism for vascular failure. Am J Physiol Heart Circul Physiol 302:H499–505
95. Vanhoutte PM, Shimokawa H, Feletou M, Tang EH (2017) Endothelial dysfunction and vascular disease—a 30th anniversary update. Acta Physiol (Oxf) 219:22–96
96. Ling S, Nheu L, Komesaroff PA (2012) Cell adhesion molecules as pharmaceutical target in atherosclerosis. Mini Rev Med Chem 12:175–183
97. Alomar F, Singh J, Jang HS et al (2016) Smooth muscle-generated methylglyoxal impairs endothelial cell-mediated vasodilatation of cerebral microvessels in type 1 diabetic rats. Br J Pharmacol 173:3307–3326
98. Jo-Watanabe A, Ohse T, Nishimatsu H et al (2014) Glyoxalase I reduces glycative and oxidative stress and prevents age-related endothelial dysfunction through modulation of endothelial nitric oxide synthase phosphorylation. Aging Cell 13:519–528
99. Tirziu D, Giordano FJ, Simons M (2010) Cell communications in the heart. Circulation 122:928–937
100. Lopaschuk GD, Jaswal JS (2010) Energy metabolic phenotype of the cardiomyocyte during development, differentiation, and postnatal maturation. J Cardiovasc Pharmacol 56:130–140
101. Eelen G, de Zeeuw P, Treps L et al (2018) Endothelial cell metabolism. Physiol Rev 98:3–58
102. Chiong M, Morales P, Torres G et al (2013) Influence of glucose metabolism on vascular smooth muscle cell proliferation. VASA Zeitschrift fur Gefasskrankheiten 42:8–16
103. Hsieh PC, Davis ME, Lisowski LK, Lee RT (2006) Endothelial-cardiomyocyte interactions in cardiac development and repair. Annu Rev Physiol 68:51–66
104. Brutsaert DL (2003) Cardiac endothelial-myocardial signaling: its role in cardiac growth, contractile performance, and rhythmicity. Physiol Rev 83:59–115

105. Tirziu D, Giordano FJ, Simons M (2010) Cell communications in the heart. Circulation 122:928–937
106. Eelen G, de Zeeuw P, Simons M, Carmeliet P (2015) Endothelial cell metabolism in normal and diseased vasculature. Circ Res 116:1231–1244
107. Zhang Z, Apse K, Pang J, Stanton RC (2000) High glucose inhibits glucose-6-phosphate dehydrogenase via camp in aortic endothelial cells. JBC. 275:40042–40047
108. Drummond GR, Sobey CG (2014) Endothelial NADPH oxidases: which NOX to target in vascular disease? Trends Endocrinol Metab 25:452–463
109. Du X, Matsumura T, Edelstein D, Rossetti L, Zsengeller Z, Szabo C, Brownlee M (2003). Inhibition of GAPDH activity by poly(ADP-ribose) polymerase activates three major pathways of hyperglycemic damage in endothelial cells. J Clin Invest 112:1049–1057
110. Rodríguez-Mortera R, Luevano-Contreras C, Solorio-Meza S et al (2018) Higher D-lactate levels are associated with higher prevalence of small dense low-density lipoprotein in obese adolescents. Clin Chem Lab Med 56(7):1100–1108
111. Maessen DE, Hanssen NM, Lips MA et al (2016) Energy restriction and Roux-en-Y gastric bypass reduce postprandial α-dicarbonyl stress in obese women with type 2 diabetes. Diabetologia 59:2013–2017
112. Hanssen NMJ, Scheijen J, Houben A, van de Waarenburg M, Berendschot T, Webers CAB, Reesink KD, van Greevenbroek MMJ, van der Kallen C, Schaper NC, Schram MT, Henry RMA, Stehouwer CDA, Schalkwijk CG (2021). Fasting and post-oralglucose- load levels of methylglyoxal are associated with microvascular, but not macrovascular, disease in individuals with and without (pre)diabetes: The Maastricht Study. Diabetes Metab, 47:101148
113. Xue M, Weickert MO, Qureshi S et al (2016) Improved glycemic control and vascular function in overweight and obese subjects by glyoxalase 1 inducer formulation. Diabetes 65:2282–2294
114. Shao CH, Capek HL, Patel KP et al (2011) Carbonylation contributes to SERCA2a activity loss and diastolic dysfunction in a rat model of type 1 diabetes. Diabetes 60(3):947–959
115. Hasim S, Hussin NA, Alomar F et al (2014) A glutathione-independent glyoxalase of the DJ-1 superfamily plays an important role in managing metabolically generated methylglyoxal in Candida albicans. J Biol Chem 289:1662–1674

Chapter 20
Marine Derived Bioactives to Combat Obesity: Potential Mechanisms of Action

Indrayani Phadtare, Hitesh Vaidya, and Sukhinder Kaur Cheema

Abstract Obesity is a complex disease caused by an interaction of genetic, dietary, lifestyle, and environmental factors. The prevalence of overweight and obesity is rising at an alarming rate worldwide, and is becoming a major public health concern. Obesity is associated with several metabolic disorders, such as diabetes, hypertension and cardiovascular diseases. Etiology of obesity involve multiple biochemical and physiological pathways such as dyslipidemia, increased oxidative stress, and inflammation; thus, a single target treatment is not beneficial towards managing this disease. Natural sources, such as marine products/foods that contain a number of bioactive molecules, may act in a synergistic way to increase bioavailability and/or action on multiple targets/organs to offer advantages over a single target treatment for obesity. Marine derived bioactives have attracted great attention in the recent years for their anti-obesity effects by targeting dyslipidemia, oxidative stress and inflammation. Thus, there is a need to develop marine based products, which are a rich source of natural bioactive molecules, as nutraceutical and functional foods, to target obesity and related complications.

Keywords Dyslipidemia · Inflammation · Cytokines · Marine bioactives · Oxidative stress · Obesity

Introduction

Obesity infers a huge economic burden on the already outstretched health systems in many countries. According to the 2016 report of the World Health Organization, more than 1.9 billion adults (18 years and older) were designated as overweight worldwide, and over 650 million of them were obese [1]. In 2018, Statistics Canada reported that 63.1% adults in Canada were obese or overweight [2], and the estimated economic burden of obesity in Canada range from \$4.6 to \$7.1 billion annually. Obesity is a

I. Phadtare · H. Vaidya · S. K. Cheema (✉)
Department of Biochemistry, Memorial University, St. John's, NL A1B 3X9, Canada
e-mail: skaur@mun.ca

© Springer Nature Switzerland AG 2021
P. S. Tappia et al. (eds.), *Cellular and Biochemical Mechanisms of Obesity*,
Advances in Biochemistry in Health and Disease 23,
https://doi.org/10.1007/978-3-030-84763-0_20

complex multifactorial chronic disease associated with excessive body fat accumulation caused by an imbalance between energy intake and expenditure. Apart from genetic factors, environmental and socio-economic factors are also major contributors of obesity. Clinical observations confirm there is not a single molecular mechanism responsible for obesity; however, majority of the molecular changes that cause obesity lead to dyslipidemia and insulin resistance (IR) as a common outcome [3–5]. Thus, the majority of the therapeutic treatments for obesity are targeted towards IR, or the causes of IR such as inflammatory cytokines, oxidative stress and dyslipidemia. Dietary habits and lifestyle are also major contributors to obesity [6]; global strategies are thus focusing on restricting calorie intake and increasing physical activity to target obesity [7]. Besides dietary modifications, intervention with natural products have shown beneficial effects under obese conditions [8, 9]. A recent study found a direct association between lower levels of omega (n)-3 polyunsaturated fatty acids (PUFA) in erythrocytes and obesity [10]. Similarly, observational and randomized controlled clinical studies have shown that seafood-based diets, or an intake of marine n-3 PUFA has beneficial effects in weight management and dyslipidemia [11]. The anti-obesity effects of marine based macroalgae as well as seaweed are also documented [12–15]. These studies highlight the potential importance of dietary inclusion of marine based foods/supplements for the prevention of obesity. The current chapter focuses on the mechanisms of action of marine derived bioactive molecules, and their potential to be utilized as functional food or nutraceutical products for the prevention and/or treatment of obesity as a safer strategy, compared to drugs.

Pathophysiology of Obesity: Dyslipidemia, Oxidative Stress, and Inflammation

Obesity is a metabolic disorder, which is associated with a number of health conditions such as IR, type-2 diabetes, hypertension, obstructive sleep apnea syndrome, non-alcoholic fatty liver disease and dyslipidemia [16]. There is not a single molecular mechanism responsible for obesity, but multiple events lead to the outcome of obesity (Fig. 20.1). Dyslipidemia is the primary observation for obesity; it has been suggested that BMI (Body Mass Index) is directly related to lipid abnormalities in obese patients [17, 18]. Approximately, 60–70% of patients who are obese are dyslipidemic, while 50–60% of patients who are overweight are dyslipidemic [19]. The hallmark of dyslipidemia in obesity is elevated fasting and postprandial circulating triglycerides (TG) levels, in combination with elevated low-density lipoprotein (LDL)-and low high-density lipoprotein (HDL)-cholesterol levels [20]. The possible reasons for an increase in circulating TG levels in obesity are either due to an increase in the release of TG from adipose tissue, or due to lipolysis of TG-rich lipoproteins within circulation [21].

Adipose tissue is a primary organ responsible for storing excess calories in the form of neutral lipids, whereas under nutrient deficit conditions, adipose tissue

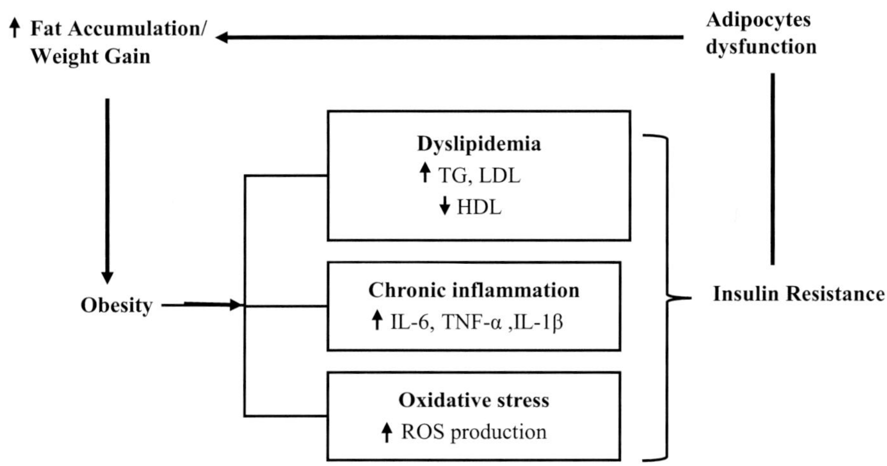

Fig. 20.1 Pathways associated with the pathophysiology of obesity. Obesity is associated with dyslipidemia, chronic inflammation and oxidative stress, causing insulin resistance and an increase in fat accumulation. TG, Triglycerides; LDL, Low-density lipoprotein; HDL, High-density lipoprotein; IL-6, Interleukin-6; TNF-α, Tumor necrosis factor-α; IL-1β, Interleukin-1β; ROS, Reactive oxygen species

supplies energy to other tissues through lipolysis [22]. Preadipocytes are converted to mature adipocytes through the process of adipogenesis and fat storage. Peroxisome proliferator-activated receptor-gamma (PPAR-γ) is the master regulator for the process of adipogenesis [23], regulates fat storage as TG, and maintains energy homeostasis [24]. Previous reports have shown that PPAR-γ gene expression is significantly reduced in obese subjects, while upregulating PPAR-γ gene expression increases adipogenesis and improves IR [25]. PPAR-γ also regulates the gene expression of adiponectin, as well as glucose transporter type-4 (GLUT-4) in adipocytes, thus is responsible for insulin sensitivity. Adiponectin is a potent adipokine, which is also known as the fat burning hormone as it increases β-oxidation of fatty acids; adiponectin levels are inversely related with BMI [26] and IR [27]. An increase in PPAR-γ gene expression also improves IR by increasing adiponectin levels [28]. GLUT-4 is a transporter protein responsible for the uptake of glucose. GLUT-4 protein expression, as well as its translocation to the surface of the cell membrane is insulin dependent, hence in the state of IR such as in obesity, the function of GLUT-4 is impaired. In obese subjects, the protein expression of GLUT-4 is decreased, which may be an early marker for obesity [29].

Adipocytes accumulate lipids via one of the two processes. Adipocytes take up dietary lipids from circulation in the form of free fatty acids, which are stored as TG with the help of the enzyme diacylglycerol O-acyltransferase (DGAT)-2 [30]. The second process involves *de novo* lipogenesis process within the adipocytes themselves [31]. The major regulator for de novo lipogenesis in adipose tissue is sterol regulatory element binding protein-1 (SREBP-1) [32]. The downstream regulator genes for lipogenesis includes acetyl-CoA carboxylase (ACC)-1, fatty acid synthase

(FAS), stearoyl-CoA desaturase (SCD) and DGAT [31]. De novo lipogenesis is increased in the adipose tissue in obese subjects [31], therefore, lipogenic genes are considered as important targets to develop treatments against obesity [33].

Adipocytes release the stored fat in energy deficient conditions as free fatty acids to the circulation by the process of lipolysis. The major regulators for this process are hormone sensitive lipase (HSL) and adipocytes triglyceride lipase (ATGL) [34]. Both enzymes are regulated by insulin signaling via the central and peripheral systems. Human studies have shown that in obese individuals, lipolysis is significantly reduced compared to lean healthy subject [35, 36]. Insulin inhibits HSL and ATGL to reduce lipolysis; these enzymes play important role in weight loss program in overweight and obese individual [37].

In response to an increase in energy intake, adipose tissue undergoes dynamic remodeling, which may include an increase in fat cell size (hypertrophy), and/or an increase in fat cell numbers (hyperplasia) or both [38]. Due to excessive fat storage and adipocytes remodeling, adipose tissue experiences an increase in oxidative stress and increase production of reactive oxidative species (ROS). Findings from obese individuals have shown that obesity is coupled with altered redox state and increased metabolic risk [39]. Karbownik-Lewinska et al. [40] reported that overweight or obese subjects have higher lipid peroxidation levels in the serum, compared to normal weight individuals. At cellular level, ROS activates PPAR-γ gene expression, which is responsible for an increase in adipogenesis and fat accumulation [38, 41, 42].

Adipose tissue also produces and releases a variety of adipokines and cytokines (anti- or pro-inflammatory cytokines), including leptin, adiponectin, resistin, and visfatin, as well as interleukin (IL)-4, interferon (IFN)-γ, tumor necrosis factor (TNF)-α, IL-6, and others [43]. An increase in oxidative stress (increase in ROS) in adipocytes is associated with decreased levels of adiponectin mRNA expression [44]. A decrease in adiponectin levels due to oxidative stress has been correlated with an increase in the secretion of inflammatory cytokines such as IL-6 and TNF-α [45, 46]. Furthermore increased hypertrophy in obesity attracts immune cells and macrophage infiltration to adipose tissue [47]. Resident immune cells and macrophages together cause a further increase in chronic inflammation via increasing inflammatory cytokine production, thereby inducing adipocytes dysfunction and IR [48, 49].

Nutrition and Obesity

Nutrition transition as a result of urbanization and affluence is considered as the major cause for the obesity epidemic [50]. Diet and dietary factors (e.g., fatty acids, antioxidants) influence dyslipidemia, oxidative stress and inflammatory state [51–54]. Nutrients are capable of directly interacting with the regulation of gene expression to impact metabolic pathways [55]. Clinical studies have shown evidence that fried food consumption predisposes individuals to obesity [56]. Moreover, replacing saturated fat with PUFA showed a significant reduction in obesity and cardiovascular

The manufacturer's authorised representative in the EU is Springer Nature Customer Service Centre GmbH, Europaplatz 3, 69115 Heidelberg, Germany. If you have any concerns regarding our products, please contact ProductSafety@springernature.com

Printed and bound by CPI Group (UK) Ltd, Croydon, CR0 4YY
30/09/2026
02225786-0001

disease [57]. Interestingly, marine foods are high in PUFA, along with the presence of other nutrients and bioactive component such as saponins, carotenoids and phytosterols that are shown to possess anti-obesity effects. The following sections focus on marine derived bioactives as potential therapeutic agents to target obesity associated dyslipidemia, oxidative stress and inflammation.

Prevention of Obesity Associated Dyslipidemia by Marine Derived Bioactives

Abnormalities in lipid metabolism are commonly observed in obese patients, showing an increase in circulating TG and a decrease in HDL levels. A major contributor to dyslipidemia in obesity is delivery of free fatty acids to the liver from adipose tissue, IR and a pro-inflammatory state. Fish oil is a rich source of n-3 PUFA, especially eicosapentaenoic acid (EPA) and docosahexaenoic acid (DHA) [58], and has acquired a significant interest to target obesity [59, 60]. Fish oil improves IR [61, 62], dyslipidemia [63], reduce pro-inflammatory mediators [64], lower body weight gain [65], and reduce obesity-related impediments. Research in humans has confirmed that supplementing n-3 PUFA significantly improved dyslipidemia by reducing TG and cholesterol levels [63]. The possible molecular mechanisms for decreasing TG levels are by inhibition of DGAT-2, increased plasma lipoprotein lipase activity, decreased hepatic lipogenesis, and increased hepatic β-oxidation [66]. Animal based studies have also confirmed that n-3 PUFA supplementation reduce fat accumulation by downregulating lipogenesis and stimulating lipid oxidation [67–69].

Other commercially valuable aquatic species such as blue mussels (BM) and sea cucumber (SC) that survive on phytoplanktons are a rich source of n-3 PUFA, and other bioactive such as polysaccharides, phytosterols and carotenoids that are known to induce several health benefits [70, 71]. We have previously shown that treating 3T3-L1 pre-adipocytes with methanolic extracts from BM and SC reduced lipogenesis and fat accumulation by inhibiting PPAR-γ gene expression [72]. Others have also shown similar anti-adipogenic effects of BM due to inhibition of PPAR-γ gene expression [73]. We further found that BM and SC methanolic extract significantly reduced TG accumulation by inhibiting lipogenic genes such as SREBP-1 and ACC-1 in 3T3-L1 cells [74]. In a follow up study using C57BL/6 mouse model, we found that supplementing BM and SC freeze dried powder in an obesogenic diet prevented weight gain, and dyslipidemia by reducing plasma TG and cholesterol levels [75–78]. Recently, Chang et al. [79] also reported that BM water extract significantly decreases abdominal adipose tissue weight and plasma cholesterol levels in mice. These findings suggest that BM and SC may target obesity related metabolic complications to provide health benefits.

Shrimp is another marine species that is rich in phospholipids, n-3 PUFA and carotenoids such as astaxanthin (ASTX) [80, 81]. Supplementing shrimp extracts in a high fat diet to LDL-receptor knockout mice was found to elicit potent anti-obesity effects by significantly reducing weight gain and decreasing plasma TG and cholesterol levels. Shrimp extract/oil is rich in ASTX (3,3'-dihydroxy-beta, beta-carotene-4,4'-dione) that possesses a unique structure compared to other carotenoids due to the presence of hydroxyl and keto groups at both ends (Fig. 20.2) [82]. This unique chemical structure brings distinctive features, such as the ability to be esterified, and a more polar configuration compared to other carotenoids [82]. ASTX exists in either free form or is conjugated with protein, or esterified with one or two fatty acids i.e., monoester or diester form, which stabilizes the molecule [83]. In recent years, ASTX gained significant amount of attention as a potential nutraceutical due to a number of suggested health benefits. Antioxidant potential of ASTX is suggested to be 10 times higher than other carotenoids such as zeaxanthin, lutein, canthaxanthin, β-carotene, and ASTX is almost 100 times more potent antioxidant than vitamin E and C [84]. Supplementing ASTX in a high-fat and high-cholesterol diet increased hepatic LDL-receptor, decreased 3-hydroxy-3-methylglutaryl CoA reductase (HMG-CoA reductase) and SREBP-2 activity, which was associated with lowering plasma cholesterol levels in mice [85]. The authors further reported a decrease in TG levels, which was due to a significant increase in the mRNA expression of carnitine palmitoyltransferase 1 (CPT1), suggesting that the TG lowering effect of ASTX might be due to increased fatty acid β-oxidation in the liver [85]. Jia et al. [86] have also shown that the possible hypolipidemic effects of ASTX is by functioning as a PPAR-γ antagonist and PPARα agonist that leads to lower plasma TG levels and improves fatty acid β-oxidation. These findings indicate that ASTX could be used as a functional food/nutraceutical to target obesity related lipid disorders.

Seaweed, also known as a microalgae, has shown benefits against dyslipidemic condition associated with obesity and IR [87]. Fucoxanthin is a major carotenoid present in seaweed. Studies have shown that potent lipid lowering effects of seaweed are due to the presence of fucoxanthin [88–90]. Research using 3T3-L1 pre-adipocytes showed anti-obesity effects of fucoxanthin by inhibiting PPAR-γ gene expression, and a significant reduction in TG accumulation in adipocytes [91]. In an animal-based study, fucoxanthin supplementation to a high fat diet significantly prevented the increase in serum TG and cholesterol levels, which was due to a decrease in SREBP-1, ACC-1 and FAS gene expression. These authors also reported that fucoxanthin increases β-oxidation by increasing CPT-1 mRNA expression [92].

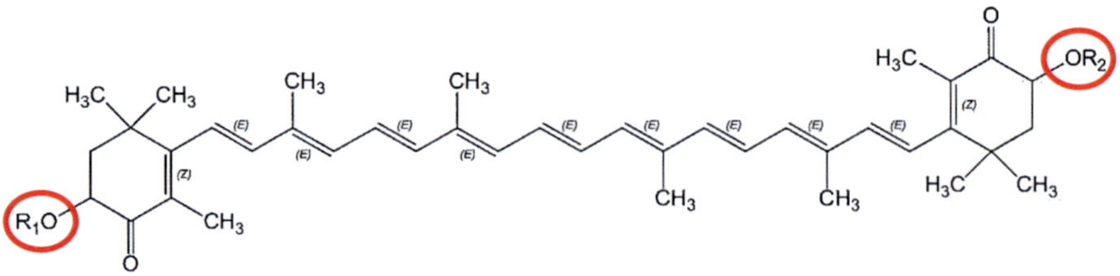

Fig. 20.2 Planar structure of Astaxanthin. R1, R2 = H, represents free ASTX; R1 or R2 = fatty acid, represents monoesterified ASTX; R1 and R2 = fatty acid, represents diesterified ASTX; ASTX, astaxanthin

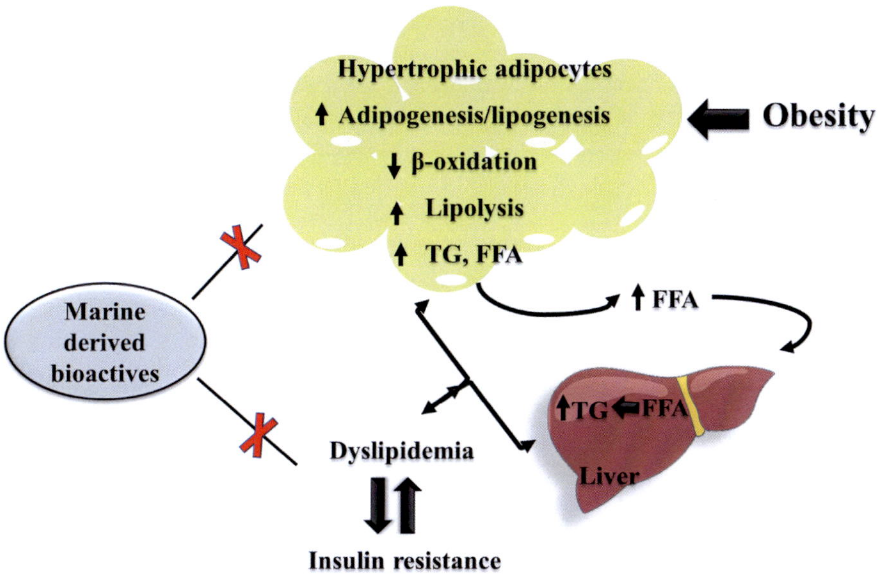

Fig. 20.3 Possible mechanisms by which marine derived bioactives target dyslipidemia associated with obesity. Marine derived bioactives target dyslipidemia by improving insulin resistance, decreasing adipogenesis, lipogenesis and lipolysis, along with increasing β-oxidation in adipocytes, as well as decreasing TG synthesis in liver. TG, Triglycerides; FFA, Free fatty acids

Thus, marine based products such as fish oil, BM, SC, shrimp/shrimp oil enriched with ASTX, and seaweed provide a clear evidence to prevent dyslipidemia by inhibiting lipogenesis and adipogenesis. The pathways/mechanisms by which marine derived bioactives target dyslipidemia are summarized in Fig. 20.3. It will be important to consider developing nutraceuticals/functional foods around these marine products and marine derived bioactives as potential therapeutic strategies to target dyslipidemia associated with obesity.

Prevention of Obesity Associated Oxidative Stress by Marine Derived Bioactives

Oxidative stress is a result of the disparity between the ROS production and the availability of antioxidants or radical scavengers. Under physiological conditions, the elevated levels of ROS can oxidize biomolecules, and in turn, can directly or indirectly damage different tissues/organs. Elevated levels of glucose and lipids, such as in obesity, can contribute as an excess source of energy substrates to metabolic pathways in adipose and non-adipose cells, that can ultimately raise ROS generation [93]. Usually, the cells are guarded from the damaging effects of ROS by intracellular and extracellular defenses, in particular antioxidant enzymes such as, superoxide dismutase (SOD), catalases, lactoperoxidases, and glutathione peroxidase (GPX). When ROS production overpowers antioxidant capacity, the resulting oxidative stress plays a part in the progression of several pathological conditions [94]. Studies have shown that biomarkers of oxidative damage are higher in obese individuals, and are directly linked to the body fat percentage, serum LDL oxidation and TG content [95]. On the other hand, antioxidant defense markers are reduced in individuals with obesity [96, 97]. Therefore, besides pharmacological strategies, other approaches such as reducing oxidative stress in obesity by dietary modifications and interventions with natural products are highly encouraged. Evidently, diets rich in certain nutrients such as n-3 PUFA, vitamins, and phytochemicals reduce oxidative stress in obese individuals [98, 99]. Other marine sources such as fish/fish oil, BM, SC, shrimp/shrimp oil have also gained a significant attention due to their potential to reduce oxidative stress and related complications [60, 100]. Marine sources are rich in naturally-occurring bioactive antioxidants that have the potential to reduce oxidative stress and exert anti-obesity effects [100].

Fish and fish oil based EPA and DHA supplements reduce oxidative stress markers, such as malondialdehyde (MDA), which is a product of the peroxidation of PUFA in individuals with obesity [101]. Previously, our laboratory has shown that BM from Newfoundland and Labrador is a valuable source of phospholipid enriched n-3 PUFAs [74]. It has been shown that BM water extract increased the activity of SOD and decreased the degree of lipid peroxidation, thereby reducing oxidative stress in high fat diet induced obese rats [102]. Besides fish and BM, SC is also a promising group of marine invertebrates that contains a diverse spectrum of bioactive compounds. SC decreases serum and hepatic ROS, free fatty acids, and reduces oxidative stress in high-fat and fructose diet fed C57BL/6 J obese mice [103]. Furthermore, these authors concluded that the antioxidant effects of SC are due to the presence of glycosphingolipids [103]. A recent study stated that Northern shrimp oil contains a high content of n-3 PUFA and antioxidants[104] that reduce oxidative stress [105]. Nair et al. [80] showed that adding shrimp oil to high fat high fructose diet reduced oxidative stress in male obese rats. Shrimp contains a significant amount of ASTX that has attracted attention in the recent years as a safe alternative to treat obesity [106]; ASTX increases the gene expression of hepatic antioxidant enzymes, and decreases myeloperoxidase and nitric oxide synthase activities, thereby reducing oxidative stress in high-fat and fructose diet induced obese mice [107]. Moreover, ASTX supplementation (5 and 20 mg once daily for 3 weeks) decreased levels of oxidative stress biomarkers by reducing lipid peroxidation and elevating the activity of the antioxidant defense system in overweight and obese individuals [108]. Seaweed is another promising natural source of antioxidants [109]. Since obesity is associated with increased oxidative stress, marine derived bioactives and marine products may provide a natural protection against oxidative stress and provide benefits under obese conditions as proposed in Fig. 20.4.

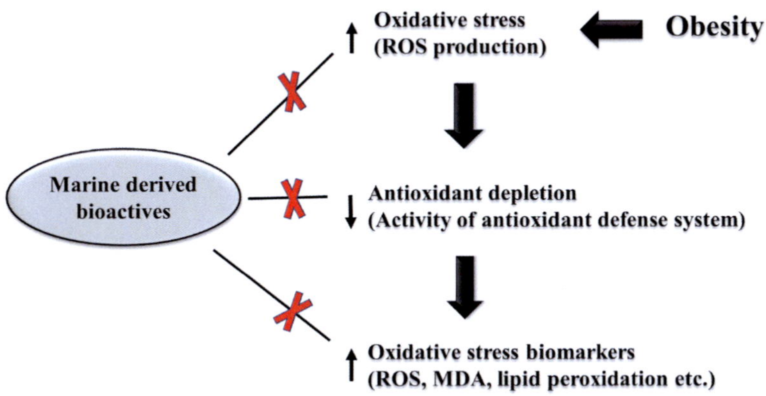

Fig. 20.4 Possible mechanisms by which marine derived bioactives target oxidative stress associated with obesity. Marine derived bioactives reduce obesity related oxidative stress by inhibiting ROS production, antioxidant depletion and oxidative stress related biomarkers. ROS, Reactive oxygen species; MDA, Malondialdehyde

Prevention of Obesity Associated Inflammation by Marine Derived Bioactives

Chronic low-grade inflammation under obese conditions results in increased levels of proinflammatory cytokines. However, it is not fully understood how obesity precisely activates inflammation. It has been proposed that excess fat in adipocytes stimulates intracellular stress, leading to the activation of inflammatory signaling cascade [110, 111]. Another approach implies that the overload of fat in adipocytes elevates the infiltration of macrophages. This progression may lead to the consequent differentiation, and activation of cytotoxic T-cells that can cause onset and development of inflammatory signaling cascades [112]. Chronic inflammation is associated with secretion of free fatty acids and inflammatory cytokines, such as IL-6, IL-1β and TNF-α [113]. Marine derived bioactives have shown the potential to reduce inflammation and associated complications arising in obesity. For example, fish oil supplements prevent obesity associated inflammation in humans [114]. Fish oil supplements showed a decrease in the mRNA expression of IL-18 and IL-1β in adipose tissue of obese individuals, and also showed a decrease in circulating IL-18 levels [114]. Moreover, plasma n-3 PUFA status inversely relates with inflammatory cytokines to show a decrease in adipocytes IL-6, IL-8 and TNF-α levels in humans with obesity [115]. Chang et al. [79] reported that BM water extract significantly decreases proinflammatory cytokine levels (TNF-α and IL-1β) in osteoarthritis rats under obesity condition [79]. Apart from fish and BM, SC have also shown protective effects with respect to obesity associated inflammation [116]. Previously, we have shown that the freeze-dried powder of SC when given along with a high fat diet significantly reduced plasma inflammatory cytokines such as IL-6 [75]. SC derived saponins and liposomes also protected adipose tissue inflammation by reducing pro-inflammatory cytokine release and macrophage infiltration in high fat diet induced C57BL/6 J obese mice [78]. Likewise, shrimp oil has the potential to prevent inflammation and associated comorbidities. Santos et al. [80] have shown that shrimp waste extract that is enriched with ASTX shows anti-inflammatory effects by inhibiting TNF-α secretion. Nair et al. [80] found that supplementing shrimp oil along with high fat high fructose diet reduced chronic inflammation in male obese rats. Additionally, reports have suggested that ASTX, a predominant carotenoid from shrimp oil, when supplemented along with high fat high fructose diet lowers TNF-α and IL-6 levels, thereby reducing inflammation in obese mice [107, 117]. Marine algae/seaweed has also shown anti-inflammatory and antioxidant properties. Seaweed supplementation mitigates chronic inflammation by reducing inflammatory cytokines in long-term high fat diet induced C57BL/6 J obese mice [118]. Park et al. [119] studied the cellular and molecular mechanism underlying the anti-inflammatory properties of fucoidan, a sulfated polysaccharide from seaweed. These authors found that the anti-inflammatory properties of this compound are related to the inhibition of cyclooxygenase-2 and pro-inflammatory cytokines (TNF-α and IL-6) [87]. Overall, the chronic inflammation in adipose tissues due to the increase in inflammatory cytokines is the primary cause of IR and adipocytes dysfunction in obesity. Marine derived bioactives and marine products have the potential as nutraceutical/functional foods to prevent obesity associated inflammation as summarized in Fig. 20.5.

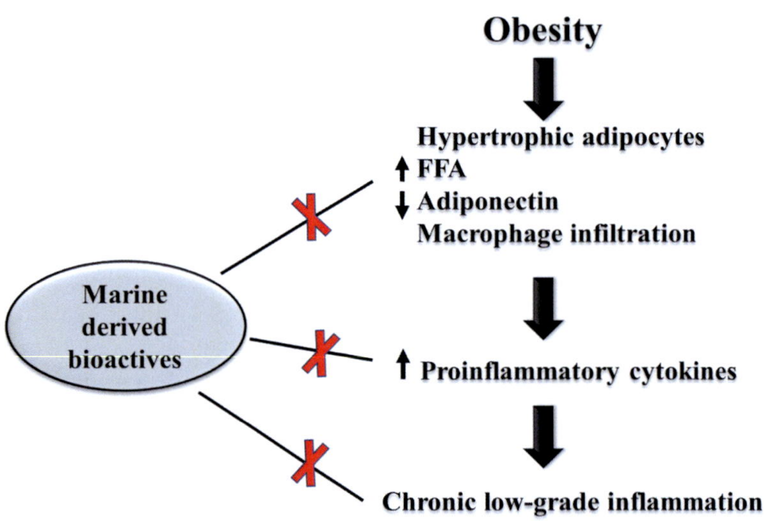

Fig. 20.5 Possible mechanisms by which marine derived bioactives target inflammation associated with obesity. Marine derived bioactives reduce obesity related low grade chronic inflammation by downregulating pro-inflammatory cytokines, as well as reducing macrophage infiltration and hypertrophy of adipocytes. FFA, Free fatty acids

Summary and Conclusions

Obesity is associated with metabolic disorders, such as chronic low-grade inflammation, IR, altered lipid metabolism and increase oxidative stress; thus, treatment for obesity requires a multi-target approach. Marine derived bioactives are attracting attention as promising sources to prevent/treat obesity. Marine species contain bioactives such as astaxanthin, fucoxanthin, saponin, phytosterol and n-3 PUFA that act through multiple pathways to control obesity. Fish oil enriched with n-3 PUFA decreases plasma TG levels, improves IR, decreases inflammatory cytokines, and prevents oxidative stress. Blue mussels and sea cucumber are also rich in n-3 PUFA, phytosterols, saponins, and other bioactives that decrease lipogenesis, inhibit inflammatory cytokines, and prevent oxidative stress. Shrimp oil contains n-3 PUFA and is a rich source of astaxanthin, which is a highly potent antioxidant. Seaweed is enriched with fucoxanthin, which has been shown to decrease inflammatory cytokines, inhibits lipogenesis and adipogenesis. Thus, there is a strong scientific evidence that marine products and/or marine derived bioactives target various metabolic disorders associated with obesity, thus have the potential to combat obesity. Nutraceutical and functional foods industry should consider developing marine based therapeutic strategies to target obesity as these are safer alternatives when compared to pharmaceuticals/drugs.

References

1. Arroyo-Johnson C, Mincey KD (2016) Obesity epidemiology worldwide. Gastroenterol Clin North Am 45:571–579
2. Obesity and overweight. https://www.who.int/news-room/fact-sheets/detail/obesity-and-overweight
3. Benomar Y, Taouis M (2019) Molecular mechanisms underlying obesity-induced hypothalamic inflammation and insulin resistance: pivotal role of resistin/tlr4 pathways. Front Endocrinol 10:140
4. Reilly SM, Saltiel AR (2017) Adapting to obesity with adipose tissue inflammation. Nat Rev Endocrinol 13:633–643
5. Gross B, Pawlak M, Lefebvre P, Staels B (2017) PPARs in obesity-induced T2DM, dyslipidaemia and NAFLD. Nat Rev Endocrinol 13:36–49
6. Leitner DR et al (2017) Obesity and type 2 diabetes: two diseases with a need for combined treatment strategies—EASO can lead the way. Obes Facts 10:483–492
7. Mohamed GA, Ibrahim SRM, Elkhayat ES, El Dine RS (2014) Natural anti-obesity agents. Bull Faculty Pharm Cairo Univ 52:269–284
8. Brown T, Moore T, Hooper L et al (2019) Interventions for preventing obesity in children. Cochrane Database Syst Rev 7(7):CD001871
9. Hays NP, Galassetti PR, Coker RH (2008) Prevention and treatment of type 2 diabetes: current role of lifestyle, natural product, and pharmacological interventions. Pharmacol Ther 118:181–191
10. Young I, Parker H, Cook R et al (2020) Association between obesity and omega-3 status in healthy young women. Nutrients 12(5):1480
11. Liaset B, Oyen J, Jacques H, Kristiansen K, Madsen L (2019) Seafood intake and the development of obesity, insulin resistance and type 2 diabetes. Nutr Res Rev 32:146–167

12. Gómez-Zorita S et al (2020) Anti-obesity effects of macroalgae. Nutrients 12:1–29
13. Kang MC, Kang N, Ko SC, Kim YB, Jeon YJ (2016) Anti-obesity effects of seaweeds of Jeju Island on the differentiation of 3T3-L1 preadipocytes and obese mice fed a high-fat diet. Food Chem Toxicol 90:36–44
14. Kang MC et al (2016) Popular edible seaweed, Gelidium amansii prevents against diet-induced obesity. Food Chem Toxicol 90:181–187
15. Sharma BR, Kim HJ, Kim MS, Park CM, Rhyu DY (2017) Caulerpa okamurae extract inhibits adipogenesis in 3T3-L1 adipocytes and prevents high-fat diet–induced obesity in C57BL/6 mice. Nutr Res 47:44–52
16. Klop B, Elte JWF, Cabezas MC (2013) Dyslipidemia in obesity: mechanisms and potential targets. Nutrients 5:1218–1240
17. Stępień A et al (2014) Assessment of the relationship between lipid parameters and obesity indices in non-diabetic obese patients: A preliminary report. Med Sci Monit 20:2683–2688
18. Oliosa PR, Zaniqueli DDA, Barbosa MCR, Mill JG (2019) Relationship between body composition and dyslipidemia in children and adolescents. Cienc. e Saude Coletiva 24:3743–3752
19. Franssen R et al (2011) Obesity and dyslipidemia. Med Clin North Am 95:893–902
20. Bays H, Toth P, Kris-Etherton P et al (2013) Obesity, adiposity, and dyslipidemia: a consensus statement from the National Lipid Association. J Clin Lipidol 7(4):304–383
21. Sobczak AIS, Blindauer CA, Stewart AJ (2019) Changes in plasma free fatty acids associated with type-2 diabetes. Nutrients 11(9):2022
22. Birsoy K, Festuccia WT, Laplante M (2013) A comparative perspective on lipid storage in animals. J Cell Sci 126:1541–1552
23. Konno T, Sasaki K, Kobayashi K, Murata T (2020) Indirubin promotes adipocyte differentiation and reduces lipid accumulation in 3T3–L1 cells via peroxisome proliferator–activated receptor γ activation. Mol Med Rep 21:1552–1560
24. Corrales P, Vidal-Puig A, Medina-Gómez G (2018) PPARS and metabolic disorders associated with challenged adipose tissue plasticity. Int J Mol Sci 19(7):2124
25. Akyürek N et al (2013) Peroxisome proliferator activated receptor (PPAR)-gamma concentrations in childhood obesity. Scand J Clin Lab Invest 73:355–360
26. Gariballa S, Alkaabi J, Yasin J, Al Essa A (2019) Total adiponectin in overweight and obese subjects and its response to visceral fat loss. BMC Endocr Disord 19:55
27. Balsan GA, Da Costa Vieira JL, De Oliveira AM, Portal VL (2015) Relationship between adiponectin, obesity and insulin resistance. Rev Assoc Med Bras 61:72–80
28. Giorgino F, Leonardini A, Laviola L, Perrini S, Natalicchio A (2009) Cross-Talk between PPARγ and insulin signaling and modulation of Insulin Sensitivity. PPAR Res. https://doi.org/10.1155/2009/818945
29. Mourelatou R et al (2019) Decreased adipocyte glucose transporter 4 (GLUT4) and aquaglyceroporin-7 (AQP7) in adults with morbid obesity: possible early markers of metabolic dysfunction. Hormones 18:297–306
30. Yu Y-H, Ginsberg H (2009) The role of acyl-CoA: diacylglycerol acyltransferase (DGAT) in energy metabolism. Ann Med 36(4):252–261
31. Song Z, Xiaoli A, Yang F (2018) Regulation and metabolic significance of De Novo lipogenesis in adipose tissues. Nutrients 10(10):1383
32. Sekiya M, Yahagi N, Matsuzaka T et al (2007) SREBP-1-independent regulation of lipogenic gene expression in adipocytes. J Lipid Res 48(7):1581–1591
33. Garrido-Sánchez L, Vendrell J, Fernández-García D et al (2012) De Novo lipogenesis in adipose tissue is associated with course of morbid obesity after bariatric surgery. PLoS One 7(2):e31280
34. Lenhard J (2011) Lipogenic enzymes as therapeutic targets for obesity and diabetes. Curr Pharm Des 17(4):325–331
35. Duncan RE, Ahmadian M, Jaworski K, Sarkadi-Nagy E, Sul HS (2007) Regulation of lipolysis in adipocytes. Annu Rev Nutr 27:79–101
36. Rydén M, Andersson DP, Bernard S, Spalding K, Arner P (2013) Adipocyte triglyceride turnover and lipolysis in lean and overweight subjects. J Lipid Res 54:2909–2913

37. Cifuentes M, Albala C, Rojas CV (2008) Differences in lipogenesis and lipolysis in obese and non-obese adult human adipocytes. Biol Res 41:197–204
38. Luglio HF, Sulistyoningrum DC, Susilowati R (2015) The role of genes involved in lipolysis on weight loss program in overweight and obese individuals. J Clinic Biochem Nutr 57:91–97
39. Lee H, Lee YJ, Choi H, Ko EH, Kim JW (2009) Reactive oxygen species facilitate adipocyte differentiation by accelerating mitotic clonal expansion. J Biol Chem 284:10601–10609
40. Warolin J et al (2014) The relationship of oxidative stress, adiposity and metabolic risk factors in healthy Black and White American youth. Pediatr Obes 9:43–52
41. Karbownik-Lewinska M et al (2012) Direct contribution of obesity to oxidative damage to macromolecules. Neuroendocrinol Lett 33:453–461
42. Kanda Y, Hinata T, Kang SW, Watanabe Y (2011) Reactive oxygen species mediate adipocyte differentiation in mesenchymal stem cells. Life Sci 89:250–258
43. Wang W, Zhang Y, Lu W, Liu K (2015) Mitochondrial reactive oxygen species regulate adipocyte differentiation of mesenchymal stem cells in hematopoietic stress induced by arabinosylcytosine. PLoS One 10(3):e0120629
44. Booth A, Magnuson A, Fouts J, Foster MT (2016) Adipose tissue: an endocrine organ playing a role in metabolic regulation. Horm Mol Biol Clin Invest 26:25–42
45. Monickaraj F et al (2013) Accelerated fat cell aging links oxidative stress and insulin resistance in adipocytes. J Biosci 38:113–122
46. Fukushima M et al (2014) Growth hormone ameliorates adipose dysfunction during oxidative stress and inflammation and improves glucose tolerance in obese mice. Horm Metab Res 46:656–662
47. Pan Y et al (2015) Losartan reduces insulin resistance by inhibiting oxidative stress and enhancing insulin signaling transduction. Exp Clin Endocrinol Diabetes 123:170–177
48. Cox AR, Chernis N, Masschelin PM, Hartig SM (2019) Immune cells gate white adipose tissue expansion. Endocrinology 160:1645–1658
49. Lu J, Zhao J, Meng H, Zhang X (2019) Adipose tissue-resident immune cells in obesity and type 2 diabetes. Front Immunol 10:1173
50. Castoldi A, De Souza CN, Saraiva Câmara NO, Moraes-Vieira PM (2016) The macrophage switch in obesity development. Front Immunol 6:1
51. Chan RSM, Woo J (2010) Prevention of overweight and obesity: How effective is the current public health approach. Int J Environ Res Public Health 7:765–783
52. Galland L (2010) Diet and inflammation. Nutr Clin Pract 25:634–640
53. Calder PC, Ahluwalia N, Brouns F et al (2011) Dietary factors and low-grade inflammation in relation to overweight and obesity. Br J Nutr Suppl 106(3):S5–78
54. Bjørklund G, Chirumbolo S (2017) Role of oxidative stress and antioxidants in daily nutrition and human health. Nutrition 33:311–321
55. Jung S, Smith-Warner S, Willett W et al. (2016) Healthy dietary patterns and oxidative stress as measured by fluorescent oxidation products in nurses' health study. Nutrients 8(9):587
56. Liu A, Ford N, Hu F et al (2017) A healthy approach to dietary fats: Understanding the science and taking action to reduce consumer confusion. Nutrition J 16:53
57. Qi Q, Chu A, Kang J et al (2014) Fried food consumption, genetic risk, and body mass index: gene-diet interaction analysis in three US cohort studies. the BMJ 348:g1610
58. Abdelhamid A, Martin N, Bridges C et al (2018) Polyunsaturated fatty acids for the primary and secondary prevention of cardiovascular disease. Cochrane Database Syst Rev 7(7):CD012345
59. Lorente-Cebrián S et al (2013) Role of omega-3 fatty acids in obesity, metabolic syndrome, and cardiovascular diseases: a review of the evidence. J Physiol Biochem 69:633–651
60. Abeywardena MY, Belobrajdic DP (2016) Long-Chain Omega-3 polyunsaturated fatty acids and obesity. Obesity 1st Edition, Springer Cham,129–44
61. Ad Lordan S, Ross RP, Stanton C (2011) Marine bioactives as functional food ingredients: potential to reduce the incidence of chronic diseases. Mar Drugs 9:1056–1100
62. Albert B, Derraik J, Brennan C et al (2014) Higher omega-3 index is associated with increased insulin sensitivity and more favourable metabolic profile in middle-aged overweight men. Sci Rep 4:6697

63. Lombardo Y, Chicco A (2006) Effects of dietary polyunsaturated n-3 fatty acids on dyslipidemia and insulin resistance in rodents and humans. A review. J Nutr Biochem 17(1):1–13
64. Skulas-Ray AC et al (2019) Omega-3 fatty acids for the management of hypertriglyceridemia: a science advisory from the American heart association. Circulation 140:E673–E691
65. Ramirez-Ramirez V et al (2013) Efficacy of fish oil on serum of TNF α, IL-1 β, and IL-6 oxidative stress markers in multiple sclerosis treated with interferon beta-1b. Oxid Med Cell Longev. https://doi.org/10.1155/2013/709493
66. Du S, Jin J, Fang W, Su Q (2015) Does fish oil have an anti-obesity effect in overweight/obese adults? A meta-analysis of randomized controlled trials. PLoS One 10:142652
67. Backes J, Anzalone D, Hilleman D, Catini J (2016) The clinical relevance of omega-3 fatty acids in the management of hypertriglyceridemia. Lipids Health Dis 15:118
68. Ruzickova J, Rossmeisl M, Prazak T et al (2004) Omega-3 PUFA of marine origin limit diet-induced obesity in mice by reducing cellularity of adipose tissue. Lipids 39:1177–1185
69. Lapillonne A, Clarke SD, Heird WC (2004) Polyunsaturated fatty acids and gene expression. Curr Opin Clin Nutr Metab Care 7:151–156
70. Azain MJ (2004) Role of fatty acids in adipocyte growth and development. J Animal Sci 82:916–924
71. Pleissner D, Eriksen NT, Lundgreen K, Riisgård HU (2012) Biomass composition of blue mussels, Mytilus edulis, is affected by living site and species of ingested microalgae. ISRN Zool. 1–12:2012
72. Heydarizadeh P et al (2013) Plastids of marine phytoplankton produce bioactive pigments and lipids. Mar Drugs 11:3425–3471
73. Vaidya H, Cheema SK (2014) Sea cucumber and blue mussel: new sources of phospholipid enriched omega-3 fatty acids with a potential role in 3T3-L1 adipocyte metabolism. Food Funct 5:3287–3295
74. Shon MS, Kim SK, Song JH, Lee SC, Kim GN (2015) Anti-adipogenic activity of blue mussel (Mytilus edulis) extract by regulation of 3T3-L1 adipogenesis through Wnt/β-catenin signaling pathway. Food Sci Biotechnol 24:315–321
75. Vaidya H, Cheema S (2014) Sea cucumber and blue mussel: new sources of phospholipid enriched omega-3 fatty acids with a potential role in 3T3-L1 adipocyte metabolism. Food Funct 5(12):3287–3295
76. Gangadaran S, Cheema SK (2017) A high fat diet enriched with sea cucumber gut powder provides cardio-protective and anti-obesity effects in C57BL/6 mice. Food Res Int 99:799–806
77. Vaidya HB, Gangadaran S, Cheema SK (2017) An obesogenic diet enriched with blue mussels protects against weight gain and lowers cholesterol levels in C57BL/6 mice. Nutr Res 46:31–37
78. Hu S, Xia G, Wang J et al (2014) Fucoidan from sea cucumber protects against high-fat high-sucrose diet-induced hyperglycaemia and insulin resistance in mice. J Funct food 10:128–138
79. Chen C et al (2018) Sea cucumber saponin liposomes ameliorate obesity-induced inflammation and insulin resistance in high-fat-diet-fed mice. Food Funct 9:861–870
80. Chang H-W et al (2020) <p>Blue Mussel (<em>Mytilus edulis</em>) water extract ameliorates inflammatory responses and oxidative stress on osteoarthritis in obese rats</p>. J Pain Res 13:1109–1119
81. Nair S, Gagnon J, Pelletier C et al (2017) Shrimp oil extracted from the shrimp processing waste reduces the development of insulin resistance and metabolic phenotypes in diet-induced obese rats.Appl Physiol Nutr Metab 42(8):841–849
82. Jiao G et al (2015) Characterization of shrimp oil from Pandalus borealis by high performance liquid chromatography and high resolution mass spectrometry. Mar Drugs 13:3849–3876
83. Higuera-Ciapara I, Félix-Valenzuela L, Goycoolea FM (2006) Astaxanthin: a review of its chemistry and applications. Crit Rev Food Sci Nutr 46:185–196
84. Hussein G, Sankawa U, Goto H, Matsumoto K, Watanabe H (2006) Astaxanthin, a carotenoid with potential in human health and nutrition. J Nat Prod 69:443–449
85. Ambati RR, Moi PS, Ravi S, Aswathanarayana RG (2014) Astaxanthin: sources, extraction, stability, biological activities and its commercial applications—a review. Mar Drugs 12:128–152

86. Yang Y et al (2011) Astaxanthin-rich extract from the green alga Haematococcus pluvialis lowers plasma lipid concentrations and enhances antioxidant defense in apolipoprotein E knockout mice. J Nutr 141:1611–1617
87. Jia Y et al (2012) The natural carotenoid astaxanthin, a PPAR-α agonist and PPAR-γ antagonist, reduces hepatic lipid accumulation by rewiring the transcriptome in lipid-loaded hepatocytes. Mol Nutr Food Res 56:878–888
88. Rosa GP, Tavares W, Sousa PM et al (2020) Seaweed secondary metabolites with beneficial health effects: an overview of successes in in vivo studies and clinical trials. Mar Drugs 18(1):8–35
89. Lange KW, Hauser J, Nakamura Y, Kanaya S (2015) Dietary seaweeds and obesity. Food Sci Human Wellness 4:87–96
90. Zhang H, Tang Y, Zhang Y et al (2015) Fucoxanthin: a promising medicinal and nutritional ingredient. Evi-Based Com Altern Med 2015:723515
91. Jeon SM et al (2010) Fucoxanthin-rich seaweed extract suppresses body weight gain and improves lipid metabolism in high-fat-fed C57BL/6J mice. Biotechnol J 5:961–969
92. Kang S, Ko H, Shin H et al (2011) Fucoxanthin exerts differing effects on 3T3-L1 cells according to differentiation stage and inhibits glucose uptake in mature adipocytes. Biochem Biophys Res Commun 409:769–774
93. Ha AW, Kim WK (2013) The effect of fucoxanthin rich power on the lipid metabolism in rats with a high fat diet. Nutr Res Pract 7:287–293
94. Matsuda M, Shimomura I (2013) Increased oxidative stress in obesity: Implications for metabolic syndrome, diabetes, hypertension, dyslipidemia, atherosclerosis, and cancer. Obes Res Clin Pract 7:e330–e341
95. Sies H (1997) Oxidative stress: oxidants and antioxidants. Exp Physiol 82:291–295
96. Pihl E et al (2006) Atherogenic inflammatory and oxidative stress markers in relation to overweight values in male former athletes. Int J Obes 30:141–146
97. Chrysohoou C et al (2007) The implication of obesity on total antioxidant capacity in apparently healthy men and women: the ATTICA study. Nutr Metab Cardiovasc Dis 17:590–597
98. Hartwich J, Goralska J, Siedlecka D et al (2007) Effect of supplementation with vitamin E and C on plasma hsCRP level and cobalt-albumin binding score as markers of plasma oxidative stress in obesity. Genes Nutr 2(1):151–154
99. Mori TA et al (2000) Effect of ω3 fatty acids on oxidative stress in humans: GC-MS measurement of urinary F2-isoprostane excretion. Redox Rep 5:45–46
100. González-Castejón M, Rodriguez-Casado A (2011) Dietary phytochemicals and their potential effects on obesity: a review. Pharmacol Res 64:438–455
101. Hu X, Tao N, Wang X, Xiao J, Wang M (2016) Marine-derived bioactive compounds with anti-obesity effect: a review. J Funct Foods 21:372–387
102. Parra D, Bandarra NM, Kiely M, Thorsdottir I, Martínez JA (2007) Impact of fish intake on oxidative stress when included into a moderate energy-restricted program to treat obesity. Eur J Nutr 46:460–467
103. Chang HW et al (2020) Blue mussel (Mytilus edulis) water extract ameliorates inflammatory responses and oxidative stress on osteoarthritis in obese rats. J Pain Res 13:1109–1119
104. Hu S, Wang J, Wang J, Xue C, Wang Y (2017) Long-chain bases from sea cucumber mitigate endoplasmic reticulum stress and inflammation in obesity mice. J Food Drug Anal 25:628–636
105. Subramanian B, Thibault M, Djaoued Y et al (2012) Chromatographic, NMR and vibrational spectroscopic investigations of astaxanthin esters: application to Astaxanthin-rich shrimp oil obtained from processing of Nordic shrimps. Analyst 140:7423–7433
106. Vincent H, Bourguignon C, Weltman A et al (2009) Effects of antioxidant supplementation on insulin sensitivity, endothelial adhesion molecules, and oxidative stress in normal-weight and overweight young adult. Metabolism 58(2):254–262
107. Maeda H, Hosokawa M, Sashima T, Funayama K, Miyashita K (2007) Effect of medium-chain triacylglycerols on anti-obesity effect of fucoxanthin. J Oleo Sci 56:615–621

108. Bhuvaneswari S, Arunkumar E, Viswanathan P, Anuradha CV (2010) Astaxanthin restricts weight gain, promotes insulin sensitivity and curtails fatty liver disease in mice fed a obesity-promoting diet. Process Biochem 45:1406–1414
109. Choi HD, Kim JH, Chang MJ, Kyu-Youn Y, Shin WG (2011) Effects of astaxanthin on oxidative stress in overweight and obese adults. Phyther. Res. 25:1813–1818
110. Sabeena Farvin KH, Jacobsen C (2015) Antioxidant activity of seaweed extracts: in vitro assays, evaluation in 5% fish oil-in-water emulsions and characterization. J Am Oil Chem Soc 92(4):571–587
111. Gregor MF, Hotamisligil GS (2011) Inflammatory mechanisms in obesity. Annu Rev Immunol 29:415–445
112. Hotamisligil GS (2006) Inflammation and metabolic disorders. Nature 444:860–867
113. Surmi BK, Hasty AH (2008) Macrophage infiltration into adipose tissue: Initiation, propagation and remodeling. Futur Lipidol 3:545–556
114. Rius B, Lopez-Vicario C, Gonzalez-Periz A et al (2012) Resolution of inflammation in obesity-induced liver disease. Front Immunol 3:257
115. Lee KR, Midgette Y, Shah R (2019) Fish oil derived omega 3 fatty acids suppress adipose NLRP3 inflammasome signaling in human obesity. J Endocr Soc 3:504–515
116. Haghiac M, Yang X, Presley L et al (2015) Dietary omega-3 fatty acid supplementation reduces inflammation in obese pregnant women: a randomized double-blind controlled clinical trial. PLoS One 10(9):e0137309
117. Kareh M et al (2018) Anti-proliferative and anti-inflammatory activities of the sea cucumber Holothuria polii aqueous extract. SAGE Open Med. 6:205031211880954
118. Arunkumar E, Bhuvaneswari S, Anuradha CV (2012) An intervention study in obese mice with astaxanthin, a marine carotenoid—effects on insulin signaling and pro-inflammatory cytokines. Food Funct 3:120–126
119. Lee Y, Oh J, Han A et al (2016) Seaweed supplementation attenuates long-term high fat diet-induced chronic inflammation and insulin resistance regardless of obesity in C57BL/6N Mice. FASEB J 30:lb217–lb217
120. Park HY et al (2011) Anti-inflammatory effects of fucoidan through inhibition of NF-κB, MAPK and Akt activation in lipopolysaccharide-induced BV2 microglia cells. Food Chem Toxicol 49:1745–1752

Chapter 21
Bitter Melon in Combination with Diet Modification and Regular Exercise Can Prevent and Treat Obesity and Hypertension Cost-Effectively

Carlin Hanoman, Jaipaul Singh, Khemraj Rupee, Sunil Rupee, Abdullah Adil Ansari, Emanuel Cummings, and Shalini Behl

Abstract Chronic diseases (CDs), including hypertension, obesity and diabetes, are responsible for a large number of global deaths annually. This is due to current life style habits, including sedentary life style, smoking, excess alcohol intake, sugar and fast-food consumption, genetic factors, stress and others. This study investigated the effect of daily consumption of bitter gourd/melon (*Momordica charantia*) combined with life style changes to reduce body weight, systolic and diastolic blood pressure (SDBP), blood glucose and lipid levels in the body. The study recruited 32 obese male (16) and female (16) subjects with an average age of 42 years (±4.5 years) and the majority of them had secondary education. They were divided into four groups (4 men and 4 women per group). Group 1 (diet only) was asked to reduce daily food intake and avoided snaking or binging for 6 weeks. Group 2 (diet and bitter melon) did the same as group 1 but combined with the consumption of 20 g of bitter melon juice (vol/weight) daily for 6 weeks. Group 3 (diet, exercise and bitter melon) did the same as group 1 but combined with daily exercise involving walking, stretching or bicycle riding or a combination for 30 min plus the consumption of 20 g of bitter melon daily for 6 weeks. Group 4 (diet and exercise) did the same as group 1 plus daily exercise involving walking, stretching or bicycle riding or a combination for 30 min Initially, at week 1 the subjects were weighed and their height and SDBP taken. Blood samples were taken for the measurements of fasting blood glucose (FBG), HBA1c, total cholesterol and triglyceride. Their BMI and blood pressure were measured weekly over 6 weeks and another blood sample for each subject was taken at the end of week 6 for analysis as in week one for comparison. The results

C. Hanoman · J. Singh
School of Natural Sciences, University of Central Lancashire, Preston, UK

C. Hanoman · K. Rupee · S. Rupee · E. Cummings
School of Medicine, College of Medical Sciences, Georgetown, Guyana

A. A. Ansari
Department of Biology, Faculty of Natural Sciences, University of Guyana, Georgetown, Guyana

S. Behl (✉)
School of Life Sciences, Manipal Academy of Higher Education, Dubai, UAE

© Springer Nature Switzerland AG 2021
P. S. Tappia et al. (eds.), *Cellular and Biochemical Mechanisms of Obesity*, Advances in Biochemistry in Health and Disease 23,
https://doi.org/10.1007/978-3-030-84763-0_21

showed that all four interventions were associated with marked decreases in BMI but with little or no change in HBA1c and FBG compared week 1 with week 6. However, significant ($p < 0.05$) decreases were observed in SDBP, total cholesterol and triglyceride comparing week 1 with week 6. It is concluded that life style changes including dieting, regular exercise and daily intake of bitter melon can help to reduce blood pressure and lipids and the weight of obese subjects leading to a better quality of life.

Keywords Obesity · Diet · Hypertension · Exercise · Bitter melon · Chronic diseases

Introduction

The World Health Organization [1] identified chronic non-communicable diseases (NCDs) or chronic diseases (CD) as cardiovascular (heart and blood vessels) diseases (CVDs), stroke, overweight and obesity, cancer, chronic respiratory diseases (CRD), cerebrovascular disease, kidney failure, dental diseases, diabetes mellitus (DM) and a few others [2]. CVDs are further classified into heart failure or cardiomyopathy, hypertension, atherosclerosis, coronary artery diseases (CAD), arrhythmias, sudden cardiac death (SCD) and others [3, 4]. Chronic diseases are often viewed to affect old people primarily but at this moment in time, they affect children as young as 12 years of age. Moreover, chronic diseases are among the most common, costly and preventable of all health problems, and they represent a growing burden for society globally [5]. They are recognized as a growing international socio-economic and public health and social care problems, accounting for over 36 million of the 57 million deaths worldwide in 2008 [6]. Currently, it is estimated that NCDs kill 41 million people each year, equivalent to 71% of all deaths globally. Moreover, 15 million people die from NCDs, especially during their working age, between 30 and 69 years and over 85% of these are premature deaths, especially in low- and middle-income countries throughout the world [1, 3, 5]. Figure 21.1 shows some of the major risk factors for obesity, a major NCD, and they include smoking, environment factors, genetics, alcohol intake, high level of stress, physical inactivity, unhealthy diets composed with excessive fats and sugar, constant snaking and binging (over eating and unhealthy eating patterns), few calories expended, family structure, socio-economic problems, no psychological intervention, some medications and health conditions including hormone imbalance, lack of knowledge in self-care, parental and birth weights not enough sleep, hypothyroidism, insulin resistance, polycystic ovary syndrome, Cushing's syndrome and others [4, 7, 8]. Detection, screening and treatment of NCDs, as well as palliative care, are key components to reduce NCDs and/or to prevent premature deaths. Metabolic risk factors for NCDs include hypertension, overweight and obesity, hyperglycaemia and hyper-lipidemia [1, 2]. The WHO predicted that the proportion of the burden of NCDs is expected to increase to 75% by the end of 2020 [6].

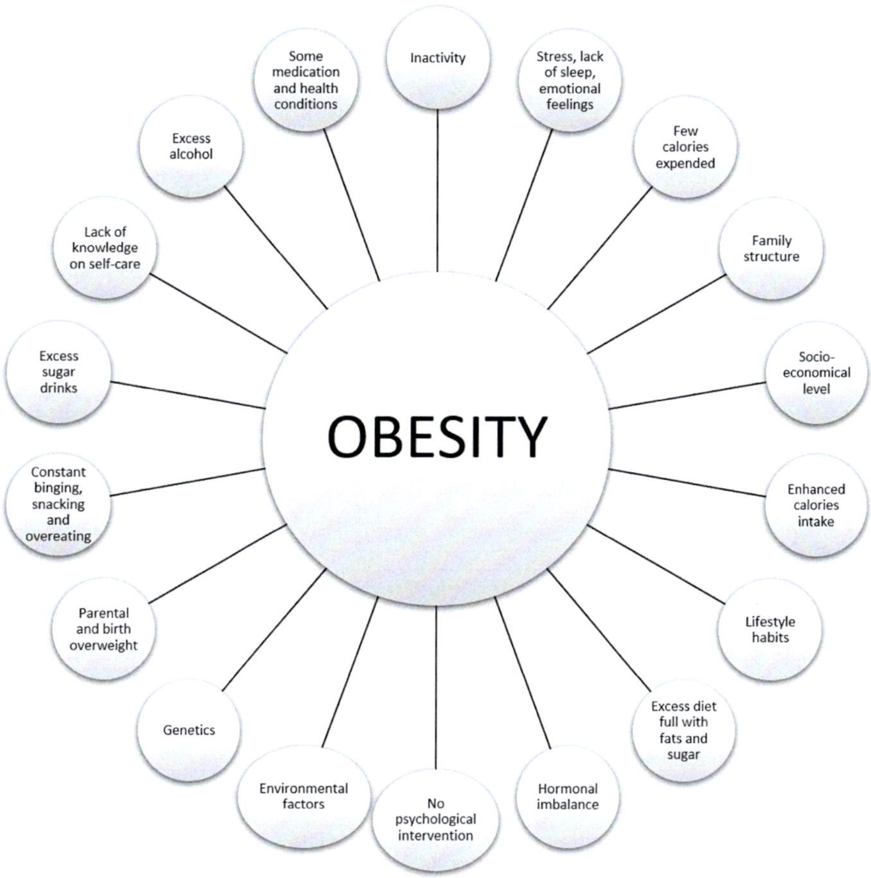

Fig. 21.1 Flow diagram illustrating some of the major risk factors in the development of overweight and subsequently, obesity in the body

Obesity has become a public global health problem. More than 13% of the world's 7.8 billion people (11% of men and 15% of women) are obese, defined as having a body-mass-index of >30 kg/m^2. In the current climate of health and nutrition, the global trend is that more than 50% of the world's population will be obese or over weight by 2030 [3, 5]. In North America, several Asian, European and Middle Eastern Countries, >30% of adults are obese. Obesity, with high fat and high sugar diets, is a risk factor for type 2 diabetes mellitus (T2DM), hypertension and early-onset heart failure with preserved ejection fraction that leads to frequent hospitalizations and in some cases SCD [9].

Obesity is also a major cause of such comorbidities as dementia, cancer, chronic pain and bullying [1, 2, 4, 5]. Similar to DM, obesity is a major NCD and it is also a major global health problem currently affecting more than 700 million adults and more than 2 billion people are overweight. Both obesity and overweight are defined as an abnormal excessive fat accumulation in the body leading to diabetes, cancer, hypertension and heart diseases over time [3, 5]. What is now worrying is that children as young as 8–12 years of age are obese due to excessive body weight? Comparing current time with 1975, overweigh has more than triple in number [7]. Obesity and over-weight are due to an imbalance between calories (fatty food) intake and expenditure combined with sedentary life style. Many people are too lazy to participate in physical activities. Obesity is defined as having a basal metabolic index (BMI) of 30 and over [3, 5]. Overweight is when the BMI is between 25 and 29 and normal weigh is when the BMI is between 20 and 24, whereas under body weight is when the BMI is less than 18 [3–5]. Obesity is now deemed as a major NCD leading to risk of chronic conditions, reduced quality of life and subsequently, to premature sudden cardiac death [1, 3, 5, 6]. Obesity and overweight vary between genders, ethnicity and economic status and both disorders are preventable depending on the individual [10].

In Guyana, there is a growing epidemic of NCDs, including obesity, diabetes and hypertension, cancer and others among its population. This is due to the effects of globalization, increased urbanization, alcoholism, smoking, physical inactivity, unhealthy diet, population ageing, behavioural, and the inadequacies of existing health and social care promotion, early disease diagnosis and prevention and management efforts. The Ministry of Public Health (MPH) reported that in 2016, NCDs accounted for over 60% of deaths amongst males and over 70% of deaths amongst females [11].

In recent years, scientists have been looking for cures of this overweight and obesity pandemic problem. Figure 21.2 shows some of the major factors which can help in reducing the weight of our body including daily exercise, eating less and more slowly, avoid constant snaking and binging, consume less alcohol and soft drinks rich in sugar, change life style habits generally, join a weight watcher's group, seek psychological intervention, educate yourself on the danger of over eating and the long-term complications of obesity-induced complications, take medication and surgery for weight loss, get hormonal treatment, browning white fat cells in the body, use of complementary medications and others. Research work in our and other laboratories have shown that bitter melon/corilla (*Momordica charantia*) can be used effectively in reducing blood sugar thereby treating diabetes [12–13] and obesity [13–14]. This study now investigates the hypothesis that bitter melon can also be used to reduce body weight gain and high blood pressure when combined with diet modification and exercise in a cost-effective manner.

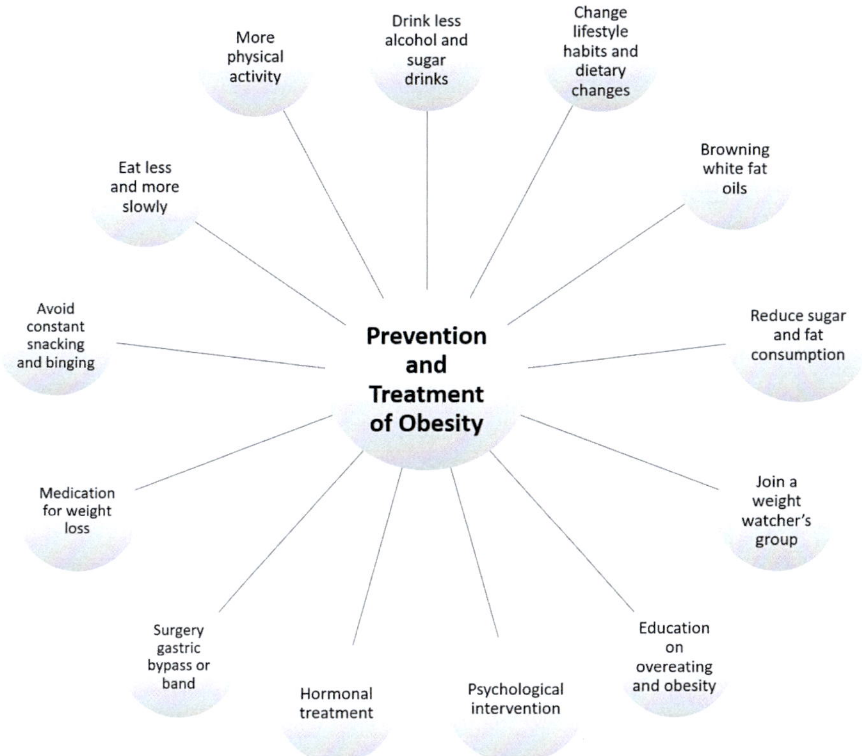

Fig. 21.2 Flow diagram showing some of the major factors which can help in reducing the weight of our body thereby either preventing the development of obesity or helping to treat the disease

Methods

Recruitment of Patients

Thirty two male and female subjects (16 males and 16 females) were recruited from newly diagnosed overweight or obese patients who had no other illness and not on any medication and who visited our clinic for treatment according to established investigative trial/study methods. The subjects signed a consent form to participate in the study at their own will and they could withdraw at any time if they wanted. Following recruitment, the subjects were explained thoroughly about their obesity or overweight and also about the experimental protocol which was designed to last over a period of six weeks. They were also given a questionnaire comprised of twenty five questions on obesity to complete based on knowledge of self-care. The thirty-two subjects were divided into four groups of eight subjects (4 males and 4 females) per group. Group 1 (diet only) was asked to reduce daily food intake, including sugar and fats, and avoided snaking and binging for a period of 6 weeks. Group 2 (Diet and bitter melon) did the same as group 1 but in addition to the consumption of 20 g of bitter melon either as a juice (vol/weight) or eat it as a salad raw once or twice daily before meals over 6 weeks. Group 3 (diet, exercise and bitter melon) did the same as group 1 but combined with daily exercise involving walking, stretching or bicycle riding or combination for 30 min daily) and consumption of 20 g of bitter melon daily either as a juice (vol/weight) or eat it raw as a salad before meals either once or twice per day for 6 weeks. Group 4 did the same as group 1 but combined with daily exercise involving walking, stretching or bicycle riding or combination for 30 min daily). At the start of the study at week 1 and up to the end of the study at week 6, all the subjects were weighed and their height taken as well as their systolic and diastolic blood pressure (SDBP) on a weekly basis. A small volume of fasting blood was taken from each subject for the measurements of fasting blood glucose (FBG), HBA1c, total cholesterol and triglyceride at the start at week 1 and at the end of the study at week 6. The BMI was calculated from the weight and height following weekly measurements. Blood pressure was measured using a digital sphygmomanometer on a weekly basis over the six weeks. Fasting blood glucose (FBG), HBA1c, total cholesterol and triglyceride were measured using our clinical laboratory auto-analyser.

Preparation of the Bitter Melon Juice

The unripe green fruits of *M. charantia* were obtained from the local supermarket or picked directly from the tree (Fig. 21.3a). They washed, cut sagitally (long-way) and subsequently cleaned of internal tissues and cut into small pieces (Fig. 21.3b, c) [15]. Approximately, one kilogram of chopped green fruit was liquidized in 1000 ml of drinking water for 5–10 min using a blender at room temperature. Once settled into a green juice, it was poured into a container and stored for use (Fig. 21.3d). Subjects were given 20 ml daily either one or twice (10 ml per time) daily (vol/weight), equivalent to 20 ml/20 g.

Statistical Analysis of Data

Statistical data analysis was done using the Statistical Package for Social Sciences (SPSS) and ANOVA. The data collected were compared according to the assigned groups. Data are expressed as mean ± standard error of the mean (SEM). A value of $p < 0.05$ is taken as statistically significant.

Fig. 21.3 Original photographs showing the bitter melon on the tree (**a**), cut long-way, cleaned of while tissues (**b**) and cut into small pieces (**c**) for blending into a green juice (**d**)

Ethical Clearance

The project had ethical clearance from the University of Guyana (UG) and UCLan Ethics Committees. A written informed consent was obtained from each participant after given written and oral information about the study.

Results

Time Course of Weight Loss

Figure 21.4 shows the time-course effect of (a) diet alone, (b) diet combined with daily intake of 20 g of bitter melon, (c) diet, exercise and daily intake of 20 g of bitter melon and (d) diet and daily exercise on BMI in obese subjects over a period

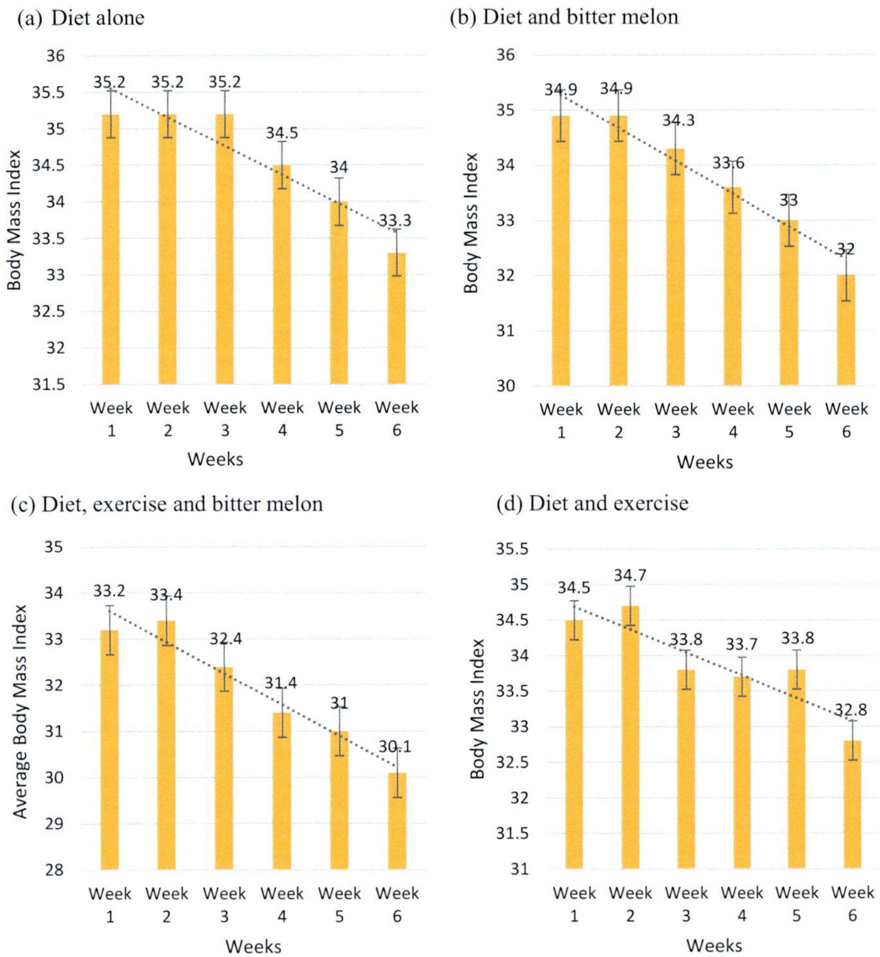

Fig. 21.4 Time-dependent effect of **a** diet alone, **b** diet and daily intake of 20 g of bitter melon daily, **c** diet, daily exercise and intake of 20 g of bitter melon and **d** diet and daily exercise on BMI over a period of 6 weeks in obese subjects. Data are mean ± SEM and n = 8 for each measured parameter. Note that all four interventions reduced BMI markedly over 6 weeks compared week 1 at the start of the experiment

of 6 weeks compared to week 1 at the start of the study. The data show that all two interventions (diet and exercise) in life style changes, combined with bitter melon, can markedly reduce body weigh gradually in a time-dependent manner comparing week 1 with week 6. Typically, body weight was reduced by 5.39%, 5.48%, 8.30%, and 9.33% for diet alone, diet and exercise, diet and bitter melon and diet, exercise and bitter melon, respectively over the six weeks of the study.

Time Course of High Blood Pressure Measurement

Systolic Blood Pressure

Figure 21.5 shows the time–course effect of (a) diet alone, (b) diet combined with daily intake of 20 g of bitter melon, (c) diet, exercise and daily intake of 20 g of bitter melon and (d) diet and daily exercise on systolic blood pressure (SBP) in obese

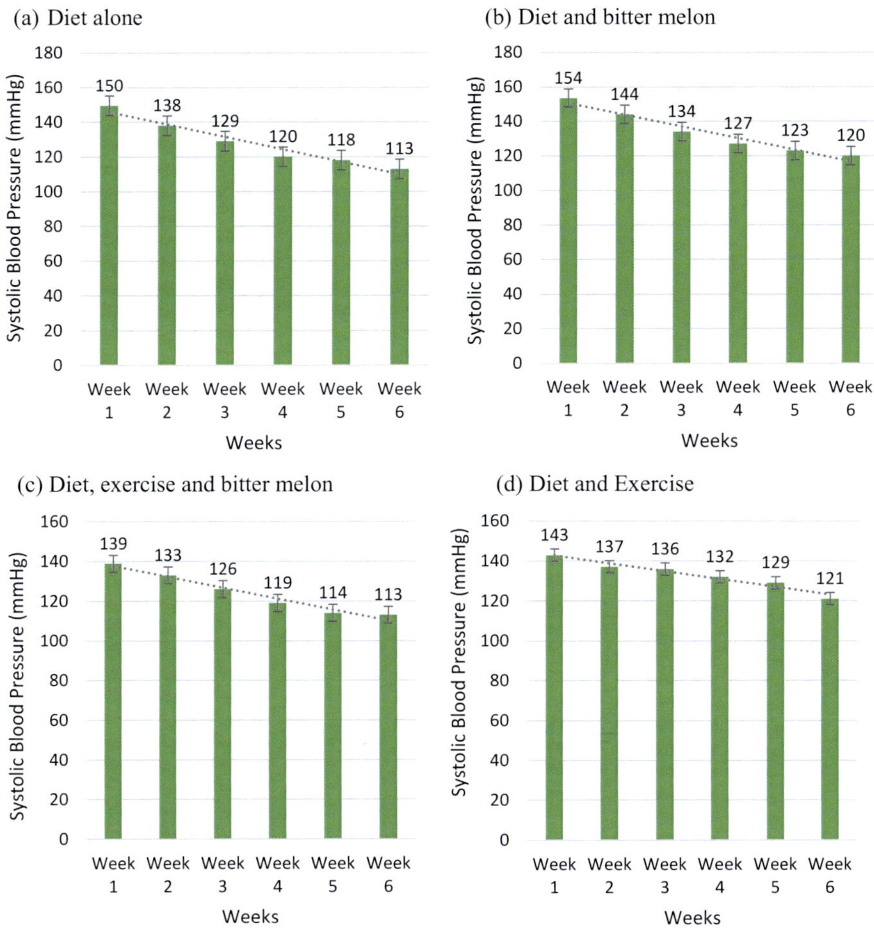

Fig. 21.5 Time-dependent effect of **a** diet alone, **b** diet and daily intake of 20 g of bitter melon, **c** diet, daily exercise and intake of 20 g of bitter melon and **d** diet and exercise on systolic blood pressure (SBP) in obese subjects over a period of 6 weeks. Data are mean ± SEM and n = 8 for each measured parameter. Note that all three interventions reduced SBP markedly over 6 weeks, compared week 1 at the start of the experiment. Note that SBP was significantly ($p < 0.05$) reduced in week 6 compared to week 1, at the start of the study

subjects over a period of 6 weeks compared to week 1 at the start of the study. The data show that all two interventions (diet and exercise) in life style changes combined with bitter melon can markedly reduce SBP gradually in a time- dependent manner from elevated levels to normal values. Typically, SBP was reduced significantly (p <0.05) by 25.38%, 18.70%, 22.665% and 24.66% with diet and exercise, diet alone, diet in combination with better melon and diet and, exercise in combination with better melon, respectively over the duration of the study, comparing week 6 with week 1.

Diastolic Blood Pressure

Figure 21.6 shows the time-course effect of (a) diet alone, (b) diet combined with daily intake of 20 g of bitter melon, (c) diet, daily exercise and 20 g intake of bitter melon and (d) diet and exercise on diastolic blood pressure (DBP) in obese subjects over a period of 6 weeks compared to week 1 at the start of the study. The data show that the two interventions (diet and exercise) in life style changes, combined with daily bitter melon intake, can markedly reduce body DBP gradually in a time-dependent manner from elevated levels to normal value. Typically, DBP was decreased significantly (p <0.05) by 14.89%, 16.66%, 17.20% and 17.52% for diet and exercise, diet alone, diet and bitter melon and diet, exercise and daily bitter melon intake, respectively over the duration of the study comparing week 6 with week 1.

Measurements of Age, Gender and Blood Biomarkers

Table 21.1 shows the mean (±SEM) age, gender and levels of such blood biomarkers such as HBA1c, fasting blood glucose (FBG), total cholesterol and triglyceride for the 8 subjects (4 males and 4 females) in each of the four different intervention groups namely, diet alone, diet and daily exercise in combination with bitter melon intake, diet and daily intake of bitter melon and diet and exercise at week 1 at the start of the study and on week 6 at the end of the study for comparison. The results show that the subjects were in their working years of age. The majority of them had secondary education. Moreover, their HBA1c and FBG only reduce very slightly for each intervention. In contrast, four interventions reduced blood total cholesterol and triglyceride, but with significance (p <0.05) in the presence of diet, exercise and bitter melon, diet with bitter melon and exercise and diet and exercise.

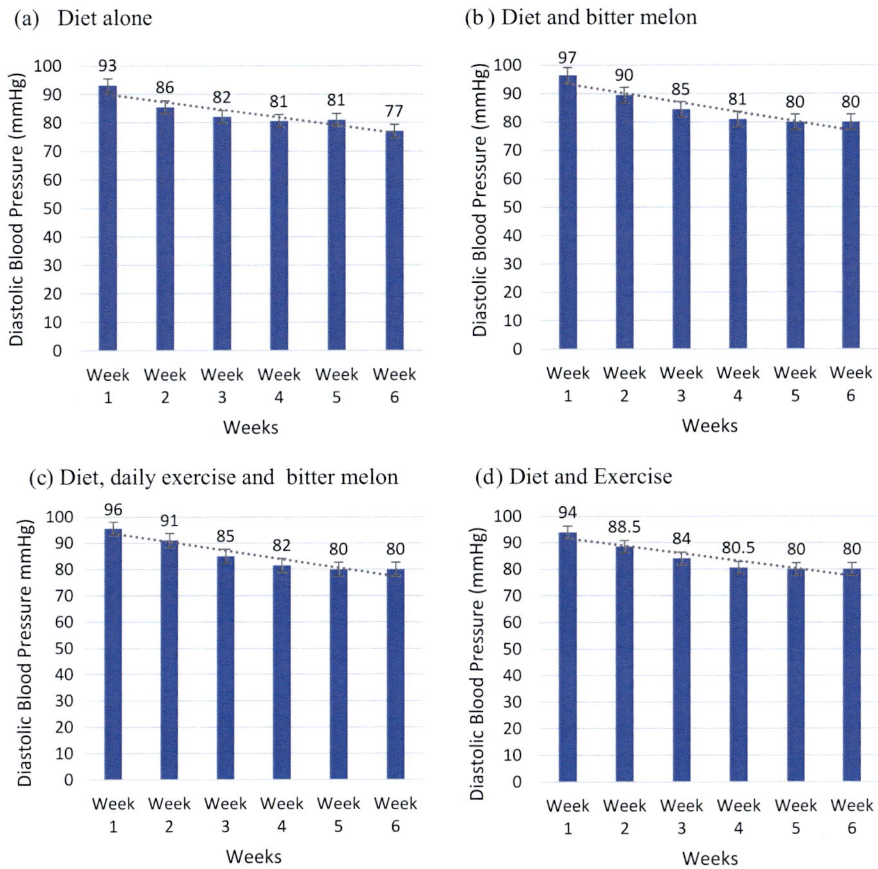

Fig. 21.6 Time–dependent effect of **a** diet alone, **b** diet and daily intake of 20 g of bitter melon, **c** diet, daily exercise and intake of 20 g of bitter melon and **d** diet and exercise on diastolic blood pressure (DBP) over a period of 6 weeks in obese subjects. Data are mean ± SEM and n = 8 for each measured parameter. Note that all four interventions reduced DBP markedly over 6 weeks compared week 1 at the start of the experiment. Note also that that DBP was significantly ($p < 0.05$) reduced in week 6 compared to week 1, at the start of the study

Discussion

Both overweight and obesity, due to an imbalance between calories intake and imbalance, are becoming major health problems affecting almost two billion people globally at present, with almost 700 million obese [3, 5]. They are responsible for abnormal or excessive fat accumulation in the body leading to impaired health. If left untreated, obesity can lead to diabetes (diabesity) which is a major cause of death from cardiovascular diseases (CVDs) [16]. Overweight, obesity, diabetes and CVDs are all classified as non-communicable diseases (NCDs) or chronic diseases [1, 6, 17]. In the

Table 21.1 Data showing age, gender, HBA1c, fasting blood glucose (FBG), total cholesterol and triglyceride in the four different groups namely (A) diet only, (B) diet, exercise and bitter melon, (C) diet and daily intake of bitter melon and (D) diet and daily exercise in obese subjects at week 1, prior to the start of this study compared to 6 weeks later at the end of the study. Data are mean ± SEM; n = 8; *p <0.05

Measured parameters	Diet only		Diet + exercise + bitter melon		Diet + bitter melon		Diet + exercise	
	Week 1	Week 6	Week 1	Week 6	Week 1	Week 6	Week 1	Week 6
Age	42 ± 4.5	42 ± 4.5	42 ± 4.5	42 ± 4.5	42 ± 4.5	42 ± 4.5	42 ± 4.5	42 ± 4.5
Male	4	4	4	4	4	4	4	4
Female	4	4	4	4	4	4	4	4
HbA1c (%)	6 ± 0.52	5.3 ± 0.34	5.8 ± 0.46	5.80.46	5.6 ± 0.52	5.6 ± 0.52	6 ± 0.0	5.8 ± 0.46
RBS (mg/dL)	126 ± 1.9	122 ± 1.4	126 ± 1.6	122 ± 2.0	124 ± 1.7	122 ± 1.8	121 ± 2.7	115 ± 3.9
Total Cholesterol (mg/dL)	239 ± 22.2	228 ± 13.9	232 ± 10.5	*186 ± 10.3	241 ± 12.9	*193 ± 7.0	254 ± 20.5	*189 ± 11.7
Triglyceride (mg/dL)	173 ± 9.3	163 ± 8.5	167 ± 7.0	*138 ± 11.0	165 ± 8.5	*139 ± 6.8	171 ± 15.8	*141 ± 8.2

olden days, it was classified as an old person disease, but recently children as young as 12 years of age are becoming obese and subsequently suffering from type 2 diabetes mellitus (T2DM) [18]. Chronic diseases such as obesity, diabetes and CVDs are very common nowadays and they are mainly due to intake of energy-dense foods rich in fats and associated with little or no physical activity. Moreover, they are very costly and preventable of all health problems representing a growing burden to mankind in both developing and developed countries worldwide and accounting for over 36 million of the 57 million deaths worldwide in 2008 [1, 3, 4, 6, 19]. The present study was designed to investigate a cost-effective and non-pharmacological way in tackling the overweight, obesity and hypertensive health epidemic problems in Guyana, a low income country in South America, by testing the impact of diet modification alone, diet modification and daily intake of 20 g of bitter melon (*Momordica charantia*), diet modification with daily exercise and intake of bitter melon (*Momordica charantia*) and diet modification and exercise in newly diagnosed obese subjects.

Over eating high calorie diets is the main cause for overweight and subsequently, obesity leading to diabesity and CVDs [16–19]. There are also a small number people who have a hormonal-induced metabolic disorder which can also cause obesity [4]. During eating, many people never feel filled and satisfied and as such they continue to eat excessively, including constant binging and snaking. The main treatment or preventative measure for overeating is to eat less or dieting in addition to the surgical intervention or bariatric surgery whereby the size of the stomach is reduced [20]. For many people without will power, this is not possible unless they educate themselves on obesity self-care and obtaining psychological intervention for adherence to food intake, just like exercise [21]. Quality of food also plays a major role in the development of obesity. It is now the norm that fast foods can induce obesity whereas the Mediterranean-style diet rich in olive oil, vegetables and fish can help in the reducing the weight of the body [22]. The results in this study have shown that daily reduction of food intake, including no snaking and binging, reduction in sugar and fast food consumption and intake of more vegetables low in calorie can markedly reduce BMI, systolic and diastolic blood pressure, total cholesterol and triglyceride comparing week 6 at the end of the experiments with week 1. Dieting had little or no significant effect on HBA1c and fasting blood glucose level but reduced cholesterol and triglyceride levels but not to significantly level through to 6 weeks of experiments compared to week 1. These data are in total agreement with other similar studies [18].

Bitter melon or corilla (*Momordica charantia*) originated from India and the plant was taken to different parts of the world including China, South East Asia, Africa, the Caribbean and South and Central America by Indians during colonization. Bitter melon has been used by many people globally as a plant remedy to treat obesity, diabetes, hypertension, heart attacks and strokes, cancer, reducing blood cholesterol and triglyceride and many other debilitating conditions [12–15]. Bitter melon is low in calories but high in fibre. It contains approximately 2 g of fibre in each 94 g. It can also promote the burning of fats in the body thereby leading to weight loss [13]. The results of this study have shown that when bitter melon was combined with diet modification there were marked decreases in BMI, systolic and diastolic blood pressure, total blood cholesterol and triglyceride levels in the subjects after 6 weeks of the intervention compared to week at the experimental period. Combination of bitter melon with diet modification had little or no effect on HBA1c and fasting blood glucose (FBG) over the experimental period. Bitter melon with dieting seems to exert a slightly better beneficial effect compared to diet modification alone. Bitter melon is rich in minerals especially calcium and sodium, vitamins, particularly C, phenolic compounds, saponins and triterpenes, glycosides, a plant–like insulin and anti-oxidants, all of which help in reducing the weight of the body by enhancing glucose and fat metabolism utilizing different mechanisms and in some cases, the different compounds act synergistically to exert their beneficial effects [16, 23, 24]. These results are in total agreement with other similar studies [25, 13, 26, 27], highlighting the evaluation of the various extracts of *M charantia* in their therapeutic potentials in obesity, diabetes and the metabolic syndrome.

Daily physical exercise is a powerful cost-effective physiological tool to prevent a number disorders and diseases in the body including overweight, obesity, diabetes, cancer, cardiovascular and respiratory diseases, stress, mental illnesses and others [21, 28, 29]. Regarding weight loss, physical activity does help to increase the number of calories the body uses for energy leading to a small reduction in body weight. It is also well known that the higher the intensity of the physical activity then there will be high weight loss due to calorific deficit more so when the obese subjects adhere to the physical activity [28]. In addition to weight loss, daily exercise can reduce stress and improve cardiorespiratory fitness especially when it is combined with dietary modification [29, 30]. The results of this study have revealed that combining diet modification with daily exercise and intake of bitter melon can help to reduce BMI, systolic and diastolic blood pressure as well as blood cholesterol and triglyceride, but with little or no effects of HBA1c and fasting blood glucose. The data further reveal that a combination of dieting with daily exercise and bitter melon intake seems to induce a small enhancing beneficial effect, similar to dieting and bitter melon intake when compared to diet modification alone. It is well established that combining diet modification with daily physical exercise is the best method for weight loss, especially in obese subjects who are insulin resistant, a serious public health problem [6, 31]. The results from the present study have now shown that the addition of bitter melon consumption to exercise and dieting can even exert a better cost-effective health benefit in treating obesity, blood pressure and blood cholesterol and triglyceride levels. Regarding exercise alone, it has to be performed at a moderate intensity such as brisk walking for at least 30 min daily to obtain the full benefit [32–33]. Moreover, further benefits can be gained from these interventions, especially if the obese subject is given some form psychological training or intervention to facilitate adherence to the measured parameters [29, 34].

It is now well known that obesity is a major contributing risk factor for cardiovascular diseases via to the release of several pro-inflammatory mediators [4]. The fluctuation in the levels of these mediators and adipocytokines released by the adipocytes further leads to cardiovascular dysfunctions [35]. The increase in the risk of CVDs and mortality caused by obesity is due to increasing levels of atherosclerotic plaques in the arteries and blood vessels leading to the heart [36]. It is also known that flavonoids are polyphenolic compounds which occur naturally in nature and they can be used in a cost-effective way to treat obesity. In this study, the small reduction in weight of the obese subjects may be due to the polyphenolic anti-oxidant compounds present in bitter melon [37].

Obesity, Hypertension and Sudden Cardiac Death (SCD)

There is a close association between obesity and such comorbidities as diabetes and hypertension which often precede the development of structural changes in the myocardium, fibrosis, apoptosis, hypertrophy, remodelling of the myocardium, diastolic and systolic dysfunctions, coronary artery diseases CAD, arrhythmias and subsequently SCD [38–39]. SCD in the young obese population normally happens in individuals without a known cardiac history [40]. Moreover, chronic obese patients have been shown to be more susceptible to increased risk of SCD. In turn, this is now becoming a major concern and challenge for clinicians and health services globally, especially since the prevalence of obesity has been increasing steadily in both developed and developing countries. In previous times, over weight and obesity used to be prevalent in high income countries such as those in North America, Europe, some parts of Asia, Australia and other parts of the world. More recently, over weight and obesity are becoming more prevalent in low income countries all over the world. The most worrying social care and health problems for National Health Services worldwide are that children as young 12–15 years of age are becoming obese. Both obesity and DM share some of the same main risk factors such inactivity, genetics, smoking, excess alcohol consumption, regular snaking and binging (over eating) and diets rich in sugar and fats [40, 41]. Most obese patients are hypertensive, pre-diabetic and diabetic and most of them tend to experience obstructive sleep apnea due to their excessive weight, increased liver size and metabolic syndrome [2, 4]. All of these pathological parameters are well-known risk factors for CVDs, arrhythmias and SCD. It is now evident that structural, functional and metabolic factors modulate and influence the risk of SCD in the obese population [41]. Obesity exerts numerous haemodynamic changes on the cardiovascular system (CVS) including enhanced cardiac output and diastolic filling pressures, both of which result in LV hypertrophy and dilatation. Moreover, obesity can elicit adverse electrical changes in the heart such as prolongation of the QRS and increase in QT intervals on the electrocardiogram (ECG), and also an enhancement in QT dispersion. Interestingly, the late potentials on signal averaged ECG are also more common in obese compared with lean individuals. Obesity, as a preventable disease, seems to induce adverse structural and electrical insults on the myocardium creating substrates that are susceptible to SCD [42].

The question which now arises is: how does obesity elicit cardiac dysfunction and how can bitter melon, combined with diet modification and exercise prevent this? Obesity-induced damage to the heart via the production of excess lipids, oxidized LDL particles and free FAs which activate the inflammatory process in the body triggering the development of cardiac dysfunction. Inflammation is responsible for the steps towards the development of atherosclerosis, from early endothelial cell dysfunction to the late atherosclerotic plaque formation causing complications. All these pathological processes are related to obesity, insulin resistance (IR) and diabetes [31]. During obesity–induced diseased processes in the heart, fatty tissues release adipocytokines which in turn induces IR, endothelial cell dysfunction, hypercoagulability and systemic inflammation. Together, these pathophysiological conditions facilitate the process of atherosclerosis in the heart. Moreover, inflammatory adipocytokine such as TNF-α rises to very high levels in visceral obesity. In turn, the heart releases excess C-reactive protein (CRP) which acts as an insult leading to enhanced risk of ischaemia, myocardial infarction and peripheral vascular disease. In turn, these pathophysiological states facilitate atherosclerosis, ischaemic heart disease, angina pectoris arrhythmias and SCD [4, 43]. Bitter melon is rich in vitamins, antioxidants, cations, proteins, fibres, phenol and several other components [44, 13, 23, 24]. Daily consumption of bitter melon as a salad or as a juice is able to reduce lipids and triglycerides in obese, diabetic and hypertensive patients [13, 37, 45]. Regarding diet modification, the less food we consume then it is less likely we will become obese unless there is a medical problem. Exercise, on the other hand, is universally known as the panacea of life which can prevent many diseases [28–31].

Prior to the start of this study, each obese subject was given a basic questionnaire based on knowledge of obesity, causes and links to diabetes and CVDs. Surprisingly, the results revealed that over 90% of the subjects had very good knowledge and understanding about obesity, it causes, long-term adverse effects to our body and obesity self-care, but they were still obese. This interesting finding suggests that people do not seem to have the will power over their eating habits irrespective that excess food intake can damage the body over time. As such, the best way forward in controlling obesity is to provide psychological intervention in order to achieve adherence to eating habits and the quality of food to consume. A similar study was done using psychological intervention to adhere to daily exercise in newly diagnosed type 2 diabetic patients with tremendous success and in delaying long–term complications due to the diabetes [21].

Summary

Figure 21.7 is a flow diagram summarising the relationship between the risk factors for obesity which can lead to sudden cardiac death. The risk factors induce increased levels of lipids and triglycerides in the body resulting in weight gain and subsequently obesity which can induce diabetes, cardiovascular diseases and sudden cardiac death. Bitter melon intake combined with dieting and regular physical activity can prevent and possible, treat the development of these adverse effects due to the risk factors. In summary, the results of this study have clearly demonstrated that daily reduction in diet intake combined with daily exercise and intake of bitter melon can help to reduce the weight of obese subjects, their blood pressure, total cholesterol and triglyceride

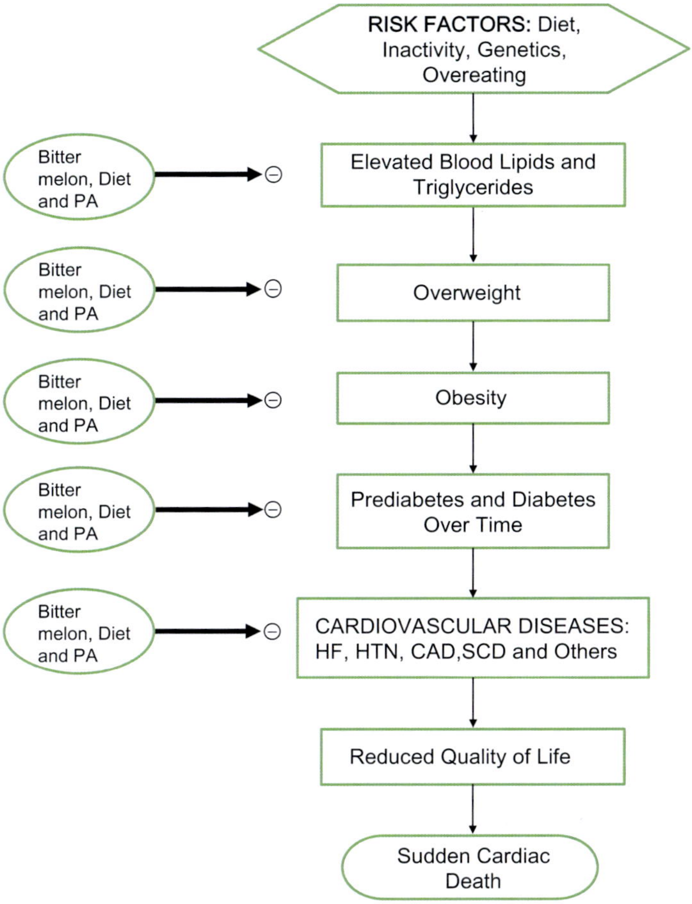

Fig. 21.7 Flow diagram summarising the relationship between the risk factors for obesity which can lead to sudden cardiac death and how bitter melon combined with diet and physical activity (PA) can help in combating the adverse effects due to the risk factors. CVD = cardiovascular diseases; CAD = coronary artery disease; HTN = hypertension; SCD = sudden cardiac death; PA = physical activity

levels at least for 6 weeks, as demonstrated in the study. These cost-effective and non-pharmacological interventions must be continued for a longer period of time in order to obtain better beneficial effects leading to a better quality of health and life for the patients.

References

1. WHO (2018) Non-communicable diseases. www.who.net
2. Carbone S, Canada JM, Billingsley HE, Siddiqui MS, Elagizi A, Lavie CJ (2019) Obesity paradox in cardiovascular disease: where do we stand? Vasc Health Risk Manag 15:89–100. https://doi.org/10.2147/VHRM.S168946
3. So I, Yadav H (2020) Obesity and its complications. Adv Biochem Health Dis 19:43–58
4. Kumar PJ Clark M (2018) Diabetes mellitus and other disorders of metabolism. In: Textbook of clinical medicine. Saunders, London, pp 1069–1121
5. Tapia PS, Defries D (2020) Prevalence, consequences, causes and management of obesity. Adv Biochem Health Dis 19:1–22
6. WHO (2010) Global status on non-communicable diseases. https://apps.who.int/iris/bitsream/10665/44579/1/98892406458
7. Riley JR, Armstrong J, Dorosty AR, Emmett PM, Ness A, Rogers I, Steer C, Sheriff A (2005) Early life risk factors for obesity in childhood: cohort study. BMJ 330(7504) (Article 1357). https://doi.org/10.1136/bmj.38470.670903.E0
8. Lopez RP (2012) Neighbourhood risk factors for obesity. Obesity 15(8):1903–2161. https://doi.org/10.1038/oby.2007.251
9. Cilia L, Saaed A, Ganga HV, Wu W-C (2019) Heart failure with preserved ejection fraction: prevention and management. Am J Lifestyle Med 13(2):182–189
10. Kanter R, Cabellaro B (2012) Gender disparities in obesity: a review. Adv Nutr 3(4):491–498. https://doi.org/10.3945/an112.002063
11. MPH, Guyana (2016) Statistics on diabetes and related diseases. Ministry of Public Health Website, Guyana
12. Ivorra MD, Paya M, Villar A (1989) A review of natural products and plants as potential anti-diabetic drugs. J Ethnopharmacol 27:243–275
13. Alam MA, Uddin R, Subhan S, Rahman MH, Jain P, Reza HM (2015) Beneficial role of bitter melon supplementation in obesity and related complications of metabolic syndrome. J Lipids 2015:496169 (18 pp). https://doi.org/10.1155/2015/496169
14. Fan M, Kim E-K, Moon S-H (2019) The role of *Momordica charantia* in resisting obesity. Int J Environ Res Public Health 16(18):3251–3268. https://doi.org/10.33390/ijerph161832251
15. Manoharan G, Cummings E, Singh J (2014) Effects of crude water-soluble extract of *Momordica charantia* on viability, caspase activity, cytochrome-c release and on cytosolic calcium levels in different cancer cell lines. Cancer Cell Microenviron 1:11 (Article e273). https://doi.org/10.14800/ccm.273
16. Zimmet P, Albert KGMM, Shaw J (2014) Global and societal implications of the diabetic epidemic. Nature 414:782–787
17. WHO (2011) Obesity and overweight. https://new.who/int/news-room-fact-sheet/details/obesity-and-overweight. Accessed Mar 2020
18. Taher S, Zaghloul H, Chagoury O, Alhadad S, Ahmed SH, El-Khatib N et al (2020) Effect of intense life style intervention on body weight and glycaemia in early type 2 diabetes (DIADEM-I): an open-label parallel group. Randomised controlled study. The Lancet 8(6):P477–P489
19. Wagner KH, Brath H (2012) A global view on the development of non communicable diseases. Prev Med 54(Suppl):S38–S411
20. Athyros VG, Tziomalos K, Karagiannis A, Mikhailidis DP (2011) Cardiovascular benefits of bariatric surgery in morbidly obese patients 12(7):515–524. https://doi.org/10.1111/j.1467-789X.2010.00831.x
21. Martinus R, Corban R, Wackerhage H, Atkins S, Singh J (2006) Effect of psychological intervention on exercise adherence in type 2 diabetic subjects. Ann N Y Acad Sci 1084:350–360
22. Singh RB, Saboo B, Mashwari A, Bharatdwaj K, Verma N, Hristova K, Ghosh S, Niaz MA, Singh J, Adeghate E, Bidasee KR, Singh M, Mishra A, Tripathi S, Singh D, Pandey S, Srivastava Jaglan P (2017) Effects of Indo-Mediterranean diet on incidence of diabetes in acute coronary syndromes. World Heart J 9(1):25–36

23. Bajpai M, Pande A, Tewari SK, Prakash D (2005) Phenolic contents and antioxidant activity of some food and medicinal plants. Int J Food Sci Nutr 56:287–291
24. Thenmozhi AJ, Subramanian P (2011) Antioxidant potential of Momordica charantia in ammonium chloride-induced hyperammonemic rats. Evid Based Complement Altern Med 11(8):1–7
25. Singh J, Cummings E, Manoharan G, Kalasz H, Adeghate E (2011) Medicinal chemistry of the anti-diabetic effect *Momordica charantia* active ingredients and modes of action. Open J Med Chem 5(Supplement 2-M2):70–77
26. Ahmed I, Cummings E, Sharma AK, Adeghate E, Singh J (2004) Beneficial effects and mechanism of action of *Momordica charantia* fruit juice in the treatment of streptozotocin-induced diabetes mellitus in rats. Mol Cell Biochem 261(1/2):63–70
27. Badhwar R, Kaur G, Popil H, Yadav D, Buttar HS (2020) Philosophy of obesity-related non-communicable chronic diseases and advancement in prevention strategies. Adv Biochem Health Dis 19:317–340
28. Paez CJ, Kravitz L (2000) Exercise vs diet in weight loss. Sports Sci Rev 28(4):165–170
29. Smail MMA, Singh RB, Bidasee KR, Howarth FC, Hanoman C, Singh J (2018) Diabetic cardiomyopathy and the role of regular exercise in preventing the disease. World Heart J 9(4):319–332
30. Chambliss HO (2005) Exercise duration and intensity in weight-loss program. Clin J Sport Med 15(2):113–115
31. Weinstock RS, Dai H, Wadden TA (1998) Diet and exercise in the treatment of obesity. Effects of three interventions on insulin resistance. Arch Intern Med 158:2477–2483
32. Ross R, Dagnone D, Jones PJH, Smith H, Paddags A, Hudson R, Janssen I (2000) Reduction in obesity and related comorbid conditions after diet-induced weight loss or exercise-induced weight loss in men: a randomized controlled trial. Ann Intern Med 133:92–103
33. Ross R, Freeman JA, Janssen I (2000) Exercise alone is an effective strategy for reducing obesity and related comorbidities. Exerc Sport Sci Rev 28:165–170
34. Barnes MS, Cassidy T (2018) Diet exercise and motivation in weight reduction: the role of psychological capital and stress. JOJ Nurs Health Care 9(5):1–6. https://doi.org/10.19080/JOJHNC.2018.09555775
35. Behll S, Salehi J, Singh J, Adem A, Jarrar M (2016) Browning the fat may reset the metabolome: role of PPAR delta network of genes in obesity and cardiovascular diseases. World Heart J 8(4):357–370
36. Ahmed Z, Pal A, Parab SB et al (2020) Biochemistry of cardiovascular dysfunction in obesity. Adv Biochem Health Dis 20:307–327. https://doi.org/10.1007/978-3-030-47336-5-16
37. Yadav BS, Yadav R, Yadav RB, Garg M (2016) Antioxidant activity of various extracts of selected gourd vegetables. J Food Sci 53(4):1823–1833
38. D'Souza AJ, Howarth CF, Adeghate E, Woods NM, Singh J (2009) Pathogenesis and pathophysiology of accelerated atherosclerosis in the diabetic heart: a review. Mol Cell Biochem 331(1/2):89–116
39. Novoa U, Arauna D, Moran M, Nuñez M, Zagmutt S, Saldivia S et al (2017) High-intensity exercise reduces cardiac fibrosis and hypertrophy but does not restore the nitroso-redox imbalance in diabetic cardiomyopathy. Oxidative Med Cell Longev 2017:1–11
40. Vishal GO et al (2020) Obesity and sudden cardiac death in the young: a nationwide retrospective study. J Am Coll Cardiol 75(11):1–10
41. Plourde B et al (2014) Sudden cardiac death and obesity. Expert Rev Cardiovasc 12(9):1099–1110
42. Adabag S et al (2025) Obesity related risk of sudden cardiac death in the atherosclerosis risk in communities study. Heart 101:215–221
43. Csige I, Ujvárosy D, Szabó Z, Lorincz I, Paragh G, Harangi M et al (2018). The impact of obesity on the cardiovascular system. J Diabetes Res 218(1):12 (Article ID: 3407306). https://doi.org/10.1155/2018/3407306
44. Joseph B, Jini D (2013) Antidiabetic effects of Momordica charantia (bitter melon) and its medicinal potency. Asian Pac J Trop Dis 3(2):93–102. https://doi.org/10.1016/S2222-1808(13)60052-3

45. Wu SJ, Ng LT (2008) Antioxidant and free radical scavenging activities of wild bitter melon (Momordica charantia Linn. var. abbreviata Ser.) in Taiwan. LWT Food Sci Technol 41:323–330
46. Platel K, Srinivasan K (1997) Effect of dietary intake of freeze-dried bitter gourd (Momordica charantia) in streptozotocin induced diabetic rats. Nahrung 39:262–268
47. Platel K, Srinivasan K (1999) Plant food in the management of diabetes mellitus: vegetables as potential hypoglycaemic agents. Nahrung 41:68–74
48. Taylor L (2002) Bitter melon (*Momordica charantia*). Herbal secrets of the rain forest, 2nd edn. Sage Press, Austin, Texas, USA, pp 1–100
49. Garau C, Cummings E, Phoenix DA, Singh J (2003) Beneficial effects and mechanism of action of *Momordica charantia* in the treatment of diabetes mellitus: a mini review. Int J Diabetes Metab 11:46–55
50. Chen QC, Li ETS (2005) Reduced adiposity in bitter melon (*Mormordica charantia*) fed rats is associated with lower tissue triglyceride and higher plasma catecholamines. Br J Nutr 99(5):747–754. https://doi.org/10.1079/BJN.20051388
51. Ross R, Janssen I (1999) Is abdominal fat preferentially reduced in response to exercise-induced weight loss? Med Sci Sports Exerc 31:S568–S572
52. D' Souza A, Howarth CF, Yanni J, Dobrzynski H, Boyett MN, Adeghate E, Bidasee KA, Singh J (2014) Chronic effect of mild hyperglycaemia on left ventricle transcriptional profile and structural remodelling in the spontaneously type 2 Goto-Kakizaki rat. Heart Fail Rev 19:65–74. https://doi.org/10.1007/s10741-013-9376-9
53. Iqbal T, Adeghate E, Howarth C, Bidasee K, Welsby PJ, Singh J (2013). Effects of diabetes-induced hyperglycemia on the heart: biochemical and structural alterations. In: Dhalla, NS, Bela T (eds) Textbook of heart failure. Springer, India. Adv Biochem Health Dis 9:77–106
54. Adeghate E, Singh J (2014). Structural changes in the myocardium during diabetes-induced cardiomyopathy. Heart Fail Rev J 19:15–23. https://doi.org/10.1007/s10741-013-9388-5

Index

© Springer Nature Switzerland AG 2021

P. S. Tappia et al. (eds.), *Cellular and Biochemical Mechanisms of Obesity*, Advances in Biochemistry in Health and Disease 23,
https://doi.org/10.1007/978-3-030-84763-0

C

D

E

T

V

W